Monica Levy Andersen • Sergio Tufik
Editors

Rodent Model as Tools in Ethical Biomedical Research

Springer

Editors
Monica Levy Andersen
Department of Psychobiology
Universidade Federal de São Paulo (UNIFESP)
São Paulo, São Paulo, Brazil

Sergio Tufik
Department of Psychobiology
Universidade Federal de São Paulo (UNIFESP)
São Paulo, São Paulo, Brazil

ISBN 978-3-319-11577-1   ISBN 978-3-319-11578-8 (eBook)
DOI 10.1007/978-3-319-11578-8

Library of Congress Control Number: 2015933848

Springer Cham Heidelberg New York Dordrecht London

Printed on acid-free paper

Springer International Publishing AG Switzerland is part of Springer Science+Business Media (www.springer.com)

Preface

The objective of this book is to concisely present information with respect to the appropriate use of experimental animals in research. The principles elaborated seek to provide knowledge of the techniques involved in both management and scientific research to all who use laboratory animals, with a focus on the well-being and ethics regarding animals. It is also within our scope to fortify the awareness of the importance of the animal as a study object and to offer orientation and assistance in conducting laboratory research, education, or tests. In this new version, we had the pleasure of working with collaborators from different universities and countries, who significantly contributed to the knowledge and quality of the book. Our most esteemed thanks to all the authors who participated in this book.

São Paulo, Brazil

Monica Levy Andersen
Sergio Tufik

Acknowledgments

To experimentation animals, who with their lives provide a notable contribution to the development of new techniques and medical treatments, significantly focusing the advancement of science and bringing benefits to human and animal health.

Contents

Contributors

Tathiana Aparecida Fernandes Alvarenga Department of Psychobiology, Universidade Federal de São Paulo (UNIFESP), São Paulo, São Paulo, Brazil

Monica Levy Andersen Department of Psychobiology, Universidade Federal de São Paulo (UNIFESP), São Paulo, São Paulo, Brazil

Isabela Beleza Antunes Master in Psychobiology, Universidade Federal de São Paulo (UNIFESP), São Paulo, São Paulo, Brazil

Heloisa Allegro Baptista Centro de Desenvolvimento de Modelos Experimentais para Medicina e Biologia CEDEME, Universidade Federal de São Paulo (UNIFESP), São Paulo, São Paulo, Brazil

Vera Baumans Division of LAS, Department of Animals in Science and Society, Utrecht University, Utrecht, The Netherlands

Maria Martha Bernardi Instituto de Ciências da Saúde, Universidade Paulista, São Paulo, São Paulo, Brazil

Leandro José Bertoglio Departamento de Farmacologia, Universidade Federal de Santa Catarina, Florianópolis, Santa Catarina, Brazil

Antônio de Pádua Carobrez Departamento de Farmacologia, Universidade Federal de Santa Catarina, Florianópolis, Santa Catarina, Brazil

Rita de Cassia Carvalho Department of Pharmacology, Universidade Federal de São Paulo (UNIFESP), São Paulo, São Paulo, Brazil

Rosaely Casalegno Helmholtz Center for Infection Research HZI, Department of Vaccinology and Applied Microbiology, Braunschweig, Germany

Nadia Carla Cheik Instituto de Ciências Biomédicas, Universidade Federal de Uberlândia, Uberlândia, Minas Gerais, Brazil

Cibele Cristina Chinen Department of Pharmacology, Universidade Federal de São Paulo (UNIFESP), São Paulo, São Paulo, Brazil

Marcos Fernando Oliveira e Costa Laboratório de Reprodução Animal, Embrapa Arroz e Feijão, Santo Antônio de Goiás, Goiás, Brazil

Renata Mázaro e Costa Department of Pharmacology, Universidade Federal de Goiás, Goiânia, Goiás, Brazil

Gabriel Cuadra Department of Pharmacology, Universidad Nacional de Córdoba, Córdoba, Argentina

Vânia D´Almeida Department of Psychobiology, Universidade Federal de São Paulo (UNIFESP), São Paulo, São Paulo, Brazil

Ana R. Dâmaso Universidade Federal de São Paulo (UNIFESP), São Paulo, São Paulo, Brazil

Michael John Decker Nursing Neuroscience Laboratory at Frances Payne Bolton School of Nursing, Case Western Reserve University, Cleveland, OH, USA

Fernanda Oliveira Duarte Departamento de Educação Física e Motricidade Humana, Universidade Federal de São Carlos, São Carlos, São Paulo, Brazil

Ana Claudia Garcia de Oliveira Duarte Departamento de Educação Física e Motricidade Humana - DEFMH, Universidade Federal de São Carlos, São Carlos, São Paulo, Brazil

Luciano Freitas Felicio Department of Pathology, Universidade de São Paulo (USP), São Paulo, São Paulo, Brazil

Armando Ferreira-Nuño Departamento de Biología de la Reproducción, Universidad Autónoma Metropolitana, Iztapalapa, Mexico

Roberto Frussa-Filho (In Memoriam) Department of Pharmacology, Universidade Federal de São Paulo (UNIFESP), São Paulo, São Paulo, Brazil

Daniela Fukue Fukushiro Department of Pharmacology, Universidade Federal de São Paulo (UNIFESP), São Paulo, Brazil

Sergé Gobaille Université Louis Pasteur Strasbourg, Faculté de Médecine, Strasbourg, Alsace, France

Patrícia Helena Zanier Gomes Professor, Universidade Federal do Triângulo Mineiro, Uberaba, Minas Gerais, Brazil

Ricardo Luis Fernandes Guerra Department of Human Movement Sciences, Universidade Federal de São Paulo (UNIFESP), São Paulo, São Paulo, Brazil

Camila Guindalini Department of Psychobiology, Universidade Federal de São Paulo (UNIFESP), São Paulo, São Paulo, Brazil

Katsumasa Hoshino Universidade Estadual Paulista, Bauru, São Paulo, Brazil

Sonia Regina Kameda Department of Psychobiology, Universidade Federal de São Paulo (UNIFESP), São Paulo, São Paulo, Brazil

Regiane Kawakami Faculty of Pharmaceutical Sciences, Universidade de São Paulo (UNIFESP), São Paulo, São Paulo, Brazil

Gui Mi Ko Department of Pharmacology, Universidade Federal de São Paulo, São Paulo, São Paulo, Brazil

Klaas Kramer Department of Health, Safety and Environment, Free University Amsterdam, Amsterdam, The Netherlands

Carlos Eduardo Antunes de Macedo Department of Psychobiology, Universidade Federal de São Paulo (UNIFESP), São Paulo, São Paulo, Brazil

Marla Simone Jovenasso Manzoni Faculty of Pharmaceutical Sciences, Universidade Estadual Paulista (UNESP), Araraquara, São Paulo, Brazil

Eduardo Ary Villela Marinho Departamento de Ciências da Saúde, Universidade Estadual de Santa Cruz, Ilhéus, Bahia, Brasil

Paulo José Forcina Martins Post-doctoral fellow in Metabolic Diseases, University of Cincinnati, Cincinnati, OH, USA

Edith Monroy-López Master in Psychobiology, Universidad Nacional Autónoma, Mexico

Luiz Eugênio Araújo de Moraes Mello Department of Physiology, Universidade Federal de São Paulo (UNIFESP), São Paulo, São Paulo, Brazil

Adriana Morales-Otal Departamento de Biología de la Reproducción, Universidad Autónoma Metropolitana, Iztapalapa, Mexico

Suzana Macedo de Oliveira Department of Biophysics, Universidade Federal de São Paulo (UNIFESP), São Paulo, São Paulo, Brazil

Ligia Assumpção Papale Department of Psychobiology, Universidade Federal de São Paulo (UNIFESP), São Paulo, São Paulo, Brazil

Camilla de Lima Patti Instituto de Genética e Erros Inatos do Metabolismo, Universidade Federal de São Paulo (UNIFESP), São Paulo, São Paulo, Brazil

Renata Pellegrino Department of Psychobiology, Universidade Federal de São Paulo (UNIFESP), São Paulo, São Paulo, Brazil

Juliana Cini Perry Technology Innovation, Curitiba, Paraná, Brazil

João Bosco Pesquero Department of Biophysics, Universidade Federal de São Paulo (UNIFESP), São Paulo, São Paulo, Brazil

Jorge Alberto Quillfeldt Department of Biophysics, Universidade Federal do Rio Grande do Sul (UFRGS), Porto Alegre, Rio Grande do Sul, Brazil

René Remie René Remie Surgical Skills Center, Flevoland, The Netherlands

Luciana Takahashi Carvalho Ribeiro Department of Pharmacology, Universidade Federal de São Paulo (UNIFESP), São Paulo, São Paulo, Brazil

Victor Proença Ricardo Universidade Paulista, Jacareí, São Paulo, Brazil

M. Jennifer Rodríguez Fisiología y Medicina del Sueño, Anáhuac Cancún, Cancún, Mexico

Elizeu Antônio Rossi Departamento de Alimentos e Nutrição, Universidade Estadual Paulista (UNESP), Araraquara, São Paulo, Brazil

Marcela Sene-Fiorese Institute of Physics of São Carlos, Universidade of São Paulo, São Carlos, São Paulo, Brazil

Andressa da Silva Department of Physiotherapy, Universidade Federal de Minas Gerais, Belo Horizonte, Minas Gerais, Brazil

Erick José Ramo da Silva Departamento de Farmacologia, Universidade Estadual Paulista (UNESP), Botucatu, São Paulo, Brazil

Christopher M. Sinton Department of Medicine, Arizona Respiratory Center, University of Arizona, Tucson, AZ, USA

Elizabeth Teodorov Centro de Matemática, Universidade Federal do ABC, Santo André, São Paulo, Brazil

Sergio Tufik Department of Psychobiology, Universidade Federal de São Paulo (UNIFESP), São Paulo, São Paulo, Brazil

Norma Lilia Anaya Vázquez Universidad Autónoma Metropolitana, Iztapalapa, Mexico

Javier Velázquez-Moctezuma Departamento de Biología de la Reproducción, Universidad Autónoma Metropolitana-Iztapalapa, Mexico

About the Editors

Monica Levy Andersen Master in Psychobiology (Universidade Federal de São Paulo – UNIFESP)
Ph.D. in Psychobiology (Universidade Federal de São Paulo – UNIFESP)
Postdoctoral Fellow – Division of Neuroscience Yerkes National Primate Research Center (Emory University), USA
Professor of Division of Sleep at Department of Psychobiology (Universidade Federal de São Paulo – UNIFESP)

Sergio Tufik Master in Physiology (Universidade Federal de São Paulo – UNIFESP)
Ph.D. in Pharmacology (Universidade Federal de São Paulo – UNIFESP)
Full professor of Division of Sleep at Department of Psychobiology (Universidade Federal de São Paulo – UNIFESP)

Part I
Ethics

Chapter 1
Introduction

Monica Levy Andersen

Humankind is, at the beginning of this millennium, in possession of a trove of scientific and technological achievements that until recently were deemed unimaginable. The multiple and profound consequences information technology has had on the scientific community and our own personal lives is remarkable. Due in part to what we now know about DNA and the entire genetic apparatus, we have seen great advances in molecular biology. Much of the developments in the field of biomedicine can be attributed to the use of laboratory animals. Because of these creatures we have been able to improve the well-being of our own species.

The close interaction between biomedical research and the use of laboratory animals occurs mainly because of the scientific knowledge we have of such animals, and the research creates this knowledge as well. The development of genetically well-defined strains of species that possess specific characteristics is particularly useful in the investigation of certain pathologies or physiological phenomena. What we have learned enables us to make better use of these animals and optimizes the work conducted by researchers, since results depend on a host of variables including how healthy the animals are, their care and handling, and their behavior. The well-being of laboratory animals must be a constant concern, and it is essential that those who make use of them be adequately trained.

Every ethical and conscientious society should be concerned with the way they handle any live species. Those who work with animals for experimentation should appreciate and value animal life and be responsible for their well-being. They should also consider these creatures as sensitive beings whose suffering should be minimized. Clinical trials and research performed on animals should be conducted in ways that minimize pain, distress, and discomfort.

M.L. Andersen (✉)
Department of Psychobiology, Chief of Sleep Division, Universidade Federal de São Paulo (UNIFESP), Rua Napoleao de Barros, 925, São Paulo, São Paulo, Brazil
e-mail: ml.andersen12@gmail.com

M.L. Andersen, S. Tufik (eds.), *Rodent Model as Tools in Ethical Biomedical Research*, DOI 10.1007/978-3-319-11578-8_1

The use of animals for experimentation must be justified. There must be a reasonable immediate or attainable benefit for mankind or for the species itself. Some criteria must be met as well: Humane treatment and protection should be dispensed to the animals, which means, in addition to minimizing pain, suffering, and discomfort, that only the necessary number of animals be used. The legislation in effect should protect not only the animals for experimentation but also those who work with them.

According to the "*Guidelines for Ethical Conduct in the Care and Use of Animals*," research should have the following objectives:

(a) To broaden the knowledge of the processes involved in the investigation, as well as to pursue better comprehension of the workings of each species;
(b) To determine whether a previous study can be reproduced;
(c) Supply results that benefit the health of humans or of other animals.

Carrying out research is not an easy task to complete, even more so when experimentation with animals is involved. Not only must the researcher master the technical aspects of the procedures, but also he or she must adhere to rigorous, self-imposed, ethical behavior, without which scientists would inevitably perpetrate irreparable damage. Each researcher must be aware of the etiology and biology of the species chosen for a given study and must be aware of the influence that his or her work and conclusions will have in the scientific community. The researcher must know that the information he or she gathered shall be used by many and that the information may help or harm people. It is therefore essential that the researcher be rigorously correct in each step of his or her work.

The ethical differences that arise when one resorts to animals for experimentation lie in the conflict between the justification of benefit for human society and the desire not to cause pain and suffering to animals. Ethical experimentation is defined as being directly beneficial to human and animal life, or, if not directly beneficial, then the research should significantly add to the understanding of the physiology and behavior of living beings.

The past 20 years have seen much debate over the validity of the use of animals for biomedical experimentation. The pros and cons against animal experimentation have been amply debated (Fig. 1.1). Opinions diverge greatly among animal protection

Fig. 1.1 Manifestato against the use of animals for experimentation (*source*: Foundation for Biomedical Research)

organizations, governmental agencies, and industries that make use of animals to test the reliability and efficacy of their products. In the course of these discussions, we must not forget that the use of animals in biomedical research is a privilege that must be carefully safeguarded so as to ensure the possibility that science will rid man and animals alike of their ailments. To ignore suffering of man and animals would be an irresponsible attitude. The vast majority of the advances in medical sciences can be attributable to the use of animals for experimentation.

According to the Nuremberg Code, established after the atrocities of World War II, all experimentation with humans "should be planned and based on results obtained from animal experimentation". The Nazis did not develop their experiments on animals. Rather, they made use of Jews and other minorities in a marginal fashion. The Helsinki Declaration, adopted in 1964 by the 18th World Medical Assembly and later revised in 1975, also emphasizes that medical research in human beings should be based on experiments using animals.

Despite the limitations the animal model faces when applications in humans are considered, it is still the best analogy researchers can find. Still, no theory can be demonstrated or refuted through mere analogy. Even though the animal model of disease is not identical to that of a human being, one must bear in mind that it was designed taking that fact into account. The animal model serves its purpose by offering grounds to investigate the applicability of a *given* procedure. Cystic fibrosis in mice does not mimic the condition encountered in humans, but it does offer insight as to what therapy would be satisfactory to improve a patient's health.

Man's curiosity has led him to pursue comprehension of the world surrounding him, which includes animals and plants. Rudimentary science gave way to objective, organized, empirical knowledge that we call scientific knowledge. Scientific discovery must tread the right paths and adopt the proper methodology. With time, as scientific data is amassed, the methodologies applied to obtain them must also adapt to the new challenges that arise along the way (Fig. 1.2).

In research, answers are sought by using models. A model is a representation of a specific reality that can be manipulated and that allows for the execution of analysis to make it better understood, as well as inference of data and testing of a hypothesis. Models can be physical, like the construction of a prototype or graphic

Fig. 1.2 Posters emphasizing the need to use animals for research (*source*: Foundation for Biomedical Research fbresearch.org)

representations such as drawings, anatomical atlases, and depictions of atomic, molecular, and cosmic structures. Models can also be mathematical when describing physical, chemical, biological, and even social phenomena. There are also biological models of live beings or parts of them, such as isolated organs or tissue or cellular culture, depending on what one wishes to investigate. Theoretically, biological models range from viruses to man himself. The results of this kind of research may be accepted as valid inasmuch as they are not contradicted by any other experimental result deemed correct. In other words, a hypothesis is accepted while the experimental data are corroborated; otherwise, it is no longer accepted and a new hypothesis must be elaborated that explains the experimental results.

In general, biological phenomena are more intricate than the physical and chemical ones. Among living beings, some phenomena can be more complex in a given species than in another. For this reason, the successful discovery of certain phenomena depends heavily on what model was adopted in that investigation.

Chapter 2
The Aspects of the Use of Rodents in Experimental Research

Vera Baumans

Mankind has long used animals for food, for transport and as companions. The use of animals in experimental research parallels the development of medicine, which had its roots in ancient Greece. Aristotle and Hippocrates laid down their knowledge on structure and function of the human body in their respective Historia Animalium and Corpus Hippocraticum, mainly based on dissections in animals. Galen (130–201 AD), physician of the Roman emperor Marcus Aurelius, performed physiological experiments on pigs, monkeys and dogs; these experiments provided the basis for medical practices in the centuries thereafter. After Galen, experimental science remained in a dormant stage until the beginning of the Renaissance when Vesalius popularized the empirical approach, starting with anatomical studies. Later on, physiological studies were performed as well. With the advent of Cartesian philosophy in the seventeenth century, experiments on animals could be performed without great moral problems. The French philosopher Rene Descartes (1596–1650) stated that living systems could be understood on pure mechanical principles. The difference between man and animals is that man has a mind, which is a prerequisite for awareness and consequently for the capability of feeling pain. Animals cannot think and are more like machines. However, Jeremy Bentham (1789) opposed Descartes' views: "The question is not, can they reason? Nor, can they talk? But can they suffer?" The discovery of anesthetics and Darwin's publication on the *Origin of Species* in 1859, defending the biological similarities between man and animal, contributed to an increase in animal experimentation. Claude Bernard published his book "*Introduction à l'étude de la médecine expérimentale*" in 1865, introducing methodology as a tool for the design of physiological experiments. The development of microbiology caused an increase in the use of animals, due to Koch's 'Postulates' where it is stated that the pathogenicity of a microorganism can

V. Baumans, D.V.M., Ph.D., Dip.E.C.L.A.M. (✉)
Division of LAS, Department of Animals in Science and Society, Utrecht University, PO Box 80166, Utrecht 3508 TD, The Netherlands
e-mail: v.baumans@uu.nl

M.L. Andersen, S. Tufik (eds.), *Rodent Model as Tools in Ethical Biomedical Research*, DOI 10.1007/978-3-319-11578-8_2

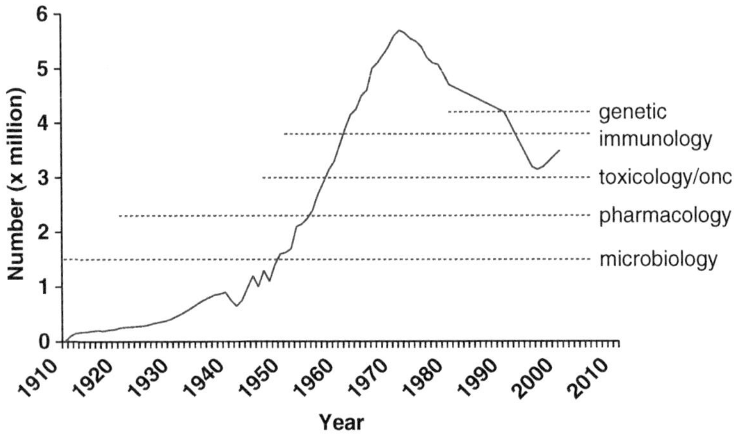

Fig. 2.1 Development of animal use in the twentieth century

be proven after successfully infecting healthy, susceptible animals (Van Zutphen 2001). The development of biomedical disciplines such as pharmacology, toxicology and immunology caused a sharp increase in the use of animals in the twentieth century. Since the early 1980s, animal experimentation has decreased due to public awareness; strict legislation regarding animal use; the development of animal ethics committees and improved animal quality (Fig. 2.1). However, in recent decades the use of animals started to increase again, mainly due to the development of genetically modified animals, which has caused a 23 % increase in the numbers of mice used each year. This increase reflects not only the animals used in research but also the large number of mice necessary to create each genetically modified line (breeding males, donor females, vasectomized males and pseudo-pregnant recipient females). Furthermore, non-transgenic and wild-type littermates may be produced that are not suitable for research or further breeding (Dennis 2002). Genetically engineered or modified mice are those with induced mutations, including mice with: transgenes; targeted mutations (knockouts); and retroviral, proviral or chemically induced mutations. Transgenic technology focuses on the introduction or exclusion (knockout) of functional genetic material in the germ-line of an animal, thus changing the genetic characteristics of an organism and its progeny. These techniques have led to the rapid development of a variety of animal models, designed for the study of gene regulation, gene expression, pathogenesis and the treatment of human and animal diseases (e.g., Alzheimer's disease, growth hormone disturbances, mastitis in cows, and poliovirus vaccine testing for eventual use in humans).

The experimental procedures involved in the process of transgenesis may compromise animal welfare. The donor animals, vasectomized males and foster mothers which are needed for the production of the transgenic offspring may experience discomfort from procedures, such as early mating (from 3 weeks onwards), anesthesia, surgery and injections.

From an ethical point of view, it can be argued that the integrity of the animal is compromised. Furthermore, concern has been expressed with respect to the patentability of transgenic animals, such as the oncomouse.

At the level of integration of the microinjected DNA into the genome, unintentional insertional mutations may occur, impacting the health of the animal. Furthermore, expression of the introduced gene may cause detrimental side-effects, e.g., the giant mouse with an overproduction of growth hormone, suffering from chronic kidney and liver dysfunction (Poole 1995).

Transgenic technology has great potential for increasing our understanding of the role of genes and may provide suitable animal models for human and animal disease. However, the welfare of transgenic animals must be carefully monitored, at least until the second generation of offspring. Score sheets can be helpful in correlating animal health problems with humane end-points in order to euthanize severely affected animals (Crawley 1998; Van der Meer et al. 2001). Data banks will be useful in providing data from existing genetically modified animals in order to predict possible health impairments in newly created genetically modified lines.

Today, 75–100 million vertebrates per year are used in research for a wide range of purposes. The major areas are drug research; testing of vaccines; and cancer research, whereas about 30 % of the animals are used for other purposes such as fundamental research, and diagnostics (Fig. 2.2).

Mice and rats are the most frequently used animal species (Fig. 2.3). In many European countries, it is mandatory by law to grade the level of discomfort experienced by the animals used in an experiment as minor, moderate and severe.

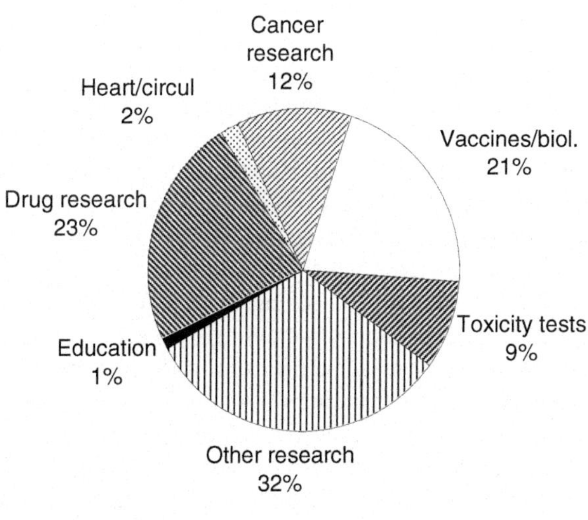

Fig. 2.2 Distribution of the purposes for animal use

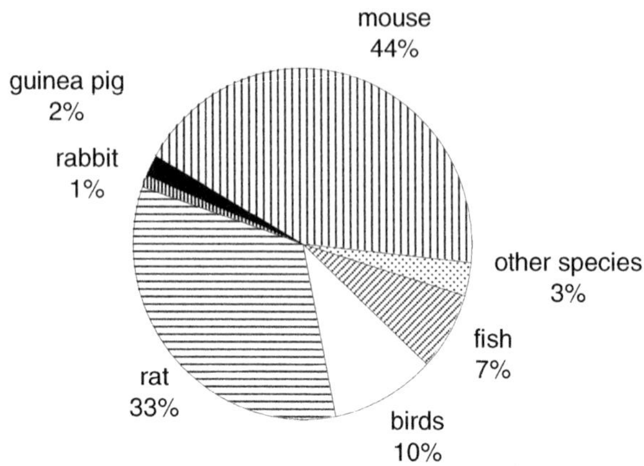

Fig. 2.3 Distribution of vertebrate animal species used for research, testing and education

On average, 50 % of the laboratory animals experience minor discomfort (e.g., single blood sampling), 30 % experience moderate discomfort (e.g., recovery from anesthesia) and 20 % experience severe discomfort (e.g., toxicity tests).

The increasing demand for high-quality animal models together with a critical view on the use of animals led to the development of Laboratory Animal Science in the fifties, a field that can be defined as a multidisciplinary branch of science, contributing to the quality of animal experiments and to the welfare of laboratory animals. Laboratory Animal Science encompasses the biology of laboratory animals and their environmental requirements, genetic and microbiological standardization, prevention and treatment of diseases, improvement of experimental techniques, anesthesia, analgesia and euthanasia, alternatives to animal experiments and ethics. The guiding principles are the three R's of Replacement, Reduction and Refinement, launched by Russell and Burch in 1959 in their book "The Principles of Humane Experimental Technique" (Russell and Burch 1959). **Replacement** means substituting the use of living animals with *in vitro* techniques (e.g., cells, tissues) or computerized models. **Reduction** means decreasing the number of animals used by standardizing the animal in terms of genotype and microbiological quality, as well as using uniform experimental procedures and environments, particularly in terms of food and climate in the animal room. Standardization of the animal cage is considered essential; laboratory practice involves a shoebox-shaped cage and standard bedding material. During the last decade, researchers have recognized the importance of an environment meeting the animal's species-specific needs, in terms of environmental enrichment and social housing, in order to improve the well-being of the animals. Furthermore, the number of animals per experimental group can be more accurately estimated by using statistics prior to the experiment, e.g., power analysis. **Refinement** means decreasing discomfort by translating the behavioral and physiological needs of the animal into adequate housing and husbandry; by providing adequate anesthesia, analgesia and care; by guaranteeing the skills of the researcher/animal staff, which can be achieved by education and training; by improving experimental procedures; and by determining a humane endpoint, when the animal can be euthanized to prevent unnecessary suffering.

Replacement: replaces the use of animals
Reduction: reduces the number of animals
Refinement: refines a method that minimizes pain and discomfort

Animal experiments have been subject to criticism ever since animals were used for research purposes. Criticism focuses in general on the ethical question of whether man has the right to use animals, as well as on the reliability and necessity of animal experiments. History has proven that reliability cannot be guaranteed. Results obtained in animals might not be reliably extrapolated to man. Despite a series of animal experiments and clinical trials in humans, the side effects of drugs may not be recognized due to extremely low incidence or may go undetected in animals (e.g., minor headaches or hallucinations). Furthermore, the necessity of certain animal experiments might be argued, such as in cosmetic testing, LD 50 tests, tests for military defense purposes, and teaching.

Undoubtedly, activities of the general public such as animal protection organizations have contributed to legislative regulations for the protection of animals used for experimental purposes. The first law, the Cruelty to Animals Act, was adopted by the UK parliament in 1876. Since then, other countries have included provisions in their laws to protect experimental animals. In Europe, two important documents controlling the use of animals in experiments were issued. In 1985, the Council of Europe (European Convention for the Protection of Vertebrate Animals used for Experimental and other Scientific Purposes) signed the *Convention for the Protection of Vertebrate Animals used for Experimental and other Scientific Purposes* (*ETS 123*). In 1986, the EU passed a European Council Directive, the *Directive for the Protection of Vertebrate Animals used for Experimental and other Scientific Purposes* (*86/609/EEC*), based on ETS 123, but more stringent. The Directive applies to vertebrate animals used in experiments likely to cause pain, suffering, distress or lasting harm, extending to the development of genetically modified animals at risk for pain and distress. The Directive contains provisions for the accommodation and care of experimental animals and for the competence of researchers and animal staff. It also lists alternatives to animal experiments, as well alternate forms of anesthesia and euthanasia. The document also contains statistics on animal experimentation and the supply of animals. In the US Animal Welfare Act (2002) the term "animal" means any live or dead dog, cat, monkey, guinea pig, hamster, rabbit, or other warm-blooded animal being used, or intended for use, for research, testing, experimentation, exhibition purposes, or as a pet. However, the term excludes birds, rats of the genus *Rattus* and mice of the genus *Mus*, bred for use in research. The guidelines for accommodation and care of laboratory animals are included in the Guide for the Care and Use of Laboratory Animals of the National Research Council, which also covers rats and mice (National Research Council 1996).

Although no specific provision in the EU Directive demands the establishment of animal ethics committees, such committees are operational in several European countries. These groups are specifically dedicated to reviewing the ethical aspects of animal experimentation. In the US, comparable Institutional Animal Care and Use Committees (IACUC's) are in operation. The task of such a committee is to perform an ethical evaluation of submitted research proposals. Animal experiments are considered acceptable only if the benefit of the proposed experiment outweighs the suffering of the animals. Ethical review of animal experiments will likely benefit the animals and improve the quality of animal-based research. As animal well-being is a prerequisite for reliable experimental results; it is of the utmost importance to seek methods and procedures that can reduce the suffering of animals and improve their well-being.

References

Crawley JN. What's wrong with my mouse? Behavioral phenotyping of transgenic and knockout mice. New York: Wiley; 1998.

Dennis Jr MB. Welfare issues of genetically modified animals. ILAR J. 2002;43(2):100–9.

European Convention for the Protection of Vertebrate Animals used for Experimental and other Scientific Purposes (ETS 123). Strasbourg: Council of Europe; 1986.

European Council Directive. Directive on the approximation of laws, regulations and administrative provisions of the member states regarding the protection of vertebrate animals used for experimental and other scientific purposes (86/609/EEC). 2006.

NRC (National Research Council). Guide for the care and use of laboratory animals. 7th ed. Washington: National Academy Press; 1996.

Poole TB. Welfare considerations with regard to transgenic animals. Anim Welf. 1995;4:81.

Russell WMS, Burch RL. The principles of humane experimental technique. London: Methuen; 1959. Reprinted by UFAW, 1992: 8 Hamilton Close, South Mimms, Potters Bar, Herts EN6 3QD England.

Van der Meer M, Rolls A, Baumans V, Olivier B, Van Zutphen LFM. Use of score sheets for welfare assessment of transgenic mice. Lab Anim. 2001;35:379.

Van Zutphen LFM. History of animal use. In: Van Zutphen LFM, Baumans V, Beynen AC, editors. Principles of laboratory animal science. Amsterdam: Elsevier; 2001. p. 2–5.

Chapter 3
The Impact of the Environment on Laboratory Animals

Vera Baumans

Housing systems for laboratory animals have often been designed on the basis of economic and ergonomic aspects (e.g., equipment, costs, space, work load, ability to observe the animals and to maintain a certain degree of hygiene) with little or no consideration for animal welfare. The environment of an animal consists of a wide range of stimuli, from the social environment (animals from the same or from other species, including humans) to the physical environment, such as the animal room (e.g., climate, illumination, noise) and the cage and its contents (Van de Weerd and Baumans 1995) (Fig. 3.1).

Environmental conditions such as housing and husbandry have a major impact on a laboratory animal throughout its life, not only during the experiment itself but also before and after the experiment. The traditional care and maintenance of laboratory animals does not usually consider the species-specific needs with regard to housing and feeding regimes. The variability in specific needs is not only different between species, but also due to variability in the genetic backgrounds among strains of the same species.

Laboratory animals have partially adapted to captive life, but still show similarities to their wild counterparts (Stauffacher 1995; Berdoy 2002; Baumans 2004). An animal's environment should cater to its innate physiological and behavioral needs, such as resting, nest building, hiding, exploration, foraging, gnawing and social contacts.

Most laboratory animal species such as rodents and rabbits—in addition to goats and sheep—are highly susceptible to predators and are likely to show strong fear responses in unfamiliar situations if they cannot shelter, as shown by attempts to flee, biting when handled or sudden immobility to avoid being detected. Careful handling from youth onward together with conditioning to experimental and

V. Baumans, D.V.M., Ph.D., Dip.E.C.L.A.M. (✉)
Division of LAS, Department of Animals in Science and Society, Utrecht University, PO Box 80166, Utrecht 3508 TD, The Netherlands
e-mail: v.baumans@uu.nl

M.L. Andersen, S. Tufik (eds.), *Rodent Model as Tools in Ethical Biomedical Research*, DOI 10.1007/978-3-319-11578-8_3

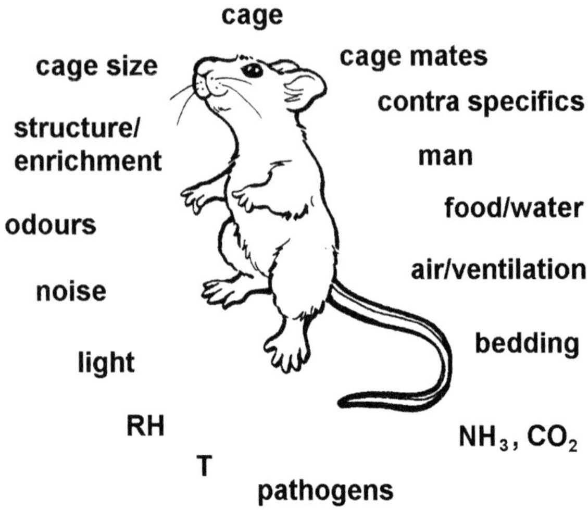

Fig. 3.1 The environment of a laboratory animal. *T* temperature, *RH* room humidity

husbandry procedures are likely to reduce stress responses considerably (Hurst 1999). For this reason, cages and pens should be provided with shelter or hiding places. Ideally, an animal should feel secure in a complex, challenging environment that it can control (Poole 1998). Security can be achieved by nestable and manipulable nesting material, hiding places and compatible cage mates.

Despite this, laboratory animals are usually housed throughout their lives in relatively barren cages and provided with *ad libitum* food, which frequently results in adverse effects on the animal's behavior and physiology, in addition to a shortened life span due to overfeeding and inactivity (Mattson et al. 2001). Standardization of environmental conditions serves to reduce individual differences within animal groups (intra-experiment variation), ultimately facilitating the detection of treatment effects, and to reduce differences between studies (inter-experiment variation), ultimately increasing the reproducibility of results across laboratories (Olsson et al. 2003), Still, it has been shown that despite rigorous efforts to equalize conditions among sites, different inbred mouse strains that originated simultaneously from three well-recommended laboratories showed significant effects resulting from their respective sites for nearly all variables examined (Crabbe et al. 1999; Wahlsten et al. 2003). Barren, restrictive and socially-deprived housing conditions interfere with the development and function of the brain and with behavior (Benefiel and Greenough 1998; Renner and Rosenzweig 1987; Würbel 2001), and restrictions such as those imposed by the standard rodent cage are potentially stressful (Ladewig 2000). Thus, the barren environment that has been designed to minimize uncontrolled environmental effects on the animals might ironically be a primary source of pathological artifacts.

Appropriate structuring of the cage/pen environment is typically more beneficial than provision of a larger floor area; however, a minimum floor area is necessary to provide a structured space. Except for locomotor activity (e.g., playing), animals do not actually use space. Instead, they use resources and structures within an area for specific behaviors. It is difficult to scientifically specify the minimal size of cages for maintaining laboratory animals, since much depends on the strain, group size and age of the animals, in addition to their familiarity with each other and their reproductive state. Cage sizes recommended in current European guidelines on accommodation for laboratory animals are preferably based on scientific evidence, but where this is lacking or insufficient, they are also based on best practice.

Environmental Enrichment

One possibility for improving the living conditions of laboratory animals is to provide opportunities for the animals to perform more species-specific behaviors by providing environmental enrichment. This can be defined as any modification in the environment of captive animals that seeks to enhance its physical and psychological well-being by providing stimuli meeting the animals' species-specific needs (Newberry 1995; Baumans 2000). Enrichment has increasingly been introduced into laboratory animal research facilities (Olsson and Dahlborn 2002). From a welfare point of view this seems to be a good development, as it is generally accepted that with the provision of environmental enrichment, an animal's well-being improves.

Beneficial effects of environmental enrichment have been described in animals with brain damage and disturbed motor function, and increased arborization of dendrites has been seen in the brain (Mohammed et al. 2002).

Enrichment of an animal's environment can focus on both the social environment (e.g., social partners, including human beings) and the physical environment, consisting of sensory stimuli (e.g., auditory, visual, olfactory and tactile) and nutritional aspects (e.g., supply and type of food). Furthermore, there is a psychological appraisal of the environment with regard to aspects such as controllability and predictability (Van de Weerd and Baumans 1995) that can be improved by structuring the cage with nest boxes, tubes, partitions and nesting material. Van de Weerd et al. (1997) showed that nesting material such as tissue was highly preferred by mice (Fig. 3.2).

For dogs and rabbits, a higher position in the enclosure such as a platform or shelf will meet their need for control of the environment (Stauffacher 2000; Hubrecht 2002), as depicted in Fig. 3.3.

Environmental enrichment should meet the animal's needs, be practical, be inexpensive, and pose no risk to humans, the animals, or the experiment. Enrichment items should be designed and evaluated based on knowledge gained in enrichment studies (Van Loo et al. 2005).

Suitable enrichment for rabbits includes at least roughage, hay blocks or chew sticks, as well as an area for withdrawal and look out such as a platform. For breeding does, nesting material and a nest box or another refuge should be provided. In floor pens for group housing, visual barriers should be provided.

Fig. 3.2 Paper nesting material as cage enrichment for laboratory mice. Photo: T.P. Rooijmans, Netherlands

Fig. 3.3 Platforms and toys as cage enrichment for laboratory dogs. Photo: Novo Nordisk A/S, Denmark

Nesting material is important for rats, mice, hamsters and gerbils as it enables them to create appropriate microenvironments for resting and breeding. Nest boxes or other refuges should be provided for guinea pigs and rats. Guinea pigs are cursorial rodents that do not burrow, but in the wild, they may live in burrows made by other animals. Refuges such as tubes or shelters should be provided within the cage or pen to allow the animal to climb onto or hide under them. Hay will satisfy the need for roughage, and wood sticks can be used for chewing and gnawing. In the wild, gerbils build extensive tunnel systems, and in the laboratory, they often develop stereotypic digging behavior unless provided with adequate facilities. For this reason, gerbils need comparatively more space in order to allow them to build or use burrows of sufficient size. They also require a thick layer of litter for digging and nesting and/or a burrow substitute, which may need to be up to 20 cm long. Nesting material (e.g., hay, straw, etc.) and wood sticks can be used for chewing and gnawing. The wild ancestors of the hamster were largely solitary, but group housing is possible. Special care should be taken in forming socially harmonious groups, and aggressive animals, especially females, should be separated. Minimum enrichment should include nesting material, a refuge area (e.g., tube, hut), roughage and gnawing objects.

In general, complexity will allow all animal species to structure their environment.

For all gregarious species, social housing should be provided and should only be denied in exceptional cases, e.g., extreme aggression in males from certain mouse strains. Being part of a compatible group provides a sense of security for the vast majority of *non-human primates*. It also provides opportunities for a whole range of species-specific social activities such as grooming, embracing, huddling, patting and kissing. Singly housed primates are particularly prone to show abnormal behavior, whereas keeping them in groups reduces the incidence of this behavior. Enclosures for primates should enable them to fully utilize the vertical dimension and should be housed in enriched environments, e.g., with swings, perches and branches, which allow them to carry out a normal behavioral repertoire. Puzzle feeders and foraging boxes are effective in reducing stereotypic behavior and increasing activity (Reinhardt 2002).

Mirrors might be useful to allow viewing events outside their cages.

Whenever possible, pigs and mini-pigs should be purchased in groups already formed of familiar or socially compatible animals. They should have permanent access to a sufficient quantity of material such as straw, hay, or sawdust to enable proper investigation and manipulation activities such as rooting. Hay is the supplement of choice for mini-pigs as it prevents gastric mucosal hyperkeratosis in addition to providing environmental enrichment. Food balls will be used for play and foraging.

Dogs should be held in socially harmonious groups or in pairs. Since chewing is an important behavior, items should be provided to meet this need. Dogs will make extensive use of such items, particularly if they are food-flavored. Proper presentation, for example by suspending chew items a few centimeters from the floor of the pen on a spring, can help to minimize cleaning and possessive aggression problems while allowing the animals to chew in a species-specific manner. Dogs extensively use platforms to play and rest and to allow easy viewing of events outside their enclosures (Hubrecht 2002).

There is concern, however, about whether or not environmental enrichment conflicts with the standardization of experiments. Standardization increases the reproducibility and comparability of experiments. It aims at reducing unwanted variation caused by animal and environmental factors and at reducing the number of animals needed in experiments. Some researchers fear that "enriched" animals show more variability in their response to experimental procedures because they show more diverse behavior. In complex environments, animals are not just responding to one stimulus in isolation but to many variable stimuli at once. This may cause increased variation among subjects (Eskola et al. 1999). On the other hand, one might argue that because an animal can perform more of its species-specific behaviors in enriched environments, it may be better able to cope with novel and unexpected changes, thus showing a more uniform response. Animals from enriched housing conditions are expected to be physiologically and psychologically more stable, therefore, they may be considered to be more refined animal models, ensuring better scientific results. Results from different studies seem to indicate that the effects of enrichment on the variability of results are dependent on the parameter, type of enrichment, strain and sex (Van de Weerd et al. 1997).

Environmental enrichment should comprise a well-designed and critically evaluated program that benefits the animals as well as the experimental outcome. It should not be a process of randomly applying objects that the staff consider attractive for the animals. Environmental enrichment should be regarded as an essential component of the overall animal care program and equally as important as nutrition and veterinary care. The key component of an enrichment program is the animal staff, whose members must be motivated and educated (Stewart and Bayne 2004). It is critically important to evaluate environmental enrichment in terms of the benefit to the animal by assessing the use of and preference for certain enrichment, the effects on behavior, on the performance of species-typical behavior, and on physiological parameters. At the same time, it is necessary to evaluate the impact on scientific outcome, how enrichment influences the scientific study, and whether and how the statistical power is affected. This will depend on the parameter measured, the type of enrichment used, and the animal strain (Van de Weerd et al. 1997; Baumans 2005).

Individually Ventilated Cage (IVC) Systems

The use of IVC systems began 30 years ago, and nowadays, these systems are frequently used, especially for housing transgenic rodents. Typically, each cage can be ventilated with 25–120 air changes per hour, with the air blown into the cage at a

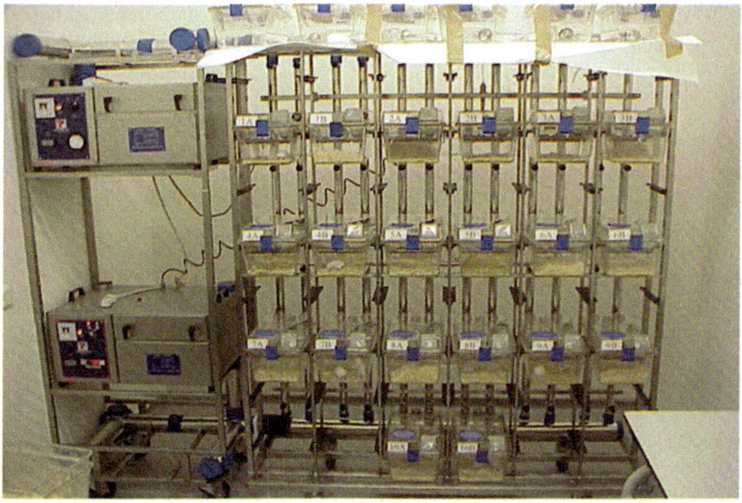

Fig. 3.4 Individually ventilated cage (IVC) rack. Photo: T.P. Rooijmans (Netherlands)

relatively high speed (Fig. 3.4). The advantages of the system are improved protection of the animals against microorganisms at the cage level, better protection of the animal staff against allergens, improved microclimate and reduced need for cage cleaning. Health monitoring and inspection of the animals may be difficult, however. Procedures and cage cleaning could also be more time-consuming, and the high intra-cage ventilation rate could induce chronic stress and heat loss by the draught (Baumans et al. 2002; Krohn 2002). It was shown that physiology and behavior of rats were not affected when the number of air changes per hour was kept below 80 [11]. In mice the location of the air supply in the cage, cage size, ventilation rate and presence of nesting material were important with regard to the impact on their well-being (Baumans et al. 2002).

References

Baumans V. Environmental enrichment: a right of rodents! In: Balls M, Van Zeller A-M, Halder ME, editors. Progress in the reduction, refinement and replacement of animal experimentation. Amsterdam: Elsevier; 2000. p. 1251–5.

Baumans V. The welfare of laboratory mice. In: Kaliste EK, editor. The welfare of laboratory animals—book series animal welfare and nutrition. Dordrecht: Kluwer Academic; 2004. p. 119–52.

Baumans V. Environmental enrichment for laboratory rodents and rabbits: requirements of rodents, rabbits and research. ILAR J. 2005;46(2):162–70.

Baumans V, Schlingmann F, Vonck M, van Lith HA. Individually ventilated cages: beneficial for mice and man? Contemp Top Lab Anim Sci. 2002;41(1):13–9.

Benefiel AC, Greenough WT. Effects of experience and environment on the developing and mature brain: implications for laboratory animal housing. ILAR J. 1998;39.

Berdoy M. The laboratory rat: a natural history. Oxford University; 2002. http://ratlife.org

Crabbe JC, Wahlsten D, Dudek BC. Genetics of mouse behavior: interactions with laboratory environment. Science. 1999;284:1670–2.

Eskola S, Lauhikari M, Voipio HM, Laitinen M, Nevalainen T. Environmental enrichment may alter the number of rats needed to achieve statistical significance. Scand J Lab Anim Sci. 1999;26:134–44.

Hubrecht RC. Comfortable quarters for dogs in research institutions. In: Reinhardt V, Reinhardt A, editors. Comfortable quarters for laboratory animals. 9th ed. Washington: Animal Welfare Institute; 2002. p. 56–64.

Hurst JL. Introduction to rodents. In: Poole T, editor. UFAW handbook on the care and management of laboratory animals. Oxford: Blackwell Science Ltd.; 1999. p. 262–73.

Krohn TC. Method developments and assessments of animal welfare in IVC systems. Thesis, The Royal Veterinary and Agricultural University, Frederiksberg; 2002.

Ladewig J. Chronic intermittent stress: a model for the study of long-term stressors. In: Moberg GP, Mench JA, editors. The biology of animal stress. Oxon: CAB International; 2000. p. 159–69.

Mattson MP, Duan W, Lee J, Guo Z. Suppression of brain aging and neurodegenerative disorders by dietary restriction and environmental enrichment: molecular mechanisms. Mech Ageing Dev. 2001;122:757.

Mohammed AH, Zhu SW, Darmopil S, Hjerling-Leffler J, Ernfors P, Winblad B, Diamond MC, et al. Environmental enrichment and the brain. In: Hofman MA, Boer GJ, Holtmaat AJGD, Van Someren EJW, Verhaagen J, Swaab DF, editors. Progress in brain research. Amsterdam: Elsevier Science BV; 2002.

Newberry RC. Environmental enrichment: increasing the biological relevance of captive environments. Appl Anim Behav Sci. 1995;44:229–43.

Olsson A, Dahlborn K. Improving housing conditions for laboratory mice: a review of "environmental enrichment". Lab Anim. 2002;3:243–70.

Olsson AS, Nevison CM, Patterson-Kane EG, Sherwin CM, Van de Weerd HA, Würbel H. Understanding behaviour: the relevance of ethological approaches in laboratory animal science. Appl Anim Behav Sci. 2003;81:245–64.

Poole TB. Meeting a mammal's psychological needs: basic principles. In: Shepherdson DJ, Mellen JD, Hutchins M, editors. Second nature: environmental enrichment for captive animals. Washington: Smithsonian Institution Press; 1998. p. 94.

Reinhardt V. In: Reinhardt V, Reinhardt A, editors. Comfortable quarters for laboratory animals. 9th ed. Animal Welfare Institute: Washington; 2002. p. 65–77.

Renner MJ, Rosenzweig MR. Enriched and impoverished environments. New York: Springer; 1987.

Stauffacher M. Environmental enrichment, fact and fiction. Scand J Lab Anim Sci. 1995;22: 39–42.

Stauffacher M. Refinement in rabbit housing and husbandry. In: Balls M, van Zeller AM, Halder M, editors. Progress in the reduction, refinement and replacement of animal experimentation, developments in animal and veterinary sciences. Amsterdam: Elsevier Science BV; 2000. p. 1269–77.

Stewart KL, Bayne K. Environmental enrichment for laboratory animals. In: Reuter JD, Suckow MA, editors. Laboratory animal medicine and management. Ithaca: International Veterinary Information Service; 2004. B2520.0404.

Van de Weerd HA, Baumans V. Environmental enrichment in rodents. In: Environmental enrichment information resources for laboratory animals. AWIC Resource Series. 1995;2: 145–9.

Van de Weerd HA, van Loo PLP, van Zutphen LFM, Koolhaas JM, Baumans V. Preferences for nesting material as environmental enrichment for laboratory mice. Lab Anim. 1997;31: 133–43.

Van Loo PLP, Blom HJM, Meijer MK, Baumans V. Assessment of the use of two commercially available environmental enrichments by laboratory mice by preference testing. Lab Anim. 2005;39:58–67.

Wahlsten D, Metten P, Phillips TJ, Boehm II SL, Burkhart-Kasch S, Dorow J, et al. Different data from different labs: lessons from studies in gene-environment interaction. J Neurobiol. 2003;54:283–311.

Würbel H. Ideal homes? Housing effects on rodent brain and behaviour. Trends Neurosci. 2001;24:207–11.

Chapter 4
Care and Maintenance of Laboratory Animals

Monica Levy Andersen, Vânia D'Almeida, Gui Mi Ko, Paulo José Forcina Martins, and Sergio Tufik

Theoretically, any organism can be used as a laboratory model for scientific investigation, but only a few species are chosen to serve in biomedical fields. To be defined as a laboratory animal, the species must be bred and raised under ideal conditions and kept in a rigorously controlled environment under constant monitoring so that all microbiologic and genetic factors are known.

Statistical surveys show predominance in the use of mice; 64 % versus 26 % for rats; thus these two species alone account for 90 % of laboratory animal usage. Inclusion of guinea pigs and rabbits brings this figure to 99 %. The remaining 1 % is composed of dogs, cats, and non-human primates, among others.

Rodents are widely used because of their low cost, high reproductive potential, short generation span, adaptability to varied environments and sociability. Much is known about rats and mice, and there are well-defined genetic strains that are free of the diversity found in wild species.

M.L. Andersen (✉) • S. Tufik
Department of Psychobiology, Universidade Federal de São Paulo (UNIFESP), Rua Napoleão de Barros, 925, São Paulo, São Paulo, Brazil
e-mail: ml.andersen12@gmail.com

V. D'Almeida, Ph.D.
Department of Psychobiology, Universidade Federal de São Paulo, Napoleão de Barros # 925, 3rd Floor, São Paulo, SP 04020-002, Brazil
e-mail: vaniadalmeida@uol.com.br

G.M. Ko
Department of Pharmacology, Universidade Federal de São Paulo, Rua Tres de Maio, 100, São Paulo, SP 04044-010, Brazil
e-mail: Guimiko.cedeme@epm.br

P.J.F. Martins, Ph.D.
Department of Internal Medicine, Vontz Center for Molecular Studies—University of Cincinnati, 3125 Eden Ave, Room 1200, Cincinnati, OH 45267, USA

M.L. Andersen, S. Tufik (eds.), *Rodent Model as Tools in Ethical Biomedical Research*, DOI 10.1007/978-3-319-11578-8_4

The use of experimental animals in the laboratory for biological investigation began with the study of "comparative pathology". Since human autopsies were forbidden, researchers studied animals. Thus, the origin and characteristics of the pathologies afflicting mankind were investigated in animals.

Experimental animals were thus considered simple "work instruments" that helped in the investigation of diagnosis and in research; sanitary and genetic considerations were disregarded. Animal centers did not have adequate facilities or trained personnel. Additionally, they lacked the appropriate chow and hygienic conditions that would allow the production of healthy animals. These conditions produced unreliable results.

Such conditions resulted in changes that took place between the end of the nineteenth century and the beginning of the twentieth century, and continue to this date. There have been enormous alterations not only in the way laboratory animals are bred and raised, but also in how specific lineages are obtained for specific purposes. Accumulated scientific knowledge, especially that of rats and mice, is now made readily available; such knowledge has made the analysis and interpretation of results more reliable. These data have become so broad and relevant that they now comprise a new, full-fledged field: Laboratory Animal Sciences. It is important to stress that knowledge of this field should not only be sought by animal facility caretakers but also by those who make use of laboratory animals in their work. It is therefore necessary to review the technological and scientific aspects of the care and maintenance of laboratory animals.

Animals are extremely sensitive to external conditions, and they modify their metabolism considerably to cope with variations in environment; this can adversely affect experiment results. Therefore, appropriate facilities, specialized equipment and adequate maintenance as well as qualified personnel are essential to maintain the well-being of the animals and the reliability of the results, as well as the safety of those involved in the study. A well-designed program of animal maintenance encompasses control over the environment and housing, and over every element allowing the animals to healthily grow and reproduce, thereby minimizing factors that might interfere with the results of the study. Good maintenance requires constant monitoring of animal health; additionally, the animal must be free to move about and if placed in the company of other species, both species must be compatible.

Another behavioral consideration receiving much attention is the requirement that humane treatment be dispensed to the animals. It must be emphasized that friendly and gentle handling of all laboratory animals is of great relevance to their well-being and good health. Animals should be respected as living beings that feel pain, hunger, thirst and fear.

Cages

The cage of a laboratory animal is more than merely a dwelling; it is the animal's world. The cage should therefore be of a size that allows the animal to move about and adopt typical postures. The material the cage is made of should be comfortable for the animal, easy to handle and have a good cost-to-benefit ratio. The most

Fig. 4.1 Conventional polypropylene cages of different sizes. The metal grid contains a slot for water and a food depository

commonly used cages are made of polypropylene. A good cage should have a sturdy structure, and internal and external surfaces should be easy to clean as well as resistant to hot water, detergents and disinfectants. In addition, the cage must be designed to facilitate maintenance and research maneuvers (Fig. 4.1). The shape of the cage has considerable influence on the concentration of ammonia, the most common and serious air contaminant in environments with high population density. Ammonia originates from the decomposition of excrement. Rigorous attention to hygiene and the prevention of over-population as well as attention to the number of cages in each room are important factors that can reduce ammonia concentrations; high levels of ammonia most typically cause respiratory infections. There is evidence that ammonia is linked to the exacerbation of respiratory complications, particularly of infections caused by *mycoplasm* in the rat.

Cages must offer comfort to animals, be appropriately ventilated, impossible to escape from, and produce little noise when being manipulated. A bedding should be spread over the floor of the cage, which should be made of wood flakes, rice straw, vegetal plywood or any other material that is not contaminated by insecticides that might ultimately alter research results. The purpose of the bedding is to absorb urine and water inside the cage in order to keep it dry. Bedding also has insulating properties that reduce the loss of body heat to the floor. The amount of bedding to be spread over the floor is an important consideration, since too little of it prevents females from being able to build nests and too much might lead to retention of excess heat. According to the Animal Facilities of the Fiocruz Laboratory, the amount of woodflake recommended for a mouse cage measuring 300 × 195 × 120 mm is about 60 g.

For rat cages measuring 410 × 340 × 175 mm, the amount is about 100 g. The bedding should be changed daily or once a week depending on the criteria of the animal facility. There are cases in which frequent bedding replacement is not advised, such as pre- and post-delivery periods.

Excessive bedding replacement may also interfere with identification and recognition among individuals. This process occurs through the innate smell of each species. The presence and production of pheromones is essential for reproduction and staking of territory. Frequent replacement of bedding stresses the animal, resulting in the production of excess pheromones and ammonia. However, not

Fig. 4.2 A cage fitted with a filter. The use of micro-isolators helps prevent animal contamination

Table 4.1 Recommended dimensions of cages holding rats and mice

Species	Dimensions (cm^2)	Minimum height (cm)
Rat		
Female with litter	800	18
In group (cm^2)	250–350	18
Mice		
Female with litter	160	13–15
In group (cm^2)	100	13–15

Source: Guide for the Care and Use of Laboratory Animals (1996)

replacing the bedding frequently enough brings the concentration of ammonia and other odors above tolerable limits, jeopardizing the study and working conditions for technicians. There has been a growing interest in the use of animals that are free of diseases. To obtain such animals, a set of physical, chemical, and working conditions must be established in the animal facility that safeguards animal health. One measure that contributes to the prevention of contamination is the installation of filters (micro-isolators) over the cages (Fig. 4.2). In this case, the cages are made of transparent polycarbonate so that the animals can be viewed. However, overpopulated animal facilities, an excessive number of animals in a single cage, and high temperature, among other factors, tend to invalidate the benefits offered by filters. Table 4.1 depicts the recommended dimensions of cages to be used for each species.

Fig. 4.3 Example of ventilated system and cage

Pressurized individual ventilation is another type of available cage that allows better housing conditions for rodents. This barrier housing has special filters to remove microorganisms and particles from the air. Pressurized individual ventilation cages protect both the housed animals and working personnel by filtering the supply and exhaust air circulation. Direct exhaust from the cage provides enhanced ventilation efficiency; as a result, temperature and humidity are controlled (Fig. 4.3). The bedding stays drier, reducing ammonia levels and extending the interval between changes without the need for air inlets at the animal level (Borrello et al. 2000).

Identification of Cages and Animals

The correct identification and registration of every event are parts of the daily work of an animal caretaker. Such reports demand great responsibility from the caretaker and researcher alike. The identification of each cage should contain all relevant information that enables the correct execution of the study. The identification tags of experimental animals should bear the name of the department, institution, number of cages and animals, birth date of the animal, species, lineage, sex, initial weight, date of experiment commencement and planned termination, and restrictions or special treatment. Moreover, the project title and researcher names should be written on the tag as well, and if needed, the name of the supplier should be provided.

In situations requiring that the animal be identified, a marking should be made on a visible part of the animal. The marking substance should not be harmful to the animal. Individual markings can be temporary or permanent.

Several substances can be utilized in concentrated solutions, including picric acid (yellow), fuchsin (purple), trypan blue (blue), and ethyl green. For application, a bundle of cotton should be soaked with the substance of choice and gently rubbed bidirectionally into the fur to reach deep into the lower layers. It is preferable to use a small quantity of a strong dye rather than a large quantity of a weak dye. Fur marking is not a long-lasting method, and the markings may have to be renewed at times. In animals with dark fur, such markings may not be easily identifiable, so it is recommended that markings be made on the abdominal region, where the fur is lighter.

Although permanent markings are still accepted in some countries, their use is restricted for ethical reasons. Ear piercing, even when performed under anesthesia, may result in bleeding and mutilation of the animals. Before using this method for animal identification, the auricular vestibule should be sterilized. In adult animals, anesthesia is recommended. The procedure entails choosing a region on the auricular vestibule that does not contain many blood vessels and piercing that patch of tissue.

The piercing instrument can be a perforator or a small straight-ending pair of scissors. Tattoos are often used for marking rabbits. The tattoos are numbers or letters made using inks such as nankeen on the inner surface of the auricular vestibule. Earrings can be made of plastic with numbers or letters pre-printed on them. Recently, electronic devices have been implanted subcutaneously for identification. Each animal is identified through an electronic detector that can also be programmed to gauge temperature. Although this is an efficient and reusable technique, its high cost is still a hindrance.

Ventilation

Ventilation should deliver an adequate supply of fresh air that in addition to providing oxygen also clears unpleasant odors, especially ammonia produced by urease-positive bacteria present in the feces and urine. Ventilation also clears the air of pollutants released from equipment and cage bedding, dilutes gaseous contaminants and helps balance temperature and humidity. Ventilation is of considerable importance in rat facilities due to this species' sensitivity to respiratory diseases. Ventilation should reach all cages and cover the entire room containing the animals. A good ventilation system should be maintained and constantly monitored for noise levels, air conditioning and exhaust. It is recommended that the air be renewed (100 % fresh air) 15–20 times an hour in animal rooms (Clough 1982). Under other circumstances, a greater flow rate may be necessary. Poor ventilation and over-population are the main factors causing respiratory problems and contamination of odors from different species.

Humidity and Temperature

The regulation of body temperature is necessary for the well-being of homeothermic organisms. Such animals control their body temperature within short spans of time by varying their metabolic rate in response to environmental conditions. Several facilities now have controlled atmospheres that maintain the environment under constant, ideal conditions. Such air-conditioning guarantees the uniform distribution of air and temperature with appropriate filters to prevent the introduction of microorganisms.

Humidity is the amount of water vapor in the atmosphere. It is expressed as a percentage, and the water contained in a certain sample of air is called the relative humidity. These factors also play an important role in animal well-being. For rats and mice kept at 20–24 °C, the recommended relative humidity is 45–55 %. It should be noted that the temperature may increase by 3–5 °C inside the cage, and humidity can be elevated as well.

Sudden variations in temperature and humidity can cause stress and greater susceptibility to infection, resulting in respiratory problems for animals kept in high humidity or low temperature conditions. Temperature and humidity variation can also alter food and water intake, jeopardizing the results of the study. The maintenance of humidity and temperature is thus essential for experimental reproducibility. Finally, relative humidity and room temperature also influence the concentration of ammonia inside the cage and affect the thermal balance of the animal.

The exposure of non-adapted animals to high or low temperatures without shelter or any other means of protection induces adaptive measures such as peripheral vasoconstriction, piloerection and enhanced metabolic activity. Such measures can have detrimental effects such as decreased reproduction. Animals that are kept under such conditions also show hunched posture, nest building initiative and augmented food intake. In special cases such as the habituation of very young animals (after weaning) or care of animals devoid of fur, the room temperature should be increased. In general, recommended temperatures for animals of larger size are closer to the minimum accepted temperature.

The sleep of animals is also altered by temperature. Schmidek et al. (1972) showed that when the room temperature was kept under 24 °C, rats slept in a curled up position. In contrast, rats tended to stretch out fully when sleeping at 30 °C (Fig. 4.4). When temperatures are over 30 °C or under 24 °C, animals tend to sleep less, and wakeful periods are increased. This reduction of sleep occurs at the expense of paradoxical sleep; the time of synchronized sleep remains reasonably constant. Acclimatizing rats to sleep to low temperatures (14 °C) and then having them sleep at higher temperatures (24 °C) showed that paradoxical sleep continued to be partially suppressed, demonstrating how thermal regulation can interfere in the programming of paradoxical sleep.

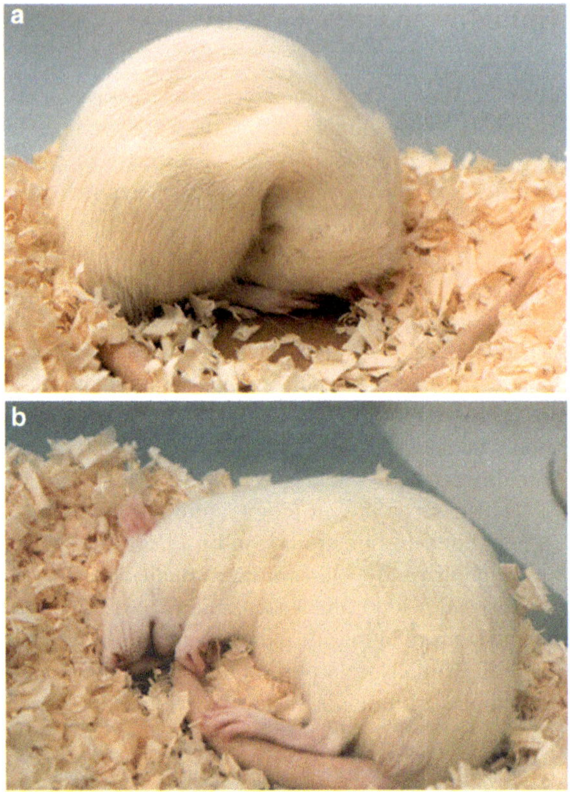

Fig. 4.4 Association between temperature and sleeping position. At low temperatures, the rat curls up to sleep (**a**). At high temperatures, the rat stretches out to sleep (**b**)

Light

Light must also be under rigorous control since it is one of the factors that most influences biological rhythms, behavior, and animal reproduction in the laboratory. Light has three important aspects: wavelength, duration of the photoperiod, and photo-intensity (expressed in Lux).

Light can affect the physiology, morphology and behavior of animals. Inappropriate light is a potential stress factor. Thus, variations in luminosity should be taken into account when building animal facilities or when preparing tools for experiments. Other factors related to light include: duration to exposure, anima background, skin complexion, hormonal condition, age, species and sex.

Albino animals are more sensitive to high intensity light even if they seem to be comfortable to the persons who handling them. Most laboratory animals cannot distinguish colors (except for cats, birds and some primates), but are adapted to see in dim light conditions. The length of light exposure is a critical factor in the regulation of the reproductive behavior of many species and may also alter body weight and food intake.

Due to their inability to see long wavelengths, red light has some advantages when there is a need to visually inspect the animals over long periods of time.

An adequate light system should allow periodic control of timers that correct for seasonal variations in luminosity. Animal facilities should not have windows, allowing control of luminosity to be entirely made through the use of artificial light. Animal adaptation to shifts in the light-dark cycle requires 2 weeks. It is notable that cold light is less irritating to animals that incandescent light.

It is recommended that the intensity of light inside the cage should not be over 60 lx, and in experimental rooms the intensity should be between 350 and 400 lx at 1 m above the floor. These values are satisfactory for routine and care of animals and do not cause phototoxic retinopathy in albino animals. Young rats prefer ambiences with low light intensity (130–325 lx). The intensity of light inside the cage is considerably influenced by its localization on the shelf. The location on the shelf where the cage is placed is positively correlated to the incidence of damage or atrophy of the retina associated with exposure to light in albino mice.

Noise

There are three sources of noise in environments inhabited by animals:

- Devices and equipment;
- The execution of tasks inside the rooms (e.g. cages being cleaned);
- The noise produced by the animals themselves.

The intensity and frequency of sound are measured in decibels (dB) and Hertz (Hz), respectively. The intensity at which the noise arrives in the human ear or in the ear of an animal is related to the distance from the source of noise.

Most animal species have superior hearing capabilities, thus it is likely that some of the noise produced in the rooms in which they are maintained will disturb them. What a rat hears is significantly relevant to its social behavior. Noise produced by animals associated with courtship, maternal care, aggression and defense are a necessary part of their communication; such communication may be thwarted by environment noise.

Animals may be able to adapt to some continuous noise. Loud, long lasting and high pitch sounds are the most disturbing. Unexpected noises or alterations in noise intensity may cause stress, metabolic alteration and decreased fertility.

The usual intensity of noise in rodent facilities is 50 dB. Environments with noise levels over 85 dB may cause increased adrenal gland weight and decreased fertility. Ideally, experiment rooms should be kept apart from the animal facility to minimize possible disturbances to the animals.

Transport

The transport of an animal from one room to another should be done with minimal disturbance to its well-being. Vibration of cages during transport bothers the animals. Cages should be properly latched so that no escape occurs. It is important to

Fig. 4.5 Transport of cages containing animals. The cage should be kept snugly between the caretaker's arms to keep it from shaking or being dropped. Observe the water spout turned upwards

remember to lift the waterspout to keep the animals and bedding from getting wet (Fig. 4.5). Animals transported in groups must have enough room in their cages to move about freely.

When transport involves moving an animal from its home cage to another cage, caution must be exercised in order to prevent exposure to daylight, rain and wind.

Food

Adequate nutrition provides laboratory animals with the means to reach their reproductive and growth potential, to enhance their longevity, and to respond to stimuli. All animals must be fed uncontaminated feed that is nutritionally balanced to the species' needs. Such care ensures that necessary nutrients will be made available in a constant manner.

The diet of laboratory animals is pelletized. This form of delivery reduces waste and facilitates handling and mastication by the animal. Feed can also be extruded, a processing method in which pellets are pressure cooked in steam. This expansion method alters the texture of the feed and makes it more palatable but requires more room in the feed container.

A well-balanced diet will include carbohydrates, protein, salts, vitamins and other essential components. The energetic needs of an animal depend on its

Table 4.2 Measured amounts of water and feed consumed daily by adult rats and mice

Species	Ingested water	Ingested feed
Rats	10–20	10–20
Mice	3–7	4–5

Source: Guide to the Care and Use of Experimental Animals (1980)

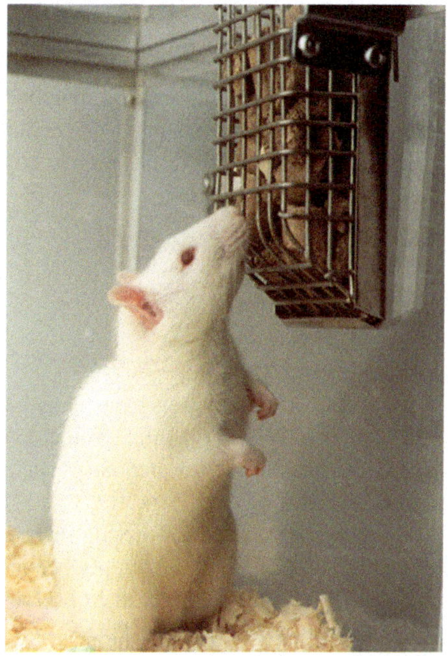

Fig. 4.6 Rat eating food. The feed container should be located in the upper part of the cage. Usually, the fed should not be spread over the floor of the cage because there is the risk of contamination and waste

physiological condition, physical activity and environmental temperature. The average daily food and water requirement of an adult rat is between 10–20 g of feed and 20–45 mL of water. Consumption (intake amount + waste) is 25 g and 50 mL, respectively. In mice, the amounts are smaller at 3–6 g of feed and 3–7 mL of water (Table 4.2).

Pregnancy, nursing, and growth augment these needs. Whenever possible, pasteurized or sterilized food purchased from trustworthy suppliers should be used. Feed supplementation is not advised since it carries not only the possibility of chemical or biological contamination, but also nutritional variation or ingestion of inadequate amounts of a given nutrient.

Storage of feed requires rigorous control. The storage quarters must be dry, cool, dark and well-ventilated so as to minimize the possibility of contamination and spoilage.

The containers in which the food is stored must be easy to wash and disinfect (Fig. 4.6). Usually, the food should not be spread out on the floor of the cage because there can be contamination and waste. The amount of nutrients and the presence of contaminants in the diet of rodents are often neglected experimental variables.

Standardized diets that are semi-purified or chemically purified are frequently used for carcinogenic studies, toxic substance assays and nutritional experiments. Whenever such diets become necessary, the use of an agar-gel base tends to promote weight loss and increases animal longevity, in addition to providing for excellent growth, minimized loss, and reduced risk of cross-contamination between the exposed caretaker and the animal. Mice and rats that are immunologically deficient or germ-free (free of pathogenic agents) must be fed sterilized diets.

The natural coprophagia of the rat may considerably affect the influence of diet on experimental results. Approximately 50–65 % of the fecal output of rats under adequate diets is re-ingested. This behavior can be enhanced by poor diet. The use of wire mesh floors does not impede coprophagia; rats will go to great lengths to have access to their feces.

Hydration

It is imperative that animals have ready access to water at all times, unless otherwise required in the experimental protocol (such changes require Ethics Committee approval). The need for water varies according to animal species, diet, and environmental condition.

It is advised to use a watering method where the dissemination of diseases or contamination of the water supply becomes unlikely. Water bottles should be clear and transparent to allow ready observation of cleanliness and water level. The watering containers should be made of a material that will tolerate sterilization and should feature a wide mouth for easy cleaning (Fig. 4.7). Watering bottles should be replaced constantly with clean, freshly-filled bottles, rather than refilling the bottles in use. Frequent bottle replacement hampers the development of microorganisms. All watering devices require frequent cleaning and periodical monitoring to prevent bacterial contamination.

Animals respond to physical, chemical and biological factors that can alter their health and well-being (Chap. 3). Thus, it is necessary to control these factors to increase the validity of experimental results.

The advent of new technologically advanced materials has allowed some structural alterations. Now, laminar flow, perfectly insulated shelves with air-flow control, constant temperature, adequate luminosity and the use of boxes or cages called microisolators are commonplace. Although these developments increase the cost of animal care, they improve the living conditions of the animals and make results more accurate and reliable while requiring fewer animals in each study. Furthermore, enforcement of the ethical use of animals has become more rigorous.

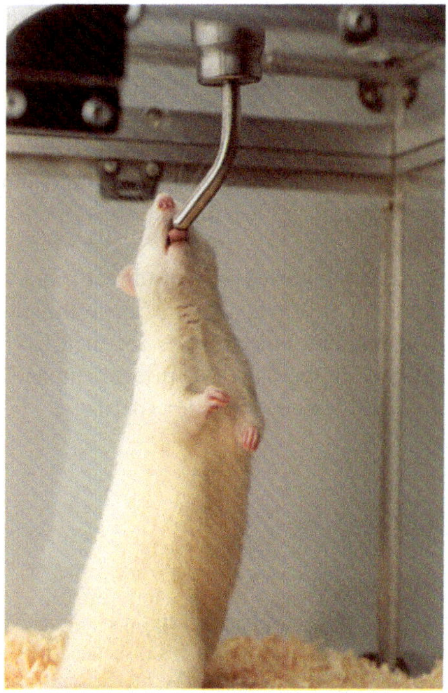

Fig. 4.7 Rat drinking water. Stainless steel should be used for the spout and cork housing

Animal Manipulation

Animals react differently to external stimuli and require a determined period to adapt themselves to environmental conditions. They should thus be handled with care, and all external factors should be taken into account in order to safeguard the well-being of the animal and the reliability of the results.

Handling should be firm and gentle so as to have a calm and docile animal. Most laboratory animals do not require special consideration for handling and seem to enjoy contact with people. Beginners under training should first use animals that have already been manipulated so that they can develop a sense of confidence and acquire a feel for the minimum strength required to safely hold and restrain a given species. Well-executed manipulation also requires that the caretaker have the ability to identify the state of the animal. Such states include being alarmed, aggressive, uncomfortable or in pain. Sudden movements and loud noise can result in respiratory and circulatory alterations as well as a state of alarm alterations.

Normally, animals will eventually become habituated to the person who cleans their cage. During the execution of an experiment, it is advised that the regular caretaker not be changed. Rodents, especially rats, enjoy being in groups. Contact with members of its species may include visual, hearing and olfactory signs.

Social contact can reduce the effects of stressful situations, abnormal behavior and cognitive stimuli. The insertion or removal of animals from their cages results in an additional effort to establish new groups. Factors such as population density, propensity for dispersion, and facility for adaptation should be considered when working with animals that live in groups. If the experiment requires complete isolation of the animal, it is recommended that other forms of distraction be provided so as to compensate for the absence of other members of the same species.

Wearing gloves prevents direct contact with urine and feces and impedes the dissemination of contaminants once the gloves are discarded after use. Door knobs, counters and other objects are then free of contamination. Such care reduces the hazards to the health of researchers and caretakers, and to all persons directly or indirectly involved in the experiment.

References

Andrade A. Animais de Laboratório: criação e experimentação/Organizado por Antenor Andrade, Sergio Correia Pinto e Rosilene Santos de Oliveira. Rio de Janeiro: Fiocruz; 2002. p. 387.

Anver MR, Cohen BJ. Lesions associated with aging. In: Baker HJ, Lindsey JR, Weisbroth SH, editors. The laboratory rat—biology and diseases. New York: Academic; 1979. p. 378–99.

Borrello P, D'Amore E, Panzini G, Valeri M, Lorenzi RN. Individually ventilated cages—microbial containment testing. Scand J Lab Anim Sci. 2000;27:142–52.

Broderson JR, Lindsey JR, Crawford JE. The role of environmental ammonia in respiratory mycoplasmosis of rats. Am J Pathol. 1976;85:115–30.

Clapp MJL, Bradbrook C. Growth and longevity of rats fed an agar-bound diet. Lab Anim. 1982;16:138–42.

Clough G. Environmental effects on animals used in biomedical research. Biol Rev Camb Philos Soc. 1982;57:487–523.

Da-Silva JB. Biossegurança em experimentação animal: um enfoque microbiológico. Niteroi: Universidade Federal Fluminense; 1998. p. 93.

De-Luca RR, Alexandre SR, Marques T, Souza NL, Merusse JL, Neves SP. Manual para técnicos em bioterismo. 2nd ed. São Paulo: Winner Graph; 1996. p. 259.

Felasa. Recommendations for the health monitoring of mouse, rat, hamster, guinea pig and rabbit breeding colonies. Lab Anim. 1994;28:1–12.

Fox GJ, Cohen BJ, Loew FM. Laboratory animal medicine. American College of Laboratory Animal Medicine series. 2nd ed. New York: Academic; 1984.

Gamble MR, Clough G. Ammonia build-up in animal boxes and its effects on rat tracheal epithelium. Lab Anim. 1976;10:93–104.

Krinke GJ, editor. The laboratory rat. San Diego: Academic; 2000. p. 634.

National Research Council—Institute of Laboratory Animal Resources, Committee on Rodents. Guide for the care and use of laboratory animals. Washington: National Academy Press; 1996. p. 125.

Nayfield KC, Besch EL. Comparative responses of rabbits and rats to elevated noise. Lab Anim Sci. 1981;31:386–90.

Neale RJ. Coprophagy in iron-deficient rats. Lab Anim. 1982;16:204–7.

Poole TB, editor. The UFAW handbook on the care and management of laboratory animals. 6th ed. London: UFAW, Churchill Livingstone; 1987. p. 635.

Saiz-Moreno L, Garcia de Osma JL, Fernández C. Animales de laboratorio: producción, manejo y control sanitario. Madrid: Instituto de Investigaciones Agrarias/Ministerio da Agricultura, Pesca y Alimentación; 1983.
Schmidek WR, Hoshino K, Schmidek M, Timo-Iaria C. Influence of environmental temperature on the sleep-wakefulness cycle in the rat. Physiol Behav. 1972;8:363–71.
Sharp PE, La Regina MC. The laboratory rat. Boca Raton: CRC Press; 1998.

Chapter 5
Euthanasia

Monica Levy Andersen and Vera Baumans

According to the International Guiding Principles for Biomedical Research Involving Animals (CIOMS 1985), because of differing legal systems and cultural backgrounds, there are varying approaches to the use of animals for research, testing, and training in different countries. Nonetheless, the use of animals should at all times be in accord with humane practices. The varying approaches in different countries to the use of animals for biomedical purposes, and the lack of relevant legislation or of formal self-regulatory mechanisms in some countries, point to the need for international guiding principles that are the result of international and interdisciplinary consultations.

The term "*euthanasia*" (Greek for good death) is used to describe the process by which the animal is killed using humane techniques. Euthanasia implies a death without suffering, fear, or anxiety. *The most important criterion in the acceptance of a method of euthanasia is the depressive effect of the central nervous system (unconsciousness) to ensure the immediate absence of sensibility to pain.* It is important to recognize that some methods of euthanasia, although they may not seem to be very elegant (such as decapitation), are considered to be compatible with the criterion above. This concept must be kept in mind when opting for a method of euthanasia. The choice should be grounded on the highest of moral standards and on refined ethics, focusing on the well-being of the animal and not the researcher's sensitivity, although this too should not be overlooked. Much effort has gone into the development of acceptable euthanasia methods. In spite of this, divergent views

M.L. Andersen (✉)
Department of Psychobiology, Chief of Sleep Division, Universidade Federal de São Paulo (UNIFESP), Rua Napoleao de Barros, 925, São Paulo, São Paulo, Brazil
e-mail: ml.andersen12@gmail.com

V. Baumans, D.V.M. Ph.D. Dip.E.C.L.A.M.
Division of LAS, Department of Animals, Science and Society, Utrecht University, PO Box 80166, Utrecht 3508 TD, The Netherlands
e-mail: v.baumans@uu.nl

M.L. Andersen, S. Tufik (eds.), *Rodent Model as Tools in Ethical Biomedical Research*, DOI 10.1007/978-3-319-11578-8_5

continue to exist between scientists and public opinion that raise the question whether these procedures are truly humane.

On June 12–13, 2004, an International Council for Laboratory Animal Science (ICLAS) Working Group on Harmonization met in Nantes, France to examine and make recommendations for the adoption of international guidelines on euthanasia. The working group agreed on general principles for euthanasia and recommended two documents as international references, namely the *2000 Report of the AVMA Panel on Euthanasia* and the *1996/1997 EC Recommendations for Euthanasia of Experimental Animals Parts 1&2*. Although there are some inconsistencies between these two documents, ethics committees can use the general principles to evaluate whether the given conditions of the situation meet the goal of assuring humane death. The working group also identified areas of inconsistency between the documents, indicating where there is insufficient knowledge. The research and animal welfare community should actively promote applied research towards performance-based standards concerning laboratory animal euthanasia methods identified under Section C. These efforts could produce much needed refinements and could contribute scientific evidence that would, in turn, support common principles to be incorporated in future versions of guidelines on euthanasia. In order to provide clarity, the working group recommends that any new or revised guidelines on euthanasia should report only the acceptable humane euthanasia methods and the conditions under which they are considered acceptable, as well as the life-stage of the concerned species.

The following guidelines are founded on generally accepted interpretations of present research in this field:

General Principles

1. Whenever an animal's life is to be taken, it should be treated with the highest degree of respect.
2. Euthanasia should place emphasis on making the animal's death painless and distress-free. Therefore, the method likely to cause the least pain and distress to the animal should be used whenever possible.
3. Euthanasia techniques should result in rapid loss of consciousness, followed by cardiac or respiratory arrest and ultimate loss of brain function.
4. Techniques should require minimum restraint of the animal and should minimize distress and anxiety experienced by the animal prior to loss of consciousness.
5. Techniques used should be appropriate for the species, age, and health of the animal.
6. Death must be verified following euthanasia and prior to disposal of the animal.

7. Personnel responsible for carrying out the euthanasia techniques should be trained:
   (a) To carry out euthanasia in the most effective and humane manner;
   (b) To recognize signs of pain, fear, and distress in relevant species;
   (c) To recognize and confirm death in relevant species.
8. Human psychological responses to euthanasia should be taken into consideration when selecting the method of euthanasia, but should not take precedence over animal welfare considerations.
9. Ethics committees should be responsible for approval of the method of euthanasia (in line with any relevant legislation) for any study involving the use of animals. This should include euthanasia as part of the experimental protocol as well as euthanasia for animals found to be experiencing unanticipated pain and distress.
10. A veterinarian experienced with the species in question should be consulted when selecting the method of euthanasia, particularly when little species-specific euthanasia research has been done.

International Reference Documents

The ICLAS working group on harmonization supports the use of the following documents as international references:

1. **2000 Report of the AVMA Panel on Euthanasia** (**2001**), ***Journal of the American Veterinary Medical Association 218*(*5*), *669–696*.**
   This document was prepared at the request of the American Veterinary Medical Association Council on Research by a Panel on Euthanasia that convened in 1999 to review and make necessary revisions to the fifth Panel Report, published in 1993. In the 2000 report, the panel updated information on euthanasia of animals in research and animal care and control facilities, expanded information on ectothermic, aquatic, and fur-bearing animals, added information on horses and wildlife, and deleted methods or agents considered to be unacceptable.
2. **Recommendations for euthanasia of experimental animals Part 1** (**1996**), ***Laboratory Animals 30*, *293–316*** **and Recommendations for euthanasia of experimental animals Part 2** (**1997**), ***Laboratory Animals 31*, *1–32*.**
   These documents were prepared for DGXI of the European Commission to be used with Directive 86/609/EEC of November 24, 1986 and discuss the approximation of laws, regulations, and administrative provisions of the Member States regarding the protection of animals used for experimental and other scientific purposes (No L 358, ISSN 0378-6978). It refers especially to Article 2(1), published by the European Commission in October 1995, which defines 'humane method of killing' as 'the killing of an animal with a minimum of physical and mental suffering, depending on the species'.

The ICLAS Working Group on Harmonization recommends that institutions carefully consider any new information on specific euthanasia techniques that may refine methods supported in the above documents in order to improve welfare outcomes of the animals.

Of note, the person in charge of the animal's killing should be apt to execute euthanasia in a safe and skillful manner. He or she must master the methods, avoiding unnecessary discomfort and distress to the animal. We must not forget that animal killing produces psychological effects in people. There is the feeling of pity, sadness, and regret, making the adoption of ethical and moral standards imperative. Promising studies are under way that will lead to the improvement of these methods in the future.

Choice of Method

It is frequently necessary to kill animals that are used in experiments. It is essential that researchers and technicians be familiar with euthanasia methods and that they be able to apply them efficiently so as not to cause suffering in respect to the animal's well-being. The choice of method will certainly be influenced by the particularities of the experimental protocols, such as the nature of the study, the animal species, the number of animals to be killed, and the speed of the method. Bear in mind that the adopted method may interfere with the final result of the study. In some cases, it might be necessary to kill each animal individually. In other instances, animals may be grouped together for mass euthanasia. Regardless of whether it is individual or mass euthanasia, the following criteria must always be met:

- death without signs of panic, pain, distress, or anguish;
- death through a method that is fast and easy to apply;
- minimal time for loss of consciousness;
- trustworthy, reproducible, and irreversible method;
- method is safe for those who execute it;
- method cause minimal emotional effect on the researchers;
- method produce minimal undesired physiological and psychological effects;
- compatible with the requirements and purpose of the study;
- minimal environmental impact through contamination occurs;
- the room where euthanasia is carried out be apart from the animal care facilities and experimental rooms;
- imposed limitations for the species to be killed.

Methods of Euthanasia

Physical Methods

The physical methods of euthanasia should cause immediate loss of consciousness.

This physical procedure is adopted when other methods may invalidate relevant data, especially those related to the biochemistry of the animal. For example, manipulation of an animal or its exposure to either chloroform or ether will cause an increase of corticosterone and catecholamines while beheading does not. Whenever chemical substances in the blood and tissue enzymes are the focus of an investigation, a physical method will often be adopted.

The physical methods of euthanasia are:

- Cervical dislocation
- Brain trauma
- Decapitation
- Electrocution
- Hypothermia (instant freezing)
- Exsanguination
- Decompression

Recently, immersion in liquid nitrogen and exposure to microwave radiation have been introduced as means of physical euthanasia for some species, but only for neonates without fur. Microwave is only used for special cases such as fixation of the brain in a certain state. It is not an easy procedure as the animal must be positioned precisely under the beam. Among the unacceptable methods are: head torsion, gaseous embolism, and the use of agents that cause hypoxia (the occurrence of convulsions before the loss of consciousness).

Cervical dislocation is carried out in mice and rats of up to 200 g. The technique consists of breaking the spinal cord by pressing the posterior base of the skull with a metal instrument against a firm surface. Simultaneously, the body of the animal is pulled along the tail axis. When breaking occurs, there may be considerable muscular activity. Blood continues to be supplied to the brain because the carotid arteries and the jugular veins are intact at this moment, but death is immediate. Studies demonstrate that corneal reflex disappears immediately after the separation of the spinal cord. Therefore, the painful stimuli cannot be transmitted and the animal has total loss of sensibility. Devices for cervical dislocation of mice, rats, other small species, and rabbits are commercially available. This method may cause the rupture of cervical and thoracic blood vessels, causing hemorrhage. Of note, the head should be destroyed by immersion in liquid nitrogen, for example.

Brain trauma (stunning) is carried out especially in rabbits, rats, and guinea-pigs. This is a method that should not be witnessed by people who are not involved in the study because it is unaesthetic and unpleasant. The method consists of striking the base of the skull with sufficient force to cause the immediate depression of the central nervous system. As mentioned earlier, those who perform the procedure must be highly capable so as to cause immediate loss of consciousness and sensibility to pain.

Exsanguination consists of either cardiac or major vessel puncturing so that total loss of blood occurs. It is normally done in small rodents and rabbits and the animals must be put under anesthesia prior to the procedure.

Fig. 5.1 Guillotine used for decapitation. Only experienced and dexterous personnel should be allowed to operate it

Decapitation is employed when the experimental protocols do not allow for the interference of other agents (Fig. 5.1). Although decapitation is aesthetically questionable, it is an extremely fast and efficient method, inducing immediate disappearance of corneal reflex and causing minimal physiological alterations in the tissues. However, the blood collected after decapitation may contain fur and salivary and respiratory secretions and animals may inhale blood into the trachea.

Chemical Methods

Non-inhalant Pharmacological Agents

The administration of non-inhalant pharmacological agents is a fast and reliable method. Those agents that are not inhaled and that are used for euthanasia include: barbituric acid derivatives, barbituric mixtures, and ketamine. The pharmaceutical agents or non-inhalant drugs are normally administered intravenously. Administration via methods other than intravenous will often result in delay of the onset of the anesthesia and prolong the time it takes to perform euthanasia. Barbiturates by intraperitoneal administration are irritating to the peritoneum, and then dilution might be better. Therefore, the dose is doubled or even tripled to ensure success. In these circumstances, it is particularly important that the animal be placed in a box to keep it form falling off of the workbench. A specialist should be consulted every time the above mentioned drugs will be administered in ways other than those recommended.

Pentobarbital sodium (200–250 mg/kg) and other barbiturate agents are those most often chosen for euthanasia for both aesthetic and scientific reasons. They must, however, be used for a reduced number of animals. Intravenous administration is the method of choice due to the rapidity with which it leads to death. However, if that method will cause discomfort to the animal, intraperitoneal administration (IP) is justifiable. Pentobarbital sodium solutions can be made 2–4 times as powerful as other agents and combinations with other agents are also possible. Unconsciousness is produced in the animals and because of the excess drug this state is deepened, ultimately leading to death. Intravenous and intraperitoneal administration of pentobarbital results in congestion and spleen swelling as well as in mild to moderate pulmonary hyperemia.

Chloral hydrate and ketamine hydrochloride are dissociate anesthetics; therefore, there is no loss of corneal reflex in the anesthetic state. In combination with other drugs, it can be used for the humane sacrifice of laboratory animals. Despite being costly, an overdose of ketamine hydrochloride is the option of choice of many researchers for the individual sacrifice of mice because it does not result in detectable, post-mortem alterations.

Use of curariform agents, which include succinylcholine, D-tubocurarine, organophosphates, strychnine, and nicotine sulfate, for euthanasia is prohibited. They do not depress the central nervous system. Rather, they act at neuro-muscular junctions and the animal dies of asphyxiation caused by paralysis of the respiratory muscles. When such substances are used, the animal may be conscious and subjected to agonizing pain until hypoxia of the brain supervenes. Hypoxia develops rapidly once the animal is paralyzed by a curariform drug.

> Excessive doses of common tranquilizers are necessary to produce a lethal dose. Many are commercially available and may even be administered subcutaneously, and intramuscularly. However, they should not be used for euthanasia as they are relatively inefficient and produce pharmacological effects only at large doses.
> Chloral hydrate is not recommended due to its painful properties.

Inhalant Pharmacological Agents

Inhalant anesthesias include ether, chloroform, halothane, methoxyflurane, and nitrous oxide and are useful agents for euthanasia. In some cases, the drawback is the possible occurrence of struggle associated with the excitement stage of anesthesia. Chloroform and ether are relatively inexpensive while halothane, methoxyflurane, and nitrous oxide are costly. Special anesthesia devices are commercially available for the use of these inhalants and, although they are not essential, they are useful and reduce leaking and contamination of the environment. Ether promotes irritation of airway mucosa, excessive salivation, and increased bronchial secretion.

Flammable agents such as ether should not be used in the presence of a flame or where electrical appliances are not discharge resistant. When chloroform and other volatile inhalant gases are used it is necessary to have adequate ventilation.

Chloroform has a toxic effect in the liver, kidneys, and testicles. Furthermore, it is considered hepatoxic and cancerous for man; therefore, it is recommended that its use be discontinued.

Non-anesthetic gases may also be used for euthanasia. These include carbon monoxide (CO), carbon dioxide (CO_2), nitrogen, and cyanide. The use of the first three listed above depends on the replacement of the available oxygen in a container, by the gas, and should be used with restrictions since they are not safe for personnel. Carbon dioxide and nitrogen are relatively innocuous to the environment but carbon monoxide and cyanide (requiring a purchase license from a law enforcement agency) are extremely toxic substances and may be fatal. If they are used, adequate air-exhaustion and special precautions should exist. The chamber used for euthanasia should be well-ventilated so as to provide for rapid replacement of air by the gas used, which is particularly necessary when CO_2 is used. It is recommended that the chamber be pre-saturated with the gas so that the method is rapid and without any visible anguish of the animal. When CO_2 is inhaled at a concentration of 40 % or over, it depresses the central nervous system and has rapid lethal properties. It is important to ensure that all animals have expired to avoid undesired situations of recovery and suffering of the animal.

Carbon dioxide has been widely used in the destruction of rodents. This is a gas that may be purchased in cylinders or in a solid state such as dry-ice. It is reasonably inexpensive, non-flammable, non-explosive, and is an essentially safe gas as long as it is used properly. Carbon dioxide poses no risk to those who work with it when used with the adequate equipment such as chambers. Carbon dioxide is heavier than air and it concentrates in the lower portion of the chamber. Rats and other animals that have the habit of burrowing will tend to keep their snouts in the lower areas where the concentration of gas is adequate to cause an irreversible narcotic effect. It is recommended that 30 % oxygen be mixed to avert the unpleasant effects of anxiety when hypoxia sets in.

Carbon monoxide at low concentrations (2 %) will quickly cause death by triggering an irreversible alteration in eritrocitary hemoglobin, with consequent paralysis of the cardiac and respiratory centers. The time needed to cause death varies according to the period of exposure and concentration. Vocalization often occurs, particularly in dogs and cats, and may last for 3–5 s before the animals fall into unconsciousness. This fact alone suggests that additional research should be conducted to ensure that animals do not suffer during that period.

Carbon monoxide from exhaust pipes of vehicles is smoky and those impurities cause irritation and discomfort. If such is the case, the exhaust gases should be filtered and cooled. Before making use of internal combustion engines to obtain a supply of carbon monoxide, researchers should ensure that the system is adequately fitted with a filtering apparatus that removes irritating substances from the exhaust gases and that the amount of carbon monoxide produced is enough to reach the desired concentration. It is also essential that protective measures be taken to prevent the exposure of other animals and people to highly toxic gases.

Pregnant women may miscarry or bear congenitally abnormal offspring if they are constantly exposed to inhalant anesthetics.

Nitrogen is an inert gas used as a rubberizing agent to remove oxygen. Hypoxia leads to unconsciousness followed by death as a consequence of the inactivity of the respiratory center and anoxic brain. However, unlike carbon dioxide, nitrogen does not produce narcosis.

Oxygen should be reduced (replaced for nitrogen) to 1.5 %, a concentration that will cause the animal loose its senses. It has been observed that cats, dogs, and rabbits vocalize at this moment and present muscular activity and struggling. The animals recover quickly if the anoxic conditions are suspended.

Cyanide is a powerful paralyzing agent that blocks the transport chain of electrons. It is one of the fastest acting poisons and there is no evidence that its effects are painful. Death is instantaneous. Its immediate action and irreversibility in causing anoxia with depression of the central nervous system guarantee fast and humane death. Still, convulsions occur before death, which is anguishing for the observer. *It is important to highlight the risk that cyanide poses to personnel and other animals of accidental exposure to this fatal gas. For this reason, cyanide is not recommended for general use in the laboratory.*

At the conclusion of euthanasia, personnel should always examine the animal to certify that it is no longer breathing and is indeed dead.

Evaluation of Pain and Distress

The evaluation of pain and distress of an animal that is being sacrificed requires that the person in charge of performing euthanasia be educated on the behavioral and physiological responses of the animal. Observation of subjective, abnormal behavioral and physiological responses in animals that demonstrate anxiety and fear must be certain. Such responses in conscious animals include anguish, restlessness, fighting, escape, defensive action or paralysis, muscular trembling, pupil dilatation, salivation, urinary and defecation reflex, increase in heart and respiratory rates, transpiration, and tachycardia.

Evaluation of Unconsciousness

Observation of the corneal reflex to help gauge unconsciousness, and hence the capacity to feel pain, is an efficient but neglected means. This reflex is evaluated by touching either the eye-lid or the cornea of the animal. The absence of blinking indicates unconsciousness and, therefore, insensitivity to pain (except when using curariform drugs that merely paralyze bone muscles or anesthesias using either chloral hydrate or ketamine hydrochloride, etc.). Chloral hydrate is typically used in euthanasia because its recovery is slow and prolonged. The absence of the corneal reflex is used to evaluate the depth of surgical anesthesia.

It is necessary that cardiac function be halted as quickly as possible so that blood stops flowing to the brain. It is imperative to verify the absence of a heartbeat before the corpse is discarded to ensure that recovery is impossible. The mere absence of muscular activity should not be used as criteria for unconsciousness or insensitivity to painful stimuli, although prolonged absence of movement indicates that the animal is dead.

Algor mortis (the cooling of a corpse), *rigor mortis* (the stiffening of a corpse), *livor mortis* (the appearance of stain patches on the corpse), ocular alterations, blood clotting, autolysis, and putrefaction are alterations that appear on the animal's corpse some time after its death and are referred to as cadaver alterations. One to 2 h after death, algor mortis settles and is followed by rigor mortis, which takes hold 3–4 h after the animal's death. Rigor mortis remains for 24 h, on average, until the first signs of putrefaction appear. Stiffness occurs firstly in the eye-lids, then, in the maxilla, neck, and finally in all of the muscles in the corpse. The disappearance of rigor mortis occurs in the reversed order of its onset.

Effects on Tissue Produced by Euthanasia Methods

In most cases, death is so fast that even electromicroscopic alterations are not observed. Researchers are concerned with the possibility of visible histologic alterations, but these do not normally happen. There are cases, however, in which euthanasia may bring about possible effects on physiological and biochemical parameters that can be gauged in assays of tissue (as, for example, the barbiturate effect on hepatic microssoma enzymes for use in cellular culture) and may have some influence on histologic or electromicroscopic findings.

Recommendations for Research Where Scientific Knowledge Is Lacking

Issues that still need further research include mass animal euthanasia, euthanasia of fetuses and neonates, euthanasia of cold-blooded animals, proper use of CO_2 according to species, decapitation with or without prior anesthesia, cervical dislocation, use of N_2 and/or argon gas.

References

Andrade A. Animais de Laboratório: criação e experimentação. Organizado por Antenor Andrade, Sergio Correia Pinto e Rosilene Santos de Oliveira. Fiocruz: Rio de Janeiro; 2002. p. 387.

Breazile JE, Kitchell RL. Euthanasia for laboratory animals. Fed Proc. 1969;28:1577–9.

Canadian Council on Animal Care (CCAC). Guide to the care and use of experimental animals, vol. 1. Ottawa: CCAC; 1984. p. 120.

De-Luca RR, Alexandre SR, Marques T, Souza NL, Merusse JL, Neves SP. Manual para técnicos em bioterismo. 2nd ed. São Paulo: Winner Graph; 1996. p. 259.

Hughes HD. Euthanasia of laboratory animals. In: Melby Jr EC, Altman NH, editors. Handbook of laboratory animal science. Cleveland: CRC Press; 1976.

Kim SU. Brain hypoxia studied in mouse central nervous system cultures I. Sequential cellular changes. Lab Invest. 1975;33:658–69.

Krinke GJ, editor. The laboratory rat. San Diego: Academic; 2000. p. 634.

Part II
Use of Rats

Chapter 6
The Health of Laboratory Animals

Monica Levy Andersen, Vânia D'Almeida, Gui Mi Ko, Paulo José Forcina Martins, and Sergio Tufik

The health of an animal comprises its physical and psychological well being, in addition to its genetic integrity. It results from the interaction of the animal with other live organisms and its physical environment. Organized and methodical daily observation of the behavior and general appearance of an animal is a way of generally evaluating its health, even without the use of deep clinical and laboratory exams.

An animal gives signs of its health and well-being. The fur should be homogeneous, shiny, and free of patches and wounds. The eyes are of fundamental importance in evaluating the health of rats and mice. Healthy animals have shiny, humidified eyes that show vivacity under normal conditions (Fig. 6.1). It is important that the airways (nose, pharynx, trachea, bronchia, bronchioles and lungs) be free of nasal drip and bleeding. The behavior displayed by the animal must also be evaluated. One must watch for any sign of discomfort, pain, length of time spent lying down, physiological signs such as the presence of pus, and increase in abdominal volume.

M.L. Andersen (✉) • S. Tufik
Department of Psychobiology, Universidade Federal de São Paulo (UNIFESP), Rua Napoleao de Barros, 925, São Paulo, SP, Brazil
e-mail: ml.andersen12@gmail.com

V. D'Almeida, Ph.D.
Department of Psychobiology, Universidade Federal de São Paulo, Napoleão de Barros # 925, 3rd Floor, São Paulo, SP 04020-002, Brazil
e-mail: vaniadalmeida@uol.com.br

G.M. Ko
Department of Pharmacology, Federal University of São Paulo, Rua Tres de Maio, 100, São Paulo, SP 04044-010, Brazil
e-mail: Guimiko.cedeme@epm.br

P.J.F. Martins, Ph.D.
Department of Internal Medicine, Vontz Center for Molecular Studies—University of Cincinnati, 3125 Eden Ave, Room 1200, Cincinnati, OH 45267, USA

M.L. Andersen, S. Tufik (eds.), *Rodent Model as Tools in Ethical Biomedical Research*, DOI 10.1007/978-3-319-11578-8_6

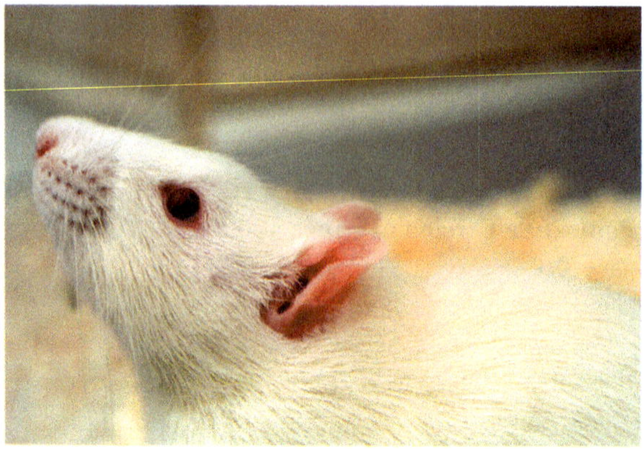

Fig. 6.1 Healthy rat. Notice the shine and the vivacity in the eyes, and the homogeneous, shiny fur

Physical Examination of Animals Includes

Body Temperature

The average body temperature of the rat is between 35.9 and 37.5 °C. Temperature gauging is done through an ear-drum thermometer or a colon or rectal thermal sensor.

Weight

Weight loss can be an indicating factor of poor health. It can also occur after experimental procedures, such as those that investigate stress and augmented motor activity.

Eyes

The occurrence of chromodacryorrhea (red tearing) is a sign of distress or suffering in rats and might be a pathologic manifestation. Because of grooming, the reddish aspect around the eyes may be transferred to the front paws. Ocular tearing may occur if irritating gases, such as ammonia, are released by hygienic products used in environmental sanitation. If the bedding is not changed frequently or if the ventilation is not sufficient, then the ammonia released by urine may cause these effects. If such substances are allowed to persist, the animal's condition tends to worsen with the formation of pus.

Chromodacryorrhea is not caused by blood but by a secretion that contains a red pigment, called porphyrin. This pigment is released through the "C" shaped Harderian glands located around the facial part of the ocular globe. When the rat is

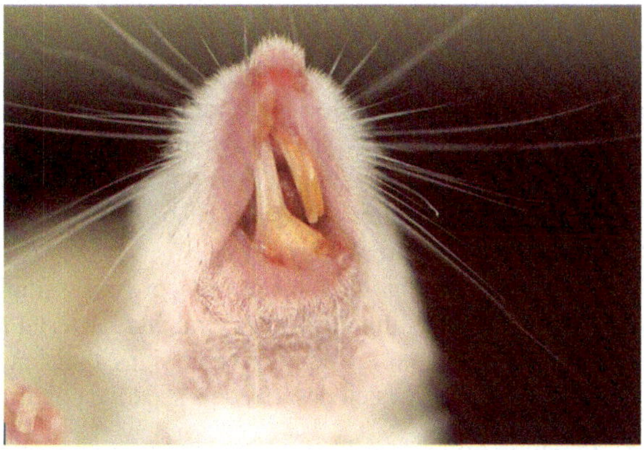

Fig. 6.2 Poor occlusion of the incisive teeth in a rat

distressed or sick, these glands excrete porphyrin, which results in a periocular and nasal secretion.

Ears

The ears should be examined for abnormalities such as swellings close to the base, which may be caused by tumors.

Teeth

The appropriate length of the incisive teeth is kept constant due to the grinding of solid foods or of rigid surfaces (such as the cage wires). Excessive growth of teeth may lead to faulty occlusion (Fig. 6.2). Poor occlusion is associated to the fracture of teeth and misalignment of the maxilla and the mandible. Clinical signs include weight loss, dehydration and deformed appearance. As it is a hereditary trait, it is suggested that these animals not be used for breeding new members of the colony. Liquid diets or diets in the form of fine flakes may lead to the overgrowth of animal teeth.

Circulatory System

The workings of the circulatory system are evaluated by an assessment of habitual behavior, signs of tiredness and unwillingness, coloring of the mucous and skin and resistance to performing physical exercises. It is possible to listen to a rat's heart and lungs with a pediatric stethoscope.

Respiratory System

Rats are very susceptible to a wide variety of respiratory disturbances. The diseases afflicting the respiratory system have their origins in either allergy or infection and are accompanied by sneezing. If infection is the cause, the bacteria become airborne encapsulated in droplets of nasal mucous and saliva, contaminating other animals and possibly the entire colony.

Mycoplasma pulmonis is the primary organism implicated in chronic respiratory infection in the rat, although other viral and bacterial agents may also be involved. The syndrome is expressed in several signs that may arise separately or jointly, some of which have been described as separate conditions, including:

- Infection of the middle ear, which induces rotation when the rat is stimulated by its tail;
- Rhinitis, with sneezing and the existence of blood stains around the nostrils;
- Pneumonia, with difficulty breathing and progressive debility.

M. pulmonis may also infect the genital tract, particularly of females. When present in this form in mating colonies, it is the primary cause of fertility reduction, reduced litter size or even complete infertility. Fortunately, *M. pulmonis* has been largely eliminated, at least in the larger mating colonies. Still, care must be exercised because *M. pulmonis* is still present in several rat colonies. Regular monitoring for *M. pulmonis* and other viral agents is an essential precaution for an animal care facility.

Urinary and Reproductive System

One of the most serious infections is that of the urinary system. It is caused by a bacterium called *Leptospira* sp., and once renal lesion occurs there is little chance for survival. Laboratory animals may become contaminated by wild rodents that find their way into the animal care facility. Besides the loss of the colony, this is one of the diseases that can be transmitted to those who handle the sick animals. Most frequent contamination occurs when the rodent's urine comes into contact with wounded human skin. This fact shows the importance of wearing gloves, washing one's hands after handling rats and mice, and neither smoking nor eating inside animal care facilities.

Nervous System

Some diseases may be located in the nervous system of animals. An example is poliomyelitis, which causes paralysis, absence of motor coordination, rotation around its own body and death.

General Conditions

Body: the swelling of any part of the animal's body may indicate the existence of a benign or malign tumor.

Skin: the skin and its annexes are the best cues to animal health. The skin should be elastic, humid, smooth, and have a pink complexion, or the typical color of its lineage or species. Rodents are very sensitive to ectoparasites that can cause fur to become less dense or less shiny, and can cause itches and wounds. Abnormal fur may also be due to distress, cold, unbalanced nutrition, allergic processes, or parasites. Whatever the case, it usually involves groups of animals, and is rarely observed in isolated animals.

Although spontaneous tumors have developed in many lineages of rats, particularly those of advanced age, the incidences across lineages are poorly documented. In the case of mice, breast tumors are the most common. In some inbred lineages, the incidence of breast tumors is about 50 %.

It is not only environmental factors that affect the well being of animals. The technician's conduct, personal hygiene and health may also alter the health of animals.

Adequate veterinary care and adherence to strict experimental procedures can generally avert the transmission of diseases from animals to humans It is important to assure that all animal caretakers be healthy and have normal resistance to infection. Special attention must be given to any allergic reactions the animals may cause. Serum, urine and other animal tissue products may trigger allergic responses.

In order to minimize the risk of problems associated to allergic reactions induced by laboratory animals it is advised to:

- Request that individuals who have a history of susceptibility to allergens take a skin test for the species involved prior to being hired. It is also helpful to minimize or avoid contact altogether by wearing personal protective suits.
- Require that purpose-made lab coats for the specific application of handling laboratory animals be worn regularly. These suits should be put on and taken off outside the holding rooms and the same suits should not be worn for experiments on other animal species.
- Wear gloves.
- Wear face masks and/or protective goggles whenever necessary.
- Wash hands before and after the manipulation of any animal. This will reduce the risk of disease dissemination and of autoinfection. To facilitate hand washing, it is very helpful to have a sink provided with soap and paper towels.
- Not to smoke, eat nor drink in any maintenance, experimental and holding rooms or any other places where there has recently been manipulation of pathogenic microorganisms.
- No person who bears an open wound or eczema may be allowed to conduct work where there is a risk of contacting pathogenic microorganisms, unless this risk is adequately addressed by the use of protective garments.

Determining the appropriate level of biosecurity for work involving a determined agent or for animal experiments depends on countless factors. The most relevant of these are the virulence, biological stability, pathogenicity, propagation, endemicity and communicability of the agent, the nature and function of the laboratory, the procedures and manipulation involving the agent, and the existence of a vaccine or effective therapeutic measures against it (Da-Silva 1998).

The cages that contain animals inoculated with infectious substances should be adequately identified. Whenever a pathogenic material is injected into an animal the caretaker should wear gloves. Gloves should also be worn when manipulating feed and removing infected animals.

- Every disposable material used in animal care and handling (gloves, face-masks, paper towel) should be discarded in purpose-made plastic bags, adequately identified and positioned for specialized sanitary collection.
- The necropsy of animals infected with highly contagious agents should be carried out in safe, ventilated quarters. The waste material from the necropsy should be conditioned into purpose-made plastic bags, and once proper identification is made, forwarded to incineration. Necropsy quarters should be adequately equipped so as to allow for refrigeration and cleansing of the room.
- In installations that are not provided with the services of a veterinarian, it is advised that contact be established with qualified professionals. These professionals should preferably be veterinarians or experienced laboratory animal caretakers who have an interest in the well being of the animals.

Several papers on animal diseases should be readily available to those who care for the health of the colony of experimental animals. It is not the purpose of this book to detail symptoms or to offer specific instruction on the diagnosis and treatment of such diseases, but it is essential for the well being of the animals and the reliability of the experimental results to detect these ills early, prevent their progression, and free the colony of the organisms that cause them are.

Good personal hygiene is another important way of protecting the health of experimental animals.
The more uniform and the homogeneous experimental animals are, the fewer animals are needed.

In order to prevent mistaken experimental results induced by environmental differences, the animals are classified according to their sanitary (or ecological) condition. The selection of mice lineages according to their microbial ecology is a relatively recent refinement in the production and use of laboratory animals. Still, it is no less important and useful than genetic definition. Ecological classification is defined as the relation of the animal with its specific environment, which includes the organisms associated to the animal and those present within the physical limits of the quarters and sanitary barriers. This body of organisms is dubbed the microbiota (viruses, bacteria, fungi and parasites).

Ecological selection is, in practice, a classification of quality, largely based on the microbiological state of the animal. The more efficient the environmental sanitary barriers are, the lower the risk of animal contamination.

According to the definition above, animals can be classified into the following groups:

- Gnotobiotic animals.
- Specific pathogen free animals—SPF animals.
- Conventional animals.

Gnotobiotic Animals

These animals possess defined microbiota and should be raised in environments that are provided with absolute sanitary barriers. The maintenance of gnotobiotics necessarily requires special equipment, such as isolators. The term "gnotobiotic" may be applied to animals free of germs and to those contaminated with one or more detectable organisms. Thus, because there is a large quantity of microbiota associated to an animal, it can be classified as either germ-free or defined-flora.

Germ-free: a germ-free animal is one that is rid of any internal, external or derivative by hysterectomy, and that is raised and kept in isolation. Germ-free animals are an additional tool in research on cancer, immunology, and nutritional diseases, among others.
Defined-flora: these are animals that have been intentionally inoculated with specific microorganisms. The first step for their production is to obtain germ-free animals. Several studies have associated selected enteric flora and growth rate with susceptibility to infection and the effect of endotoxins.

Specific Pathogen Free Animals: SPF Animals

The term SPF means that the animal is free of a variety of specific microorganisms but not necessarily of other non-specific[1] organisms. SPF animals originate from germ-free animals and do not present disease causing microbiota, but only non-pathogenic microorganisms. Only high quality microbiological animals can vouch for an experimental result that is free of the interference of other variables. The use of these animals has been on the rise, as researchers need more trustworthy results from their experiments.

Conventional Animals

Most experimental animals in use are still raised in conventional mating colonies. These animals have undefined microbiota, since they are kept in environments that are not provided with rigorous sanitary barriers.

[1] Nomenclature recommended by the International Laboratory Animal Committee (1964).

The selection of the ecological class of the animals to be used will depend on the experimental objectives. There are many applications in which a good quality conventional animal is perfectly suitable. There are, however, obvious advantages to having a precise definition of the microbial nature of animals from environments provided with sanitary barriers. In some cases, such as in the maintenance of athymic mice, barrier conditions are essential.

Clearly, the availability of genetically and ecologically (sanitarily) defined animals, combined with current advances in the fields of molecular biology and genetic engineering, provide the researcher with a highly sophisticated system that enables the development of new models and the improvement of existing ones.

References

Andrade A. Animais de Laboratório: criação e experimentação. Organizado por Antenor Andrade, Sergio Correia Pinto e Rosilene Santos de Oliveira. Rio de Janeiro: Fiocruz; 2002. p. 387.

Canadian Council on Animal Care (CCAC). Guide to the care and use of experimental animals, vol. 1. Ottawa: CCAC; 1984. p. 120.

Cassell GH, Lindsey JR, Baker HJ, Davis JK. Mycoplasmal and rickettsial diseases. In: Baker HJ, Lindsey JR, Weisbroth SH, editors. The laboratory rat, biology and diseases. New York: Academic; 1979. p. 235–73.

Da-Silva JB. Biossegurança em experimentação animal: um enfoque microbiológico. Niteroi, Rio de Janeiro: Universidade Federal Fluminense; 1998. p. 93.

De-Luca RR, Alexandre SR, Marques T, Souza NL, Merusse JL, Neves SP. Manual para técnicos em bioterismo. 2nd ed. São Paulo: Winner Graph; 1996. p. 259.

Krinke GJ, editor. The laboratory rat. San Diego: Academic; 2000. p. 634.

National Research Council—Institute of Laboratory Animal Resources, Committee on Rodents. Guide for the Care and Use of Laboratory Animals. Washington: National Academy Press; 1996. p. 125.

Poole TB, editor. The UFAW handbook on the care and management of laboratory animals. 6th ed. London: UFAW, Churchill Livingstone; 1987. p. 635.

Chapter 7
Rats

Monica Levy Andersen, Renata Mázaro e Costa, and Marcos Fernando Oliveira e Costa

The rat is a mammal and belongs to the order of Rodentia that encompasses several species, two of which are the ubiquitous brown rat (*Rattus norvegicus*) and the black rat (*Rattus rattus*). The rat that is employed in scientific research worldwide belongs to an albino variety called *Rattus norvegicus albinus* and is characterized by the complete absence of the pigment melamine in its fur and iris.

Its origin dates back to 1905 when researcher Henry Donaldson was transferred to the Wistar Institute to become its first scientific director. He brought four mating pairs of albino rats to that colony. These were the animals that gave rise to the famous strain that is used worldwide, including Brazil. The importance of this subspecies lies in the fact that Dr. Donaldson carried out, along with the team working at the Wistar Institute, countless studies that obtained the fundamental data of this strain mainly growth curves of the animal, skull, skeleton, and several organs, individually. All of these results were compiled in a famous book entitled "The Rat: Data and References Tables for the Albino Rat (*Mus norvegicus albinus*) and the Norway rat (*Mus norvergicus*)," the first edition of which was published in 1915 and the second in 1924. All of these data were fundamental in the characterization of the Wistar lineage and in the dissemination of its use in research.

Another two important commercial subspecies of rats were produced, called Long-Evans, obtained between 1915 and 1920 by two professors of the University

M.L. Andersen (✉)
Department of Psychobiology, Chief of Sleep Division, Universidade Federal de São Paulo (UNIFESP), Rua Napoleao de Barros, 925, São Paulo, São Paulo, Brazil
e-mail: ml.andersen12@gmail.com

R.M. e Costa
Department of Pharmacology, Universidade Federal de Goiás, Goiânia, Goiás, Brazil

M.F.O. e Costa, D.Sc.
Laboratório de Reprodução Animal, Embrapa Arroz e Feijão, Rua S-05, N. 700, Apto 1702, Res Asturias, Setor Bela Vista, Santo Antônio de Goiás, Goias 74.823-460, Brazil
e-mail: mfocosta@hotmail.com

M.L. Andersen, S. Tufik (eds.), *Rodent Model as Tools in Ethical Biomedical Research*, DOI 10.1007/978-3-319-11578-8_7

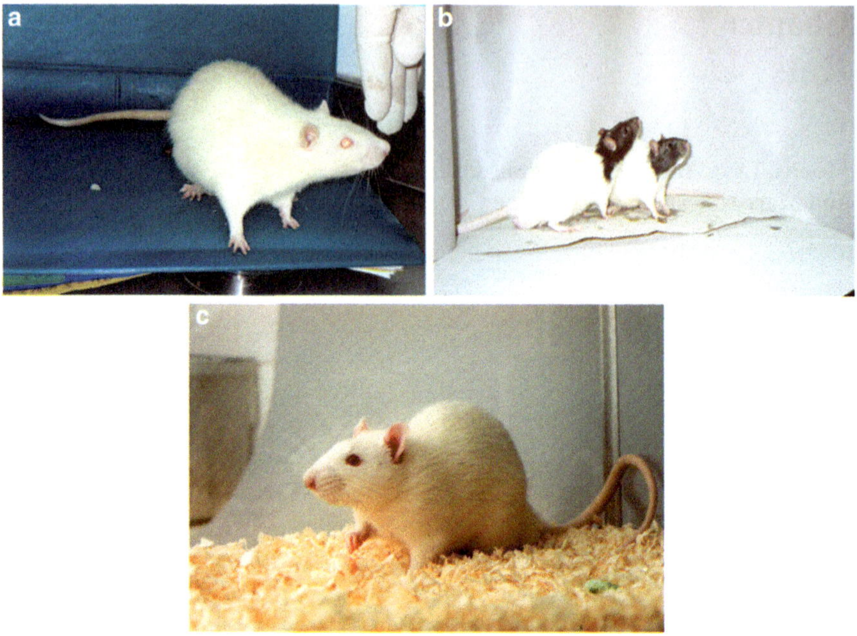

Fig. 7.1 (**a**) Wistar rats. (**b**) Sprague–Dawley rats. (**c**) Long Evans

of California, Joseph Long and Herbert Evans. This lineage was obtained from the crossing of females (probably Wistar) and males with *agouti* fur that were captured in the wild. The descendents that formed the new subspecies featured either black or gray fur. Dr. Long and Dr. Evans went on to become authors of a great number of papers on rats, especially on their physiology and reproduction.

A second form, also very important, is the Sprague–Dawley produced by Robert Dawley through the crossing of hybrid males and female Wistar albinos. The male had black fur on its head running along its body (the so-called hood-rats), and these were very large and vigorous rats. The descendants were cross-bred until they were completely white, and vigorous but docile offspring were obtained. This lineage has been greatly used in research.

In regards to these subspecies, there are some differences that can easily be recognized: Sprague–Dawley rats have a wider head and a larger body while Long-Evans and Wistar and other varieties are smaller (Fig. 7.1). Propensity for disease and aggressive behavior also vary in these lineages, but, in practice, these characteristics seem to vary more in substrains selected from other forms.

Use in Research

It is important to remember that one of the fundamental requirements in scientific research is that data obtained from experimentation must be reproducible. Such results are only possible when experimentation is performed on the same model,

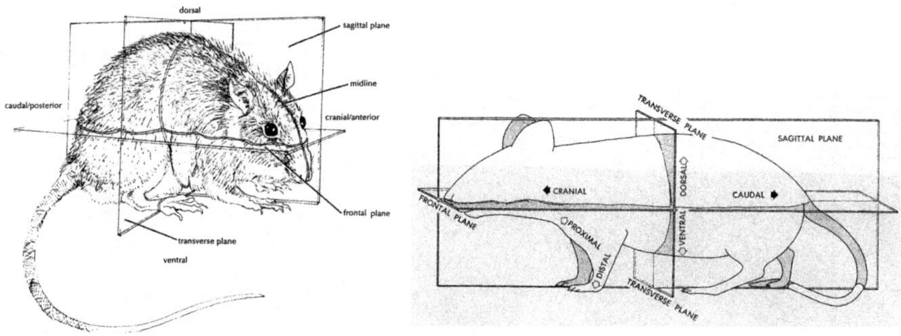

Fig. 7.2 Lateral view of the Wistar rat illustrating planes and directions used in anatomical terminology

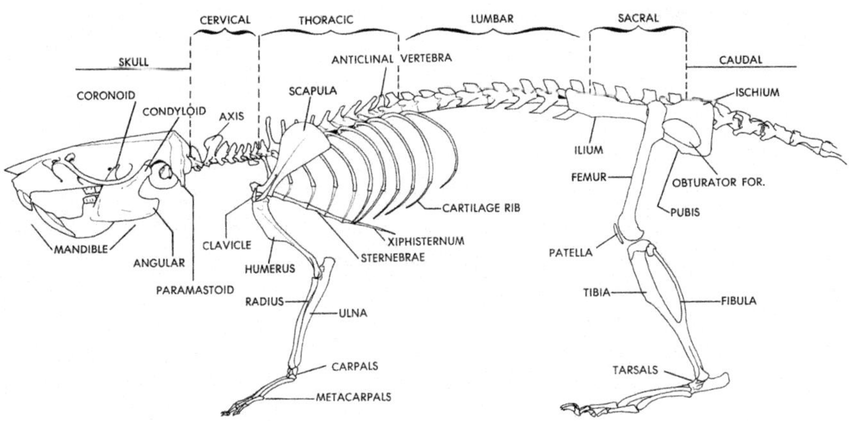

Fig. 7.3 Lateral view of a rat's skeleton

which is why biomedical research worldwide is performed on very few subspecies of rats and mice, such as the Wistar rat and the Swiss mouse. In the case of special subspecies used for determined ends, researchers generally use animals of the very same lineage so that their data can be reliably compared to others Figs. 7.2 and 7.3.

At the same time, however, it is necessary to remain watchful because although subspecies from different laboratories have the same origin and name, they eventually accumulate genetic mutations that set them apart (Fig. 7.4). These differences may not necessarily interfere with the experimental results.

After the mouse, the rat is the most common laboratory mammal to be used. The rat is the most used animal in behavioral investigation, for which mice are not suitable. The rat is traditionally the animal of choice in nutrition research, although its convenience for this purpose may be limited by its coprophagia habits.

In the past 80 years, rats have been used for investigation in nearly all fields of biomedical and behavioral research and tests. Mutations and genetic selection

Fig. 7.4 Comparison of same-age Wistar rats. One is normal and the other is a product of spontaneous genetic mutation, probably in growth-related genes

produce models of great value for research, the examples of which will be discussed later. Rats are widely used and are particularly advantageous in biomedical research in the fields of toxicology, gerontology, cardiology, dentistry, immunology, reproduction, neurosciences, and parasitology.

Several factors contributed to the adoption of this species for research, namely short intervals between litters, numerous individuals in each litter, adaptation to being caged, low maintenance cost, docile behavior, and easy handling, in addition to the copious yield of biologic material such as blood and tissue, which reduces the need for large experimental groups. Also, the rat reaches puberty within 50 days and sexual maturity within only 90 days, and their mating and pregnancy take 30 days with a 120 day interval between generations. This is quite a short period that can hardly be matched by any other species except for mice, which yield a much lower amount of biological material.

Genetic Selection

In the first half of the twentieth century, through successive breeding between males and females that featured specific characteristics, several subspecies used for research were created. Thanks to the discovery of some of these forms, today there are animal models that make it possible to satisfactorily investigate several clinical entities. In the past few decades, for example, several subspecies of rats with arterial hypertension and genetically transmittable epilepsy were created and these have been of great value in the investigation of hypertension and epilepsy in

Fig. 7.5 Obese Zucker rat (fat) and its phenotypically thin (lean)

Fig. 7.6 Phenotype of the subspecies of the Brattleboro rat that presents *diabetes insipidus*

humans. Although rat subspecies are genetically well-characterized, they are less numerous and are often less defined than subspecies of mice.

A great number of less frequent and more defined strains (see Figs. 7.5, 7.6 and 7.7) is established by means of selection or may arise through mutation, in addition to co-sanguineous subspecies. These are perpetuated due to the fact that their

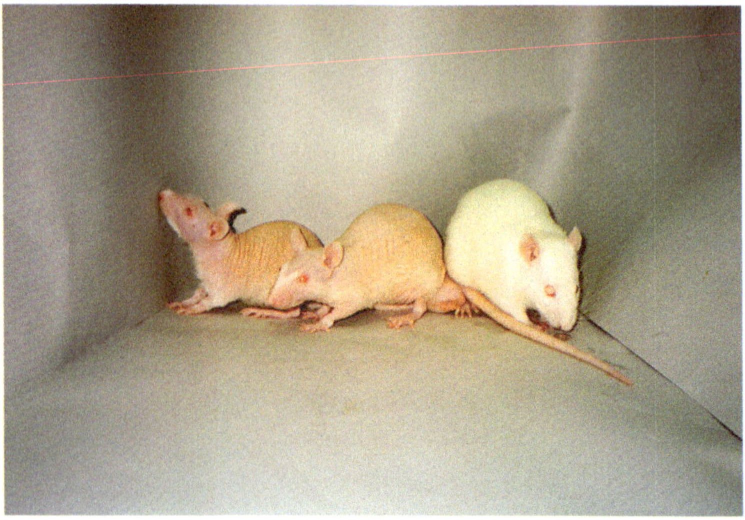

Fig. 7.7 Strains of rat that are completely devoid of fur (except for the whiskers—nude rats) next to a Wistar rat

genotype is particularly adapted to biomedical and specific behavioral studies. Some of the best known forms are:

- ACI: urogenital congenital abnormalities.
- CAR<CAS: resistance to dental cavities.
- SHR: spontaneously hypertensive rat.
- AA<ANA: acceptance or non-acceptance of alcohol.
- Brattleboro: *diabetes insipidus* (absence of vasopressin).
- BB: spontaneous diabetes.
- WAR: audiogenic-induced convulsion rat.
- Zucker: obesity

In one of the most typical systems, called *outbreeding* (non-co-sanguineous or heterogenic), the lineage presents a high degree of heterozygosis (99 %), that is to say, genetic heterogeneity. This makes highly genetically diversified colonies possible. The *outbreeding* system is produced by chance crossing between descendants of several other different crossings. As opposed to this, the *inbreeding* system (co-sanguineous or isogenic) is used to produce special forms and results in high degree of *inbreeding* or homozygosis by systematic crossings between a single pair of siblings for generations. Maintenance of this form should be very rigorous to assure minimal variations in subsequent generations. According to theoretical studies, it takes 20 successive generations to create an animal holding 99 % of the loci genes, all of which are homozygous, making these animals very similar indeed. These forms are extremely important in cancer research because they enable rejection-free tumor transplants to be carried out, and they are instrumental in the investigation of the existence of genetic factors that may influence the heredity of cancer.

Such subspecies were independently produced in 1909 by two researchers, Clarence Little, who worked with mice, and Helen King from the Wistar Institute, who worked with rats. One of these forms is the King Albino, known today as PA. Another form that is often utilized and that was produced by King around 1932 is the Brown Norway (BN).

Time and the intense manipulation of these rats brought on mutations that were observed and special forms of these mutants were selected for research. Many of them constitute extremely useful models in the investigation of specific abnormalities.

Characteristics

Ontogenetic Development

The pregnancy of the rat lasts about 20–23 days. However, the period between the moment of fertilization and birth may extend to 30 days or more due to delayed implantation followed by post-delivery coupling. This delay tends to be proportional to the number of offspring the female nurses. Post-delivery estrus occurs 48 h after delivery, and couplings in this period are successful in more than 50 % of the cases. The female that does not become pregnant in this period will delay mating until 2 or 4 days after the litter weans.

The pregnancy of the rat may be divided into three periods:

1. Pre-implantation, which encompasses the first to fifth day of pregnancy. This is an important period for the implantation and initial development of the embryo.
2. Organogenetic, from the 6th to the 15th day into pregnancy, when organs are formed.
3. Fetal, from the 16th day to the 21st day, when the organs are already developed and the fetus begins its intra-uterine growth.

Knowledge of these periods is fundamental for research in reproduction toxicology to assess embryofetotoxicity. Normal delivery occurs during the day period of pregnancy (14 h light and 10 h dark cycle), and 37 % of the animals are born on the 21st day, 20 % during the night, and 42 % in the light phase of the 22nd day. There are two periods in which there is a concentration of deliveries, from the 13th to the 15th h of the 21st day and from the 9th to 11th h of the 22nd day. Pregnant females that are housed in the company of a male present the "*Bruce effect*," in which there is a delay in the implantation of the embryo in the uterus's horn.

The size of the litter varies according to the subspecies, but normally there are 8–15 offspring from the Wistar rat (Fig. 7.8). When they are born, the offspring weigh about 5 g and are blind, very active, and depend on care and maternal nursing to survive until the weaning phase. During the rat's first 2 weeks of life, the nervous center is maturing and the hypothalamic-hypophysis-adrenal axis is hyperresponsive to stress. In this phase, external interference with maternal behavior (alterations in environmental conditions, tampering with the nest during bedding replacement, inappropriate handling of the mother—Fig. 7.9, for more information see Chap. 11),

Fig. 7.8 Ten day old rats

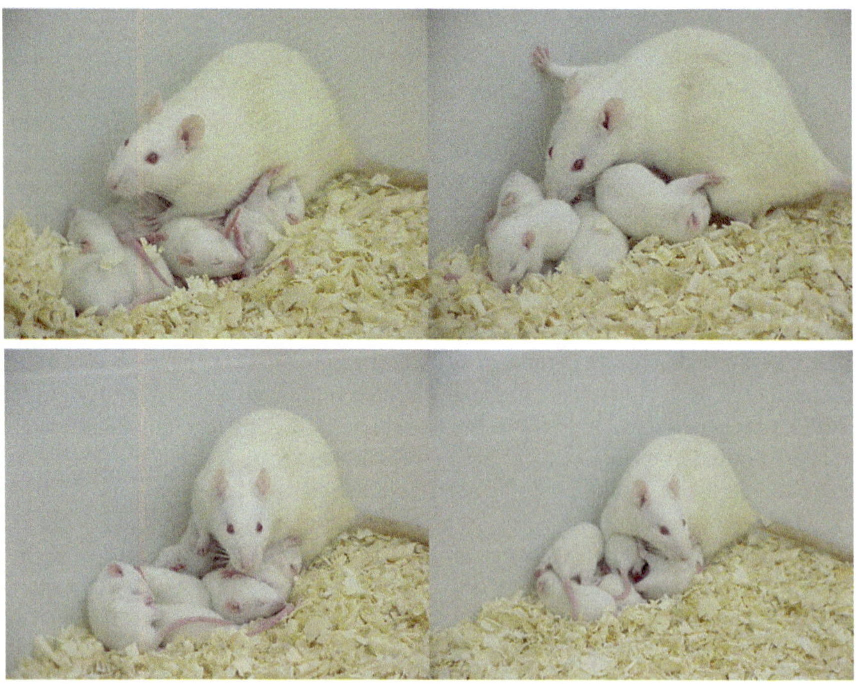

Fig. 7.9 The female is always on the alert for possible threats to her offspring

and manipulation of the offspring may cause permanent alterations in the behavior of the animals in puberty and when they reach adulthood (Levine 2001; Mazaro and Lamano-Carvalho 2006).

Several factors may alter the relationship between a mother and her offspring. When studying the consequences that co-habitation with an arthritic litter of rats had on the mother, in terms of maternal behavior, Roizenblatt et al. (2010) observed that it increases maternal behavior, anxiety, reduction in pain thresholds, and sleep disturbances.

The determination of gender can be made easily in new-born offspring by comparing the anogenital distance of each offspring (Figs. 7.10 and 7.11). The distance in the male is twice that of the female. Besides, the genital papilla of the male is larger. The nipples of the female offspring are visible within a week and the testes in the male can be seen in the scrotum at about 3 weeks of age when the offspring is examined in an upright position.

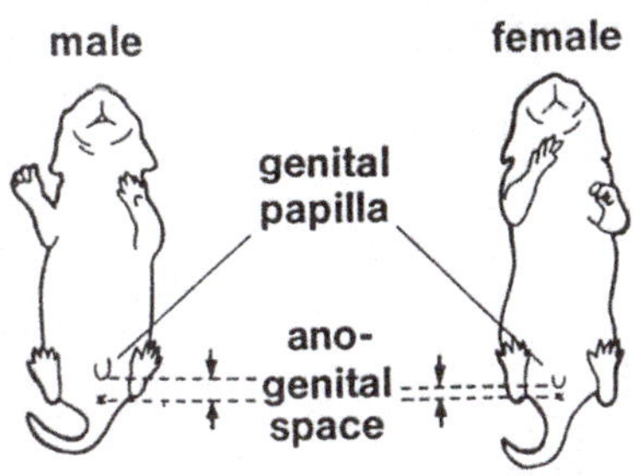

Fig. 7.10 Illustrative diagram of the structures used in the determination of sex in new-born offspring

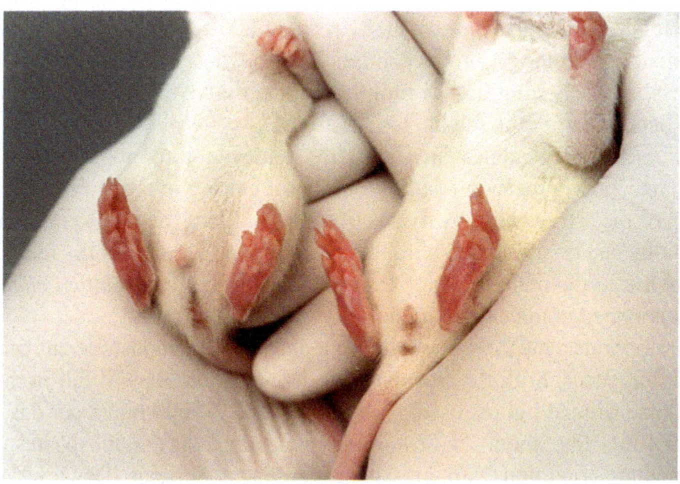

Fig. 7.11 Sexual differentiation in new-born offspring should be made by simultaneous comparison of the anogenital distance between males and females. The distance in the male is twice that of the female

When handling offspring to determine their sex, it is advised that some measures be taken. Before touching the offspring, rub the gloves you are wearing in the bedding material of the cage (careful not to ruin the nest) and manipulate some chow. Only after you have done this should the offspring be gently handled. This will prevent the mother from rejecting them on account of alien odors.

Weaning occurs when offspring replace breast suckling for solid food. There is a diminution in the synthesis of prolactin in the mother followed by the consequent end of the nursing period. This happens close to the 21st day after birth. At this point in time, the offspring weigh 40–50 g and should be separated by sex to avoid precocious mating.

At puberty, the secretion pattern of GnRH (releasing hormone of gonadotrophins) goes from tonic to pulsing, influencing spermatic maturation in the male rat and giving rise to the female estrus cycle. Puberty of the male rat begins when it is around 50 days old. That is when mature spermatids appear in the testes and the first spermatozoids can be seen in the cauda epididymis (Robaire et al. 1988; Zanato et al. 1994). The descent of the testicles, however, occurs well before puberty, around the weaning period, and the rat's testicles are retractable. Young males should not be used for mating until the age of three months, when they weigh 300–450 g. Puberty in females begins in the same period and is associated with the opening of the vagina and the occurrence of the first proestrus. Vaginal opening normally occurs about 33–42 days after birth and the regular estrus cycle begins around 1 week after vaginal opening.

The age at which animals reach puberty varies among species, and in each species there are several variables, including environment, nutrition, temperature, humidity, photoperiodicity, and co-habitation of both sexes that influence the initiation of puberty.

Mating may occur at any time after vaginal opening, but it should be avoided until the female is 90 days old, when she reaches 200–275 g, depending on the species. Females continue to produce litters until advanced ages, although progressively less regularly after 12–15 months. Productivity (the size of the litter) will begin to decrease after the female is 1 year old.

Programmed pregnancies are normally achieved through nocturnal mating. Greater conception precision results from 2 h of pairing in the morning of the confirmation of the estrus phase by means of rubbing. Estrus may be synchronized and its beginning can be stimulated with the introduction of a male in the female's cage. The response to the male pheromones (smell) is called the "*Whitten effect*" and is more pronounced in mice than in rats.

Laboratory rats mate throughout the year and do not demonstrate any meaningful seasonal influences. Still, the frequency of delivering litters will fall in the wintertime, unless artificial light is used to produce about 14 light hours per day.

Healthy rats live about 24–36 months depending on the subspecies, sex, and environmental conditions, among other variables. Sprague–Dawley rodents of defined microbiota have relatively short life spans. The males live for about 2 years while male ACI live for approximately 31 months. The fur of the albino rat is soft and white when they are young but progressively becomes coarse and faded (yellowish-gray) with age.

Fig. 7.12 Rat sleeping in different positions

Sleep-Wake Cycle

In animals that have nocturnal habits, wakefulness predominates at night and sleep during the day (Fig. 7.12). However, sleep also occurs during the night, although at a lesser proportion than during the day. Chapter 11 depicts the distribution of the sleep states of the rat along the 24 h of the day. Slow-wave sleep as well as paradoxical sleep (or REM—*Rapid Eye Movement*) last for shorter periods of time at night than during the daytime. According to Van Luijtelaar and Coenen (1983), the rat sleeps about 62 % of the daytime and 33 % of nighttime. The rat spends most of its time at night eating, prowling and exploring the environment, interacting with other rats, reproducing, and nest building. If the periods of sleep are totaled, it is verified that the rat may sleep up to 47.5 % of the 24 h, 17.1 % of which is light sleep, 21 % deep sleep, and 9.4 % paradoxical sleep.

Investigation of the sleep-wake cycle requires the registry of cortical and subcortical electrophysiological potentials. The graphic recording of the fluctuations in electrical potentials of the nervous system is generically dubbed oscillogram. Its best known form is the electroencephalogram (EEG) obtained in animals by implanting electrodes directly over the cortical surface (electrocorticogram—ECoG) or in the interior of the encephala. The simultaneous registry of oscillograms from several different regions of the central nervous system is of incalculable value in the investigation of the mechanisms of the processes implicated in wakefulness and sleep (Timo-Laria 1985; Andersen et al. 2001).

Anatomy

Nervous System and Sense Organs

The nervous system of the rat is formed by peripheral nerves, nervous ganglia, spinal cord, encephalic trunk, brain, and cerebellum (Figs. 7.13 and 7.14). Figure 7.15 shows different positions of the skull of the rat. Observe the bregma and lambda.

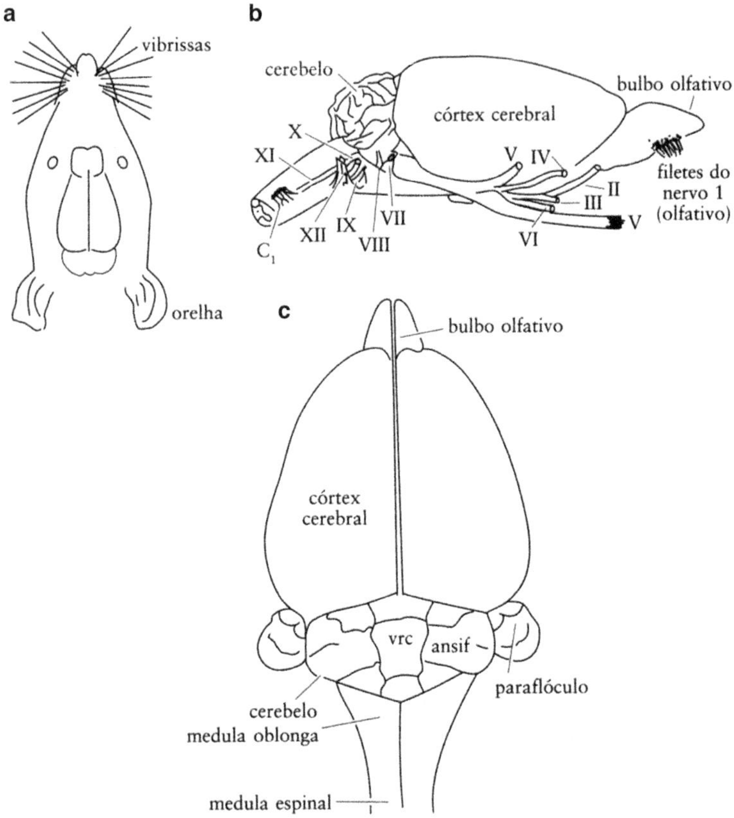

Fig. 7.13 (**a**) Dorsal view of a rat's head, showing the relative position of the brain in the skull. (**b**) Lateral view of the brain, highlighting the cranial nerves (modified from Greene 1955). (**c**) Dorsal view of the brain, showing relative positions of the olfactory bulbs, cerebral cortex, and cerebellum (Andersen et al. 2001)

Some diseases centered in the nervous system of laboratory animals can be exemplified in poliomyelitis of mice (Threiler's disease), in which paralysis, lack of motor coordination, rotation around the animal's own body, and death are observed. Lymphocytic choriomeningitis affects several animals and may affect man. Affected animals are hunched and present dull fur, convulsions, torpor, and generally die.

The rat is a macrosmatic animal that has high olfactory sensitivity and a large olfactory bulb (Fig. 7.16). This is the most developed sense of rats. Therefore, the researcher should abstain from wearing perfume when working with these animals and should avoid actions that alter environmental smell, not only in the experimental room but also in the animal care facility (for example: volatile anesthesia like ether, presence of fresh blood on the apron and/or on the counters).

Figure 7.17 shows the large arterial (in red) and venous vessels (in blue) of the dorsal and ventral surface of the encephala, while Figs. 7.18, 7.19, and 7.20 show

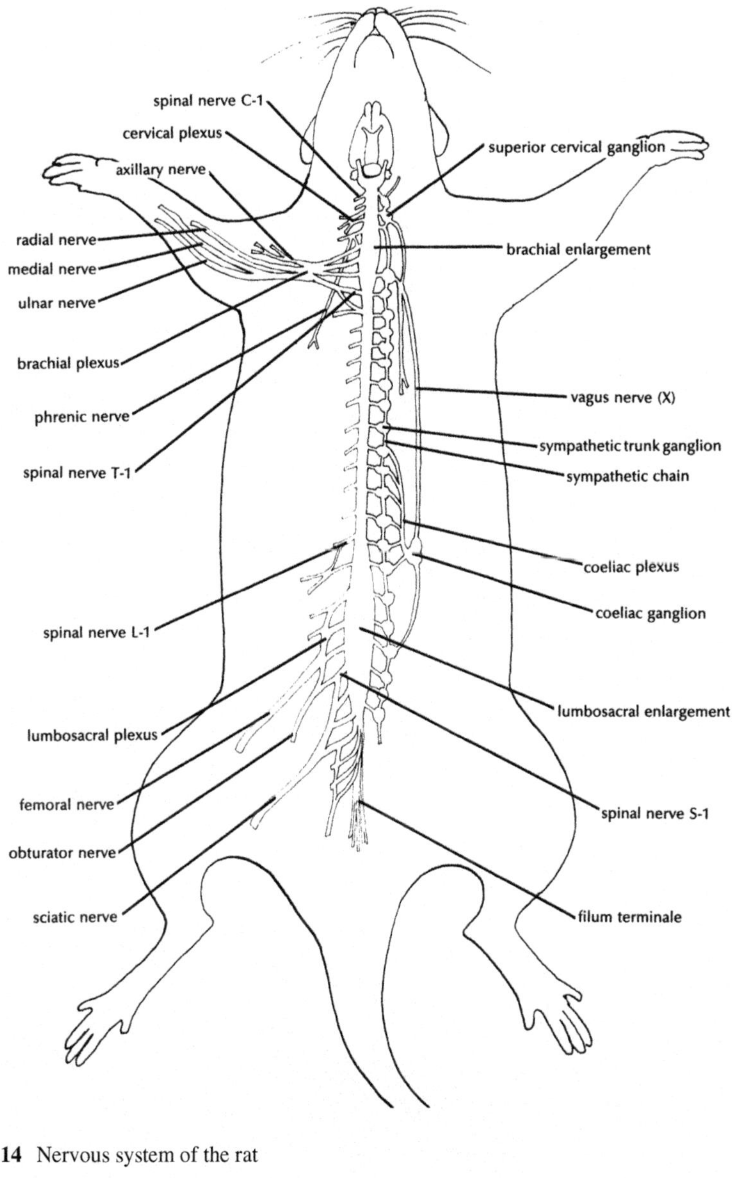

Fig. 7.14 Nervous system of the rat

the main arteries (Fig. 7.18), veins (Fig. 7.19), and lymphatic system (Fig. 7.20) of the rat. Notice the simplicity of the vascular structure of the rat's encephala compared to the primate's. During surgery, a puncture may occur that causes a local hemorrhage. When heavy bleeding occurs, it is advised that the animal be euthanized because damage to the cerebral cortex may be inflicted. Still, the animal's brain is resilient to manipulation provided that hygienic and aseptic conditions are adequate for the surgery.

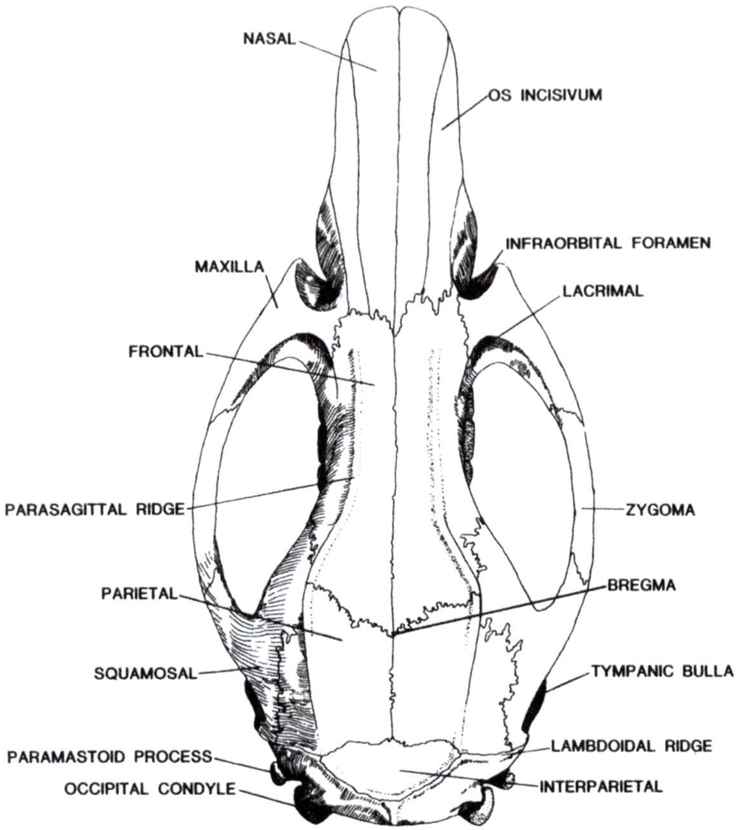

Fig. 7.15 Dorsal view of the skull (modified from Chiasson 1994; Wingerd 1988)

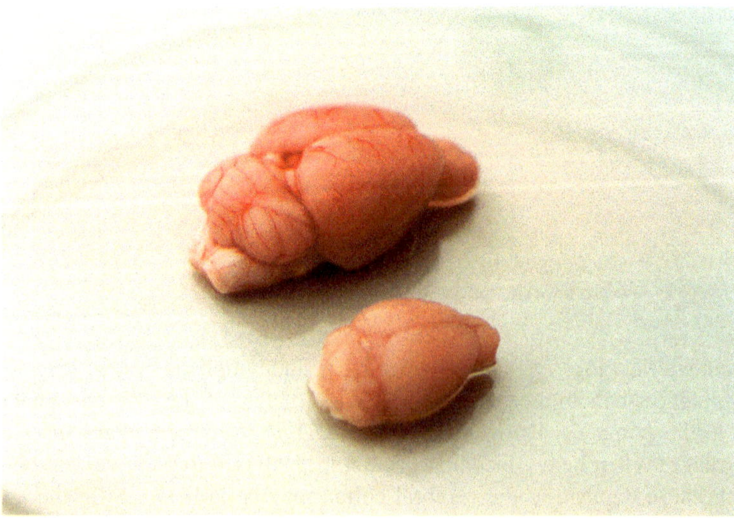

Fig. 7.16 Comparison between the brain of a rat and that of a mouse. Lateral view. The large bulging volume of the olfactory bulb is observable

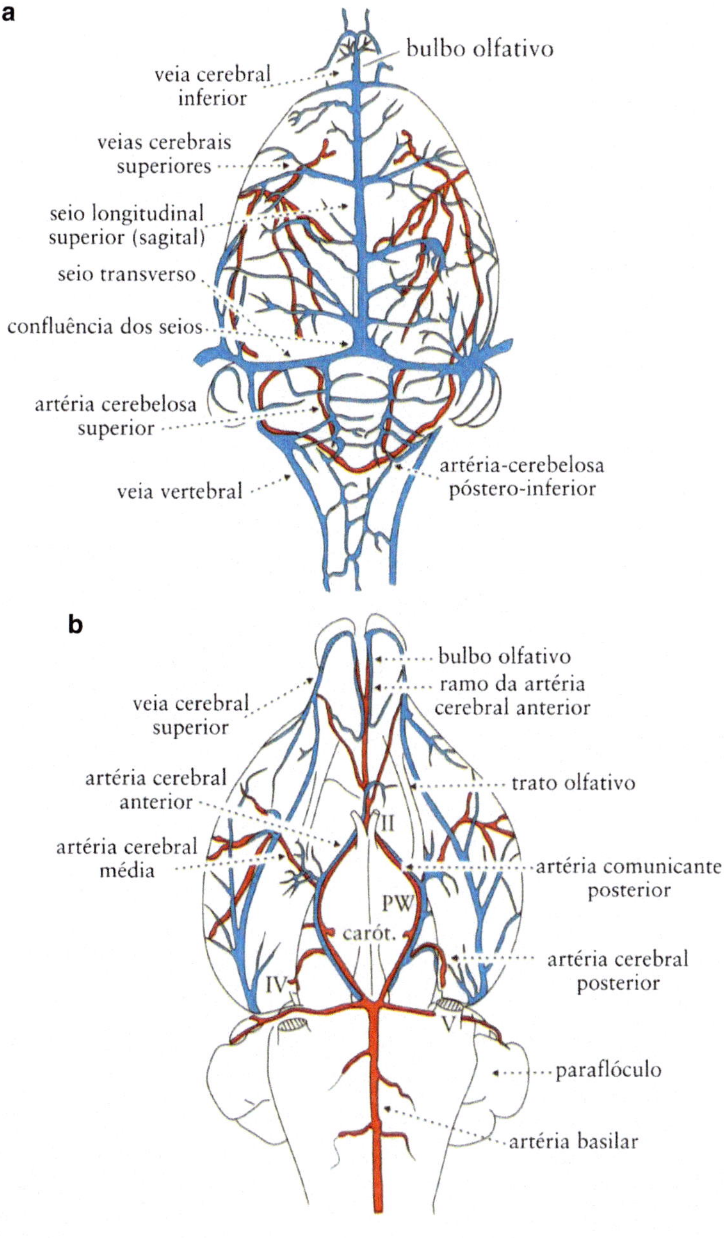

Fig. 7.17 Superficial veins of the rat's encephala. (**a**) Veins (in blue) and arteries (in red) of the dorsal surface. (**b**) Veins and arteries of the ventral surface (modified from Greene 1955)

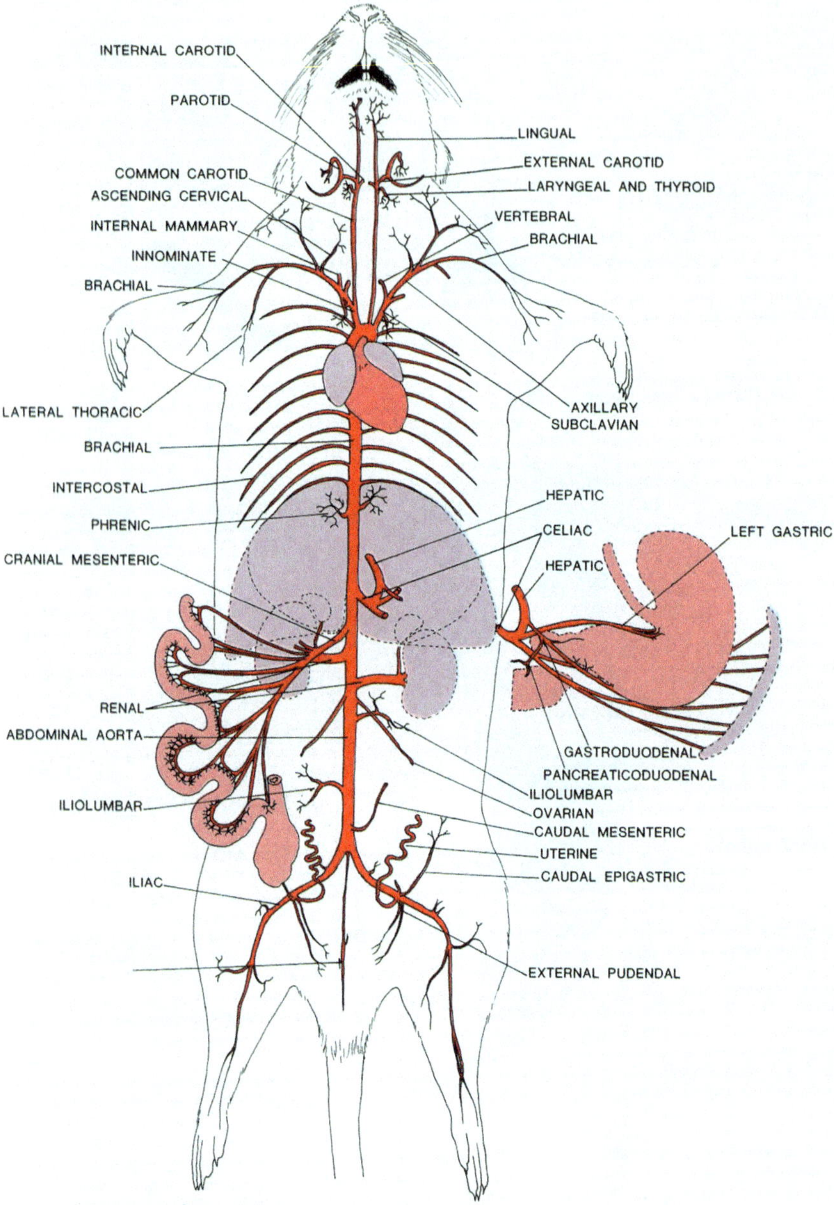

Fig. 7.18 Ventral view of the main arteries of the rat (modified from Chiasson 1994)

The rat features long vibrissae; approximately 1.5 cm on each side, close to the snout (see Fig. 7.21). These are rooted deeply and linked to mechanical receptors (tactile) that are highly sensitive to slight movements of each whisker. As the rat is a nocturnal animal, the whiskers detect the presence of objects in the vicinity even

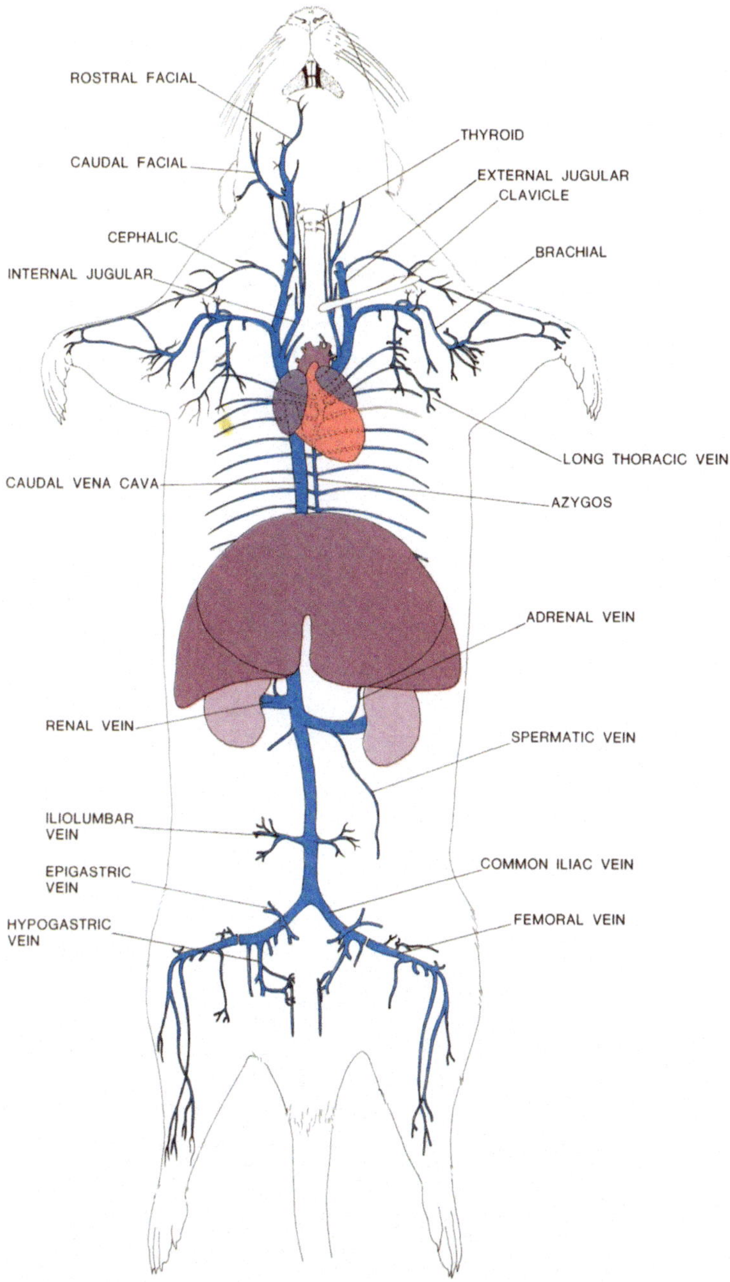

Fig. 7.19 Ventral view of the main veins of the rat (modified from Chiasson 1994)

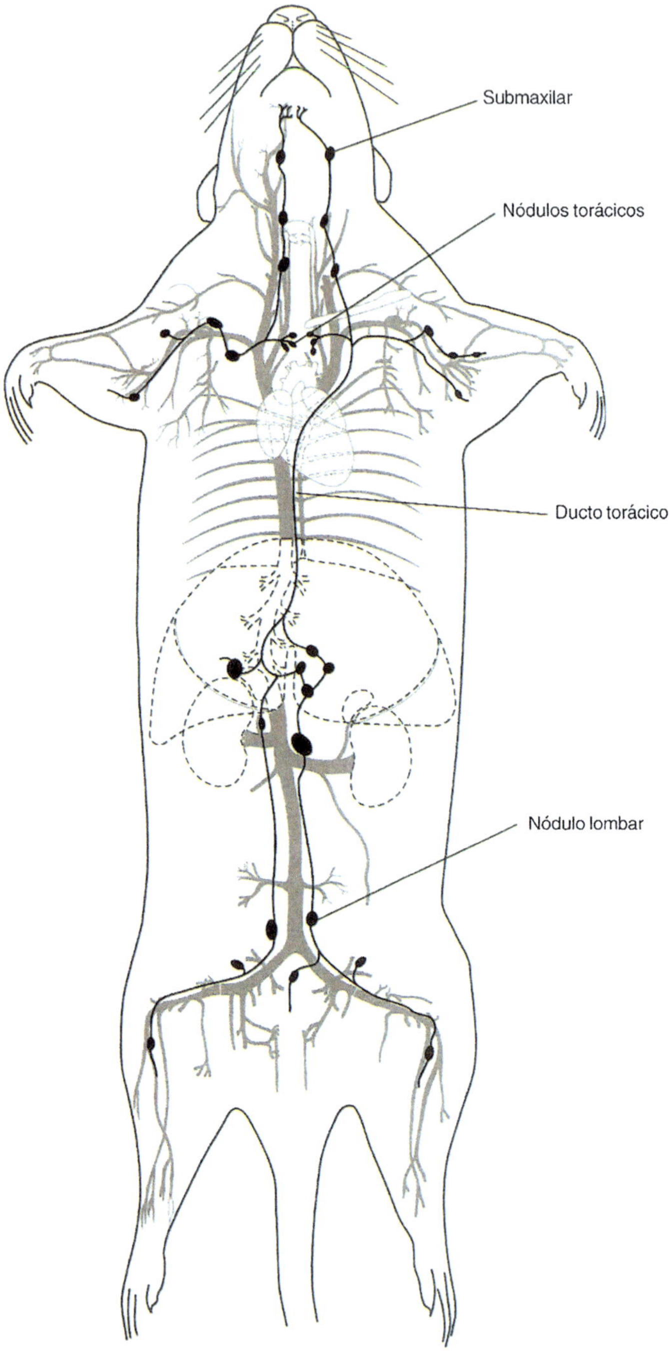

Fig. 7.20 Lymphatic system of the rat (modified from Chiasson 1994)

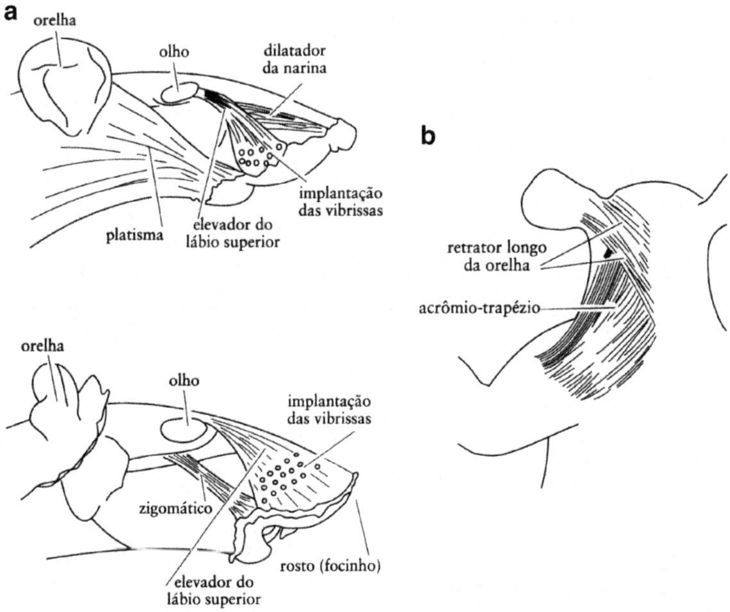

Fig. 7.21 Lateral view (**a**, **b**) and dorsal view (**c**) of the head of the rat, showing the position of the muscles that move the face, whiskers, and ears. The small orifices that are visible on the face are the anchorage entrances of the whiskers (Andersen et al. 2001)

if they are odorless, enabling the rat to safely guide itself through darkness. The rat relies on olfactory sensitivity, made even sharper by nostril movement, and on the tactile sensitivity provided by the whiskers as its main senses, which it uses in the exploration of the environment, search of food, and detection of the opposite sex and predators.

Denture, Appendix, and Energetic Reserves

The rat's denture is typical of a *Muridae*, with incisive pairs in continuous growth and enamel only on the cutting edges. The three pairs of molar teeth are present only as permanent teeth (non-deciduous denture); these possess open cuspidate and are devoid of enamel. The dental formula is: 2 (incisive 1/1 canine 0/0 pre-molars 0/0 M 3/3) = 16.

The rat features five digits on each of its four limbs, a long tail that represents approximately 85 % of the total length of the body and that is relevant in the animal's thermal regulation. Blood circulation in the tail increases when the animal needs to cool down, and circulation decreases when heat must be conserved. Vasodilatation dissipates heat and vasoconstriction is essential in maintaining body heat. There are no developed thermal-regulating mechanisms in newborn animals until the end of their first week of life.

The multilocular adipose tissue consists of cells that are filled with multiple droplets of lipids that carry a brown pigmentation and that do not mix as lipids in the common adipose cell "ring with bell" (unilocular adipose tissue). Multilocular adipose is diffused and distributed to the dorsal sides, lateral and ventral sides of the neck, as well as in the retroperitoneal region, particularly in the pelvis and kidney. The prominent accumulation of this tissue in the intercapsular region appears as glands and has been referred to as a gland at rest. Its true significance is not completely understood, although it is critical to a rat's life. The features listed above make the rat the most common model for cold adaptation studies.

Urinary and Digestive System

The esophagus extends from a small curvature of the stomach until it reaches a fold in the limiting groove of the stomach. The rat's stomach possesses a large aglandular portion, or anterior stomach, that constitutes one-third of the total gastric mucosa and is rich in gastric cells that produce histamine, and a glandular portion with pyloric glands that are restricted to the antrum. These two regions of the stomach are separated by a limiting fold that prevents the rat from being able to vomit. The ileocecal appendix helps in the digestion of cellulose. The rat does not have a gallbladder.

> The rat is an excellent model for antiulcerogenic drugs due to the ability to observe lesions of its gastric mucosa.

The presence of superficial nephrons in the renal cortex of the rat makes it an excellent model, widely used in investigations that require micro-puncture to evaluate tubular function in vivo.

The urethral *plug* was recently described as a normal characteristic, present in the proximal urethra of all male *Muridae* and *Cavidae*, and its absence may be associated with health problems. The *plug* is chemically similar to secretions of the seminal vesicle. In females, there is the copulation *plug* of the vagina that forms within 3–8 h after mating and remains in place for approximately 24 h. Specifically, it forms from secretion of the vesicular glands and lodges itself in the reproductive tract of the female, from the vulva to the cervix.

Care to be Taken with the Environment

The success of an experimental procedure is directly linked to the quality of the animals. The less external interference, the better it will be for the well-being of the animals, and this will impact directly on the reliability of the analyses. The longer the experiment, the more careful one should be so as to preserve the animal's health.

The light/dark cycle should be under control as the rat is highly active in the dark period, whereas in the light period, it tends to rest. Thus, any noise should be blocked from reaching the room, and animal manipulation should be kept to a minimum and done by as few people as possible.

Animals should be isolated only when necessary. Under normal conditions, the rats should be placed together in groups within cages. Abnormal behavior or alteration in the behavioral pattern may call for the animal to be removed from the cage or even for the experiment to be suspended altogether until the causes of the alteration of behavior are identified and corrected.

Any manipulation should be done adequately and in the shortest period of time possible so as not to aggravate stress. One should refrain from handling pregnant rats, but if it is necessary, they should be manipulated gently and not jolted or squeezed. The change of the box of nursing females must be done with great care without touching the nest or offspring. The wood flake bedding should be delicately removed on the sides of the nest, but the nest itself should remain untouched. All care must be taken to avoid any interference. The offspring should only be touched or removed from the nest if such actions are part of the experimental procedure.

Frequent manipulation of the offspring or an increase of stress in the female may lead to cannibalism or total or partial elimination of the litter by the mother. The presence of alien odors, blood, or any abnormality in the offspring may lead to abandonment, precocious weaning, or even their death by the mother.

The execution of experimental procedures on newborns should be done with the utmost care, especially the surgical procedures, which can lead to bleeding or extravasations of bodily fluids from the offspring that may promote anatomical alteration or taint them with other odors.

The quality of the bedding may interfere directly in experimental investigations. The bedding should be soft, odorless, and with great absorption properties. It would be ideal to produce and sterilize the bedding on-site in order to achieve control over its quality. However, because of storage and cost limitations, there are few resources for production, drying, and storage of bedding. Still, satisfactory material may be obtained from carpentry shops, and such materials should at least be sterilized. Light-colored wood shavings are preferred as darker woods tend to be rich in tannins that ooze odors that are potentially irritable for animals.

Wood shavings of intermediate size should be used for bedding purposes. They should not be too big, which may make nesting unfeasible, or too pulverized, which will create an atmosphere laden with particulates. A high concentration of airborne particles will worsen the quality of the air within the cage and increase the occurrence of problems in the rats' sensitive respiratory system. Also, it is highly advisable to learn more about the origins of the wood shavings and whether or not pets, livestock, or other animals had access to the shavings during production and storage as their odor will certainly compromise the quality of the material. If it is impossible to know such things, one should proceed to sterilize the shavings by physical or chemical means in order to minimize contamination of fleas, mites, ticks, and other pests. If sterilization is not possible, an entire suspicious batch of wood shavings should be discarded.

The daily portions of chow that are served to the rats should remain in the 10–20 g range. More than that heightens the risk of obesity and requires daily monitoring. Also, there is the waste consideration as the rat will often resort to the chow pellets to satisfy its natural need to simply gnaw on something, an activity that has much to do with controlling the growth of incisive teeth.

Feeding smaller amounts of chow may limit the growth of the rats. The preparation of special diets that either remove or add some form of nourishment should take into consideration the nutritional demands of the animals so as not to exceed those needs or fall short of them.

Chow should preferably be fed in pellets, which are more accepted than the pulverized form. If pellets cannot be made available, then particulated forms might be possible, in which case an adaptation period must be considered. For long term studies, alternative methods must be devised so that the rats have something to gnaw at and control the growth of the incisive teeth.

Tracking water ingestion is a means of verifying thermal comfort as an excessively hot environment will increase water intake and, thus, a bias as the animal must cope with thermal discomfort.

Reproduction

Male Reproductive System

In males, the reproductive system is formed by a penis, a pair of testicles, accessory sexual glands (ampoule, seminal vesicles, prostate, and bulbourethral, clotting and preputial glands), and a set of seminal ducts (epididymis and vas deferens) through which the spermatozoids and seminal fluid are eliminated at the moment of ejaculation.

The testicles are constituted by tubular loops separated by interstitial tissue that contains blood and lymphatic vessels, nerves, Leydig cells (synthesis and secretion of testosterone), and a considerable number of macrophages and mastocytes. It is in the tubular loops that the spermatozoids develop, influenced by the action of steroidal sexual hormones and by hypophysis gonadotrophins among other paracrine and autocrine factors. In the rat, spermatogenesis occurs in accordance with the spermatogenic cycle, which is composed of 14 distinct stages and defined by the arrangement of germ cells (spermatogonia, primary and secondary spermatocytes, and round and enlogated spermatids) in testis (Fig. 7.22). The cells that exist in the

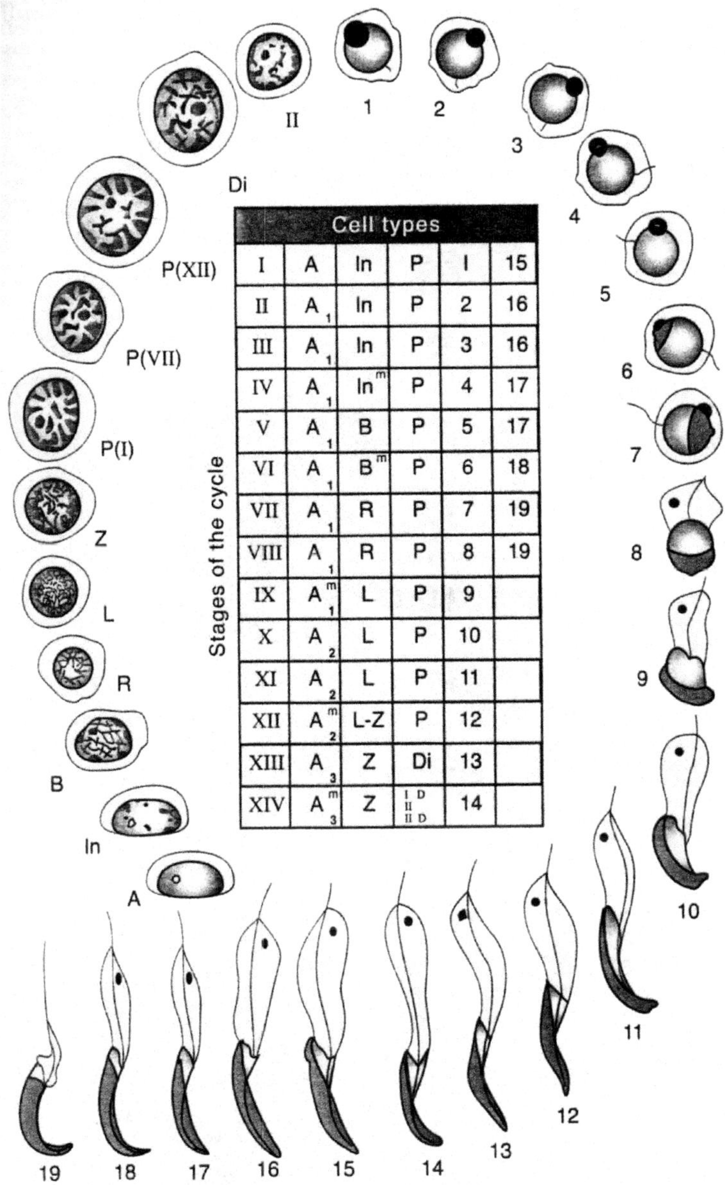

Stages of the cycle	Cell types				
I	A	In	P	1	15
II	A_1	In	P	2	16
III	A_1	In	P	3	16
IV	A_1	In^m	P	4	17
V	A_1	B	P	5	17
VI	A_1	B^m	P	6	18
VII	A_1	R	P	7	19
VIII	A_1	R	P	8	19
IX	A_1^m	L	P	9	
X	A_2	L	P	10	
XI	A_2	L	P	11	
XII	A_2^m	L-Z	P	12	
XIII	A_3	Z	Di	13	
XIV	A_3^m	Z	I D II II D	14	

Fig. 7.22 Stage I–XIV of the cell cycle in the spermatogenesis of the rat. Type A spermatogonium (A1–A4), intermediate type spermatogonium (In) and B; Type B spermatogonium; *R* resting primary spermatocyte, *L* leptotene spermatocyte, *Z* zygotene spermatocyte; *PI, PII, PIII* early, mid, and late (respectively) pachytene spermatocytes. The numerals indicate steps 1–19 of spermatogenesis, *Di* diplotene, *II* secondary spermatocyte. The table shows cellular composition of the stages of the cycle of the seminiferous epithelium (modified from Dym and Clermont 1970)

testes are called mature spermatids and have neither locomotion nor fertilization capabilities, which are acquired through spermatic maturation when mature spermatid travel along the epididymis.

The epididymis is an organ that consists of an expanded extremity, the head, followed by a longer, narrower portion, the body. The opposing extremity is constituted by a single, long, and intricately coiled duct (cauda), internally lined with a pseudostratified epithelia containing different kinds of cells whose morphology and abundance vary from region to region (Fig. 7.23). Spermatic maturation begins in

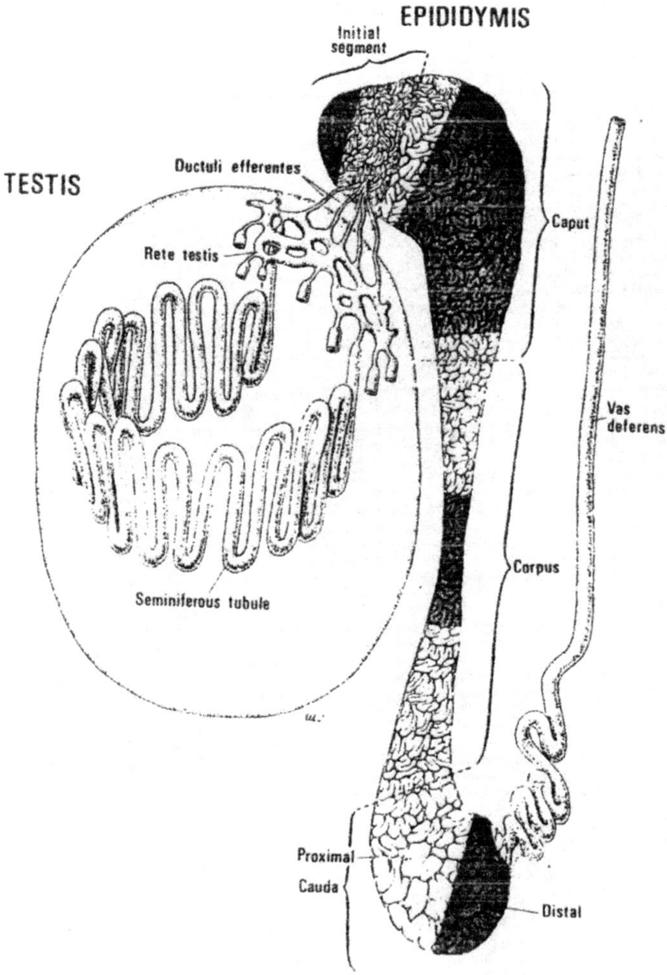

Fig. 7.23 Diagram representation of the testis, showing a semineferous tubule and testicular net, vas efferens, the epididymis, and the vas deferens. The shaded areas indicate the different segments of the epididymis, for example: the initial segment, head, body, and proximal and distal cauda (modified from Robaire et al. 1988)

the head of the epididymis and continues through the body to the proximal cauda. This trajectory allows spermatozoids to acquire their potential for progressive motility, survival, and success in fertilization. During passage through the epididymis, the fluid to which the spermatozoids are exposed undergoes substantial alterations in composition, including changes in osmolarity, ionic proportion, energetic reserves, and proteic types. Many of the alterations in the membranes of spermatozoids that coincide with spermatic maturing are mediated by proteins secreted in restricted regions of the epididymal duct (some of which are androgen-dependent) and by subsequent associations of these proteins with the spermatozoids. Spermatic transit through the epididymis takes 8–10 days; 3–4 days are needed to go through the head and body of the epididymis, and 5–6 days for transit through the cauda epididymis.

Female Reproductive System

In females, the reproductive system is formed by the vagina, a pair of ovaries, and a set of sexual ducts [oviducts (Fallopian tube), uterine horn, corpus of the uterus, the cervix, where several steps take place: the transport of oocytes, the development of embryos after fertilization, and the exit of the fetus at delivery. Females present six pairs of mammary glands: three in the thoracic region, one in the abdominal region, and two in the inguinal region (Fig. 7.24).

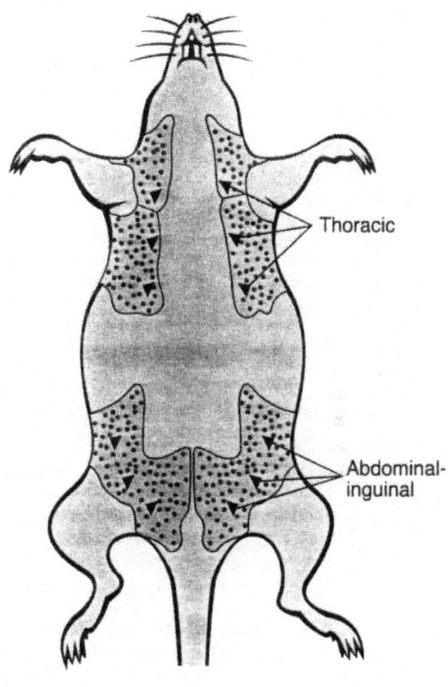

Fig. 7.24 Representation of the female mammary glands

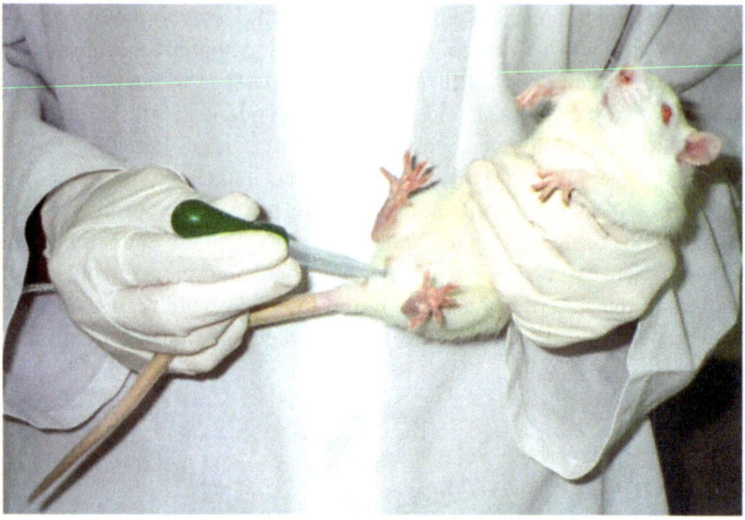

Fig. 7.25 Technique of vaginal smearing. Pipette (glass or plastic) containing distilled water is delicately inserted in the vagina and the liquid containing epithelial cells is collected and placed on a slide for microscopic visualization

Females are polycystic animals, presenting spontaneous ovulation and estrus occurs for a period of 12–14 h every 4–5 days.

Activation of the corpus luteum is not spontaneous and depends on the neural input of mechanical stimuli of the uterine cervix near the moment of ovulation. Such stimuli is typically associated with coitus but can be mimicked in the laboratory by stimulating the cervix with a glass rod or during vaginal rubbing or vaginal smearing (Fig. 7.25), which results in a condition known as pseudo-pregnancy in which the female constantly remains in the diestrus 1 phase (metestrus). This condition exhibits the same endocrine mechanisms of the initial phase of normal pregnancy and lasts about 12 h, delaying the next estrus cycle.

The use of saline solution for vaginal rubbing is not recommended because crystals may form as the liquid dries up, compromising the identification of the estrus cycle, especially diestrus phase II. In the case of vaginal washes, saline solution does not interfere with the interpretation of results.

Vaginal smear easily identify cycle stages cytologically (Chap. 8). Because of its rapidity and regularity, the estrus cycle of the rat rarely needs to be monitored cytologically (for the reproduction of the rat colony in animal care facilities), except when programmed pregnancies are needed. Mating success can be confirmed by observing the copulation *plug* (vaginal) in the female (more easily visible in mice) or by identification of spermatozoids in the vaginal rubbing (present for at least 12 h).

Mating System

The mating system is the manner in which the males are kept in the company of females, taking into consideration aspects related to the proportion of females for each male, in addition to those related to permanence periods and characteristic of each species. Among the most adopted systems are:

1. Monogamy: the system in which the female remains with the same male throughout her entire reproductive life. There is the exploitation of the post-delivery estrus since the female nurses a litter at the same time that she is expecting the next. In addition to better and more detailed registries, the rates are precise since the animals are identified individually and the litters are more homogeneous. Consequently, this is the preferred system in small colonies of co-sanguineous animals. Monogamy does require a greater number of cages and more room for animal reproduction.
2. Polygamy (harem): the reproductive family is formed by one male and a number of females that varies between two and six. When several females find themselves in the same environment, it is advisable to remove those that are pregnant before delivery. By doing so, the offspring are free of male interference, and losses due to overpopulation and the mixing of litters are avoided.

Factors that Alter Fertility and Reproduction

Genetic—a great number of variations in reproductive performance have been described in different species and subspecies of laboratory rats. Much of this variation will obviously be a reflex of the multifactorial differences that accidentally occur within one genotype during the selection of other characteristics. An example of this could be the increase of fertility in senile male ACI rats compared to male Sprague–Dawley rats of the same age, a characteristic that is related to a 6-month increase in the life expectancy of artificial subspecies.

Environmental—the temperatures and the relative humidity values recommended for rat colonies have been described in Chap. 3. The most critical factor is avoiding excessive fluctuations, keeping them within a range of 2 °C.

The kind of cage, size, and type of material used for bedding, overpopulation, and cleaning frequency seem to exert an influence on mating performance. These parameters and photoperiodicity are essential during pregnancy.

The increase in population density causes hypertrophy of the supra-renal gland and gonadal atrophy proportional to population growth, leading to reduced fertility. The pheromones secreted by the male, one urinary and the other plantar (secreted by the perspiration glands of the plantar region), are responsible for this effect (the *Christian effect*) that is observed in rats, rabbits, and mice.

Behavior

Laboratory rats are generally curious, docile, and, if often handled gently, become domesticated and trained. They display intense exploratory behavior and great curiosity when novelties are introduced into the environment. Rats are easier to handle than mice, which are generally jumpier. Both become aggressive when threatened; therefore, a calm posture and gentle handling deter aggression and bites.

Aggressive behavior is rarely observed. However, males fight more amongst each other than do females (Fig. 7.26 Rats live and raise their young in a community, sharing care and duties. If there is overpopulation, territorial fights can be observed among the members of the group. These behavioral patterns vary slightly according to species and vary more specifically according to the choice within a subspecies.

Extreme behavior is rare, but very quiet or excessively active animals may signal some pathology or environmental stress. Aberrations in behavior are not common although coprophagia is relatively common in this species. Additionally, something should be said here about cannibalism. Cannibalism also occurs and is a sign that either the housing conditions are inappropriate or that the animals are undernourished, in which case an examination of the chow is required. However, cannibalism may also be an indication of stress, mishandling of the system, and poor quality of the water and bedding. Females will commonly resort to cannibalism immediately after delivery if the offspring are too small. Within a few days after consuming her offspring, the female will go into estrus. Cannibalism of offspring at later time points after delivery may be an indication of an environmental alteration that befell either the female or the offspring that is strong enough to lead the female to discard

Fig. 7.26 Fighting behavior between two male rats

the possibility of successfully nursing her litter. The causative agent of such behavior should then be identified and corrected.

Rats are intelligent animals that demonstrate a broad variety of behavioral traits that are relevant in biomedical research. Furthermore, they adapt very well to studies in psychobiology and pharmacology and recover fairly well from surgery.

The rat keeps its fur clean and shiny through its grooming behavior, which redistributes the fur's natural oily secretion that is produced by skin glands (Fig. 7.27).

Yawning is a common physiological event that occurs in rats. It is an involuntary movement triggered by reduced oxygenation in the brain. During yawning, there is great inhalation and a considerable amount of oxygen is brought into the body, replenishing its ideal concentration (Fig. 7.28). In rats, it is possible to induce yawning with the administration of certain drugs such as apomorphine (dopaminergic agonist) and pilocarpine (muscarinic agonist).

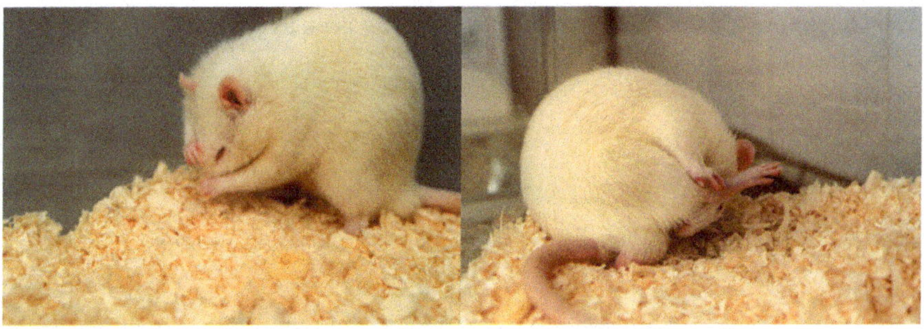

Fig. 7.27 Typical grooming behavior of rats

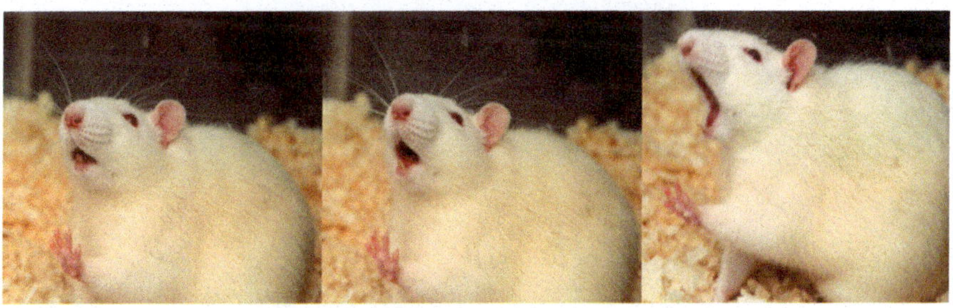

Fig. 7.28 Sequence of yawning movements

Handling and Physical Restraint

The rat easily adapts to the heat of the latex-gloved hand and quickly ceases its escape and aggressive behavior and becomes progressively more docile in subsequent manipulations. In general, handling consists of holding the animal by the base of its tail with one hand while the other is enclosed around its back and ribs so as to have the thumb, index, and middle finger behind the rat's elbows, pushing these forward. An alternative maneuver is to have the thumb and index finger positioned on the back of the neck behind the mandible and pinch the skin (Fig. 7.29), taking in the slack tissue to ear level so that the head becomes immobile, preventing accidental bites. When manipulated correctly, the rat is unable to bite even if it so desires. The use of a flannel cloth to restrain the animal is a widespread technique and reduces the likelihood of someone being bitten on the hand.

The manipulation of pregnant females carries the risk of precipitating cannibalistic behavior in view of the threat of traumatic effects on the offspring.

Several specific points should be considered when one manipulates an adult laboratory rat for the first time:

(a) Do not hold the rat too strongly or make any sudden or jolting movements.
(b) Do not be afraid; fear is contagious.
(c) Do not press the thorax or the area around the neck as this will make breathing difficult and prompt the rat to struggle.

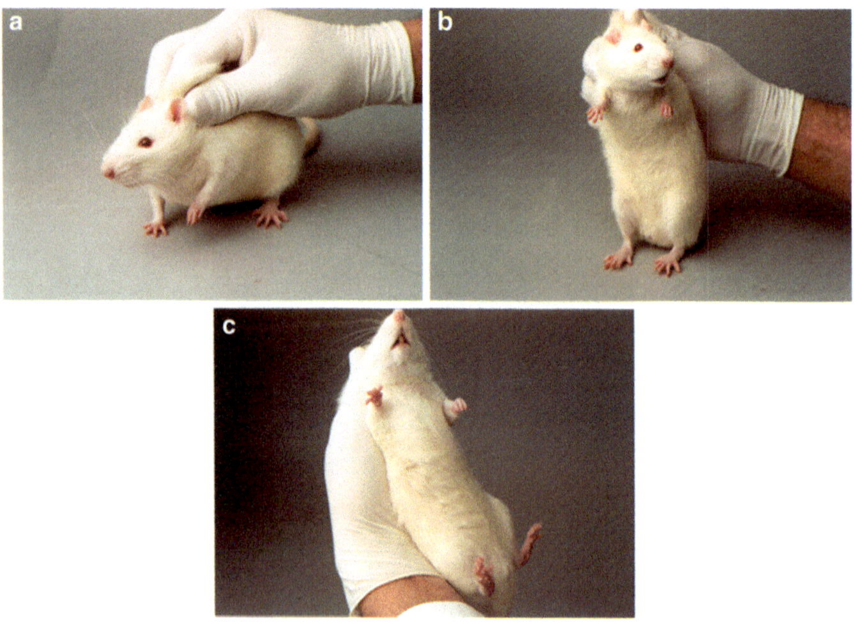

Fig. 7.29 Sequence of movements that immobilize a rat (**a**, **b**). In (**c**), the animal is completely immobile in the researcher's hand

(d) Do not hold a larger rat by the tail and do not cover large distances holding small rats and mice this way. The animal will perceive it is in a perilous situation and will begin to struggle. Skin can then detach from the tail and the animal will fall.
(e) Smaller rats and mice should be held by the base of the tail and not by its distal extremity because they can curl onto themselves, climb their own tails, and bite the startled researcher.
(f) Be sure that all of the animals in the experimental group are handled at the same frequency as control animals. This ensures results under equal conditions of stress.

Collection of Material

Blood

The choice of collection procedure will typically depend on the amount of blood needed and the frequency of the sampling. Frequently, the same technique used in the administration of intravenous injections is used in the collection of blood, although for some ends, arterial blood may be needed. Other considerations are the effects that the collection technique and anesthesia may have on the on the properties of the blood.

An animal that is to remain alive after having its blood collected should first be weighed. In a rat, 10 % of its weight represents its volume of blood. Up to 2 % of this volume can be collected while constantly monitoring the animal's condition. It is advised that blood should not be collected from animals that are debilitated or that have unsatisfactory appearances.

For small and repetitive samplings, bleeding of the retroorbital plexus is the most recommended method of collection. This is normally done under anesthesia, recovery is fast and complete, and the procedure may be repeated in a few days. It is a simple and apparently non-stressing procedure and can be repeated without any visible damage to the eye, due to the almost immediate repair of the simple endothelial wall cells of ruptured vessels. This commonly used technique is also applied to mice. Generally, blood collection is performed with either a Pasteur pipette or a glass capillary that has a rounded end so as not to cause any lesion to the conjunctiva when it is gently penetrated, breaking the venous plexus at eye Large amounts of blood can be collected in this way; 4–6 mL can be collected from rats that weigh 115–130 g if a larger pipette (13 × 100 mm) or even com.mercial heparinized tubes are used. In the first attempt to bleed the orbital bell, it is essential that the rat be under anesthesia and that the researcher be knowledgeable of the anatomy of that region. Note that if this procedure is used repeatedly, lesions will form in the tissue of the Harderian gland.

Cardiac puncture is commonly used in rats for the fast collection of volumes of blood over 5 mL. This is a procedure that should only be conducted on animals under anesthesia. A size 24 G needle may be used to pierce through the thoracic wall. The entrance angle should be 45° between the 5th and 6th rib to the left of the breast-plate (sternum). If there is penetration of the left ventricle, blood will gush into the syringe.

There are variations of the cardiac puncturing technique for its use in smaller rats or newborns as well as derivatives of other venopuncturing procedures.

Discrepancies in the results of a study may be caused by factors that may be considered irrelevant, such as excessive moving in the cage or inadequate collection of samples. Studies show that removing 1 mL of blood every 2 h, an acceptable volume that does not modify most hematological values, will cause a persistent decrease in weight gain. Many variables may be difficult to control or even impossible to be removed from the protocol of an experiment; however, they should be considered and efforts should be made to minimize their effects.

In many cases, blood is collected after euthanizing the animal. Decapitation (see Chap. 29) provides samples that may be compromised with fur and precocious clotting of the blood. The collection of biological material, such as feces, urine, and blood, from mice follow the same recommendations as those from rats.

Urine and Feces

The use of metabolic cages is the easiest way to collect samples of urine and feces separately. Commercial metabolic cages are constructed to separate urine from feces without admitting contaminants such as food, water, and fur. These cages may be made of either plastic or metal (Fig. 7.30). Several laboratories manufacture

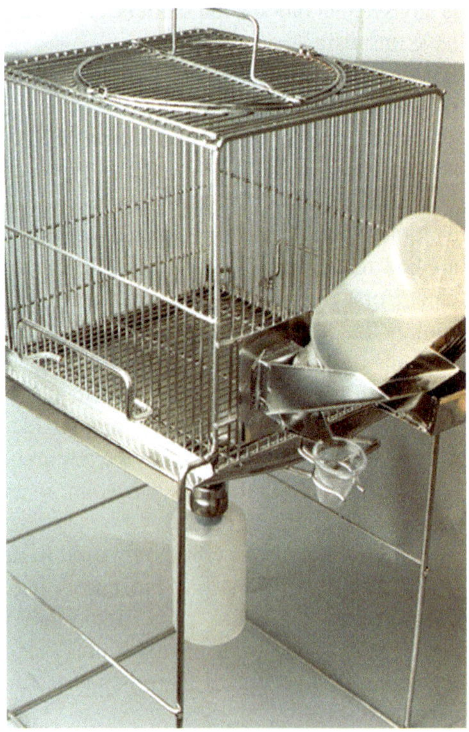

Fig. 7.30 Stainless-steel metabolic cage. (**1**) Cage compartment. (**2**) Wire-mesh floor. (**3**) Urine collector. (**4**) Drinking bottle. (**5**) Feed container. (**6**) Feces collector. (**7**) Metal tray

metabolic cages and other means of collection. The urine of rats may be collected on an aluminum plate placed under a wire-mesh floor; 1 mL of urine may be collected over a period of 1 h in this way. The collection of contaminant-free feces may be done by using sterilized containers, which are more efficient in collecting feces from male rats due to the fact that their anogenital distance is twice that of females.

References

Alexander G. Cold thermogenesis. Int Rev Physiol. 1979;20:43–155.

Andersen ML, Valle AC, Timo-Iaria C, Tufik S. Registro de Potenciais. In: Implantação de elétrodos para o estudo eletrofisiológico do ciclo vigília-sono do rato. São Paulo: Universidade Federal de São Paulo; 2001. p. 35–59.

Andersen ML, Bignotto M, Machado RB, Tufik S. Does sleep deprivation and cocaine induce penile erection and ejaculation in old rats? Addict Biol. 2002;7:285–90.

Andersen ML, Bignotto M, Machado RB, Tufik S. Effects of chronic stress on steroid hormones secretion in male rats. Braz J Med Biol Res. 2004a;37:791–7.

Andersen ML, Bignotto M, Tufik S. Hormone treatment facilitates penile erection in castrated rats after sleep deprivation and cocaine. J Neuroendocrinol. 2004b;16:154–9.

Andersen ML, Martins PJF, D'Almeida V, Santos RF, Bignotto M, Tufik S. Effects of paradoxical sleep deprivation on blood parameters associated with cardiovascular risk in aged rats. Exp Gerontol. 2004c;39:817–24.

Andersen ML, Martins PJ, D'Almeida V, Bignotto M, Tufik S. Endocrinological and catecholaminergic alterations during sleep deprivation and recovery in male rats. J Sleep Res. 2005; 14:83–90.

Andrade A. Animais de Laboratório: criação e experimentação. Organizado por Antenor Andrade, Sergio Correia Pinto e Rosilene Santos de Oliveira, Rio de Janeiro: Fiocruz; 2002. p. 387.

Baker DEJ. Reproduction and breeding. In: Baker HJ, Lindsey JR, Weishbroth SH, editors. The laboratory rat. Biology and diseases. New York: Academic; 1979. p. 154–68.

Barnett SA. The rat: a study in behavior. Chicago: Aldine Publications; 1964.

Bertholet JY. Mating method to produce accurate timed pregnancies in rats. Lab Anim Sci. 1981;31:180–2.

Black WD, Claxton MJ. A simple, reliable and inexpensive method for the collection of rat urine. Lab Anim Sci. 1979;29:253–4.

Breazile JE, Kitchell RL. Euthanasia for laboratory animals. Fed Proc. 1969;28:1577–9.

Cameron TP, Lattuada CP, Kornreich MR, Tarone RE. Longevity and reproductive comparisons for male ACI and Sprague-Dawley rat aging colonies. Lab Anim Sci. 1982;32:495–9.

Canadian Council on Animal Care. Guide to the care and use of experimental animals. Ottawa: Canadian Council on Animal Care (CCAC); 1984. p. 120.

Cardy RH, Warner JW. Effect of sequential bleeding on body weight gain in rats. Lab Anim Sci. 1979;29:179–81.

Chiasson RB. Laboratory anatomy of the white rat. New York: WCB McGraw-Hill; 1994.

D'Almeida V, Lobo LL, Hipolide DC, De Oliveira AC, Nobrega JN, Tufik S. Sleep deprivation induces brain region-specific decreases in glutathione levels. Neuroreport. 1998;9:2853–6.

Davis DR, Yeary RA. Impaired fertility in the jaundiced female (Gunn) rat. Lab Anim Sci. 1979;29:739–43.

De Oliveira AC, D'Almeida V, Hipolide DC, Nobrega JN, Tufik S. Sleep deprivation reduces total plasma homocysteine levels in rats. Can J Physiol Pharmacol. 2002;80:193–7.

De-Luca RR, Alexandre SR, Marques T, Souza NL, Merusse JL, Neves SP. Manual para técnicos em bioterismo. 2nd ed. São Paulo: Winner Graph; 1996. p. 259.

Dym M, Clermont Y. Role of spermatogonia in the repair of the seminiferous epithelium following X-irradiation of the rat testis. Am J Anat. 1970;128:265–82.
Ehrensreund D. Use of rodents in behavioral research. In: Gay WI, editor. Methods of animal experimentation. New York: Academic; 1968. p. 1–25.
Gartner K, Buttner D, Dohler K, Friedel R, Lindena J, Trautschold I. Stress response of rats to handling and experimental procedures. Lab Anim. 1980;14:267–74.
Harkness JE, Wagner JE. The biology and medicine of rabbits and rodents. 2nd ed. Philadelphia: Lea & Febiger; 1983.
Hughes HD. Euthanasia of laboratory animals. In: Melby Jr EC, Altman NH, editors. Handbook of laboratory animal science. Cleveland: CRC; 1976. p. 3.
Jaenisch R. Transgenic animals. Science. 1988;240:1468–74.
Kim SU. Brain hypoxia studied in mouse central nervous system cultures I. Sequential cellular changes. Lab Invest. 1975;33:658–69.
Krinke GJ, ed. The laboratory rat. San Diego: Academic; 2000. 634p.
Lane-Petter W. The laboratory animal: principles and practice. New York: Academic; 1971.
Levine S. Primary social relationships influence the development of the hypothalamic-pituitary-adrenal axis in the rat. Physiol Behav. 2001;73:255–60.
Lindsey JR. Historical foundations. In: Baker HJ, Lindsey JR, Weisbroth SH, editors. The laboratory rat. Biology and diseases. New York: Academic; 1979. p. 2–36.
Mazaro R, Lamano-Carvalho TL. Prolonged deleterious effects of neonatal handling on reproductive parameters of pubertal male rats. Reprod Fertil Dev. 2006;18:497–500.
Morse I, Herbert C. The laboratory mouse. A historical perspective. In: Foster HL, Small D, Fox J, editors. The mouse in biomedical research. History, genetics and wild mice. New York: Academy; 1981. p. 1–17.
Palma BD, Gabriel Jr A, Bignotto M, Tufik S. Paradoxical sleep deprivation increases plasma endothelin levels. Braz J Med Biol Res. 2002;35:75–9.
Pesquero JB. Animais transgênicos. Biotechnol Cien e Desenvol. 2002;27:52–6.
Randall D, Burgeren WW, Eckert R, French K. Animal physiology: mechanisms and adaptations. 5th ed. New York: WH Freeman; 2002.
Roizenblatt S, Andersen ML, Bignotto M, Smith AK, Tufik S. The impact of arthritics offspring on the lactating dam's behaviour, pain threshold and sleep. J Sleep Res. 2002;11:195–6.
Roizenblatt S, Andersen ML, Bignotto M, D'Almeida V, Martins PJ, Tufik S. Neonatal arthritis disturbs sleep and behaviour of adult rat offspring and their dams. Eur J Pain. 2010;14:985–91
Sharpe RM, Morris A, Wyatt AC. The effect of the sex of littermates on the subsequent behaviour and breeding performance of cross-fostered rats. Lab Anim. 1973;7:51–9.
Suter P, Luetkemeier H, Zakova N. Lifespan studies on male and female mice and rats under SPF-laboratory conditions. Arch Toxicol Suppl. 1979;2:403–7.
Windhager EE. Micropuncture techniques and nephron function. New York: Appleton-Century-Crofts; 1968.
Wingerd BD. Rat dissection manual. Baltimore: The Johns Hopkins University Press; 1988. p. 68.

Chapter 8
The Female Rat

Isabela Beleza Antunes, Andressa da Silva, Regiane Kawakami, and Monica Levy Andersen

Estrous Cycle

The rat is one of the most ubiquitous experimental animals used in the investigation of reproductive physiology. This specie shows sexually dimorphic behaviors, which are those that differ due to gender and may also vary depending on the physiologic state of the animals involved (Kustritz 2005). Like in humans, the reproductive cycle of the female rat acts in tandem with the cycle of sexual hormones.

The reproductive system consists of the vagina, the uterus, the fallopian tubes, and the ovaries. The female rat possesses six pairs of mammary glands distributed in the following regions: three pairs in the thoracic region, one in the abdominal region, and two in the inguinal region (Fig. 8.1). After birth, the nipples of the females become visible within 1 week.

Female rats experience a hormonal cyclicity called the estrous cycle that usually begins at 35–45 days of age and ceases between 10 and 15 months of life (Meites et al. 1980). The length of a rat's estrous cycle may vary from 4 to 5 days and con-

I.B. Antunes
Universidade Federal de São Paulo (UNIFESP), São Paulo, São Paulo, Brazil

A. da Silva, Ph.D. (✉)
Department of Physiotherapy, Universidade Federal de Minas Gerais, Rua Antonio Auusto de Carvalho, 97/105, Belo Horizonte, Minas Gerais 031340-020, Brazil
e-mail: silvadressa@gmail.com

R. Kawakami
Faculty of Pharmaceutical Sciences, Universidade de São Paulo (UNIFESP), São Paulo, São Paulo, Brazil

M.L. Andersen
Department of Psychobiology, Chief of Sleep Division, Universidade Federal de São Paulo (UNIFESP), Rua Napoleao de Barros, 925, São Paulo, São Paulo, Brazil
e-mail: ml.andersen12@gmail.com

M.L. Andersen, S. Tufik (eds.), *Rodent Model as Tools in Ethical Biomedical Research*, DOI 10.1007/978-3-319-11578-8_8

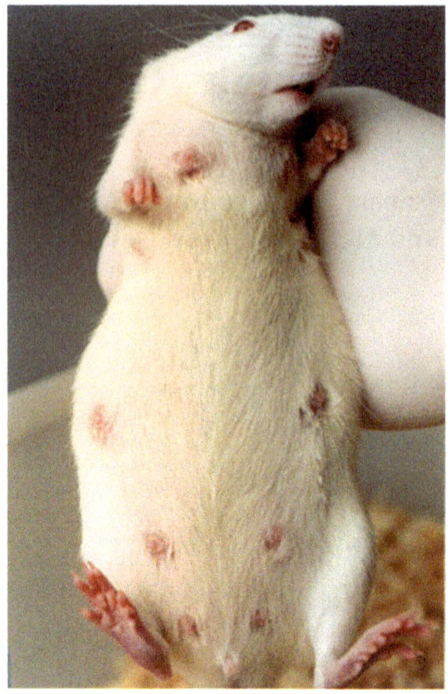

Fig. 8.1 Image of the mammary glands of the adult female rat

sists of four phases: proestrus, estrus, metestrus (or diestrus I), and diestrus (or diestrus II), as can be observed in Fig. 8.2.

The four phases of the estrous cycle are described below:

(a) **Proestrus:** the phase of the estrous cycle that is characterized by the presence of epithelial cells with predominantly large and nucleated cells and occasional cornified cells. It lasts 12–14 h and precedes estrus. During this phase, the estrogenic action over the vaginal epithelium becomes evident, and the highest concentrations of estradiol occur with follicular development.
(b) **Estrus:** this phase is evidenced by the predominance of cornified cells. This phase begins at night and lasts about 25–27 h. During this phase, ovulation occurs spontaneously, and females become receptive to males and exhibit behavioral alterations that increase the probability of mating.
(c) **Metestrus:** this phase is often called diestrus I and is characterized by the presence of few leukocytes, mucus, and few cornified epithelial (nucleated) cells. It lasts about 6–8 h, during which vulva tumefaction and caseous mass occur in the vagina. This is the luteinic phase that corresponds to the formation of *corpora lutea* on the surface of the ovary. The *corpora lutea* produces the hormone progesterone, which impedes the maturation of new follicles and allows the maintenance of pregnancy.
(d) **Diestrus:** is also recognized as diestrus II. The predominant cells are leukocytes, there are few epithelial cells, and there is abundant mucus in the vaginal smear. This is the longest phase and typically lasts about 55–57 h. This period corresponds to the action of progesterone in the impediment of new follicle maturation. The duration of the phase depends on the time it takes for *corpora lutea* regression to occur, signaling the cessation of the action of progesterone.

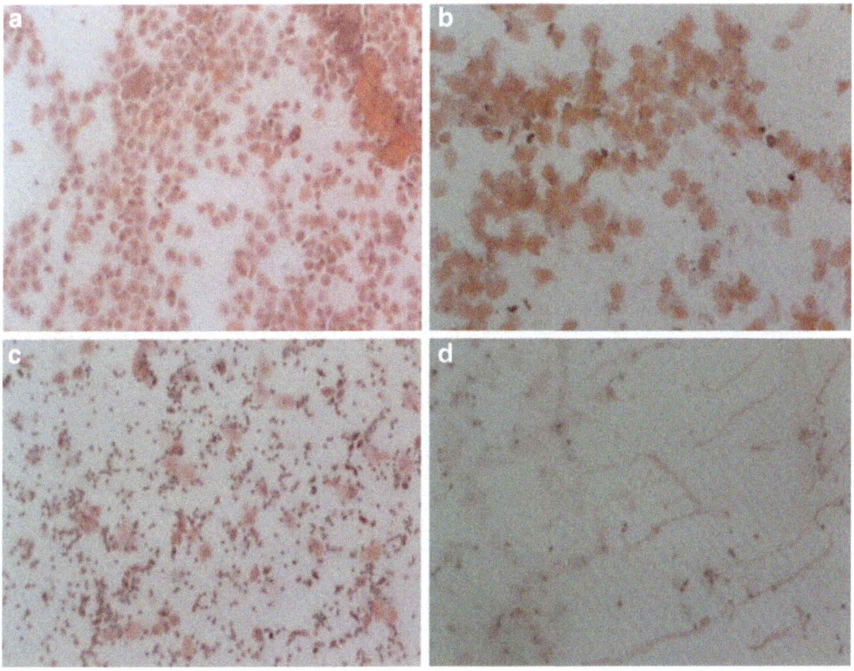

Fig. 8.2 Predominant cells of the four phases: proestrus (**a**), estrus (**b**), metestrus (**c**), and diestrus (**d**)

Technique to Obtain Vaginal Smears

The vaginal smear is the most popular and reliable method to identify the phase of the estrous cycle (Fig. 8.3). The cycle stages may be determined according to the predominant cell types observed in the vaginal smear.

A glass or plastic pipette containing distilled water is gently inserted into the vagina, and a small amount of the liquid is flushed and drawn back into the pipette. The collected cells are smeared onto a slide, stained using different methods, and examined under a microscope. The use of saline solution to obtain cells is not recommended because crystals may form as the liquid dries up, compromising the identification of the estrous cycle.

In some instances, anestrus can be observed. This phase takes place when ovaries become inactive without follicular maturation, and is observed during pregnancy or at the end of reproductive life.

Sexual maturity occurs between 6 and 8 weeks, and although rats ovulate spontaneously, the activation of the *corpora lutea* depends on the neural input originating from mechanical stimulation of the uterine cervix. Such stimulation is usually associated with coitus but can be mimicked in the laboratory by cervical stimulation with a pipette during collection of vaginal cells. This accidental stimulation results in a hormonal and physiological condition known as pseudopregnancy and exhibits

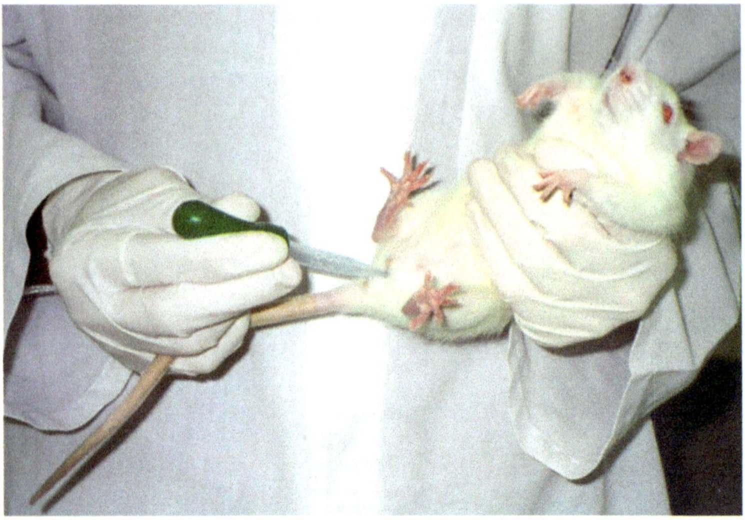

Fig. 8.3 Female rat prepared for a vaginal smear procedure

the same endocrine mechanisms of the initial phase of normal pregnancy. This condition lasts about 12 h, resulting in a delay of the next estrous cycle.

The vaginal smears require a plastic pipette used for sampling with a small amount of distilled water put into the vagina. The cells are subsequently sucked up into the pipette. These cells are placed onto a slide and observed through a microscope (usually ×100).

Hormonal Regulation of Estrous Cycle

Between puberty and aging, the ovary has a cyclical function under the control of the hypothalamus-pituitary axis. Indeed, the hypothalamus-pituitary-ovary axis promotes the endocrine reproductive control over the estrous cycle by negative and positive feedback. Ovarian function is considered a continuing phenomenon, in which the ovary has its own activity and local regulation, whereas the hypothalamus-pituitary axis controls and modulates the final steps of follicular growth and ovulation under the influence of the alternating tides of ovarian signals (estradiol and progesterone) (Buffet et al. 1998). Ovulation occurs from the beginning of proestrus to the end of estrus (Marcondes et al. 2001).

The hypothalamus, anterior pituitary, gonads, and reproductive tract regulate reproductive functions. The hypothalamus is essential for the secretion of two hormones: gonadotrophin-regulating peptide hormone and luteinizing-hormone-releasing hormone (LHRH). LHRH controls the secretion of the follicle-stimulating hormone (FSH) and luteinizing-hormone (LH) released by the pituitary gland. This gland is essential for prolactin release, another hormone involved in the hormonal cycle. Specifically, the medial pre-optical area and medial septal area are neuron groups that synthesize and secrete gonadotrophin-release-hormone GnRH in the central nervous system (CNS). This hormone stimulates the pituitary to release gonadotrophins, LH and FSH. The release of gonadotrophin stimulates female gonads to synthesize and release steroid hormones, mainly estradiol and progesterone. These steroids participate in the control of cellular and behavior modification during the estrous cycle.

FSH regulates growth of follicles in the ovaries, and LH induces ovulation and formation of the *corpora lutea*. These two gonadotrophins also stimulate estrogen and progesterone secretion by the ovaries, which in turn promote development of the female reproductive tract and feedback to the hypothalamus-pituitary system to regulate gonadotrophin release (Meites et al. 1980).

Both the menstrual and the estrous cycles are divided into a follicular phase and a luteal phase. Both are dominant expressions dictated by the predominance of estrogen in the former and progesterone in the latter (Meites et al. 1980). As such, the proestrous phase of the female rat can be considered the follicular phase and the diestrous phase, the luteal phase. During the estrous cycle, the estrogen produces many changes in the vaginal epithelium.

Hormonal evaluation is necessary to determine the endocrine reproductive function. This evaluation is made possible by the assessment of protein hormones (LH, FSH) or steroids (estradiol, progesterone) using radioimmunoassay or chemiluminescence techniques in serum or in homogenized tissue (steroidogenesis evaluation). By punch and microdialysis, it is possible to measure GnRH and other hormones (arginine vasopressin [AVP], corticotrophin-releasing hormone [CRH]) or neurotransmitters (e.g., dopamine) that control the estrous cycle.

The follicular phase is characterized by the development of the follicle and a rise in estrogen secretion shortly before ovulation. The rise in estrogen stimulates the hypothalamus to release LHRH and sensitizes the pituitary to its action, followed by a surge of LH and FSH release from the pituitary. The surge of gonadotrophins evokes ovulation and the formation of the *corpora lutea*.

During the luteal phase, progesterone and estrogen secretion from the *corpora lutea* inhibits further surges of LHRH and gonadotrophin release. If fertilization fails to occur, the *corpora lutea* degenerates and progesterone and estrogen secretion fall, allowing for the reinitiating of the cycle. At the end of the luteal phase, there is also some loss of uterine and vaginal epithelium, but no bleeding occurs (Meites et al. 1980).

Estradiol serum concentration is low between the estrous phase and the diestrous morning and begins to rise in the diestrous afternoon. Estradiol serum concentration reaches its peak in the midday of the proestrous phase and decreases to baseline values towards the end of the night of the estrous phase. This pattern of secretion is a trigger for pre-ovulatory peaks of gonadotrophins and prolactin (PRL).

Progesterone serum concentrations begin to surge at the end of the estrous phase, and the maximum value is reached in tandem with the LH pre-ovulatory peak at proestrus and begins to fall in the estrous morning. At midday of diestrus I, a second peak occurs, returning to baseline values in the diestrus II morning. This hormone promotes negative feedback on FSH and LH secretion.

LH concentration shows one peak in proestrous mid-afternoon that induces ovulation. The baseline values are obtained at estrus and maintained in diestrus I and II. FSH plasmatic concentration is similar to LH, but this hormone presents a secondary peak at the final night of estrus. With respect to prolactin secretion, two peaks can be observed: one in the middle of the proestrous afternoon, accompanied by a pre-ovulatory LH peak, and another at post-ovulatory time.

The periovulatory interval is defined as the onset of an ovulatory gonadotrophin stimulus until follicle rupture. In rats, this interval lasts approximately 12–16 h, during which time there are rapid alterations in steroid synthesis, increased expression of proteolytic enzymes, as well as the production of angiogenic factors (Murphy 2000; Richards et al. 2002). These hormonal oscillations (Fig. 8.4) and other functional cascades work in concert to ensure the extrusion of a fertilizable oocyte and formation of a *corpora lutea*. It has been assumed that luteinization represents the terminal differentiation of granulosa into luteal cells (Rao et al. 1978).

There is divergence among authors about the age of the first overt sign of reproductive aging in female rats. Some speculate it occurs around 10–15 months of age (Meites et al. 1980), and others report 8–12 months of age when the preovulatory LH surge becomes attenuated and delayed (Wise 1982). It remains to be determined what causes the attenuation of LH surges at middle age, although it most likely results from a decrease in response of the hypothalamic-pituitary axis to the positive feedback action of estradiol. The regular estrous cycle tends to become irregular, usually characterized by an increase in the number of days in estrus or diestrus. By 15–24 months of age, most female rats show a constant estrous syndrome or persistent estrus, characterized by many well-developed and even cystic follicles, but no evidence of ovulation or formation of *corpora lutea* (Meites et al. 1980). This state of acyclicity is termed *persistent estrus* and is characterized by persistent vaginal cornification, which is indicative of sustained follicular development and estradiol secretion in the absence of LH surges and ovulation (Meites and Huang 1976; Lu et al. 1979).

The oldest rats, 2–3 years of age, often exhibit an anestrous state characterized by atrophic ovaries. Hormonal secretory patterns in the aging female, constant estrous or "anestrous", differ from those of the mature regular estrous cycling rats, and the difference is mainly that gonadotrophins, estradiol, progesterone, and prolactin show no cyclic surges (Meites et al. 1980). There is some evidence that the pituitary and ovaries of aging female rats may be less responsive to stimulation by LHRH and gonadotrophins, respectively, than mature regular estrous cycling rats (Meites et al. 1976, 1978).

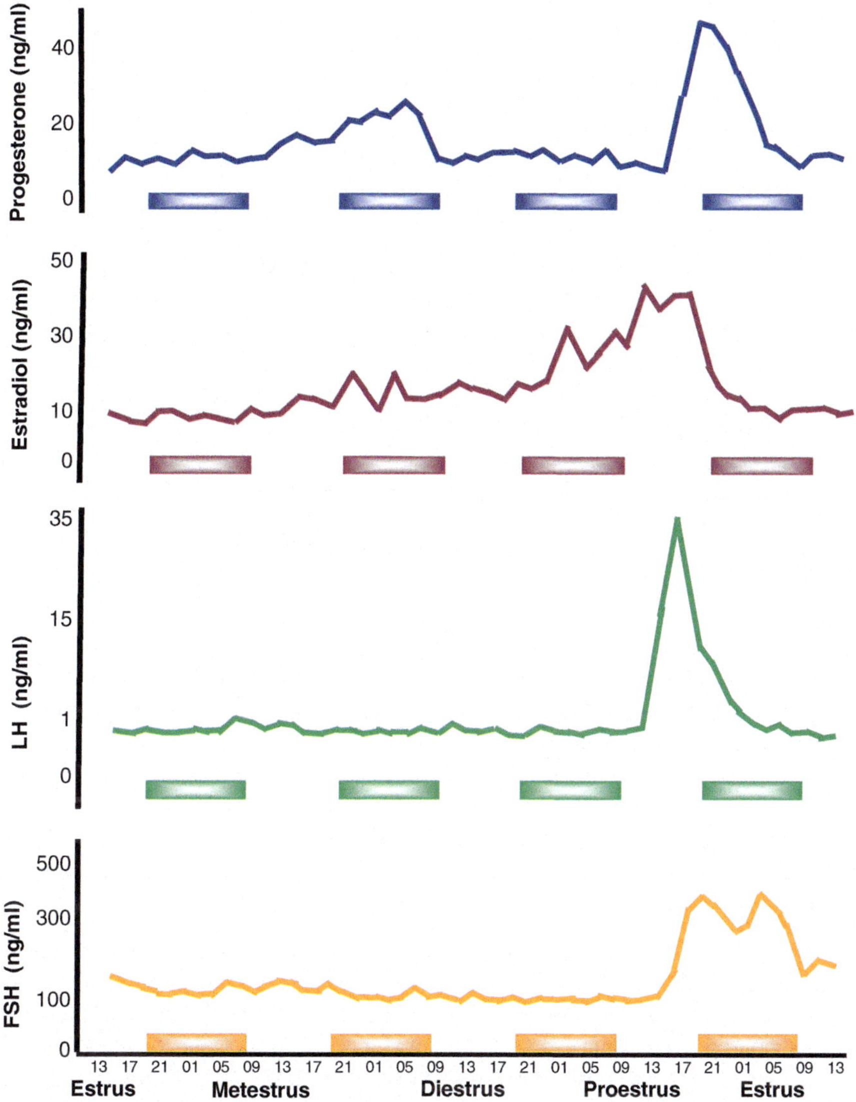

Fig. 8.4 Hormones concentrations (progesterone, estradiol, luteinizing hormone—LH, and follicle-stimulating hormone—FSH) across the estrous cycle of rats. Modified from Smith et al. (1975). For review see Krinke (2000)

Pregnancy Evaluation

Mating is usually nocturnal and a copulatory plug forms from secretions of the seminal vesicle and coagulating glands following copulation. This copulatory plug fills the female reproductive tract from vulva to cervix, and remains for a few hours

Fig. 8.5 Interactions of mother and her pups

following copulation. The presence of copulatory plugs in the cage is an indicative of successful mating. Alternatively, the presence of sperm in the vaginal smear in the morning following cohabitation can also be used to confirm mating.

The pregnancy lasts about 20–23 days. Post delivery estrus occurs 48 h after delivery, and matings in this period are more than 50 % successful. For practical purposes, pregnancy may be divided into three periods. Knowledge of these periods is fundamental for reproductive research, especially in reproductive toxicology to assess embryofetotoxicity:

- preimplantation: lasts from the beginning of pregnancy until embryo implantation in the uterus (day 5 of gestation), and is characterized by the presence of totipotent cells. A delay in embryo implantation can occur when pregnant females are housed in the same cage as males, and this is known as the "Bruce Effect";
- organogenesis: lasts from implantation until the time when closure of the hard palate is complete (from day 6 to day 15 of gestation), and during this period the germ layers develop into the internal organs of the forming organism. This period is characterized by differentiation and individualization of organs and systems.
- fetal: extends from day 16 until delivery, characterized by fetal growth, and maturation and development of organ systems formed during embryogenesis.

The rat has a discoidal placenta and hemochorial placentation. Normal delivery occurs during daytime on day 21–22 of pregnancy. The litter size varies according to the subspecies but normally is 8–15 pups for Wistar rats. Newborns weigh about 5 g at birth and are blind, very active, and depend on care and maternal nursing to survive until weaning (Fig. 8.5).

Weaning occurs at about 21 days after birth and is characterized by the replacement of breast suckling for solid food by the pups. There is a decrease in prolactine synthesis by the dam followed by the end of the nursing period.

During the weaning time the experimenter needs to devote much care to the litter, because it is in this phase that postnatal neural maturation of rats occurs. The relationship between dams and pups needs to be preserved. The manipulation of pups can be done with minimal disturbance to the mother-pup interaction because they form one biologic unit, and disruption in this chemistry may cause behavioral, physiological, and reproductive alterations at puberty and adulthood.

Experimental Methods in Reproductive and Developmental Toxicity

There are different standardized methods for assessment of adverse effects on reproduction and development caused by chemicals, and comprehensive guidelines have been published by various governmental agencies and international organizations.

Prenatal Developmental Toxicity Test

The prenatal developmental toxicity test is used to assess the potential a chemical substance has to affect the organism during critical periods of growth and development. Formerly, this test was often referred to as teratology. Pregnant animals are treated from implantation until the day before the expected birth, and the fetuses are assessed for toxic effects. Rats, mice, and rabbits are the species of choice for this test.

The following procedures are used in rats and must be adapted to the other species. Mating and pregnancy diagnostic: for mating, the animals should be healthy, nulliparous, and sexually mature (about 10–12 weeks old). In the late afternoon, 1–3 females are placed in the same cage with a male, and in the subsequent morning females are examined for evidence of copulation. The presence of a vaginal plug on the bottom of the cage or in the vagina, or the presence of sperm in a vaginal smear, is indicative of mating, and this day is considered day 0 of gestation.

Treatment and maternal evaluation: At least 20 mated females per group are assigned to the experimental groups (usually four groups, including control). These animals are treated daily with the chemical to be tested from implantation (day 6 of gestation) until one day before expected birth (day 20 of gestation). The chemical is usually administered orally, by gavage, using a gastric probe. During the gestation period, dams are observed for clinical signs of toxicity, and body weight and food consumption are recorded at regular intervals.

Necropsy: On day 20 of gestation, the dams are euthanized by carbon dioxide inhalation and a caesarian section is performed. The dams are evaluated macroscopically, the ovaries removed, and *corpora lutea* (round and brownish pink aspect) counted. The uterus is removed and its contents evaluated. The conditions of the placenta, the umbilical cords, the fetal membranes, and fluids are examined. The number and location of viable and dead fetuses, early (only decidual or placental remnants are visible) and late (embryonic and fetal tissues visible), are recorded as well as the number of implantation sites. The uteri of apparently non-pregnant animals are immersed in 0.5 % ammonium sulfide solution for approximately 10 min, and the presence of implantation sites (stained blue-black) examined. Viable fetuses and respective placentas are weighed.

External Evaluation: After weighing, the fetuses are examined for external abnormalities in an orderly manner from head to tail. Each fetus is sexed based on the anogenital distance. The fetuses are euthanized, and the visceral and skeletal examinations performed.

Visceral Evaluation: Visceral or soft tissue examination may be performed using the fetal sectioning method of Wilson (1965) or the micro dissection method (Barrow and Taylor 1969). For the fetal sectioning method, one half of the fetuses are fixed in Bouin's solution after external examination, preserved in 70 % alcohol, and free-hand razor blade sections of head, thorax and abdomen are harvested. All sections are examined in the appropriate order under a stereoscope or a dissection microscope. For the microdissection method, abdominal and thoracic organs of all fetuses are macroscopically examined in situ and organ positions carefully observed. In addition, the heads of one half of the fetuses are removed, fixed in Bouin's solution, and examined using the serial sectioning method.

Skeletal Evaluation: For skeletal examination, the fetuses are eviscerated, fixed in 95 % alcohol or acetone, placed in potassium hydroxide (KOH), stained with Alizarin Red S (stains calcified bone), and cleared and stored in glycerin. Another technique is double staining with alizarin and alcian blue, which stains calcified bone and cartilage, respectively.

Single and Multigenerational Reproduction Tests

These tests involve continuous exposure to a chemical for one or more generations. The objective is to detect effects on the integrity and performance of male and female reproductive systems, including gonadal function, the estrous cycle, mating behavior, conception, gestation, parturition, lactation, and weaning, as well as on the growth and development of the offspring.

Usually, the chemical is added to the diet that is continuously offered to one or two generations. The treatment starts when the animals are approximately 7 weeks old (P parental animals) and ends when the animals are euthanized. At least 20 animals of each sex are assigned to each experimental group (usually four groups including control), and these animals constitute P parental generation.

Premating Period: This period lasts about 70 days. During this period, body weight and food consumption of P parental males and females are recorded weekly, and the animals are observed for clinical signs of toxicity. The estrous cycle is monitored for the last 3 weeks of premating period by daily vaginal smears. The stage of cycle is determined to evaluate possible effects on normal cycling.

Mating Period: At the end of the premating period, females are cohabited with an assigned male (1 female:1 male) from the same experimental group until evidence of copulation is observed, or either three estrous periods or 2 weeks have elapsed. Vaginal smears are collected daily during the mating period and examined for the presence of sperm. Day 0 of gestation is defined as the day sperm is found in the vaginal smear.

Gestation Period: During this period, body weight and food consumption of pregnant animals are recorded as well as clinical signs of toxicity. From day 20 of gestation, the animals are examined at least two times per day for signs of parturition. At birth, pups are weighed, counted, sexed, and examined for external abnormalities.

Lactation Period: At day 4 of lactation, litters can be culled to eight pups per litter, whenever possible four males and four females. Each pup is weighed at intervals of about 7 days. In order to evaluate physical development, the day of occurrence of eyes opening, incisor eruption, and unfolding of the external ear are recorded. Functional investigations (reflex ontogeny), such as righting reflex, grip reflex, and negative geotaxis are carried out in the pups. In studies of one generation, all parental animals and all pups are killed at weaning (day 21 of lactation), ending in-life phase. In multigenerational studies, parental animals are killed, and one pup/sex/litter is selected to constitute the F1 parental animals. The day of occurrence of vaginal opening and preputial separation is recorded on the selected animals. Nonselected pups are killed on day 21 postnatal (weaning).

F1 and F2 generation: All P generation procedures during premating, mating, gestation, and lactation periods are similar for the F1 generation, and the offspring of these animals are called F2 generation. The study ends at the weaning of F2 pups.

Necropsy: All parental animals and pups are examined macroscopically for any structural abnormalities or pathological changes. Reproductive organs of parental animals are weighed and evaluated microscopically. Sperm evaluation, including sperm motility and morphology assessment and enumeration of homogenization-resistant spermatids and cauda epididymal sperm reserves, is performed in parental animals.

Stress and Female Behavior

Stress is revealed by the inability of an animal to cope with its environment, a phenomenon that is often reflected in a failure to achieve genetic potential, e.g. for growth rate, milk yield, disease resistance, or fertility (Dobson and Smith 2000). The complex nature of some stressors in the modern environment simultaneously exposes animals to several different stimuli (Dobson and Smith 2000).

It is well known that stress interferes with reproductive function in humans and animals (Rivier and Rivest 1991; Dobson and Smith 1995; Magiakou et al. 1997). In fact, a large amount of work has been devoted to understanding the mechanisms by which stress inhibits reproductive function (Kam et al. 2002). The activation of the hypothalamic-pituitary-adrenal (HPA) axis during stress is known to affect the hypothalamic-pituitary-gonadal axis, especially by modulating the secretion and pulse frequency of gonadotrophin (Calogero et al. 1998). The suppression of reproductive function following activation of the HPA axis also appears to be species-specific (Jeong et al. 1999).

It is common knowledge that the duration and nature of the stressor are important determinants (Kloet 2003). CRH has been proposed to negatively regulate GnRH secretion. An imbalance of central CRH has been implicated in the suppression of reproductive function in humans under stressful conditions (Jeong et al. 1999).

Ovarian cyclicity has profound effects on most, if not all, of the nervous system. In rats, as well as other species, these effects produce transient changes in behavior. For example, reproductive behavior, food intake, fluid intake, and locomotor activity vary markedly with the stage of the estrous cycle in female rats (Long and Evans 1922; Freeman 1994; Eckel et al. 2000). Some behavioral studies with female rats have investigated the differences in experimental designs associated with the estrous cycle. The cyclicity of the ovarian hormones is the main difference between male and female rats. All ovarian hormones show fluctuations throughout the estrous cycle, and these fluctuations may contribute to alterations in response to stress depending of the phase of the estrous cycle. In cyclic rats, the preovulatory LH surge occurs on the afternoon of proestrus as a consequence of increased hypothalamic GnRH secretion (Roozendaal et al. 1997).

In general, female rats are more vulnerable to chronic mild-stress, showing disruption of sacarose intake, decreases in open filed activity, and increased corticosterone concentrations when compared with males (Dalla et al. 2005). All these alterations can indicate the abnormal modulation of the HPA axis and the sex hormones. Other stress protocols commonly used are forced swimming (Axelson 1987) and forced running (Carlberg and Fregley 1985), which are responsible for the disruption of the estrous cycle showing anestrous periods for 19 days (Axelson 1987).

The disruption of the estrous cycle can be evidenced in female rats submitted to physical stress (food deprivation, forced swim test, electrical shocks, etc.) within the first week of a 14-day treatment (Rodríguez-Echandía et al. 1988). Otherwise, immobilization tress during 28 days showed no significant difference in average of the estrous cycle length between the stress group and the control group (Bowman et al. 2001). Except for the immobilization stress, all physical stress caused a disruption in the estrous cycle. It was verified that emotional stress (blood odor, individual housing, exposure to darkness, etc.) treatment did not affect the estrous cycle (Rodríguez-Echandía et al. 1988).

It is not only the estrous cycle that has been observed in various studies using a stress protocol; some behaviors during the estrous cycle are also investigated. It is well know that the variation in motor activity and sleep is observed in the course of the estrous cycle (Yokoyama et al. 1966; Kleinlogel 1983; Fang and Fishbein 1996). A disruption in the estrous cycle of female rats submitted to sleep deprivation in diestrous phase for 96 h was observed. These female rats showed an anestrous period during the first week of the rebound period, demonstrating the relevance of scrutinizing the estrous cycle during and after a stress protocol (Antunes et al. 2006) as well as sleep alterations (Andersen et al. 2008).

In many cases, fluctuation of the ovarian hormones such as estrogen and progesterone has been associated with changes in mood (Hiroi and Neumaier 2006). In rodents, the open field evaluation showed a significant alteration in outer sector crossings, with stressed females making fewer overall visits than controls, although without significant alterations in grooming, rearing or inside sector crossing (Bowman et al. 2001). Furthermore, daily restraint stress for 21 days led to a small albeit significant difference in performance on the radial arm maze. In contrast, stress enhanced performance as measured by the total number of visits required to complete the task. Stressed females completed the task in fewer visits than controls (Bowman et al. 2001). Interestingly, the effect of neonatal isolation in food-restricted female rats in their estrous cycle compared with the non-handling group was that the greatest percentage of time was spent in diestrus compared with the time spent in estrus or proestrus (Kosten et al. 2005). Such experimental studies using a stress protocol may contribute to the understanding of the stress-response mechanism and the influence of the sex hormones in some behaviors.

Endocrine systems appear to be an ideal way of coordinating this regulation throughout the whole body. In order to unravel the complexity of stress-induced subfertility, it is necessary to study the reaction to stressors of repeatable severity, firstly by examining responses to clearly defined stimuli, and then by investigating the influence on reproductive mechanisms (Dobson and Smith 2000).

Sleep and Estrous Cycle

Results of studies focusing on the relationship between sleep and the estrous cycle in rats have offered a comprehensive view of the alterations observed in the sleep pattern during the female hormonal cycle. The greatest alterations in the sleep pattern occur during proestrus, when there is a surge in wakefulness time and a consequent reduction of sleep time and slow wave and paradoxical sleep (Colvin et al. 1969). There is evidence of an increase of sleep and a reduction of wakefulness during metestrus (Kleinlogel 1975). According to these authors, female rats presented an increase of total wakefulness time and a reduction of slow wave sleep as well as of paradoxical sleep during the night of the estrous phase. In contrast, Fang and Fishbein (1996) and Kimura and colleagues (1996) reported an absence of any alteration in slow wave sleep throughout the estrous cycle. A more recent investigation has confirmed a meaningful reduction in slow wave sleep and of paradoxical sleep during the night period, and an increase in the number of arousals in the proestrous phase as well as an increase in the total sleep time during the light phase of estrus when compared to the other phases (Schwierin et al. 1998). No differences in sleep patterns of female rats across the estrous cycle during the baseline period were observed by Andersen and colleagues (2008). These authors observed that paradoxical sleep deprivation-induced alterations in sleep patterns during the rebound period in male and female rats in different phases of the estrous cycle. According to the findings, males and females submitted to PSD in estrus and diestrus presented an increase in sleep efficiency and paradoxical sleep in the first day of the light rebound period. Although all females returned to paradoxical sleep baseline values on the third day, males and cycling diestrus females returned to normal on the second day of the dark rebound period. These findings demonstrate that females and males displayed distinct coping resources when sleep was disrupted.

On this basis, more in-depth knowledge of the functional significance of sleep deprivation in females should be obtained to assist current endeavors to find ways of preserving and promoting better sleep throughout the lifespan.

References

Andersen ML, Antunes IB, Silva A, Alvarenga TA, Baracat EC, Tufik S. Effects of sleep loss on sleep architecture in Wistar rats: gender-specific rebound sleep. Prog Neuropsychopharmacol Biol Psychiatry. 2008;32:975–83.

Antunes IB, Andersen ML, Baracat EC, Tufik S. The effects of paradoxical sleep deprivation on estrous cycles of the female rats. Horm Behav. 2006;49:433–40.

Axelson JF. Forced swimming alters vaginal estrus cycles, body composition, and steroid levels without disrupting lordosis behavior or fertility in rats. Physiol Behav. 1987;41:471–9.

Barrow MV, Taylor WJ. A rapid method for detecting malformations in rat fetuses. J Morphol. 1969;127:291–306.

Bowman RE, Zrull MC, Luine VN. Chronic restraint enhances radial arm maze performance in female rats. Brain Res. 2001;904:279–89.

Buffet NC, Djakoure C, Maitre SC, Bouchard P. Regulation of the human menstrual cycle. Front Neuroendocrinol. 1998;19:151–86.

Calogero AE, Bagdy G, D'Agata R. Mechanisms of stress on reproduction. Evidence for a complex intra-hypothalamic circuit. Ann N Y Acad Sci. 1998;851:364–70.

Carlberg KA, Fregley MF. Disruption of estrous cycles in exercise-trained rats. Proc Soc Exp Biol Med. 1985;179:21–4.

Colvin GB, Whitmoyer DI, Sawyer CH. Circadian sleep-wakefulness patterns in rats after ovariectomy and treatment with estrogen. Exp Neurol. 1969;25:616–25.

Dalla C, Antoniou K, Drossopoulou G, Xagoraris M, Kokras N, Sfikakis A, Papadopoulou-Daifoti Z. Chronic mild stress impact: are females more vulnerable? Neuroscience. 2005;135:703–14.

Dobson H, Smith RF. Stress and reproduction in farm animals. J Reprod Fertil Suppl. 1995;49: 451–61.

Dobson H, Smith RF. What is stress, and how does it affect reproduction? Anim Reprod Sci. 2000;60–61:743–52.

Eckel L, Houpt T, Geary N. Spontaneous meal patterns in female rats with and without access to running wheels. Physiol Behav. 2000;70:397–405.

Fang J, Fishbein W. Sex differences in paradoxical sleep: influences of estrus cycle and ovariectomy. Brain Res. 1996;734:275–85.

Freeman ME. The neuroendocrine control of the ovarian cycle of the rat. In: Knobil E, Neill J, editors. The physiology of reproduction. 2nd ed. New York: Raven; 1994. p. 613–58.

Hiroi R, Neumaier JF. Differential effects of ovarian steroids on anxiety versus fear as measured by open field test and fear-potentiated startle. Behav Brain Res. 2006;166:93–100.

Jeong KH, Jacobson L, Widmaier EP, Majzoub JA. Normal suppression of the reproductive axis following stress in corticotropin-releasing hormone-deficient mice. Endocrinology. 1999;140: 1702–8.

Kam KY, Park YB, Cheon MS, Kang SS, Kim K, Ryu K. Influence of GnRH agonist and neural antagonists on stress-blockade of LH and prolactin surges induced by 17beta-estradiol in ovariectomized rats. Yonsei Med J. 2002;43:482–90.
Kimura M, Zhang SQ, Inoue S. Pregnancy-associated sleep changes in the rat. Am J Physiol. 1996;271:1063–9.
Kleinlogel H. The rat's sleep in oestrous cycle. Experientia. 1975;31:712–3.
Kleinlogel H. The female rat's sleep during oestrous cycle. Neuropsychobiology. 1983;10:228–37.
Kloet ER. Hormones, brain and stress. Endocr Regul. 2003;37:51–68.
Kosten TA, Sanchez H, Jatlow PI, Kehoe P. Neonatal isolation alters the estrous cycle interactions on the acute behavioral effects of cocaine. Psychoneuroendocrinology. 2005;30:753–61.
Kustritz MVR. Reproductive behavior of small animals. Theriogenology. 2005;64:734–46.
Long J, Evans H. The oestrus cycle in the rat and its associated phenomena. In: Leuschner A, editor. Memoirs of the University of California. Berkeley: University of California Press; 1922. p. 1–149.
Lu KH, Hopper BR, Vargo TM, Yen SSC. Chronological changes in sex steroid, gonadotropin and prolactin secretion in aging female rats displaying different reproductive states. Biol Reprod. 1979;21:193–203.
Magiakou MA, Mastorakos G, Webster E, Chrousos GP. The hypothalamic-pituitary-adrenal axis and the female reproductive system. Ann N Y Acad Sci. 1997;816:42–56.
Marcondes FK, Miguel KJ, Melo LL, Spadari-Bratfisch RC. Estrous cycle influences the response of female rats in the elevated plus-maze test. Physiol Behav. 2001;74:435–40.
Meites J, Huang HH. Relation of neuroendocrine system to loss of reproductive function in aging rats. In: Kumar A, editor. Neuroendocrine regulation of fertility. Basel: Karger; 1976. p. 246–58.
Meites J, Huang HH, Riegel GD. Relation of the hypothalamo-pituitary-gonadal system to decline of reproductive functions in aging female rats. In: Labrie F, Meites J, Pelletier G, editors. Hypothalamus and endocrine function. New York: Plenum; 1976. p. 3–20.
Meites J, Huang HH, Simpkins JW. Recent studies on neuroendocrine control of reproductive senescence in rats. In: Schneider EL, editor. The aging reproductive system. New York: Raven; 1978. p. 213–36.
Meites J, Steger RW, Huang HHH. Relation of neuroendocrine system to the reproductive decline in aging rats and human subjects. Fed Proc. 1980;39:3168–72.
Murphy BD. Models of luteinization. Biol Reprod. 2000;63:2–11.
Rao MC, Midgley Jr AR, Richards JS. Hormonal regulation of ovarian cellular proliferation. Cell. 1978;14:71–8.
Richards JS, Russell DL, Ochsner S, Espey LL. Ovulation: new dimensions and new regulators of the inflammatory-like response. Annu Rev Physiol. 2002;64:69–92.
Rivier C, Rivest S. Effect of stress on the activity of the hypothalamic-pituitary-gonadal axis: peripheral and central mechanisms. Biol Reprod. 1991;45:523–32.
Rodríguez-Echandía EL, Gonzalez AS, Cabrera R, Fracchia LN. A further analysis of behavioral and endocrine effects of unpredictable chronic stress. Physiol Behav. 1988;43:789–95.
Roozendaal MM, De Kruijf HF, Reuling RJ, Threels A, Swarts JJ, Wiegant VM, Mattheij JA. Inhibition of the LH surge by restraint stress in cyclic rats: studies on the role of GABAA and GABAB receptors. Stress. 1997;1:241–8.
Schwierin B, Borbély AA, Tobler I. Sleep homeostasis in the female rat during the estrous cycle. Brain Res. 1998;811:96–104.
Smith MS, Freeman ME, Neill JD. The control of progesterone secretion during the estrous cycle and early pseudopregnancy in the rat: prolactin, gonadotropin and steroid levels associated with rescue of corpus luteum pseudopregnancy. Endocrinology. 1975;96:219–26.
Wilson JG. Methods for administering agents and detecting malformations in experimental animals. In: Wilson JG, Warkany J, editors. Teratology: principles and techniques. Chicago: University of Chicago Press; 1965. p. 262–77.
Wise PM. Alterations in proestrus LH, FSH, and prolactin surges in middle-aged rats. Proc Soc Exp Biol Med. 1982;169:348–54.
Yokoyama A, Ramirez VD, Sawyer CH. Sleep and wakefulness in female rats under various hormonal and physiological conditions. Gen Comp Endocrinol. 1966;7:10–7.

Part III
General Procedures and Methods in Rodents Research

Chapter 9
Anaesthesia in Laboratory Animals

René Remie

Introduction

In the biomedical sciences, many animal experiments are performed with the animal under anaesthesia. Therefore, understanding the proper use of anaesthesia and the rationale for using particular forms of anaesthesia are important in animal research. Anaesthesia is used during surgery and other painful procedures. In addition, anaesthesia is used to immobilize experimental animals in procedures that would otherwise cause the animal distress or require physical restraint.

Definitions

Anaesthesia (from Greek αν- an- "without" + αἴσθησις aisthesis "sensation") has traditionally meant the condition of having pain and other sensations blocked. The word was coined by Oliver Wendell Holmes, Sr. in 1846. An alternative definition is a state of controllable, reversible insensibility in which sensory perception and motor responses are both markedly depressed. This state may include a total lack of awareness (general anaesthesia) or a lack of awareness of a part of the body, such as with a spinal anaesthetic or any nerve block.

Anaesthesia aims to eliminate pain and immobilize the animal so that surgical and experiment procedures can be carried out safely, humanely and accurately.

R. Remie (✉)
René Remie Surgical Skills Center,
Madernastraat 21, Almere, Flevoland 1323 HE, The Netherlands
e-mail: r.remie@rrssc.eu

M.L. Andersen, S. Tufik (eds.), *Rodent Model as Tools in Ethical Biomedical Research*, DOI 10.1007/978-3-319-11578-8_9

Today, the term "general anaesthesia" in its most basic form may include:

- Hypnosis: producing unconsciousness;
- Analgesia: blocking the conscious sensation of pain;
- Relaxation: preventing unwanted movement or muscle tone;
- Obtundation of reflexes: preventing exaggerated autonomic reflexes (e.g. the vagus nerve).

Anaesthetics can also cause amnesia for the purpose of preventing memory formation.

- Sedatives produce drowsiness and appear to reduce fear and apprehension in animals.
- Tranquilisers produce a calm state without sedation, while high doses cause ataxia (unsteady, uncoordinated movement). While various anaesthetics have different mechanisms of action, there is considerable overlap in the action of many agents. Furthermore, different animal species often respond differently to various anaesthetics.
- Muscle relaxants produce neuromuscular blockade, which results in paralysis of skeletal muscles.

Anaesthesia can be administered by injection or inhalation of drugs. In some instances, only a single anaesthetic agent is used (mono-anaesthesia), but more frequently a combination of drugs is administered that have combined effects to produce anaesthesia. Importantly, combinations of drugs at low rates of administration (i.e., lower doses with time) potentiate each other to produce balanced anaesthesia.

Anaesthetic Regime

Depending on the type of surgery or procedure, the particular combination of anaesthetics may differ. For example, for procedures involving only X-ray images of a healing bone, sleep and muscle relaxation are required. Because this procedure is not painful, there is no need for pain relief. In contrast, other procedures, such as invasive abdominal surgery that would be painful for the animal, require additional pain relief (analgesia). Vagal reflexes should be inhibited when using volatile anaesthetics that are inhaled because they are known to irritate the mucus membranes of the airways. Furthermore, parasympatholytics should be administered to protect the heart from a dangerously low heart rate evoked by the release of acetylcholine, the transmitter of the parasympathetic nervous system.

> Adjust your anaesthetic regime for the type of procedure you want to perform

The choice of a particular technique depends on a variety of factors:

- Species
- Strain
- Depth of anesthesia required
- Duration
- Availability of equipment (anaesthetic delivery system)
- Experience and skill level of staff

For an experiment requiring anaesthesia, it is important to select an anaesthetic technique/regime that would interfere minimally with the particular experiment.

Anaesthetic Depth

When administering anaesthetics to an animal, ataxia and resistance are the initial responses. The animal then shows signs of excitation before finally losing consciousness. At this point the animal is lightly anesthetized, and this phase of anaesthesia allows for only minor non-invasive procedures. The animal must reach the next phase of anaesthesia (3.2) before beginning surgery, which can be assessed by testing the animal's reaction to a painful stimulus, such as a foot-or tail pinch. In the event of overdose, the animal enters deep anaesthesia and respiration becomes slow and sometimes irregular. For many anaesthetics, side effects on the respiratory and the cardiovascular system are expected. In cases of significant over dosage, these side effects will lead to death of the animal.

Four phases of anaesthesia

1. Ataxia, resistance
2. Excitation

3.1. Light anaesthesia

- For minor, non-invasive procedures

3.2. Surgical anaesthesia

- Sufficient for most procedures

3.3. Deep anaesthesia

- All reflexes are lost
- Intercostal muscles become paralyzed

3.4. Overly deep anaesthesia

- Intercostal muscles become paralyzed
- Significant cardiopulmonary depression

4. Deepest phase of anaesthesia (far to deep)
   - Cessation of respiration
   - Cardiac arrest
   - Death

Factors Influencing Anaesthetic Depth

When choosing a proper anaesthetic dosage for your animal, the sensitivity of the specific strain to the anaesthetic combination must first be determined. Furthermore, many factors influence the anaesthetic depth and must be taken into account. The following list gives an overview of these factors.

- Body temperature
- Nutrition
- Disease
- Parasite load
- Infection
- Estrus
- Pregnancy
- Lactation
- Metabolism

Special attention is needed in monitoring body temperature. Once the animal is anesthetized, the thermo-regulating center in the brain is blocked. As a consequence, the animal will become poikilothermic (behave like a cold-blooded animal) and will thus be dependent on environmental temperature to regulate its body temperature. The environmental temperature is usually below 37–38 °C, so the animal's body temperature will decrease. Importantly, most anaesthetics are more potent at a lower body temperature. Therefore, if the animal cools, there is a greater probability that it will overdose and possibly die.

Injectable Anaesthetic Agents

A number of different injectable anaesthetic agents are used in laboratory animals. Recommended doses for common species are listed at the end of this chapter. Injectable anaesthetics can be administered by a variety of routes, but intravenous administration is usually preferable because it produces the most predictable and rapid onset of action. This consistency allows for the drug to be administered 'to effect' so that a sufficient dose can be administered to reach the desired depth of anaesthesia. In many smaller laboratory species, practical considerations, such as the absence of suitable superficial veins or difficulty providing adequate restraint of the animal, may limit the use of the intravenous route. Administration by intraperitoneal (i.p.), subcutaneous (s.c.) or intramuscular (i.m.), injection is relatively simple in most species, but the rate of drug absorption, and thus its anaesthetic effects, may vary considerably.

As mentioned above, significant variation in response to anaesthetics occurs among different strains, ages, and sex of animals. When first using a particular anaesthetic regime, it is essential to assess the effects of the anaesthetic in a small group of animals before working with a larger group. The recommended doses should then be adjusted based on the findings from the smaller group in order to optimize dosing for the particular animals being used.

Other factors that should be taken into account when selecting anaesthetics for i.m., i.p., or s.c. administration are a wide safety margin and possible irritation cause by the drug that could result in unnecessary pain or discomfort to the animal. This is a particular concern with i.m. injection in small rodents. For example, ketamine is known to induce tissue necroses. Pain on injection can also occur when using a large injection volume.

If the i.v. route can be used, a number of short-acting anaesthetics can be used to provide 5–10 min periods of anaesthesia (e.g. propofol, thiopental, methohexital).

Neuroleptanalgesic Combinations

A combination of a neuroleptic and an analgesic drug are often used in anaesthetizing rodents and rabbits.

Fentanyl/fluanisone (Hypnorm®), when administered alone, produces sedation and sufficient analgesia for superficial surgery in most small rodents, rabbits and dogs. However, muscle relaxation is poor, and high doses that are needed for more major surgery produce marked respiratory depression. In small rodents and rabbits, combining this regimen with the benzodiazepine midazolam (Dormicum®) produces surgical anaesthesia with only moderate respiratory depression. This combination has the advantage of being partially reversed with a mixed opioid agonist/antagonist, such as butorphanol, or a partial agonist, such as buprenorphine (Temgesic®). As a result, respiratory depression caused by fentanyl is reversed, but post-operative analgesia is maintained. The benzodiazepine antagonist flumazenil can be used to speed recovery. In small rodents, a mixture of midazolam and Hypnorm, the so-called FFM mixture, can be given as a single i.p. injection.

Fentanyl/droperidol (Innovar-Vet®) produces effects similar to Hypnorm, however in combination with midazolam the effects are unpredictable. Fentanyl/droperidol should be used only for minor procedures.

Ketamine

Ketamine produces immobility in most species but only limited analgesia.

In combination with acepromazine or midazolam, ketamine produces light to moderate surgical anaesthesia in non-human primates, cats, sheep, pigs, ferrets and rabbits. In small rodents, the effects of these combinations are less predictable, and usually only light planes of anaesthesia are produced. However, ketamine, in combination with medetomidine or xylazine, results in surgical anaesthesia in many species. In guinea pigs and pigs, the effects are less reproducible, and some animals may not reach a sufficiently deep plane of anaesthesia for major surgery. Additional analgesia may be provided with inhalation anaesthetics or local analgesics. Since ketamine alone has limited effects in many species, reversal of medetomidine or xylazine with the alpha-2 antagonist atipamezole (Antisedan®) speeds recovery. In dogs, ketamine should not be given alone, because it can produce seizures.

Barbiturates

Pentobarbital has long been used as a general anaesthetic. While pentobarbital induces sleep, it has little analgesic power. Invasive surgery can be performed only with high doses. Importantly, in many species this anaesthetic dose is close to the lethal dose. When given i.v., the dose can be finely adjusted. However, in rabbit, even fine delivery of a diluted solution (6 mg/mL) is hazardous, and respiratory arrest may occur before surgical planes of anaesthesia are attained.

Thiopental, thiamylal and methohexital can all be used to produce short periods of anaesthesia when administered i.v. This approach can be used for short surgical procedures and allows intubation followed by anaesthesia maintenance with volatile agents.

Propofol

Propofol may be administered i.v. to provide short periods of surgical anaesthesia in most species. With continuous infusion additional doses can be given to prolong the period of anaesthesia. In rabbits, propofol can provide sufficient depth of anaesthesia for intubation, but respiratory arrest usually occurs before the onset of surgical anaesthesia. In sheep, relatively high doses are needed to induce and maintain anaesthesia.

Local Analgesics

Although local anaesthesia is most often used in larger species, it can be used to support the general anaesthesia in smaller animals. Local anaesthetics, such as lidocaine, can be infused into or dripped upon the area to be made insensitive.

By injecting around a nerve trunk larger areas can be anaesthetized. Administration into the epidural space or into the cerebrospinal fluid produces anaesthesia of the hind limbs and lower abdomen.

When using these drugs in small animals, care must be taken to avoid overdose that will result in death of the animal.

Inhalation Anaesthesia

Anaesthetics that are given by inhalation are usually volatile liquids that are vaporized before delivery and carried into the animal by a gas or a gas mixture (e.g. isoflurane in nitrous oxide and oxygen, or oxygen and air (carrier gases). The minimal concentration of oxygen in the carrier gas should be 30 % (Fig. 9.1).

Halothane, the most potent inhalation anaesthetic, is no longer used due to side effects, such as liver necrosis and hepatitis. In laboratory animal anaesthesia Halothane has been replaced with isoflurane. Inhalation anaesthetics are liquids at room temperature. They are extremely potent, and must therefore be delivered in a controlled manner, typically using a calibrated vaporizer.

Sevo- and desflurane have a very low blood-to-gas coefficient, which means that they rapidly produce anaesthesia and the animal rapidly recovers upon removal of the volatile anaesthetics. Rapid recovery may not be optimal when working with small rodents because time is needed for surgical preparation, such as shaving and disinfecting the area, before the animal regains consciousness.

Usually, anaesthetic vapor of a certain concentration is produced by leading the carrier gas through a vaporizer. The vaporizer setting can be adjusted easily to provide an appropriate concentration for induction of anaesthesia, or for maintenance (see Table 9.1). The depth of anaesthesia can be quickly and easily regulated by changing the vaporiser setting.

The depth of anaesthesia can be assessed fairly easily by checking for the presence of withdrawal responses. As anaesthesia deepens, these responses are lost and respiration becomes progressively shallower. If the plane of anaesthesia becomes dangerously deep, respiration may stop or gasping respiratory movements may occur.

Anaesthetic Delivery Systems

Volatile anaesthetics are preferably delivered by an anaesthetic delivery system. This device consists of a source of compressed oxygen, air and/or nitrous oxide, either from cylinders or piped from a central supply (nitrous oxide has very little anaesthetic effect in animals, so it can only be used in combination with other anaesthetics). The cylinder pressure is reduced by a regulator in the system, and the gasses are delivered to a flow-meter that controls the amount of anaesthetic flow to the animal. A vaporizer adds a preset concentration of the anaesthetic to the carrier gas.

Fig. 9.1 Images taken from Laboratory Animal Anaesthesia, by Paul Flecknell (1987)

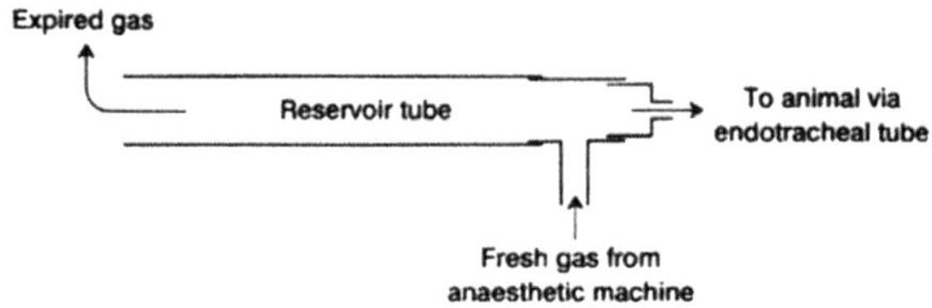

Ayre's T-piece.

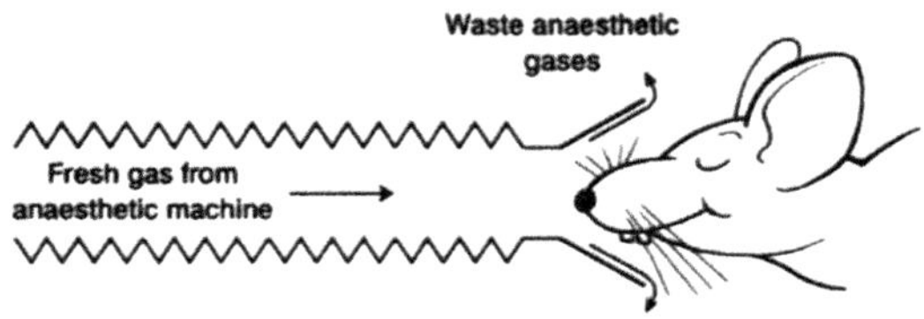

Open anaesthetic circuit with delivery of anaesthetic gases by means of mask. Expired gases pass around the mask.

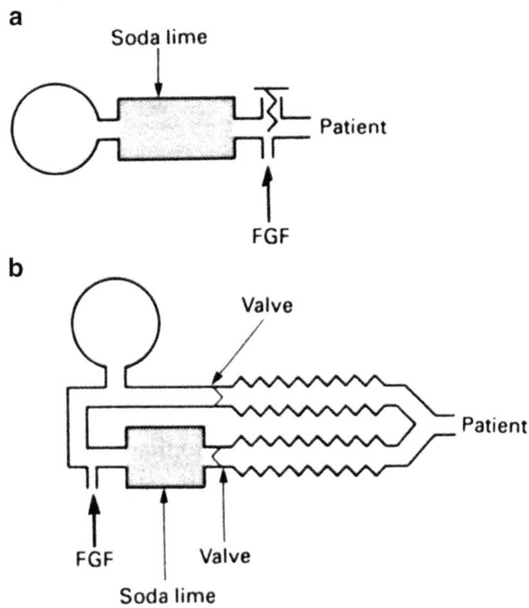

To-and-fro (a) and circle absorber systems (b) (FGF = fresh gas flow).

Endotracheal Tubes

Endotracheal tubes may be plain tubes, or may have an inflatable cuff that seals the gap between the wall of the tube and the trachea. The cuff can be carefully inflated with a syringe. Take care not to overinflate the cuff as this will lead to necrosis of the tracheal wall. During inflation, one must listen for the sound of escaping gas during inspiration. Once this sound can no longer be heard, the cuff is sufficiently inflated.

Table 9.1 Concentrations of volatile anaesthetics for induction and maintenance of anaesthesia. Adapted from Flecknell (1992)

Anaesthetic	Induction concentration (%)	Maintenance concentration (%)
Enflurane	3–5	1–3
Isoflurane	3–4	1.5–3
Methoxyflurane	3–3.5	0.4–1
Sevoflurane	3–4	2–3
Desflurane	10–14	7–9

Delivery of Anaesthetics by Inhalation

In small rodents, the anaesthetic can be delivered easily using an anaesthetic chamber. Animals first become ataxic and may then go through a period of excitement. They will next lose the righting reflex and become immobile. The animal should be watched during induction of anaesthesia to determine when the animal is down but not allow it to be overly exposed to the anaesthetic. Therefore, the induction chamber should be transparent (polycarbonate or Perspex). The inlet of the anaesthetic gases should be at the base of the chamber, and the chamber should be placed in a properly ventilated hood. The chamber should always be filled with anaesthetic mixture before introducing the animal. Once the animal is unconscious, it can be removed from the chamber, and administration may be continued (at a lower concentration, e.g. 2 % isoflurane) using a face-mask. In larger animals, it is often preferable to use an injectable agent to sedate the animal. Next, the animal is put to sleep by i.v. anaesthetic injection, and the animal can remain intubated and maintained under inhalation anaesthesia until recovery.

Anaesthetic Circuits

Anaesthetic gases can be delivered through a face mask or an endotracheal tube. For endotracheal tube delivery, several different delivery systems are available, and they are divided into non-rebreathing and rebreathing. Examples of the delivery systems include the Ayres T-piece, the to-and fro circuit, and the circle system.

As a general rule, the amount of gas delivered to an animal using a face mask should be three times the animal's minute volume (the volume of gas inhaled in 1 min). Minute volume is calculated by multiplying the volume of one breath (the tidal volume) by the respiratory rate.

Pre-anaesthetic Preparations

Before anaesthetising an animal, all equipment should be checked for proper functioning. In species that readily vomit during induction of anaesthesia, it is advisable to withhold food for 8–12 h before induction. Dogs, cats, ferrets, and non-human primates may vomit on induction, but most rodents do not. If the particular surgery involved the intestinal tract in rodents it is advisable to withhold food for a period of 6 h before surgery. However, recall that food withdrawal changes an animal's physiology and will therefore respond slightly differently to the anesthesia.

Pre-anaesthetic Medication

Before anaesthetizing the animal, pre-anaesthetic medication may be administered to the animal. For example, the use of sedatives and tranquilisers can reduce aggression and fear, while the use of analgesics can reduce pain and provide 'pre-emptive analgesia'. Atropine or glycopyrrolate can be used to reduce bronchial and salivary secretions, while also protecting the heart from vagal inhibition.

Sedatives, tranquilisers and analgesics can reduce the amount of anaesthetic needed to produce the desired level of anaesthesia (anaesthetic sparing).

Peri-anaesthetic Care

Animals under anaesthesia are unable to control their body temperature, so they become poikilothermic and assume the temperature of the environment. During surgery, the body temperature of the animal must remain in the range of 36–38 °C. If the body temperature falls too low the animal may suffer hypothermia, and anaesthetics become more potent. As a result, the animal may die from hypothermia or from side effects of the drug compromising the heart and the respiratory tract.

Fluid balance should also be monitored. While under anesthesia, animals lose fluids via the respiratory tract, the surgical area, and from blood loss. Lost fluids can be replaced by warm sterile saline injection. Blood loss should be compensated for by transfusion or by plasma replacement therapy.

Post-anaesthetic Care

A separate recovery area should be available for post-surgical animals. This recovery area, which would allow proper attention to each individual animal, should include appropriate environmental conditions, such as increased environmental temperature. During recovery from anaesthesia, care should be taken to keep the respiratory tract unobstructed and to keep the animals warm. In most instances, small animals can be allowed to recover in their normal (home) cages or inside an incubator. Adult animals require and environmental temperature of 25–30 °C, and neonates require a temperature of 35–37 °C. If an incubator is unavailable, heating pads and lamps should be provided. Although warmth is needed, care must also be taken not to overheat the animal.

Following invasive surgical procedures, animals may experience post-operative pain that must be controlled with analgesics.

Remember that most animals we use in our research are prey animals that do not show overt pain or illness. If they did, predators would immediately spot the animal and catch it.

Pain Relief Agents

Analgesics can be broadly divided into two groups: the opioids (major or narcotic analgesics) and the minor analgesics, and the nonsteroidal anti-inflammatory drugs (NSAID's), such as aspirin. Local analgesics can also be used for post-operative pain relief by blocking all sensation from the affected area.

Pre-emptive Analgesia

During anaesthesia, nerve impulses from the surgical site reach the central nervous system (CNS). These nerve impulses trigger changes in the CNS that increase the degree of pain (hyperalgesia) that is perceived when the animal regains consciousness. If analgesics are administered before any potentially painful stimuli are encountered, then there analgesics are generally more effective in preventing post-operative pain. For this reason, it is now widely recommended that analgesics be administered pre-operatively to provide more effective pain relief and possibly reduce the dose of anaesthetic required (Tables 9.2, 9.3, 9.4, 9.5, 9.6, 9.7, 9.8).

Table 9.2 Anaesthesia and analgesia in mice

Agent	Dose	Duration of surgical anaesthesia
Anaesthesia in mice		
Pentobarbital	50 mg/kg IP	20–40 min
Tribromoethanol (avertin)	240 mg/kg IP	15–45 min
Metomidate/fentanyl	60 mg/kg + 0.06 mg/kg SC	20–30 min
Ketamine/xylazine	80–100 mg/kg + 10 mg/kg IP	20–30 min
Ketamine/medetomidine	80–100 mg/kg+	
Agent	Dose	
Analgesia in mice		
Buprenorphine	0.05–0.1 mg/kg SQ BID	
Butorphanol	1–5 mg/kg SQ QID	

Table 9.3 Anaesthesia and analgesia in rats

Agent	Dose	Onset	Duration
Parenteral anaesthesia in rats			
Pentobarbital	30 mg/kg IP	5	90 min
Pentobarbital	40 mg/kg IP	5	120 min
Ketamine-xylazine	40 mg/kg; 5 mg/kg IP	5	80 min
Ketamine-xylazine	60 mg/kg; 7.5 mg/kg	2	115 min
Ketamine-medetomidine	75 mg/kg; 0.5 mg/kg	1	20–30 min
Agent	Dose		
Analgesia in rats			
Buprenorphine	0.01–0.05 mg/kg SQ or IV TID to BID		
	0.1–0.25 mg/kg PO, BID to TID		
Butorphanol	2 mg/kg SQ every 4 h		

Table 9.4 Anaesthesia and analgesia in rabbits

Agent	Dose (mg/kg)	Duration of surgical anaesthesia
Parenteral anaesthesia in rabbits		
Ketamine	35 mg/kg IM	25–40 min
Xylazine	5 mg/kg IM	
Ketamine	10 mg/kg IV	20–30 min
Xylazine	3 mg/kg IV	
Ketamine	35 mg/kg IM	
Xylazine	5 mg/kg IM	100 min
Acepromazine	0.75 mg/kg IM	
Ketamine	35 mg/kg IM	
Xylazine	5 mg/kg IM	60–90 min
Butorphanol	0.1 mg/kg IM	
Pentobarbital	Induce with 10 mg/mL IV and administer at increments of 2–10 mg until a satisfactory level of anaesthesia is reached	20–30 min
Agent	Dose	
Analgesia in rabbits		
Morphine sulphate	2–4 mg/kg SC every 2–4 h	
Buprenorphine	0.02–0.05 mg/kg SQ or IV TID or BID	
Butorphanol	0.1–0.5 mg/kg IV every 4 h	

Table 9.5 Anaesthesia and analgesia in guinea pigs

Agent	Dose (mg/kg)	Duration
Parenteral anaesthesia in guinea pigs		
Ketamine-xylazine	30+5 IM	30–45 min
Ketamine-xylazine	44+5 IM	74 min
Pentobarbital	15 mg/kg IP +	60 min
Fentanyl-droperdiol	0.4 mL/kg IM	
Pentobarbital[a]	37 mg/kg IP	60–90 min
Agent	Dose (mg/kg)	
Analgesia in guinea pigs		
Morphine sulphate	2–5 mg/kg SC every 4 h	
Buprenorphine	0.05 mg/kg SQ TID to BID	
Indomethacin	2.5–8.8 mg/kg PO	

[a]Not suitable for procedures requiring deep anaesthesia

Table 9.6 Anaesthesia and analgesia in dogs

Agent	Dose (mg/kg)	Duration
Parenteral anaesthesia in dogs		
Ketamine-xylazine	5; 1–2 IV or IM	30–65 min
Propofol	5–7.5 mg/kg IV; maintain with 0.2–0.4 mg/kg/min	5–10 min unless a continuous infusion
Thiopental[a]	10–18 mg/kg IV	5–15 min
Pentobarbital[a,b]	20–30 mg/kg IV	30–40 min
Analgesia in dogs		
Agent	Dose (mg/kg)	
Butorphanol	0.2–0.4 mg/kg SQ every 2–5 h	
Buprenorphine	0.005–02 mg/kg IM, IV or SQ QID to BID	
Morphine sulphate	0.5–5 mg/kg SC every 4 h	
Carprofen	4 mg/kg IV, SC SID or 1–2 mg/kg PO BID for 7 days	
Ketoprofen	2 mg/kg SC, IM, IV SID for maximum 3 days or 1 mg/kg PO SID for 5 days	

[a]Use of acepromazine (0.05–1 mg/kg IM) as a pre-anesthetic will reduce the amount of barbiturate required for anaesthesia up to 50 %

[b]Repeated doses of thiopental or pentobarbital will greatly prolong the recovery from anaesthesia

Table 9.7 Anaesthesia and analgesia in cats

Agent	Dose (mg/kg)	Duration
Parenteral anaesthesia in cats		
Ketamine	22 mg/kg IM	20–30 min
Xylazine	1.1 mg/kg IM	
Ketamine	20 mg/kg IM	20–30 min
Acepromazine	0.11 mg/kg IM	
Thiopental[a,b]	10–18 mg/kg IV	5–15 min
Pentobarbital [a,b]	20–30 mg/kg IV	60–90 min
Agent	Dose (mg/kg)	
Analgesia in cats		
Butorphanol	0.4 mg/kg SQ every 3–4 h	
Buprenorphine	0.005–0.01 mg/kg IV or SQ TID to BID	
Morphine sulphate	0.1 mg/kg SC every 4 h	
Carprofen	4 mg/kg SC or IV	
Ketoprofen	1 mg/kg SQ SID for up to 3 days or 1 mg/kg PO SID for up to 5 days	

[a]Use of acepromazine (0.05–1 mg/kg IM) as a pre-anaesthetic will reduce the amount of barbiturate required for anaesthesia up to 50 %.
[b]Repeated doses of thiopental or pentobarbital will greatly prolong the recovery from anaesthesia

Table 9.8 Anaesthesia and analgesia in macaques

Agent	Dose (mg/kg)	Duration
Parenteral anaesthesia in macaques		
Ketamine	10 mg/kg IM	30–40 min
Xylazine	0.5 mg/kg IM	
Thiopental[a]	15–20 mg/kg IV	5–10 min
Pentobarbital[a]	25–35 mg/kg IV	30–60 min
Agent	Dose (mg/kg)	
Analgesia in macaques		
Butorphanol	0.01 mg/kg IV every 3–4 h	
Buprenorphine	0.005–0.01 mg/kg IV or IM QID to BID	
Morphine sulphate	1–2 mg/kg SC or IM every 4 h	
Flunixin	2–4 mg/kg SC SID	

[a]Repeated doses of thiopental or pentobarbital will greatly prolong the recovery from anaesthesia

References

Brown JN, Thorne PR, Nuttall AL. Blood pressure and other physiological responses in awake and anesthetized guinea pigs. Lab Anim Sci. 1989;39:142–8.
Flecknell PA. Laboratory animal anaesthesia: an introduction for research workers and technicians. New York: Academic; 1987.
Flecknell PA. Laboratory animal anaesthesia. 3rd ed. New York: Academic; 2009.

Frisk CS, Herman MD, Senta KE. Guinea pig anaesthesia using various combinations and concentrations of ketamine, xylazine, and/or acepromazine. Lab Anim Sci. 1982;32:434.
Jenkins WL. Pharmacologic aspects of analgesic drugs in animals: an overview. J Am Vet Med Assoc. 1987;191:1231–40.
Lipman NS, Marini RP, Erdman SE. A comparison of ketamine/xylazine and ketamine/xylazine/acepromazine anaesthesia in the rabbit. Lab Anim Sci. 1990;40:395–8.
Manning PJ, Ringer DH, Newcomer CE, editors. The biology of the laboratory rabbit. New York: Academic; 1994.
Morgan WW, Morlan SL, Krupp JH, Rosenkrantz JG. Pentobarbital anaesthesia in the rabbit. Am J Vet Res. 1966;27:11–33.
Papaioannou VE, Fox JG. Efficacy of tribromoethanol anaesthesia in mice. Lab Anim Sci. 1993;43:189–92.
Principles and Practice of Veterinary Anaesthesia. Short CE; 1987.
Svendsen P, Hau J. Handbook of laboratory animal science. Boca Raton: CRC Press; 1994.
Wixson SK, White WJ, et al. A comparison of pentobarbital, fentanyl-droperidol, ketamine-xylazine and ketamine-diazepam anaesthesia in adult male rats. Lab Anim Sci. 1987;6:726–30.

Chapter 10
Various Surgical Procedures in Rodents

René Remie

Principles of Surgery

Surgery in laboratory animals, regardless of species or size, is governed by the same principles as surgery in human beings. A basic surgical tenet is Halstead's rule, which cautions the surgeon to not do harm to the tissue. However, this is only one of a set of interrelated principles regarding tissue handling, exposure, asepsis and homeostasis.

- Tissue handling
  Remember that every time you pick up tissue with your instruments, you kill cells. Try to kill as few cells as possible. Be goal-oriented in your approach and remember that sharp dissection is generally less traumatic than blunt dissection.
- Exposure
  Make sure your view is unobstructed, with proper illumination and physical access. This means that the wound you make should be sufficient in size and certainly not too small. Do not worry about the healing of the wound, as it is less likely to be affected by size than by appropriate approximation of the wound edges.
- Asepsis
  Conditions favorable to bacterial growth must be avoided. First, a meticulous sterile technique must be followed. Second, and equally important, dead tissue and foreign materials should be removed, together with blood or serum residue.

R. Remie (✉)
Rene Remie Surgical Skills Center,
Madernastraat 21, Almere, Flevoland 1323 HE, The Netherlands
e-mail: r.remie@rrssc.eu

M.L. Andersen, S. Tufik (eds.), *Rodent Model as Tools in Ethical Biomedical Research*, DOI 10.1007/978-3-319-11578-8_10

- Hemostasis
  Blood loss can have serious consequences, especially in small laboratory animals, and may eventually result in the untimely death of your animal. Clamping or applying light pressure can stop almost all bleeding. Other methods are coagulation (mono- or bi-polar), cauterization, ligation and chemical treatment with collagen and/or ADP-containing hemostatics.

The Surgeon

From the beginning, the study of surgical and especially microsurgical techniques makes many mental and physical demands. Attention to detail and a high level of concentration are very important when learning new techniques. The PVC rat (Fig. 10.1) can provide an ideal setting for training for advanced microsurgical and various experimental techniques. The number of live animals needed for training is drastically reduced with the use of this non-animal training device.

A new surgeon must also contend with the difficulties inherent in working under a microscope. Here, you do not have direct eye-to-hand contact and must develop an awareness of the fact that even simple movements in a significantly smaller world are more complex than they are at the normal perspective. Under the operating microscope, not only are objects enlarged, but your movements are also magnified between 8 and 40 times. The physical ability required to coordinate one's movements decreases in proportion to the magnification used.

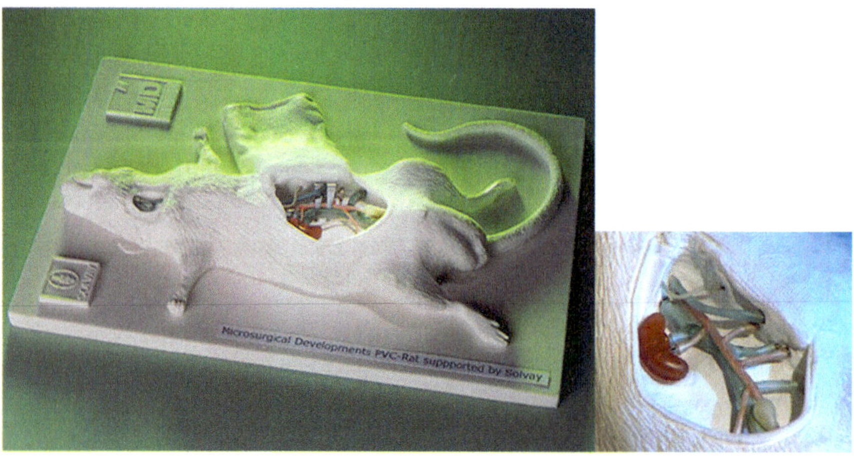

Fig. 10.1 The PVC rat is developed to master skills in microsurgery

Individual Preparation

Listed below are some suggestions to improve your performance.

- Make sure that the environment is quiet. A dedicated operating theater is ideal.
- Try to avoid any mental stress.
- Make sure that your table is at the right height, giving adequate support to your arms. Sit upright to avoid strain injury to your shoulders, neck and back.
- Schedule your exercises carefully. You need to be able to devote all of your attention to the exercise at hand. Make sure that you have no conflicting appointments and that you cannot be disturbed by telephone calls, etc. It is unrealistic to expect to perform well when your attention is divided or if you are forced to act hastily.
- Try to avoid heavy physical exertion during the 24 h preceding the surgical exercise, as this will interfere with your fine muscular control and will probably increase your tremor. It was recently shown that experienced surgeons recover very quickly (within 4 h) from heavy exercise.
- Do not change any habits relating to your intake of coffee, as a radical increase or decrease will increase your tremor.
- Do not become discouraged when something has gone wrong. If you encounter a difficulty, evaluate and try to correct it before you continue. Do not let frustration become your greatest enemy.
- Do not work for too long at a stretch. Surgery is very fatiguing. If possible, take a 10 min break every hour, as otherwise you will lose concentration and your coordination and learning ability will be reduced.

Planning

When a surgical procedure unfamiliar to the investigator is to be carried out, it is good practice for the work to be carefully planned in advance. This will allow the investigator to feel and act calmly and confidently, while the animal gets the best possible treatment. Planning and preparation for an operation are quite important, although often underestimated. You should preferably commence operations in the morning in order to provide optimal post-operative care. Surgery at the end of the week should also be avoided unless proper attention can be provided over the weekend.

Before you start an operation, make sure that you have everything you need at your disposal. It is helpful to prepare a checklist of requirements that must be completed before surgery is begun. It is particularly frustrating when you have to leave the operating theater to search for things you urgently require, not to mention the break it causes in the aseptic technique. For detailed instructions, the reader is referred to Acland and Remie et al.

Anatomy

A thorough knowledge of the animal's anatomy is essential, as it will reassure you during the performance of an operation. Do not start an operation until you are familiar with all the structures in the area of interest. Animals euthanized after being used in an experiment make ideal learning materials. Use the operation microscope to look in detail at all kinds of tissues and learn how to handle them with your instruments. For example, try dissecting blood vessels of different sizes to see how much tension they can withstand. As you will see, the majority of microdissection activities are performed by slight pronation and supination movements of the fingers and forearm. Small spreading movements parallel to the blood vessel (or other structures) will prevent tearing of branches. Try to use atraumatic techniques even when dissecting dead animals, i.e., do not grab the whole thickness of the blood vessel between the jaws of your forceps. Pick it up only by its outer layer, the adventitia. The less damage you cause to the blood vessels and the surrounding tissues the better. For the study of rat and mouse anatomy, the work of Greene, Hebel and Stromberg and Iwaki and Hayakawa are highly recommended.

Spatial Relationships

In anatomy, all spatial relationships are defined with respect to the imaginary median plane, which runs from the head to the tail (Fig. 10.2). This plane divides the body into two equal (right and left) halves. Another imaginary plane is the sagittal plane, which runs parallel to the median plane. The sagittal plane also divides the animal into two (right and left) sections, though these are not necessary equal. Thus, there can be many sagittal planes, but by definition only one median plane. The transverse plane lies perpendicular to the median plane; it divides the animal into a rostral half,

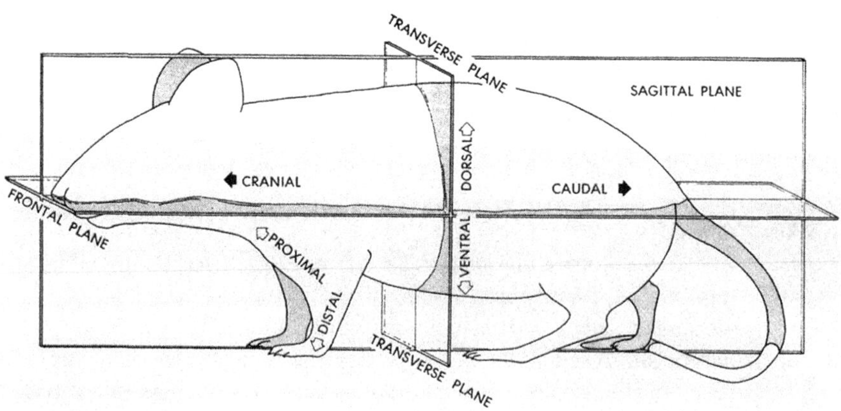

Fig. 10.2 This plane divides the body into two equal (*right* and *left*) halves

comprising those structures that lie in the direction of the head, and a caudal half, which includes those structures situated in the direction of the tail. The transverse planes are traversed in turn by the coronal planes; these divide the body into a ventral and a dorsal component. The terms "proximal" and "distal" respectively indicate towards and away from the center, the median line or point of attachment or origin. On the head itself, we use the terms rostral and caudal, while in the rest of the body we use cranial and caudal. In the brain, we use the term anterior instead of rostral and posterior instead of caudal.

Basic Instrumentation

Surgical Instruments

There are three main categories of surgical instruments: those that cut, grasp or retract tissue. Always choose the instrument of proper size and strength to do the job. Try to get your own set of instruments. These should be of good quality and must be well maintained. Do not begin with old, worn out or obsolete equipment. Once you have acquired your set, do not lend it out. Your set should consist of:

Macro instruments

- Anatomical or dissecting forceps (straight)
- Anatomical or dissecting forceps (90°-angled)
- A pair of fine-toothed forceps
- A pair of ring-handled scissors (straight, sharp/sharp or sharp/blunt)
- A needle holder (Matthieu or Castroviejo)
- A pair of artery forceps, a baby-Mosquito or a micro-Halstead clamp
- Towel clips (Backhaus)
- Baby Dieffenbach Serrefines (bulldog haemostatic clamps)
- An instrument case

Micro instruments

- Two pairs of jeweler's forceps (No. 5, Dumont)
- One pair of jeweler's forceps (45°-angled, Dumont)
- A micro needle holder (Barraquer-type) without a lock
- A pair of ring-handled dissecting scissors with gently curved blades
- A pair of fine 45°-angled iridectomy scissors (Wecker-type)
- Vessel clamps including a clamp applicator
- Two 12 mm single clamps (Acland, Biemer or Heifetz)

Micro instruments should not be too long; 11–12 cm is a suitable length. An advantage of the Barraquer needle holder is that the round grips allow easy rotating movements. Clamps can be divided into venous and arterial types, the latter having a higher shutting strength. One must be careful with clips that have a very high shutting strength, as they will damage the tissue. Always clean your instruments immediately after surgery with tap water. Every now and then, clean your instruments by placing them in an ultrasonicator. Prior to storage, the instruments must be dried and oiled using standard instrument oil. Some operations will require additional specialized instruments (Figs. 10.3, 10.4, 10.5, 10.6).

The Operating Area

The physical environment in which surgery is to be performed can vary from a specially designed and sophisticated surgical suite to a small area on a laboratory bench. What is important is that the area is easily kept clean and uncluttered and is distant from general human traffic. The liberal use of a disinfectant on all surfaces can helpfully reduce the room's population of unwanted microorganisms and dust.

Operating Table

Specially designed operating tables, or boards, made from metal, plastic or Silicon rubber are available and considerably aid surgery of small animals. Such tables may incorporate a heating facility that can be used to maintain the animal's body temperature. Other helpful accessories for surgery of small animals, such as small skin and tissue retractors, are available from commercial sources (Lone Star Medical Products, Inc.). For simple operations on small animals, a silicon plate (25 × 30 cm) may suffice. Silicon can be easily kept clean and may even be sterilized.

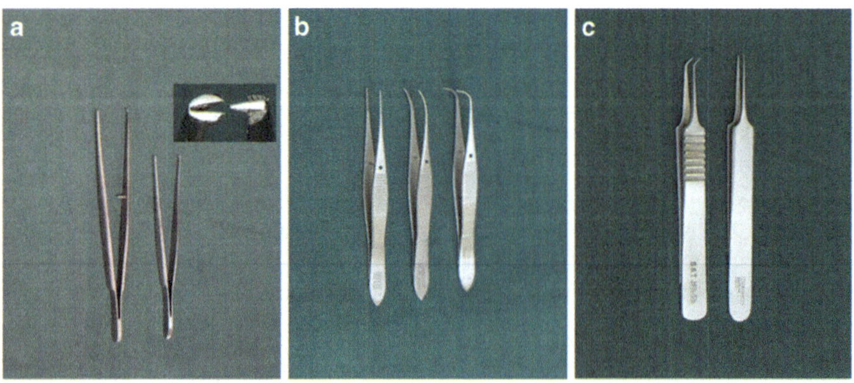

Fig. 10.3 (**a**) Surgical forceps; inset shows the 'rat teeth'. (**b**) Anatomical forceps. (**c**) Jeweler's forceps

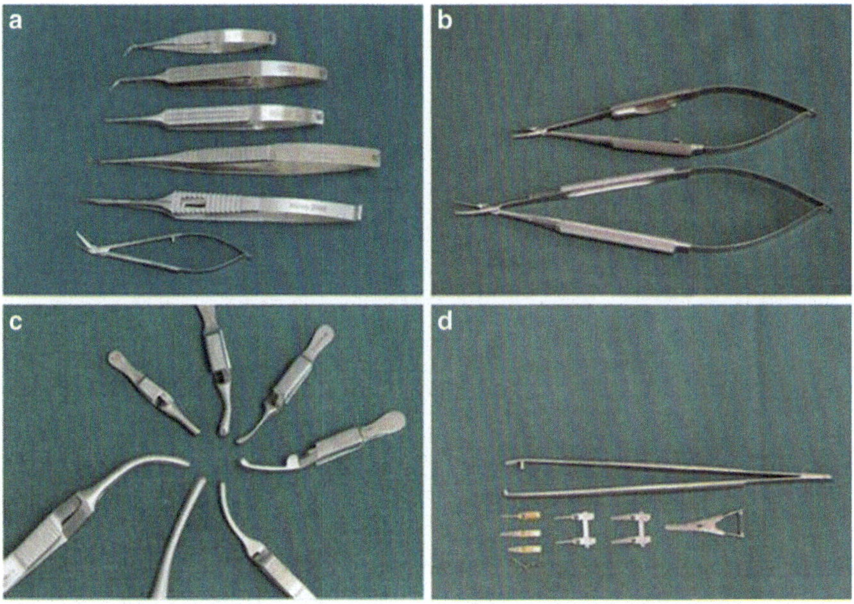

Fig. 10.4 (**a**) Micro scissors. (**b**) Barraquer needle holders (with and without lock). (**c**) Bulldog clips. (**d**) Microvascular clips with applying forceps

Lighting

A portable cold-illumination lighting system that can be regulated to give diffuse or pinpointed lighting is very effective. Proper lighting is of enormous help in all types of surgery and the use of special surgical theater lights is highly recommended.

Other Apparatus and Accessories

The apparatus required for surgery will depend on the type of surgery being carried out, the species used and whether or not the surgery is an active part of an experiment. Similarly, surgical accessories such as gauze, drapes, cotton wool, etc., will be variously required. All apparatus should be clean, free of dust and preferably wiped down with a detergent or disinfectant. Some apparatus and accessories must be sterilized, which can be achieved by autoclaving or, in the case of glass or metal apparatus, by dry heat in a 160 °C oven for at least 1 h. Plastic items can only satisfactorily be sterilized by irradiation, but in many cases it is acceptable to use them after they have been disinfected in, for example, 0.5 % aqueous chlorhexidine. Plastic cannulas disinfected in this way and rinsed with sterile saline have been kept in various parts of the body of both large and small animals for long periods of time without producing infection (Figs. 10.7 and 10.8).

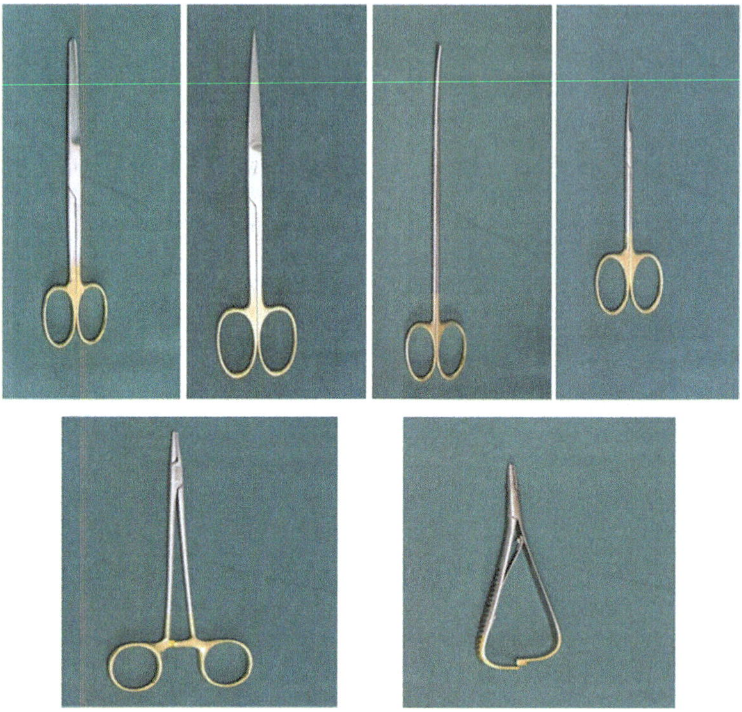

Fig. 10.5 Scissors. From left to right: Straight blunt-blunt acc. to Sims, straight sharp-sharp, bent blunt-blunt acc. to Metzenbaum, straight sharp-sharp (small). Needle holders. *Left*: De Bakey. *Right*: automatic needle holder acc. to Matthieu

Fig. 10.6 The three-point grip. A thumb and ring finger grip while the index finger supports the instrument

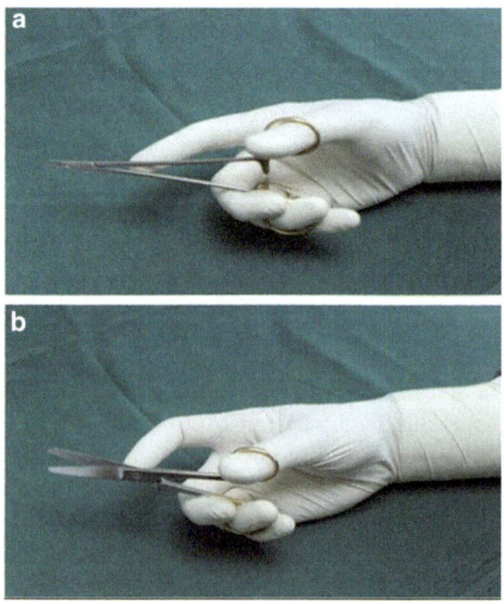

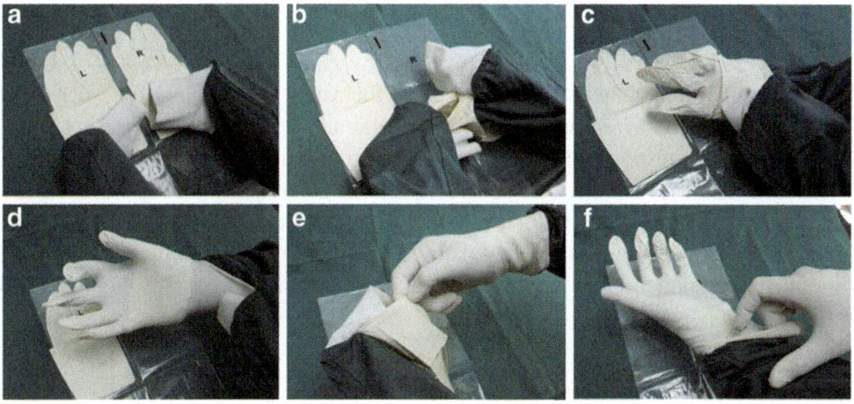

Fig. 10.7 Closed gloving technique. Gloves lie opened with the sleeve turned to the surgeon. The right sleeve is grasped with the "closed" left hand. The thumb of the glove is positioned over the thumb of the right hand, with the fingers pointing in the direction of the surgeon. The right hand holds the base of the inner sleeve of the glove. The left hand pulls the outer sleeve over the right hand and covers the sleeve of the gown. Make sure all fingers are in the correct positions. Using the right hand (with glove), the left hand glove is put over the left hand (thumb on thumb). The base of the glove is grasped. With the right hand, the sleeve is pulled over the left hand until it covers the sleeve of the gown

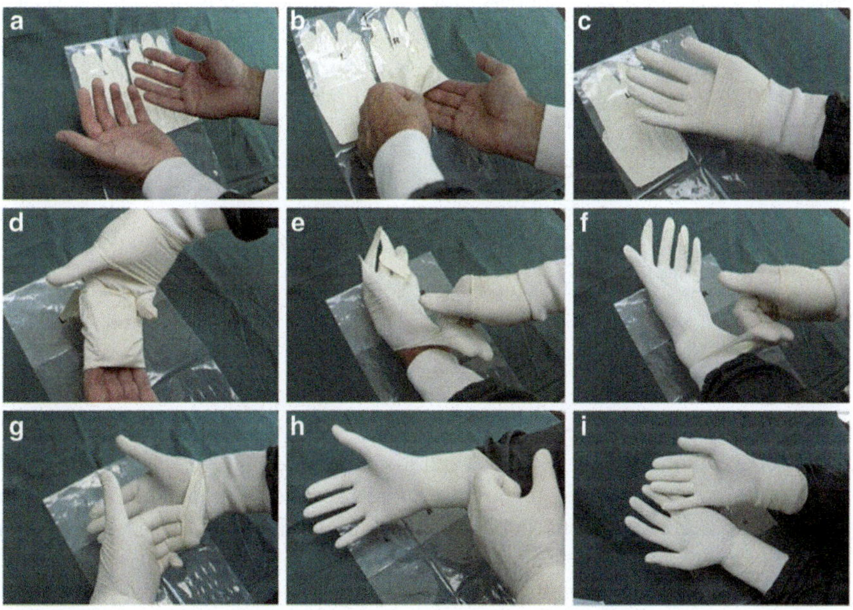

Fig. 10.8 Open gloving technique. (**a**) Gloves are lying opened with the sleeve turned to the surgeon. (**b**) With the "open" left hand, the right hand glove is grasped at its sleeve and the right hand is put inside. (**c**) The sleeve of the glove is pulled over the sleeve of the gown. (**d**) The middle, index, and ring fingers of the right hand are put inside the outside of the left hand glove. The left hand is pushed into the glove. (**e, f**) The right hand pulls the sleeve of the glove over the gown. (**g**) Thumb, index and ring fingers of the left hand are brought into the outside of the inner sleeve. (**h**) The left hand pulls the sleeve over the gown. (**i**) Final result

The Animals

Animals should be allowed to acclimatize to their new environment for a period of 7–14 days. This is to make sure that metabolic and hormonal changes that result from stressful transport are eliminated. It is a good habit to keep records of body weight as well as food and water consumption. A severe loss of body weight and a reduced intake of food and water are strong indications that the animal is in pain and that adequate measures should be taken. Given the fact that small rodents do not vomit during induction of anesthesia, it is not necessary to deprive them of food prior to the operation.

Preparation of the Incision Site

It is necessary to remove hair from the site of incision and from a sufficient surrounding area so that hair does not get into the wound. Apart from the possibility that associated bacteria might cause infection, hair can irritate the wound and delay healing. However, since short hair can be adequately disinfected, it is not always necessary to remove it for simple operations in rats and mice. Removal of hair should be carried out with clippers, after which a razor can be used if necessary, though clipping suffices in most cases. The incision site must then be wiped liberally with an antiseptic such as 70 % ethyl or isopropyl alcohol, or 0.5 % chlorhexidine or 0.1 % benzalconium chloride in 70 % alcohol. Also, tincture of iodine containing at least 1 % free iodine and 10 % povidone-iodine solution is very commonly used for larger animals in particular.

Don't use cream to remove the hair of the anesthetized animals, since it contains toxic components.

The Incision

The incision is made with a scalpel and a sharp blade, though scissors are more frequently employed for small animals. Some knowledge of anatomy is useful in determining where incisions into skin and muscle should be made, e.g., when obtaining access to the trachea or associated structures such as the carotid arteries. These lie beneath the sternohyoid muscle, which is divided into two halves kept together by overlying connective tissue. Merely tearing the connective tissue will allow separation of the two muscle halves without damaging them and thus permit an approach to the underlying tissues.

Temperature

Animals undergoing surgery lose heat due to suppression of the thermoregulating center in the hypothalamus. The smaller the animal, the faster the heat loss, and for the mouse this rate can be quite alarming, as a 7 °C drop in body temperature within the first 10 min is possible, depending on the environmental temperature. Methods of monitoring body temperature vary from sophisticated automatic recording apparatus to the simple thermometer placed in the rectum. Any of these methods can be used for any of the common laboratory species. Since a substantial drop in body temperature puts the animal at risk (anesthetics are more potent at lower temperatures), the temperature should be maintained near normal during the operation. This can be done using special thermostatically-controlled heating pads on which the animal lies.

Fluid Balance

Animals may lose substantial amounts of water and minerals during surgery via respiration, evaporation of tissue fluid and loss of blood. Dehydration puts the animal at considerable risk, and if much fluid is lost it must be promptly replaced. Maintenance of fluid balance is a complex matter, but an interim measure is to provide warm sterile saline. Any route of administration can be used, and although the intravenous route is preferred, the fluid can be given intraperitoneally or subcutaneously, as these routes are more accessible in small animals. Drying of tissues from prolonged exposure during surgery must be avoided. This effect can be prevented by applying warm saline or a saline-soaked gauze pad to the tissue.

Hemorrhage

Some bleeding during surgery is inevitable, but blood loss must be minimized at all times as even mild loss can result in the animal showing signs of shock. Persistent bleeding must be stopped promptly. There are several methods to effect hemostasis, including clamping, ligation, electrocautery and bipolar coagulation.

The bipolar coagulator is an indispensable surgical instrument that contributes to high-grade hemostasis and increases the speed of the process, leading to a higher efficiency of operation. Learn to use the coagulator properly with an appropriate pair of clean, oiled bipolar forceps. Oiling reduces sticking of the forceps to the tissue. If bleeding has occurred, excess blood must be removed before the wound is closed; otherwise, it acts as an ideal medium for bacterial growth.

Suture Materials

The primary function of a suture is to hold tissue together. In the first week of wound healing in the rat, the strength of the wound is only 10–15 % of the original strength of the intact tissue. This means that you have to adjust your choice of suture to the rate of tissue healing. Another important property is the tissue reaction caused by the suture. Generally, absorbable sutures cause less reaction as compared to non-absorbable sutures. Sutures should be removed after about 7 days.

Try to avoid excessively tight sutures, as this allows bacteria to be protected in tissues made ischemic by pressure. Also avoid using an excessive number of sutures, for this will induce larger ischemic areas in the tissue that are likely to give rise to infection. Today, both absorbable (polydioxanone, polyglycolic acid, and polylactic acid or a combination thereof) and non-absorbable materials are used. Most suture materials are of high-quality, pull easily through tissue and can be knotted securely. Mechanical performance of knotted sutures is generally measured by knot break load, minimum numbers of throws required for knot security, knot rundown force, first throw holding force, and tissue drag force (7.8). As a rule of thumb, always use absorbable material. Only in cases where you want to fixate a foreign object such as a cannula or electrode should you use non-absorbable sutures.

In rats, the maximum suture size is 4-0 for closure of the abdominal cavity, though in mice a 5-0 or 6-0 suture will suffice. Inside the animal, for ligation of small vessels, etc., we only use 6-0 and 7-0 sutures (Table 10.1, Figs. 10.9, 10.10, 10.11, 10.12).

Tying Knots

In small animal surgery, only instrument knots are tied. Knot tying starts with picking up the suture with the forceps. A loop (throw) is made on the tip of the needle holder, which is placed just above the wound edges. Depending on the tension between the wound edges, single or double throws (to increase the friction in the first half of the knot) are made around the tip of the needle holder. Next, the short end of the suture is grasped with the needle holder and the loop is pulled off, while moving each hand to the other side of the wound. Gently tighten to the point at which the wound edges are just approximated. This is the first half of the knot. Do not let go of the suture held with the forceps, but immediately make a second loop around the needle holder, which again should be just above the wound edges in the middle of the "V" formed by the drawstrings of the suture. Pull off the loop and tighten the knot. You will see that the first half of the knot is progressively tightened during this procedure. It is essential to maintain equal tension on the strands during knot tying. If you keep only one strand under tension while the loop is pulled off, the knot will tumble and slip.

Table 10.1 USP standard, nonabsorbable materials and synthetic absorbable materials

	Diameter				Minimum limit for knot pull strength			
	Min.		Max.		Class I[a]		Class III[b]	
Size	(in.)	(mm)	(in.)	(mm)	(lbs.)	(kp)	(lbs.)	(kp)
10/0	0.0005	0.013	0.0010	0.025	0.09	0.04	0.11	0.05
9/0	0.0010	0.025	0.0015	0.038	0.11	0.05	0.13	0.06
8/0	0.0015	0.038	0.0020	0.051	0.20	0.09	0.24	0.11
7/0	0.0020	0.051	0.0030	0.076	0.31	0.14	0.35	0.16
6/0	0.0030	0.076	0.0040	0.102	0.51	0.23	0.60	0.27
5/0	0.0040	0.102	0.0060	0.152	0.99	0.45	1.20	0.54
4/0	0.0060	0.152	0.0080	0.203	1.50	0.68	1.80	0.82
3/0	0.0080	0.203	0.0100	0.254	2.50	1.13	3.00	1.36
2/0	0.0100	0.254	0.0130	0.330	3.50	1.59	4.00	1.80
0	0.0130	0.330	0.0160	0.406	5.00	2.27	7.50	3.40
1	0.0160	0.406	0.0190	0.483	7.00	3.17	10.50	4.76
2	0.0190	0.483	0.0220	0.559	7.40	3.85	13.00	5.90
3	0.0220	0.559	0.0250	0.635	10.00	4.54	16.00	7.26
4	0.0250	0.635	0.0280	0.711	12.50	5.67	20.00	9.11
5	0.0280	0.711	0.0320	0.813	16.00	7.26	25.00	11.4
6	0.0320	0.813	0.0360	0.914	20.00	9.1	30.00	13.6
7	0.0360	0.914	0.0400	1.016	25.00	11.3	35.00	15.9

[a]Class I = silk, synthetic absorbable and nonabsorbable sutures

[b]Class III = stainless steel

Knot configuration can be classified into two general types by the relationship between the ears of the knot and the loop. When the right ear and loop of two throws exit on the same side of the knot (parallel to each other), the knot is judged to be square or reef. When the right ear and loop exit on or cross different sides of the knot, it is called a granny knot. The above described knot with two throws to overcome tension on the wound edges is called a friction or surgeon's knot. Tera and Åberg devised a simple description of a knot's configuration. The number of wraps for each throw is indicated by the appropriate Arabic numeral. The relationship between each throw, either cross or parallel, is given by the symbols × or =, respectively. In accordance with this code, the square knot is designated 1 = 1, and the granny knot as 1 × 1, while the simple surgeon's knot is 2 = 1 (this knot should not be used). A surgeon's knot with an extra half knot for security is represented by 2 = 1 = 1. By following the sequence described above, you will always tie 1 = 1 or 2 = 1 = 1 knots.

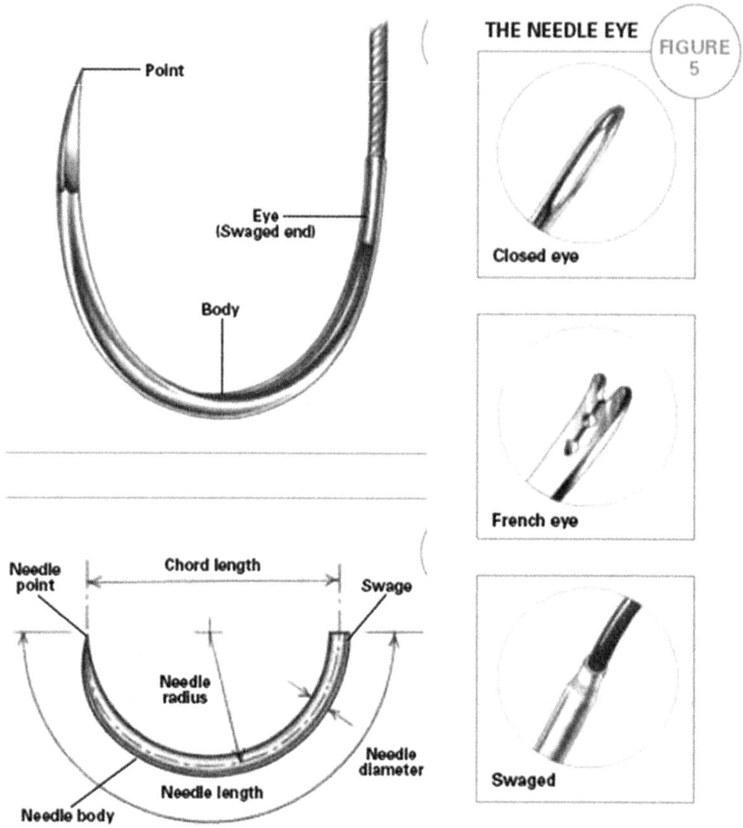

Fig. 10.9 The needle eye

Wound Closure and Care

Gentle tissue handling combined with careful hemostasis, thorough debridement and careful aseptic techniques, decrease the risk of infections and promote healing (Figs. 10.13, 10.14, 10.15).

Sterilization of Instruments

Surgical Asepsis

The opinion that rodents in particular are especially resistant to surgical infection is widespread amongst scientists, but evidence from work on intentional infection of these animals with pathogenic bacteria suggests that this may not be true.

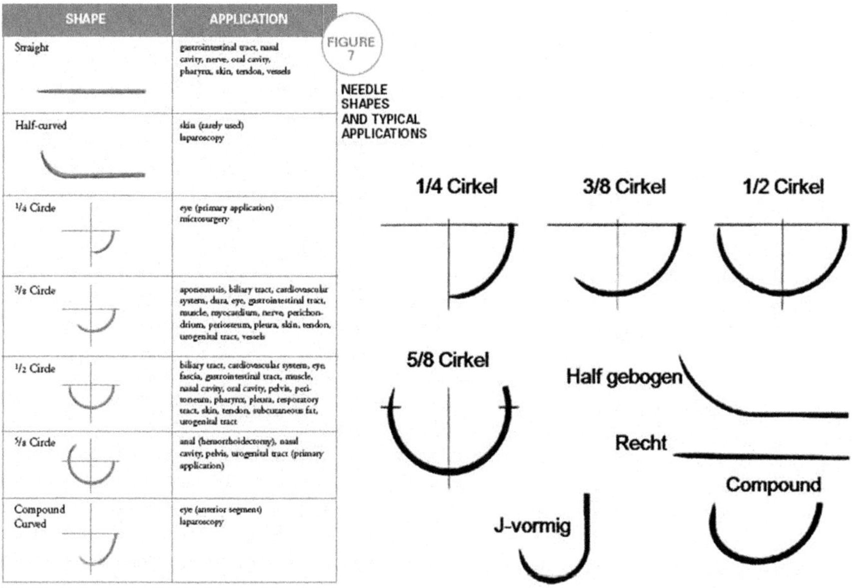

Fig. 10.10 Needle shapes, sizes and applications

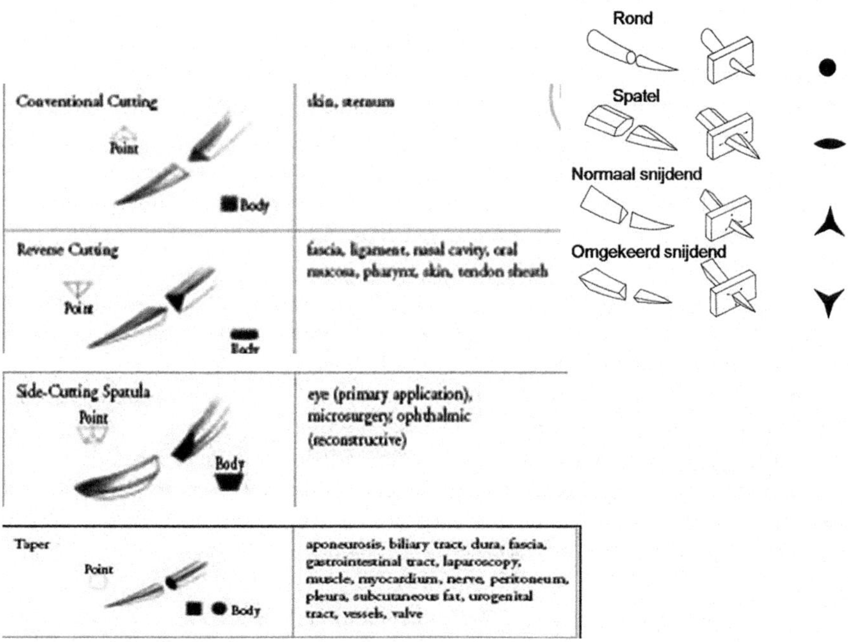

Fig. 10.11 Needle profile

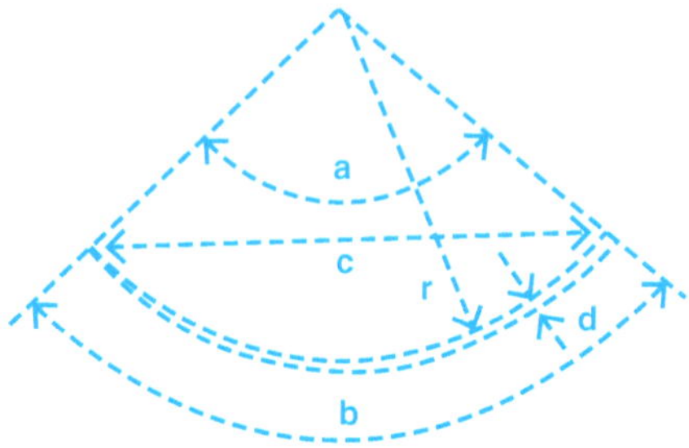

Fig. 10.12 Anatomy of the needle; a = angle, b = arch, c = chord, d = diameter, r = radius

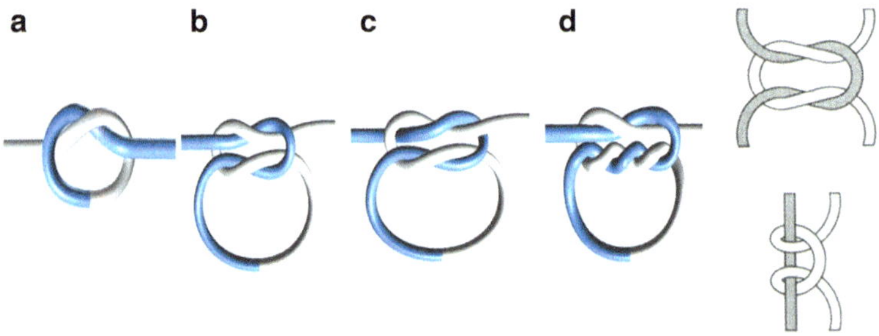

Fig. 10.13 *Right*, (**a**) single throw. (**b**) Square knot (two throws). (**c**) Granny knot. (**d**) Surgical knot. *Left*, square knot (*top*) and a tumbled knot (*bottom*)

Considering the ubiquity of microorganisms, it is fortunate that only a small percentage is capable of causing diseases. It is against these pathogens that numerous methods of sanitization, disinfection and sterilization have been developed. Although the rat demonstrates a remarkable resistance to wound infection, it is careless to neglect aseptic technique for surgical procedures in this animal. Surgery in the laboratory rat should therefore be governed by the same basic principles as those that address surgery on human beings. Surgical asepsis can be defined as the body of techniques that are designed to maintain an object or area in a condition as free of microorganisms as possible. Unfortunately, the opinion that aseptic technique is a waste of time and money is still widespread amongst researchers.

Surgery on laboratory animals is often characterized by the use of so-called "sanitary clean technique," meaning that the principles of aseptic surgery are completely neglected during procedures on small rodents. Most researchers gloss over these shortcomings by saying that a high percentage of animals remain alive after the operation. However, the argument should not be over whether or not the animals stay alive, but instead whether they can also be used as reliable models that give useful results.

Asepsis during surgery seems far more important than is generally acknowledged, though the literature about this specific topic is unfortunately limited. This is probably due to the fact that the untimely death of an animal is seldom connected to the conditions during surgery. As a consequence, more animals than strictly necessary are often used. Furthermore, the consequences on the welfare aspects of the

Fig. 10.14 The knot formula

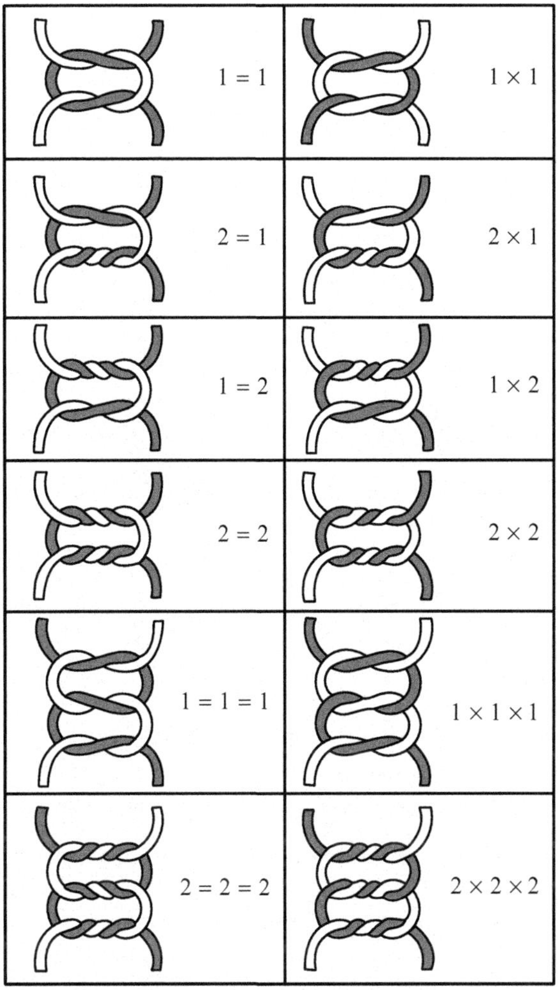

animals are unknown. The work of Baker on natural pathogens of laboratory animals and their effects on research gives additional arguments in favor of application of aseptic technique (Figs. 10.16 and 10.17).

Another outcome of bad aseptic technique is the reduction in long-term patency rates of inserted catheters. In our hands, permanent jugular vein catheters can remain patent for up to 6 months (and sometimes even longer) when both catheters and surgical instruments are sterilized before use, in addition to some standard precautions (see below). When using non-sterile catheters and instruments, patency is drastically reduced to between 1 and 2 weeks. Not only is patency affected, but the recovery of the animal is also delayed and the time needed to return to the pre-operative weight is extended.

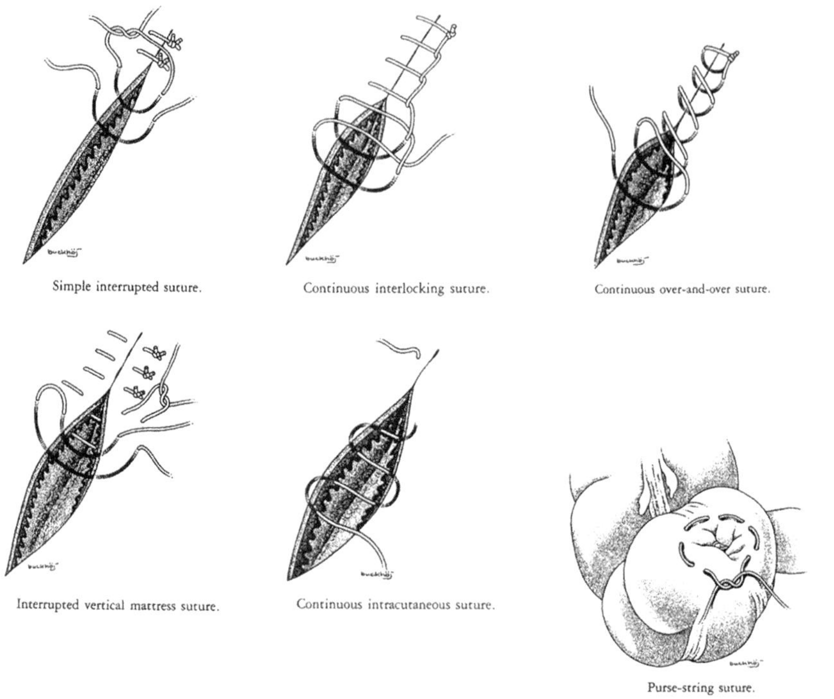

Fig. 10.15 Examples of sutures

A high level of aseptic technique can be achieved using the following procedures:

- Sterilization (generally achieved by autoclaving) of surgical instruments and all materials that are permanently implanted in the animal.
- Ask an assistant to help you during surgery.
- The use of proper scrub when washing hands, e.g., Betadine-scrub, which is a polyvinylpyrolidone-iodine solution or Hibi-scrub, which is a chlorhexidine solution.
- The use of rubber gloves (this was an important element in the transition from antiseptic surgery, in which microorganisms that are already present are killed, to aseptic surgery, where the environment is kept free from microorganisms from the outset.
- Subsequent disinfection with Sterillium® (isopropanol, n-propanol and ethylhexadecyl-dimethyl-ammoniumethylsulphate).
- The use of sterile suture materials, hypodermic needles and syringes, as well as sterile solutions (saline and heparin).
- A clean operating area, including a silicon rubber plate (30 × 25 × 1 cm), which can easily be sterilized or disinfected using 0.5 % of an aqueous chlorhexidine solution or 70 % ethanol.
- During surgery, try to avoid talking, sneezing, coughing, and unnecessary body movements.

Furthermore you should know what is sterile and what may be contaminated. Needless to say, only sterile objects may be touched by the surgeon. In case of doubt, consider the object as contaminated. For a wealth of information concerning

a

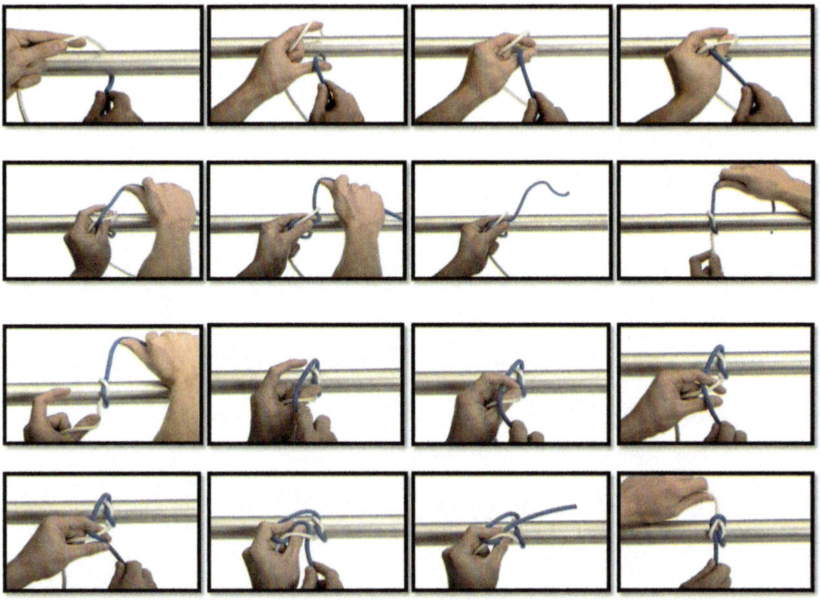

b

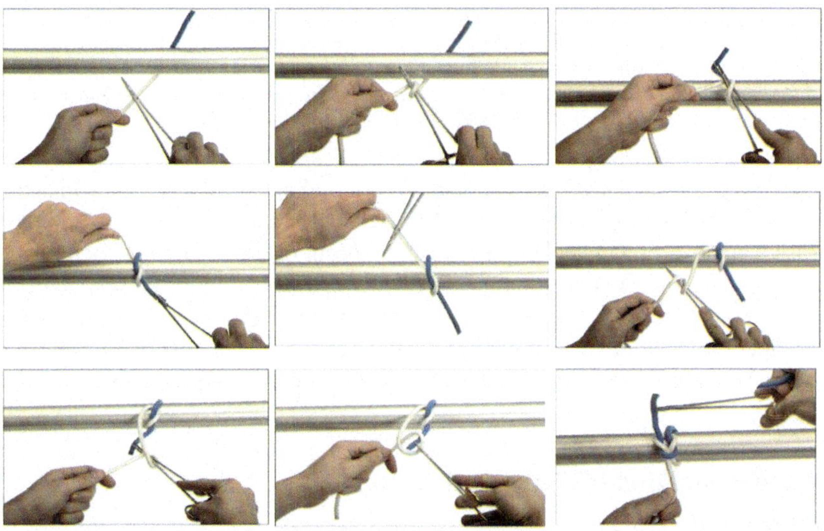

Fig. 10.16 (**a**) Two-hands tying technique. *Top left*, the white thread lies over the index finger while the dark thread is held by the right hand. The dark thread is brought between the thumb and index finger of the left hand. The left hand turns inside while the thumb turns underneath the white thread. The dark thread is brought over the white thread. Thumb and index finger turn underneath the white thread. Both threads are pulled through in the horizontal plane. The first throw is finished. The white thread is over the thumb. The dark thread goes over the white while thumb and index finger are opposed. The left hand continues to turn and grasps the dark thread between thumb and index finger. The dark thread is brought through the loop and held with the right hand. The knot is formed with both threads pulled in a horizontal plane. (**b**) Instrument tying technique. *Top left*, the white thread in the left hand, the needle holder on the white thread. The white thread is turned around the needle holder. The dark thread is grasped by the needle holder and pulled downwards over the wound, while the white thread is pulled upwards. The first throw is finished. The needle holder is replaced on the white thread, which is then turned around the instrument. The needle holder grasps the dark thread and pulls it upwards, while the white thread is pulled downwards. With both threads in the horizontal plane, the knot is tied

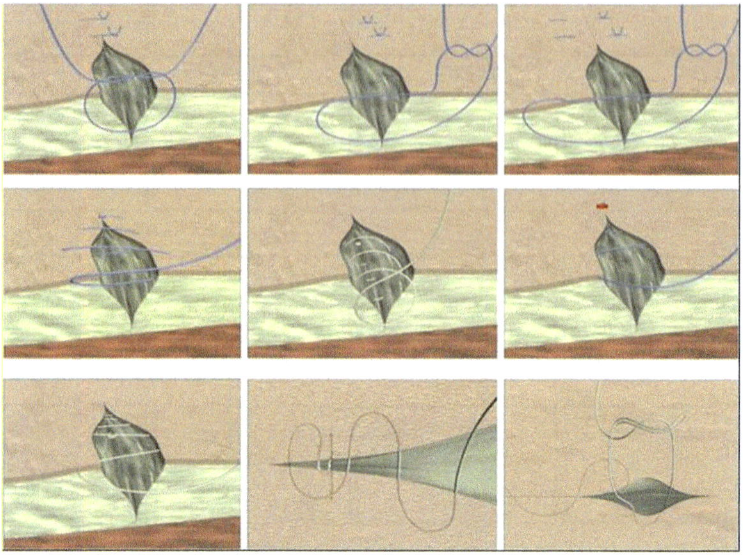

Fig. 10.17 Suture patterns. (**1**) Single knots. (**2**) Allgöwer suture. (**3**) Donati suture. (**4**) Simple running suture. (**5**) Running subcutaneous suture. (**6**) Intracutaneous suture using non-absorbable suture material. (**7**) Intracutaneous suture using absorbable suture material. (**8**) Top view of the intracutaneous suture (**7**). (**9**) Tying the knot of the intracutaneous suture (**7**)

good surgical practice (GSP), the reader is referred to the work of Tracy. In addition to the above mentioned precautions, animals could be administered an antibiotic in case a break of aseptic technique is likely to occur or when the total operating time exceeds 180 min. A single dosage of 150 mg/kg amoxycillin given subcutaneously (s.c.) or ampicillin (150 mg/kg s.c.) provided 10 min before the operation will give adequate protection against possible infections. Alternatively, Baytril® (Enrofloxacin, Bayer) could be administered subcutaneously at a dose of 4 mg/kg body weight.

Catheterization of Veins and Arteries

Blood Sampling

Introduction

Blood samples are often needed in biomedical research. There are numerous ways in which blood can be sampled from laboratory animals. Techniques can be divided into three categories, namely acute without surgery, acute with surgery (anesthetized animals) and chronicle models.

Without surgery:

- puncture of the ophthalmic venous plexus
- puncture of the heart

- cutting the tail vein
- decapitation

In all of these methods, the animal is anesthetized, handled or restrained. This treatment causes adverse reactions in the animal, such as a rise in a.o. glucose, prolactin, catecholamine and corticosteroid levels. Needless to say, these changes may interfere with the results of the experiment. Cutting of the tail vein can be done without anesthesia and seems to give reliable results.

With surgery, while the animal remains anesthetized during the complete experiment:

- catheterization of the jugular vein;
- catheterization of the carotid artery;
- catheterization of the femoral vein;
- catheterization of the femoral artery.

In these acute models, the anesthetic has a marked influence on a number of physiological parameters. These effects can be avoided by taking blood samples from surgically prepared animal models with permanent catheters. These techniques allow blood samples to be taken from freely moving animals without disturbing them. Several techniques have been proposed:

With surgery (permanent catheters):

- catheterization of the jugular vein;
- catheterization of the femoral vein;
- catheterization of the carotid artery;
- catheterization of the femoral artery.

Permanent catheterization of the jugular vein was described in the rat and the ground squirrel by Popovic et al. (1963). Catheterization of the jugular vein in combination with a head attachment apparatus allowing easy connection of catheters was first introduced by Steffens. These techniques enable continuous blood sampling from the general circulation and even infusion of fluids in the freely moving rat. During sampling or infusion, the animal remains undisturbed, which is of vital importance in experiments monitoring behavior or where stress factors are expected to influence results. Several modifications have been introduced, including those of Brown and Hedge, who introduced the L-shaped adapter, Nicolaidis et al., who used an additional stainless steel head bolt, and Dons and Havlik, who used a multilayered catheter. A nice overview of methods for vascular access and collection of body fluids from the laboratory rat has been given by Cocchetto and Bjornsson.

Catheters and Their Preparation

Catheters can be made of different materials, and thus can have different biocompatibility and thrombogenic properties. For acute catheterization, polyethylene (PE), polyvinyl chloride (PVC) or polyurethane (PU) catheters are used.

The catheters should be cut under 45°, as a beveled end allows for easy insertion. Permanent catheters used in rats and mice are preferably made of silicon rubber. Silicon tubing is rather flexible, causes hardly any immunity problems, and can be sterilized easily. We routinely use Silastic® or Silclear™ Medical-Grade tubing from Dow Corning or Degania Silicone, respectively. These tubes are available in various sizes. Alternatively, you might use commercially available catheters such as those provided by Instech. Normally, these catheters are cut in a rectangular manner, ensuring minimal intima damage. To enhance biocompatibility and reduce thrombogenicity, catheters can be coated (e.g., CBAS or Hydromer).

Acute Catheterizations

Catheterization of the Femoral Vessels (Vein or Artery)

Since this technique is almost the same for the artery and vein, it will only be described once. A skin incision of approximately 1.5 cm is made starting at the point where the femoral vessels leave the abdominal cavity towards the hind leg. The underlying fat is carefully dissected using small forceps to expose the femoral vessels. Regular flushing with warm sterile saline is carried out throughout the procedure. At this point, you should separate the vein from the artery, leaving the saphenous nerve untouched. The vessels are manipulated by their adventitia using jeweler's forceps. Any residual fat and connective tissue is removed to a distance of approximately 12 mm. At this point, it should be easy to raise the vessels by inserting the tip of the forceps beneath each one. Subsequently, two ligatures (6-0 silk) are placed around the vessel: a distal one that will tie off the vein or artery and a proximal one to tie the vessel around the catheter once it is inserted. Catheterization is facilitated by clamping the distal ligature close to the vessel and putting it under slight tension, thus stabilizing and lifting the vessel. Before cutting a V-shaped hole in the vessel close to the distal ligature using microvascular scissors, a clamp should be placed on the vessel just above the proximal ligature. The catheter, previously primed with sterile heparinized saline (50 IU/mL), is then gently inserted into the vein using straight anatomical forceps. To secure the catheter, the drawstring of the distal ligature should be tied around the catheter. Slightly open the clamp to check for leaking and patency.

Alternatively to cutting a hole in the vessel, a sterile 23-gauge needle fitted to a 1 mL syringe with the needle-tip bent to a 90° angle can be used. This method works best for the artery. Using it on the vein may result in an uncontrolled longitudinal hole.

Catheterization of the Jugular Vein

A 1.5 cm incision should be made just above the right clavicle. This place can easily be found by drawing imaginary lines between the animal's right ear and its left armpit, and between its right armpit and the chin. Using two or three spreading movements with sharp scissors, connective and adipose tissues are pushed aside and the external jugular vein is exposed and mobilized over a distance of about 1 cm.

Small artery forceps (Micro Halstead or baby-Mosquito) are used to clamp the vessel as far rostral as possible. The vein should then be ligated rostral to the clamp using 6-0 silk. A second ligature should be put loosely around the vessel. Using iridectomy scissors, a V-shaped hole should be cut in the vein just proximal to the clamp. Prior to its insertion into the vessel, the catheter must be connected via a 23-gauge needle to a 1 mL syringe filled with a sterile heparinized saline solution (50 IU/mL). After insertion, the second ligature must be tied, the artery forceps removed and the first ligature tied to secure the catheter.

Chronic Catheterization

Catheterization of the Jugular Vein

The technique described below is one of the most simple and reliable, and should be regarded as one of the basic techniques for catheterization of blood vessels. The operation can be divided into four parts, including:

- Preparation of the crown of the head
- Catheterization of the jugular vein
- Subcutaneous tunneling of the catheter
- Fixation of the catheter

After the neck of the animal has been shaven on the right side, the skin should be disinfected with chlorhexidine solution. An incision is made just above the right clavicle. This place can be easily found by drawing imaginary lines between the animal's right ear and its left armpit, and between the chin and its right armpit. Using two or three spreading movements with sharp scissors, connective and adipose tissue are pushed aside and the jugular vein is exposed. The division of the external jugular into the maxillary vein and the linguofacial vein should be identified. This bifurcation is recognizable by the presence of a small lymph node. At this point, one should choose the largest vessel for catheterization and mobilize it over a distance of about 5 mm.

Small artery forceps (micro-Halstead or baby-Mosquito) are used to clamp the vessel 3 mm rostral from the bifurcation. The vein should then be ligated rostral to the clamp using 6-0 silk. A second ligature is put loosely around the vessel. Using iridectomy scissors, a V-shaped hole is cut in the vein 2 mm rostral from the bifurcation. Prior to its insertion into the vessel, the sterile catheter must be connected via a 23-gauge needle to a 1 mL syringe filled with a heparinized saline solution (50 IU/mL). No air bubbles should be left in the catheter. Using sharp pointed jeweler's forceps to dilate the vessel, the catheter should be slid between the legs of the forceps and gently pushed into the vessel until the silicon ring reaches the V-shaped hole. Sometimes while trying to push the catheter gently into the vessel, it bounces on release.

In this case, the catheter has almost certainly entered the subclavicular vein. This can be remedied by pulling the catheter back, though still leaving it inside the vessel, then pushing it back in gently while lifting the animal's chest by the skin. When the silicon ring has reached the vessel, the tip of the catheter should now be at the level of the right atrium. This can be checked by removing the needle and the syringe and looking at the fluid in the free end of the catheter. During inspiration, fluid should be sucked into the catheter, while during expiration, the fluid should be pushed back (intra-thoracic pressure). Moreover, the heart frequency should be superimposed on the respiratory-induced fluid movements. The catheter positioning should be further checked by aspirating some blood, after which the catheter should be flushed gently with heparinized saline solution. The artery forceps are now removed, the caudal ligature gently tied, and the rostral ligature used to anchor the catheter to the vessel. Subsequently, the syringe is removed and the catheter gently clamped using a small microvascular clamp. To ensure that the catheter cannot move, one drawstring of each ligature should be cross-tied.

Subcutaneous Tunneling

Small artery forceps are pushed in longitudinally under the skin in a caudal direction to a distance of about 3 cm. Subsequently, the forceps should be turned anticlockwise through an angle of 90°, and pushed in the direction of the incision in the neck. The catheter should be grasped by the forceps and pulled back. Always make sure that the catheter makes a smooth curve, ensuring that the animal can move freely without tearing the catheter. The forceps should then be removed and replaced by a small microvascular clamp. Subsequently, the catheter is flushed with sterile saline, connected to an L-shaped adapter, and filled with a catheter lock solution. Finally the wound in the neck of the animal should be closed with two absorbable sutures (for details see Remie et al. 1990).

Fixation of Catheters

Head Attachment

The head of the animal should be shaven and disinfected with a chlorhexidine solution. An incision of approximately 2 cm should be made in the crown of the head. This will provide enough space to mount three stainless steel screws (1.1 mm in diameter and 4.2 mm in length) in the crown of the skull, which are used for additional anchoring. The membranous tissue should be removed using curved jeweler's forceps, and the bregma (the point on the top of the skull where the coronal and sagittal sutures meet) is then exposed. With a 3/0-round dental drill, three holes should be made, two on the left and one on the right of the bregma. The stainless steel dental drill should be loosely held between the thumb and forefinger to allow rotating movements. To ensure that underlying tissues like the dura, the rostral sagittal sinus or the transverse sinus are not punctured, the conical end of the drill is covered with a piece of polyethylene (PE) tubing, leaving about 2 mm of the tip uncovered. If you have access to a mechanical drill, this procedure can be carried out very quickly. Make sure that you drill the holes with a sharp drill bit rotating at low speed, as high speed drilling will dehydrate the bone structure and lead to necrosis. The screws are fitted into the holes using specially prepared surgical forceps and a small screwdriver. They are tightened to such an extent that approximately 2 mm is left between the skull and the head of the screws. Prior to fixation, the catheter has to be slid over the short end of the L-shaped adapter. To facilitate this action, you may use some diethyl-ether to make the Silicon tubing more supple so that the catheter will slide smoothly over the stainless steel tubing. Catheters placed in blood vessels are next flushed with saline (0.5 mL) and filled with a catheter lock solution. The long end of the L-shaped stainless steel adapter should be closed with a piece of heat sealed PE-tubing (PE-cap). Next, the catheter together with the L-shaped adapter should be fixed to the skull with suitable glue. Make sure the glue flows properly under the heads of the screws and that it is wrapped around the vertical part of the adapter, as this will prevent any movement.

Different Glues

Traditionally, glues used to fixate catheters or electrodes on the crown of the head were taken from the dental praxis. The two component acrylic glue was the standard over an extended period of time. However, due to the exothermic reaction that occurs during curing, the temperature of 1 g of glue could rise to 80 °C. Other problems with this glue were its moderate binding to wet surfaces and the relatively long curing time (5 min). Today, we prefer to use glasionomeric glue such as the Fuji CG. This glue binds to moist surfaces and develops less exothermic heat. Spoelstra has tested different pretreatments of the bony surface of the skull (personal communication). Etching with diluted acid gel or mechanical roughening of the surface enhanced binding. When the glue achieves optimal binding, it is not necessary to use additional anchoring screws. On their website, Alzet advocates the use of Cyanoacrylate Gel (CAG) (Loctide 454 Adhesive) in place of dental cement. The results indicate that the use of CAG significantly reduced surgical time for the preparation of ICV implants in rats and had no adverse effects on the postoperative period as compared to the dental cement technique. Additional advantages of the CAG technique were the reduction of the number of screws used for the implant, resulting in less trauma to the skull, and reduction in the size of the headcap. In addition, the transparent nature of the gel may allow early visual detection of inflammation and infection under the headcap.

Tethering Button

A button can be used instead of the head attachment. There are a number of different types of implantable buttons available. Tethering buttons are generally attached by suturing in place under the skin or on top of the skin in the scapular area. The button allows the catheter to be externalized and attached to a spring tether. The button may also be held in place with a rodent jacket without suturing the button to the rat. The type of button used depends on the needs of the research. Stainless steel buttons have been used for many years. They can be autoclaved and reused virtually without wear. Stainless steel buttons appear to be best suited for short-term studies lasting less than 7 days. Longer use of the stainless steel buttons can cause adverse tissue reactions.

Polysulphone buttons are rigid buttons that can be autoclaved and reused. Another type of button is the Dacron mesh button. This button has a Dacron mesh screen base and a soft plastic tip. The mesh base allows the tissue to infiltrate the material during longer studies. This produces an excellent attachment between the button and the animal. The flexible, soft plastic tip creates a simple yet effective method to attach the button to the spring tether. Dacron mesh buttons are not reusable, but are relatively inexpensive.

Catheter Lock Solutions

The patency life of a catheter can be affected by many factors, including flushing regimen, catheter material and lock solution used to fill the lumen of the catheter. Luo et al. studied the abilities of several different lock solutions to maintain patency of un-manipulated, indwelling polyurethane vascular catheters in rats, as the use of a vascular flushing regimen would give unwanted variation.

Heparinized saline: Sodium heparin (10,000 IU/mL) is added to physiological saline (0.9 %) to achieve a final concentration of 500 IU/mL.
Heparinized Dextrose: Sodium heparin (10,000 IU/mL) is added to 50 % dextrose solution to achieve a final concentration of 500 IU/mL.
Heparinized PVP: Sodium heparin (10,000 IU/mL) and polyvinylpyrrolidone (SIGMA PVP-40) are added to physiological saline (0.9 %) to achieve a final concentration of 500 IU heparin/mL and 1 g PVP/mL.
Heparinized Glycerol: Sodium heparin (10,000 IU/mL) is added to a glycerol solution (1.26 g/mL) to achieve a final concentration of 500 IU/mL of heparin.

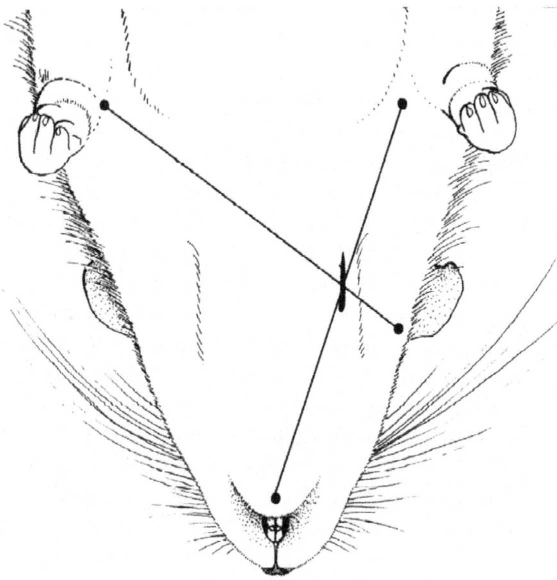

Fig. 10.18 Place of incision. The imaginary lines between the animal's right ear and its left armpit and between the chin and the right armpit are used for the determination of the incision site

Recovery After Surgery

Before you start using your animals, you should allow them to recover completely. In the past, we used the return to pre-operative weight as a sign that the animal had recovered completely. Today, however, using telemetric devices we know that it takes at least 7–9 days before the circadian rhythm has returned to normal. This is long after the animal has returned to its pre-operative weight (Figs. 10.18, 10.19, 10.20).

Catheterization of the Carotid Artery

The ventral neck area from the lower mandible to the sternum should be shaven. This area is then disinfected using chlorhexidine solution. A longitudinal incision is made along the ventral midline. The membrane overlying the glands is opened and the glands pushed aside with curved jeweler's forceps. The right carotid artery can be found lateral to the muscles overlying the trachea. The carotid artery should be mobilized for a distance of 12 mm in the caudal direction. Make sure that the vagus nerve, which lies next to the artery, is carefully dissected from the carotid artery without being damaged. The artery should then be ligated with 6-0 silk. This ligature should be tightened and put under minimal tension in the rostral direction.

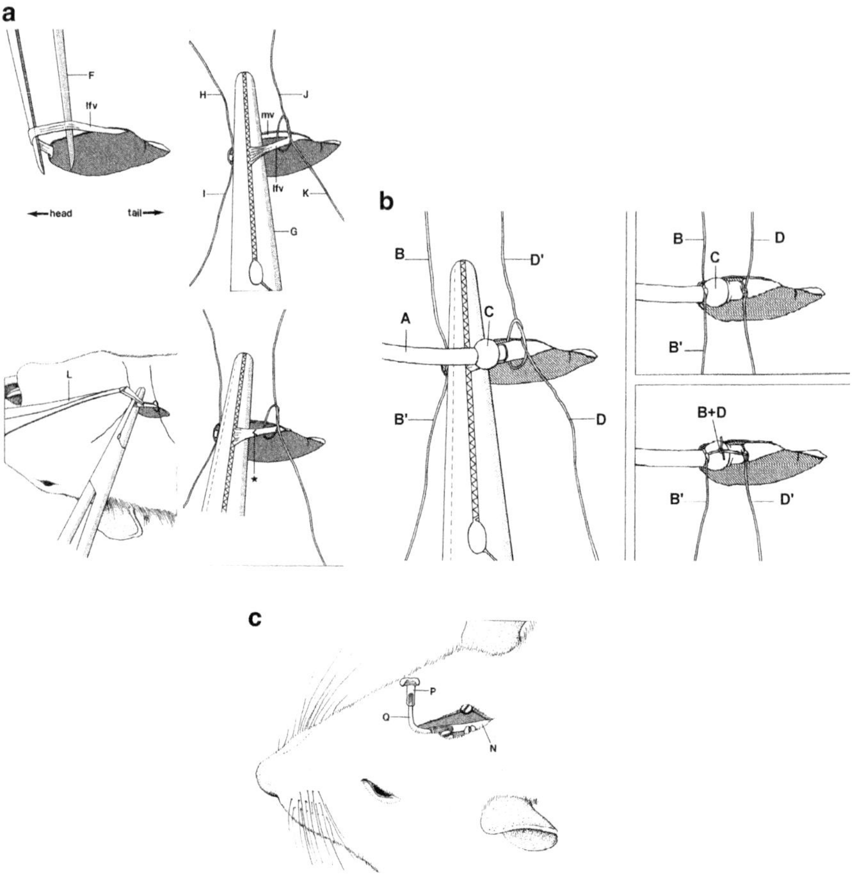

Fig. 10.19 (**a**) The linguofacial vein is mobilized using blunt dissection technique (F. anatomical forceps, lfv, the linguofacial vein). *Top right*: Clamping and ligation of the vessel. The bifurcation of the external jugular vein into the maxillary vein (mv) and the linguofacial vein (lfv) is located. Small artery forceps (G) are used to clamp one of the two vessels. The cranial ligature (H/I) is tightened and the caudal ligature (J/K) is put loosely around the vessel. *Below left*: Position of the iridectomy scissors (L) while cutting a V-shaped hole in the vein. *Below right*: detail of the vein with the V-shaped hole (*). (**b**) *Left*: The catheter is inserted into the vessel until the silicon ring (C) has reached the V-shaped hole, after which the strings D and D′ are tightened. *Top right*: The artery forceps are removed and the strings B and B′ are tightened behind the silicon ring. *Below right*: Strings B and D are tightened across the silicon ring and all remaining strings are cut to size. (**c**) Situation on the crown of the head. Three stainless steel screws have been inserted, and the L-shaped adapter (Q) connected to the catheter (N) is sealed with a polyethylene cap (P)

A vessel clamp should be placed 10 mm caudal to the bifurcation. This will prevent extensive bleeding when the vessel is cut. A V-shaped hole is cut 1 mm caudal from the bifurcation using micro-scissors. Prior to its insertion into the vessel, the catheter must be connected to a 1 mL syringe filled with sterile heparinized saline solution (50 IU/mL). As described before, a modified 25-gauge needle with its tip bent

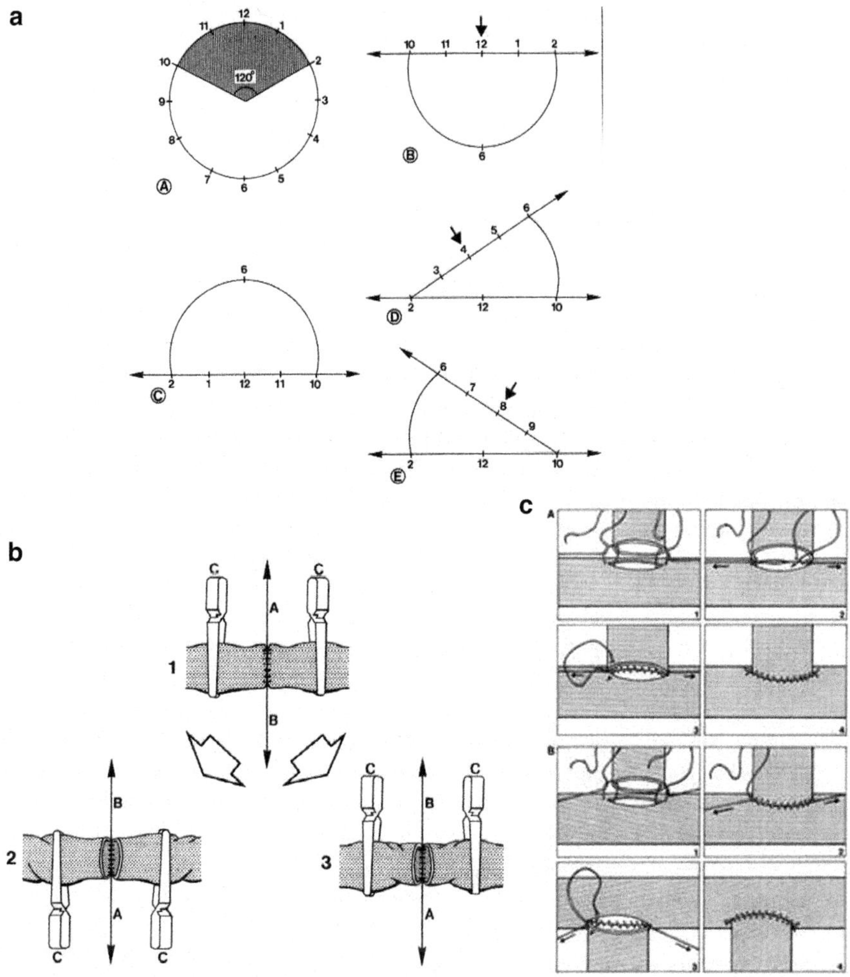

Fig. 10.20 Schematic drawing of the steps in an end-to-end anastomosis. Panel **a**: The corner sutures are placed at angle of 120° ('2 o'clock' and '10 o'clock'). (b) Light tension is put on the corner sutures, and the next three sutures are placed. The arrow indicates the first suture to be placed. (c) The anastomosis is rotated, and the '6 o'clock' suture is placed. (d, e) The '6 o'clock' suture is retracted to the right, and the next three stitches are placed. (d) The '6 o'clock' suture is retracted to the left, and the anastomosis is completed. Panel **b**: Rotation of the anastomosis: (1) The anterior wall has been sutured. (2) Clamp rotation: The clamps (C) are rotated and the corner sutures (A and B) are repositioned. (3) Suture rotation: The corner sutures are repositioned, while the clamps retain their position, resulting in slight torsion of the vessel. Panel **c**: Inside suture technique. (b) Normal suture technique. Both use a running suture

to a 90° angle can also be used to puncture and subsequently catheterize the vessel. The catheter is pushed in until its tip reaches the vessel clamp. The proximal ligature should be tied with the first half of a square knot. This will prevent the cannula from being pushed out by blood pressure when the vessel clamp is released, while making it still possible to advance the catheter into the artery. To be sure that the cannula is not pushed out, the inserted part of the cannula must be held in place with the jeweler's forceps while the cannula is pushed further using anatomical forceps. The second half of the square knot should now be tied and the distal ligature is tied to secure the catheter.

Craniectomy and Craniotomy

Craniectomy

Craniectomy is defined as the removal of the brain. In most experimental work on the brain, it is necessary to perfuse and remove the brain after the experiment(s). In this way, one can see the exact locations of electrodes, lesions, infusion cannulae, dialysis probes, etc. Therefore, we have to perfuse the brain with saline and formaldehyde fixative (10 %). Perfusion can be performed by using two reservoirs, one filled with saline and the other with 10 % formaldehyde. Using a 3-way stopcock, both fluids may be perfused through the same needle (19G).

Cardiac Perfusion-Fixation and Removal of the Brain

The animal should be deeply anesthetized using twice the amount of barbiturate needed for surgical anesthesia. Open the abdomen via a medial laparotomy, and open the thorax by cutting the ribs. Insert a perfusion needle into the left ventricle of the heart. Keep the needle in place by clamping it with hemostatic forceps at the point where the needle enters the heart. Next, cut a hole in the right atrium to allow blood to be drained from the head. Flush with saline solution until the effluent is clear and then switch to formaldehyde perfusion (250 mL). The brain can then be removed.

Make an incision from behind the ears to the nose. Cut the skin from the skull down to the nose. Remove the temporal muscles from the skull and the first two vertebrae. Next, remove the cranial bone using small bone cutters until the olfactory bulbs are visible. Cut the dura and lift the brain, starting with the anterior part. Cut the cranial nerves, lift the rest of the brain and cut the upper part of the spinal cord. Continue the fixation process by placing the brain in formaldehyde (10 %) for 3 days. At this point, the tissue will be ready for sectioning and staining.

Craniotomy

Introduction

Craniotomy is defined as an operation on the skull or an incision into the skull. Surgery on the head is easier than surgery on other parts of the body because there is little risk of excessive hemorrhage and the area is easily exposed.

Stereotaxic brain surgery is one of the techniques used to apply mono- or bipolar needle-shaped electrodes, infusion catheters or sample catheters for the collection of cerebrospinal fluid. In this section, some basic principles for the use of stereotaxic equipment will be discussed. Some frequently used techniques will also be described.

Anatomy

As with all surgery, a thorough knowledge of the anatomy in the area of interest is of vital importance. Fortunately, there are a number of atlases available describing different aspects of the rat brain. More information on stereotaxic atlases can be found at http://www.kopfinstruments.com/Atlas/Rat.htm. Look at the skull diagram of a 290 g Wistar rat from dorsal and lateral views. Important landmarks are the bregma, lambda and the interaural line. Note that lambda is 0.3 mm anterior to the coronal plane that passes through the interaural line.

The principle of stereotaxic surgery is based on the constant relationship between these landmarks on the skull and parts of the brain. A system of three coordinates is used to determine a specific location in the brain relative to one of these landmarks:

- Anterior-posterior (A-P)
- Dorsal-ventral (D-V)
- Lateral (Lat)

Stereotaxic Brain Surgery

Preparation of the Animal

Animals (adult Wistar rats of either sex) should preferably weigh between 250 and 350 g. There may be some differences in craniometric and stereotaxic data for rats of different strains, sex and weight; however, no substantial stereotaxic error will occur when 290 g rats of a different strain and sex are used (see Paxinos and Watson for details). For the use of stereotaxis in newborn rats, the reader is referred to Cunningham and McKay.

After the animal has been anesthetized, the head of the animal is shaven and disinfected. The rat is placed in a Kopf small animal stereotaxic instrument. The head will be fixed in three places, at the two bony ear canals and the upper jaw. Start by inserting the ear bars into each ear canal. The head should pivot freely about the interaural axis and should have little lateral movement. Move the incisor bar under the upper incisors. Place the nose clamp over the nose, gently retract the incisor bar anteriorly and tighten it. Next, the incisor bar should be adjusted vertically until the heights of lambda and bregma are equal (both in a coronal plane), resulting in a flat-skull position. For the implantation procedure, it is imperative that all of the periosteum is removed from the crown of the head, ensuring that the glue will properly adhere to the skull. After positioning of the implantable device, three additional holes may be drilled in the skull as described previously.

Positioning of Implantable Devices

To determine the exact position of an implantable device (electrode, catheter, microdialysis probe, etc.) you must use a stereotaxic atlas. The Paxinos and Watson atlas (26) comes highly recommended. After the animal has been placed in the stereotaxic instrument, the implantable device is clamped into the electrode carrier. Make sure that the device is straight and is set at a 90° angle to the coronal plane. Next, the tip of the device is adjusted directly above the bregma and the reading for the A-P and the Lat zero points are taken. Calculate the readings after the distances given in the stereotaxic atlas are added or subtracted from the A-P and Lat zero readings. The device should now be moved to the newly calculated A-P and Lat position. The device is lowered until it just touches the skull, giving the vertical zero reading. Calculate what the final reading must be in order for the device-tip to penetrate the brain to the specific depth. When the device is in the newly calculated position, it should be slightly raised and the place on the skull where the hole is to be drilled should be marked with a sharp pencil. Move the electrode aside and drill a hole of sufficient diameter in the skull at the pencil mark. The device is then positioned once more in the A-P and Lat planes according to the previous calculation. Now the device can be lowered over the required distance to the reading of the calculated vertical placement (D-V value).

Lesion or Stimulation Using Electrodes

Monopolar electrodes can be used to damage or stimulate brain areas by passing an electrical current between the relatively small surface area on the tip of the electrode to a relatively large amount of body tissue, which acts as a ground terminal. Electrodes are preferably made of platinum and should be insulated. A two-component epoxy resin can be used for insulation. Simply dip the electrode once or twice and let the resin cure. Next, 0.5 mm of the tip is freed of its insulation using a scalpel blade. Always connect the positive terminal (anode) of the current source to the electrode, thus preventing the formation of hydrogen and oxygen. The cathode should be clamped onto a muscular layer of the head. After the lesion has been made, the electrode is removed and the hole in the skull is closed with bone wax. The skin is then closed using the interrupted suture technique. Bipolar electrodes are very similar to the monopolar electrodes. They are commercially available, but you can also build them yourself. Before implantation of the bipolar electrode, it should be soldered to a miniature connector (e.g., an IC-foot).

Technically, the bipolar electrode is implanted in the same way as the monopolar electrode. After the electrode has been placed in a specific brain region, according to the calculated A-P, Lat and D-V values, the electrode together with the small IC-foot connector are glued to the skull using acrylic dental cement. Note that the electrode carrier is still holding the electrode during this process. After the dental cement has cured, the skin should be closed around the connector using absorbable sutures (4-0).

Electroencephalography

Electroencephalography recording (EEG) requires specialized equipment and techniques. The signals picked up from the cortex are of very low voltage and should be derived as close to the source as possible. Electrode placement is therefore of vital importance. We routinely use 1 mm stainless steel screws. The screws are equipped with 2 cm long small-diameter non-insulated stainless steel wires, which are point-welded to the top of the screw. The screws are placed on the crown of the head: the first 2 mm posterior and 2 mm lateral to bregma and the second 2 mm anterior and 2 mm lateral to lambda. A third screw is used for additional anchoring and can also be used as a ground electrode. The wires of all three screws are soldered in situ to a small connector.

Signals may be registered via a multi-channel electrical swivel while the animal is kept in a slowly rotating cage (to keep it awake) placed in a cage of Faraday. More sophisticated is the use of a telemetric device that can be magnetically switched on and off, ensuring a long lifespan of the battery (approximately 1 year).

Central Nervous System Injections and Infusions

Several techniques have been described for the placement of a catheter in one of the ventricles of the rat for injection, sampling and infusion. Most techniques are rather simple and use a combination of an external guide cannula and an internal cannula. A very reliable and inexpensive cannula and injection system for local chemical brain stimulation with small volumes of fluids has been described by Strubbe. This system, which will be described below, has several advantages over others mentioned in the literature. Its construction is very simple, it is inexpensive and therefore disposable, it allows bilateral infusions without disturbing the animal and it gives you a direct visual determination of the rate at which fluid enters the brain.

The system consists of a permanent guide (outer) cannula, made from a disposable hypodermic 23-gauge needle. The colored plastic is removed with the exception of the white fixed inner ring. The thin plastic layer above the ring is removed using a sharp knife. The needle is cut to a length determined by the depth of the brain area to be stimulated. After disinfection in chlorhexidine solution, the cannula tip is placed stereotaxically as described above. The plastic ring is just above the surface of the skull and is fixed using acrylic dental cement and three screws to anchor the cement to the skull. A polythene cap (ID 0.58 mm, OD 0.96 mm) is placed on the outer cannula. In case bilateral cannulae are wanted, they can be mounted in a brass bar with holes at the required distance and can be glued together with dental cement.

The inner cannula consists of a stainless steel tube (ID 0.1 mm, OD 0.29 mm) and should exactly be 3 mm longer than the guide cannula. Over these a 3 mm polythene tube (PE) is slipped (ID 0.29 mm, OD 0.61 mm). A silicon cuff (ID 0.5 mm, OD 1.0 mm) is slid over this connection. On the other end of the PE tube, a silicon cuff (ID 0.5 mm, OD 1.0 mm) is placed. Over this cuff, a second silicon tube (ID 1.0 mm, OD 3.0 mm) is slipped. A small nail with a head of suitable size is pushed into the silicon tube to act as a plunger, after which the injection system is ready. The injection system is first filled completely with methylene blue (1 % in saline). Subsequently, the nail is pushed down and slightly pulled back again until an air bubble can be barely seen above the inner cannula. Now, the tip of the inner cannula is placed in the fluid to be injected. The nail is pulled up so that the infusion tube is filled with that fluid. The inner cannula is then placed into the outer cannula. The silicon cuff on the lower end of the injection tube serves to attach it firmly to the guide cannula. The injection can now be made by pushing down on the nail. The air bubble, separating methylene blue from the injection fluid, may serve as a marker to read the volume administered. This volume can be calculated from the diameter of the PE tubing.

Microdialysis

Chemical interplay between cells occurs in the extracellular fluid, a compartment usually overlooked due to the fact that it is hard to be accessed experimentally. Many experimental approaches have been proposed to acquire information about the extracellular environment of the intact brain, for example, ventricular perfusion, cortical cup perfusion and push-pull cannulae. The introduction of a dialysis membrane into the tissue has provided the first generally applicable way of interacting with the extracellular compartment. Brain dialysis is a nice technique for the investigation of in vivo release of neurotransmitters and amino acids. It has some advantages over other techniques like push-pull perfusion, which risks doing mechanical damage to the tissue, and in vivo voltammetry (Gonon et al. 1980; Ewing et al. 1982), where the chemical identity of the detected material is often uncertain. Today, a range of different probes is used, e.g., the trans-striatal, the U-shaped, the I-shaped and the commercially available Carnegie cannulae.

Four different types of intracerebral microdialysis probes were characterized by Santiago and Westerink. They concluded that every type of dialysis probe causes a certain amount of damage when implanted into the brain. In acute experiments, the best results were obtained using the I-shaped probe. In chronic conditions (24 h after implantation), all probes performed well. During the second day after implantation, conditions were optimal to carry out dialysis experiments. After 48 h, certain restrictions are to be expected due to elevated K^+ levels in the neuronal tissue. The trans-striatal probe displayed a high output given its capacity to perfuse bilateral brain structures, but from an animal welfare perspective, much discomfort is produced due to damage done to the temporal muscles. Details about perfusion fluids and other practical aspects can be found in the work of T.E. Robinson and J.B. Justice Jr.

Noninvasive and Invasive Measurement of Blood Pressure in Rodents

Experiments using acutely prepared, anesthetized or immobilized laboratory animals have been the standard for many years. The majority of physiological and pharmacological knowledge on blood pressure (BP), its regulation and the influence of drugs on the cardiovascular system has been derived from these experiments. The site of measurement is traditionally either the carotid artery or the femoral artery. Although valuable, this information is obscured by direct effects of anesthesia and surgical intervention. General anesthesia, for example, has a considerable impact on baseline blood flow, myocardial function (pressure), respiration, etc. Therefore, if subtle effects are to be measured, the need for refined methods becomes apparent.

The two most commonly used techniques for measuring BP in small laboratory animals such as rats and mice are the indirect tail-cuff method and the direct measurement using an exteriorized catheter connected to a pressure transducer. Disadvantages of these methods are reviewed in the papers of Brockway and Hasler and Kramer and Kinter.

Radio-telemetry with an implantable transmitter, which minimizes exposure to stress, provides a way to obtain accurate and reliable measurements from awake and freely moving animals. BP measurements via radio-telemetry have been described for small laboratory animals from rabbits to mice. From an animal welfare point of view, telemetry has another advantage, as it allows the possibility of group-housing the animals. This is very complicated in animals that have exteriorized catheters, as cage mates will gnaw on the stoppers that close the catheters, leading to the untimely death of the animal. In this section, the surgical aspects of the radio-telemetry techniques currently used to monitor and measure blood pressure in awake animals will be described.

Tail Cuff

A common technique currently employed for monitoring BP in conscious rats and mice is the tail-cuff device. The tail-cuff method has the advantage of being noninvasive. The technique is dependent on maintaining a minimum amount of blood flow in the tail, so any physiological, pharmacological or environmental factor that affects tail blood flow will affect the BP measurement. In addition, mice are subject to the stress of handling, heating and restraint, to which they do not seem to adapt, even after considerable training.

However, considerable measurement errors have been found when using the tail-cuff technique in rats. Bunag et al. reported that systolic arterial pressure measurements taken with this method in conditioned Sprague–Dawley rats may deviate from measurements obtained by simultaneous aortic cannulation by as much as 37 mmHg. Such errors result in additional variability in the data. In light of the fact that only a few momentary measurements are possible in rats subjected to stimuli such as handling, restraint and heating, this high variability may interfere with the detection of small chronic changes in pressure.

Exteriorized Catheters

Another common technique currently employed for monitoring BP in conscious rats and mice is the use of an exteriorized catheter that refers pressure to a nearby transducer. This technique involves placement of an open-lumen catheter in an artery and exteriorizing it at a site inaccessible to the animal, usually in the nape of the neck or at the crown of the head. Measurements of pulsatile arterial pressure and heart rate (HR) can thus be obtained. However, accurate measurements of systolic and diastolic pressure are difficult to obtain. There is often a poor dynamic response due to the small size of the catheters used for cannulation of mouse arteries, significant catheter compliance, and a relatively long length of conduit from the point of cannulation to the transducer. The resulting damping renders measurements of systolic and diastolic pressure unreliable.

Probably the most significant drawback of this technique is that catheter patency is short. A carefully conducted study on BP measurement in mice using exteriorized catheters showed a failure rate of 50 % by the end of the fourth week post-implantation. There appear to be no documented scientific studies where the use of exteriorized catheters in mice results in stress artifacts. However, in rats, significantly higher levels of systolic and mean BP (both +17 mmHg) have been found when recorded by indwelling and exteriorized catheters vs. implanted radio-telemetry devices. The authors hypothesized that the differences in BP and HR were due to the stress associated with performing cardiovascular measurements through externalized catheters.

Intracranial Pressure Measurement

One of the most elegant ways of measuring intracranial pressure is by radiotelemetry. The best results are achieved when the tip of the pressure catheter is placed in the Cisterna Magna.

Placement of the Catheter Tip

Following the same procedure as described previously, the skin is opened on the crown of the head and a 1 mm trepan drill is used to make a hole in the skull 2 mm from the occipital bone. In rats weighing between 150 and 300 g, the catheter is placed in the Cisterna Magna over a distance of 9–11 mm measured from the top of the skull. The angle of insertion should approximately be 60° to the flat skull position, with the tip pointing in caudal direction. Once inserted, the catheter is fixed using tissue glue to cover the hole and Glasionomeric glue to further fixate the catheter in its 60° position. The transmitter body is subsequently placed subcutaneously and sutured to the skin. Take care to not kink the catheter and make sure the whole curve of the catheter is covered with glue. Finally, the skin is closed over the glue using absorbable sutures.

Recording of Body, Brain, and Muscle Temperature

Introduction

Body temperature in the rat is usually recorded via the rectum. Today, small electronic thermometers come with a variety of sizes of sensor probes (thermistors, thermocouples). However, handling the animal during measurement (even for 15 s) will inevitably lead to a rise in body temperature. Measuring more animals in a row will lead to stress-induced hyperthermia. Ideally, cerebral ischemia studies should measure brain temperature, as even slight hypothermia conveys protection against cerebral ischemia. It is well known that brain and body temperatures can dissociate under anesthesia and especially during ischemia. Under normal circumstances, however, brain and body temperatures correlate reasonably well.

The preferred method of continuous temperature measurement in rodents is via implanted telemetry probes. Telemetry probes have several advantages over conventional methods of measuring temperature (e.g., rectal probe).

- Less stress to the animal
- Rectal temperature measurements can result in a stress-induced fever (peak of ~1 °C, duration of ~1 h)
- Allows continuous and automated data collection and storage
- Allows better feedback regulation of temperature, resulting in more consistent and valid data
- Less animal usage

Implantation

Probes are available to measure core temperature (abdominal implant) or brain temperature. These probes are mounted on the skull. A core temperature probe (model TA10TA-F40) is manufactured by DataSciences, Int. It has been sterilized and can be implanted into the abdominal cavity. For mice and gerbils, smaller models are available. A brain temperature probe (model XMFH-BP) is manufactured by Mini-Mitter Co. The probe with glued-on base assembly (cannula and nut) can be glued to the skull using glasionomeric glue. The probe tip is sterilized prior to use and the cannula is implanted under general anesthesia. Depending on the area of interest, the tip of the probe can be implanted in different brain areas.

Muscular Temperature

Muscular temperature can easily be measured by injection of a probe into the muscle of interest. The IPTT-200 programmable temperature transponder (dimensions 2.2×14 mm) is one of the leading transponders (http://www.plexx.nl/UK/index.htm). These pre-coded or programmable injectable transponders for electronic animal identification and body temperature monitoring can be programmed with 32 characters of choice. The temperature sensor has an accuracy of 0.1 °C. It is encapsulated in biocompatible glass and has an anti-migration feature. Special equipment is required to read these transponders.

References

Agterberg MJH. Het rokende brein. Biotechniek. 2001;40:139–45.

Baker DG. Natural pathogens of laboratory animals: their effects on research. Washington, DC: American Society for Microbiology Press; 2003.

Bouman HJ, Wimersma Greidanus TB. A rapid and simple cannulation technique for repeated sampling of cerebrospinal fluid in freely moving rats. Brain Res Bull. 1979;4:575–7.

Bradfield JF, Schachtman TR, McLaughlin RM, Steffen EK. Behavioral and physiologic effects of inapparent wound infection in rats. Lab Anim Sci. 1992;42:572–8.

Brown MR, Hedge GA. Thyroid secretion in the unanesthetized, stress-free rat and its suppression by pentobarbital. Neuroendocrinology. 1972;9:158–74.

Clement JG, Mills P, Brockway B. Use of telemetry to record body temperature and activity in mice. J Pharmacol Methods. 1989;21:129–40.

Cocchetto DM, Bjornsson TD. Methods for vascular access and collection of body fluids from the laboratory rat. J Pharm Sci. 1983;72:465–92.

Colbourne F, Sutherland G, Auer RN. An automated system for regulating brain temperature in awake and freely-moving rodents. J Neurosci Methods. 1996;67:185–90.

Criado A. Use of cyanoacrylate gel as a substitute for dental cement in intracerebroventricular cannulations in rats. Contemp Top Lab Anim Sci. 2003;42:13–6.

Cunningham MG, McKay RD. A hypothermic miniaturized stereotaxic instrument for surgery in newborn rats. J Neurosci Methods. 1993;47:105–14.

DeBow S, Colbourne F. Brain temperature measurement and regulation in awake and freely moving rodents. Methods. 2003;30:167–71.

Dons RF, Havlik R. A multilayered cannula for long-term blood sampling of unrestrained rats. Lab Anim Sci. 1986;36:544–7.

Ewing AG, Wightman RM, Dayton MA. In vivo voltammetry with electrodes that discriminate between dopamine and ascorbate. Brain Res. 1982;249:361–70.

Faulkner BC, Gear AJL, Hellewell TB, Watkins FH, Edlich RF. Biomechanical performance of a braided absorbable suture. J Long Term Eff Med Implants. 1996a;6:169–79.
Faulkner BC, Tribble CG, Thacker JG, Rodeheaver GT, Edlich RF. Knot performance of polypropylene sutures. J Biomed Mater Res. 1996b;33:187–92.
Fluttert M, Dalm S, Oitzl MS. A refined method for sequential blood sampling by tail incision in rats. Lab Anim. 2000;34:372–8.
Gonon F, Buda M, Cespuglio R, Jouvet M, Pujol JF. In vivo electrochemical detection of catechols in the neostriatum of anaesthetized rats: dopamine or DOPAC? Nature. 1980;286:902–4.
Greene EC. Anatomy of the rat. New York: Hafner; 1963.
Hebel R, Stromberg MW. Anatomy and embryology of the laboratory rat. Wörthsee: BioMed Verlag; 1986.
Hsu PA, Cooley BC. Effect of exercise on microsurgical hand tremor. Microsurgery. 2003; 23(4):323–7.
Imperato A, Di Chiara G. Trans-striatal dialysis coupled to reverse phase high performance liquid chromatography with electrochemical detection: a new method for the study of the in vivo release of endogenous dopamine and metabolites. J Neurosci. 1984;4:966–77.
Iwaki T, Hayakawa T. A color atlas of sectional anatomy of the mouse. Tokya: Adthree; 2001. ISBN 4-900659-58-4.
Kline J, Reid KH. Variability of bregma in 300 gram Long-Evans and Sprague-Dawley rats. Physiol Behav. 1984;33:301–3.
Luo YS, Luo YL, Ashford EB, Morin RR, White WJ, Fisher TF. Comparison of catheter lock solutions in rats. In: Proceedings AALAS 2000, San Diego. www.criver.com/pdf/cath_lock_solutions.pdf
Nicolaidis S, Rowland N, Meile MJ, Jallat PM, Pesez A. Brief communication. A flexible technique for long term infusions in unrestrained rats. Pharmacol Biochem Behav. 1974;2:131–6.
Paxinos G, Watson C. The rat brain in stereotaxic coordinates (deluxe edition). 4th ed. New York: Academic; 1998.
Paxinos G, Watson C, Pennisi M, Topple A. Bregma, lambda and the interaural midpoint in stereotaxic surgery with rats of different sex, strain and weight. J Neurosci Methods. 1985;13:139–43.
Popovic V, Kent KM, Popovic P. Technique of permanent cannulation of the right ventricle in rats and ground squirrels. Exp Biol Med (Maywood). 1963;113:599–602.
Popp MB, Brennan MF. Long-term vascular access in the rat: importance of asepsis. Am J Physiol. 1981;241:H606–12.
Remie R, Rensema JW, van Wunnik GHJ, van Dongen JJ. General principles of microsurgery. In: van Dongen JJ, Remie R, Rensema JW, van Wunnik GHJ, editors. Manual of microsurgery on the laboratory rat. Amsterdam: Elsevier; 1990. p. 11–5.
Remie R, van Dongen JJ, Rensema JW. Permanent cannulation of the jugular vein (acc. to Steffens). In: van Dongen JJ, Remie R, Rensema JW, van Wunnik GHJ, editors. Manual of microsurgery on the laboratory rat. Amsterdam: Elsevier; 1990. p. 159–70.
Robinson TE, Justice Jr JB, editors. Microdialysis in the neurosciences. Amsterdam: Elsevier; 1991.
Rosenberg RS, Bergmann BM, Rechtschaffen A. Variations in slow wave activity during sleep in the rat. Physiol Behav. 1976;17:931–8.
Santiago M, Westerink BH. Characterization of the in vivo release of dopamine as recorded by different types of intracerebral microdialysis probes. Naunyn Schmiedebergs Arch Pharmacol. 1990;342:407–14.
Steffens AB. A method for frequent sampling of blood and continuous infusion of fluids in the rat without disturbing the animal. Physiol Behav. 1969;4:833–6.
Strubbe JH. Insulin, glucose and feeding behaviour in the rat: a reappraisal of the glucostatic theory. Thesis, University of Groningen; 1975. p. 8–12.
Tera H, Aberg C. Tensile strengths of twelve types of knot employed in surgery, using different suture materials. Acta Chir Scand. 1976;142:1–7.
Tracy DL, editor. Mosby's fundamentals of veterinary technology: small animal surgical nursing. St. Louis: Mosby Year Book; 1994.

Ungerstedt U. Measurement of neurotransmitter release in vivo by intracranial dialysis. In: Marsden CA, editor. Measurement of neurotransmitter release in vivo. New York: Wiley; 1984. p. 81–105.

Ungerstedt U. Introduction to intracerebral microdialysis. In: Robinson TE, Justice Jr JB, editors. Microdialysis in the neurosciences. Amsterdam: Elsevier; 1991. p. 3–22.

Van der Heyden JAM, Zethof TJJ, Olivier B. Stress-induced hyperthermia in singly housed mice. Physiol Behav. 1997;62:463–70.

Waynforth HB. Standards of surgery for rodents: do we need to change? Scand Lab Anim Sci. 1993;2:43–6.

Westerink BH, De Vries JB. Characterization of in vivo dopamine release as determined by brain microdialysis after acute and subchronic implantations: methodological aspects. J Neurochem. 1988;51:683–7.

Westerink BHC, Justice Jr JB. Microdialysis compared with other in vivo release models. In: Robinson TE, Justice Jr JB, editors. Microdialysis in the neurosciences. Amsterdam: Elsevier; 1991. p. 23–43.

Zetterstrom T, Sharp T, Marsden CA, Ungerstedt U. In vivo measurement of dopamine and its metabolites by intracerebral dialysis: changes after d-amphetamine. J Neurochem. 1983; 41:1769–73.

Chapter 11
Transgenic Animals: Principles, Methods and Applications

Suzana Macedo de Oliveira, Heloisa Allegro Baptista, and João Bosco Pesquero

This chapter focuses on the basics of technology, methods, application and future prospects of transgenic animals in medicine, agriculture and industry. Transgenic technologies and the ability to introduce or delete functional genes into animals have revolutionized our ability to address complex biomedical and biological questions.

What Is a Transgenic Animal?

There are various definitions for the term *transgenic animal*. The Federation of European Laboratory Animal Associations (FELASA) defines the term as an animal whose genome, the genetic makeup of an organism responsible for inherited characteristics, was deliberately modified.

In general, the term *transgenic* designates an animal that has had its genetic content altered by the introduction, modification or inactivation of a gene. This manipulation is performed *in vitro* and leads to an alteration of the genome of all cells, including the germ cells (eggs and sperm), which enables the mutation to be transmitted to the litters after breeding (Fukamizu 1993). The introduced gene is termed a 'transgene' and the process is designated transgenesis.

S.M. de Oliveira (✉)
Department of Biophysics, Universidade Federal de São Paulo (UNIFESP), Rua Pedro de Toledo, 669, 90 andar, São Paulo, SP 04039-032, Brazil
e-mail: clavedessol@gmail.com

H.A. Baptista
Centro de Desenvolvimento de Modelos Experimentais para Medicina e Biologia CEDEME, Universidade Federal de São Paulo (UNIFESP), São Paulo, São Paulo, Brazil

J.B. Pesquero
Department of Biophysics, Universidade Federal de São Paulo (UNIFESP), São Paulo, São Paulo, Brazil

M.L. Andersen, S. Tufik (eds.), *Rodent Model as Tools in Ethical Biomedical Research*, DOI 10.1007/978-3-319-11578-8_11

The term 'transgenic' encompasses different types of animal models, as follows:

- overexpression model: when one or more copies of a gene (endogenous or from another species) are added to the genome;
- *knockout* and *knockin*: when an endogenous gene is inactivated (*knockout*) or modified previously and exchanged in the genome (*knockin*).

Introduction and History

Before the development of the modern techniques of molecular biology were applied to genetics, the only available models to study gene regulation and function were the spontaneous mutants or mutants obtained by chemical agents, like N-ethyl-N-nitrosourea (Justice et al. 1999). These spontaneous mutations allow for the identification of a gene by the analysis of a pathological phenotype in the altered animal, different from the non-mutated; for instance, differences in coat color and pigmentation. In this manner, genes have been cloned and identified as responsible for diabetes and obesity in experimental animals (Zhang et al. 1994). The development of new techniques in molecular and cellular biology applied to genetics has made it possible to alter the genome in a controlled manner. Therefore, these novel tools to generate transgenic models are the most promising technologies for commercial purposes and also for basic and clinic research.

At the end of 1960s, the first publication on transgenesis described the technique used to introduce DNA fragments in procaryotic and eukaryotic somatic cells *in vitro* and the induction of the expression of genes in cells from the nervous system and blood (Gavrilova 1967; Loyter et al. 1975). In 1968, Richard L. Gardner isolated cells from a mouse embryo and injected them into other embryos. These embryos were later transferred to the uterus of foster mothers, which generated mice with cells from both embryos, called chimeric animals.

The techniques to generate transgenic animal models by integration of foreign DNA into the genome or by the deletion of DNA sequences from the genome of mammals were established long time ago. The first work describing the introduction of exogenous genetic material into a mouse embryo was published in 1974, before the development of recombinant DNA techniques. Rudolf Jaenisch and Beatriz Mintz observed the presence of exogenous viral DNA in several mice tissues after injecting SV40 virus in the blastocell. However, in these trials there was no integration of the exogenous genes in the germ line cells and no transmission to the litters (Jaenisch and Mintz 1974). Later, the same authors demonstrated with success DNA integration in germ line cells and transmission after exposing the embryos to the Moloney murine leukemia virus (Jaenisch 1976, 1977). In 1980, Gordon et al. described for the first time the pronuclear microinjection technique with a viral DNA construct in mice embryos; however, no germ line transmission was observed (Gordon et al. 1980). In 1982, the first transgenic mouse line was obtained using the pronuclear microinjection technique by Richard Palmiter from University of Pennsylvania and Ralph Brinster and colleagues from University of Washington In this new study, the mice were generated by the injection of a DNA construction containing a rat growth hormone gene. From the 21 animals born, six had the exogenous DNA inserted in their genome and were twice bigger than the control animals, which was a hallmark in the development of transgenic technology (Palmiter et al. 1982).

Parallel to the development of transgenesis, in 1981 totipotent embryonic stem (ES) cells were isolated for the first time (Evans and Kaufman 1981). In 1986, Gossler and colleagues generated the first chimeric mice obtained from ES cells, definitively proving that these cells were able to generate a complete organism (Gossler et al. 1986). Some years later, in 1982, the conditions for cultivating mice ES cells were established, and therefore, the manipulation of embryo cells *in vitro* and the inactivation of specific genes was made possible (Evans and Kaufman 1981).

In 1987, using undifferentiated ES cells, two independent groups successfully obtained site directed mutagenesis by homologous recombination, making it possible to delete a specific gene in the genome (Doetschman et al. 1987; Thomas and Capecchi 1987). Based on these findings, the first knockout model in mice (Schwartzberg et al. 1989) and later the model of knockin, with a punctuated modification of a gene, were generated (Shastry 1998).

Presently, it is also possible to generate conditional knockouts based in a CRE/loxP recombination system, developed by Orban et al. (1992). This system allows the deletion or activation of an endogenous or exogenous gene in a time- and organ/cell-specific manner (Orban et al. 1992; Gu et al. 1994).

Most of the experiments in transgenesis have been performed in mice, the animal that pioneered the technology. However, today experiments in transgenesis research has been extended to other species such as the rat, rabbit, swine, fish, poultry, sheep, cows and non-human primates. Mice continue to be the most used models in science and research because of their small size, low maintenance costs and the possibility of obtaining suitable embryos for manipulation. Moreover, murine ES cells can be kept undifferentiated in culture, enabling the generation of knockout/knockin models (Hammer et al. 1985, 1990; Mullins et al. 1990; Wall et al. 1991; Pursel et al. 1989; Hogan et al. 1986). Another important factor that supports the use of mice in research is the genetic similarity to men, as recently demonstrated by the sequencing of its genome (Gregory et al. 2002).

Overexpression Transgenic Models

Overexpression transgenic animals are those that have one or more copies of a transgene (endogenous or exogenous) inserted into their genome. The presence of multiple copies in the genome leads to an overexpression of the added gene and consequently to an increase in the expression of the coded protein and function.

One important feature of this technique is the random integration of the transgene into the genome, which might lead to an inactivation of an interrupted gene. This insertion can be lethal to the embryo or to the adult animal, if the interrupted gene is essential to the development of the embryo, or can lead to a phenotype not related to the inserted transgene. Therefore, different transgenic lines must be generated and present the same phenotype in order for the transgene to be linked to the phenotype. The precise molecular mechanism responsible for the integration of the transgene into the genome is still not totally clear. Some characteristics of the integration into the host genome are:

- the frequency of the integration is around 10–30 % of the cells receiving the DNA into their nuclei;
- the integration event generally occurs in only one, and rarely in more than one *locus* in the chromosome.

DNA is microinjected in the linear form and is integrated into the genome in an array of multicopies known as concatamers, which consist of head-to-tail DNA associations, indicating the existence of events like extrachromosomal homologous recombination before integration.

The number of copies of the transgene integrated into the genome cannot be controlled, as well as its expression level, which is dependent on the promoter sequence used to generate the construct, the number of copies inserted and the insertion site. However, the random characteristic of the technique can be of great value, since the insertional site can be spotted and the inactivated gene identified and characterized (Jaenisch 1988; Gridley 1991; Meisler 1992).

Several techniques have been used to generate overexpression in transgenic animals, like pronuclear microinjection (Gordon et al. 1980), embryo infection by retrovirus vectors (Nagano et al. 2001), aggregation of genetically modified ES cells (Gossler et al. 1986; Wood et al. 1993), transfer of chromosome segments (Richa and Lo 1989) and sperm-mediated DNA transfer (Lavitrano et al. 1989).

Pronuclear Microinjection

Pronuclear microinjection is the most commonly used technique to generate transgenic overexpression animal models, in which about 100 copies of a DNA construct are injected into the pronucleus of a zygote. This method involves the direct microinjection of a chosen gene construct (a single gene or a combination of genes) from the same or from different species into the pronucleus of a fertilized ovum using a micromanipulator attached to a high resolution inverted microscope. This technique allows the insertion of long DNA sequences into the genome of different species, producing high expression levels and stable integration of the transgene into the genome of germ line cells. The general microinjection scheme is shown in Fig. 11.1. After microinjection, the transgene is integrated randomly into the genome of all cells of the animal, mostly at a single site as tandem repeats of 1–100 copies by a

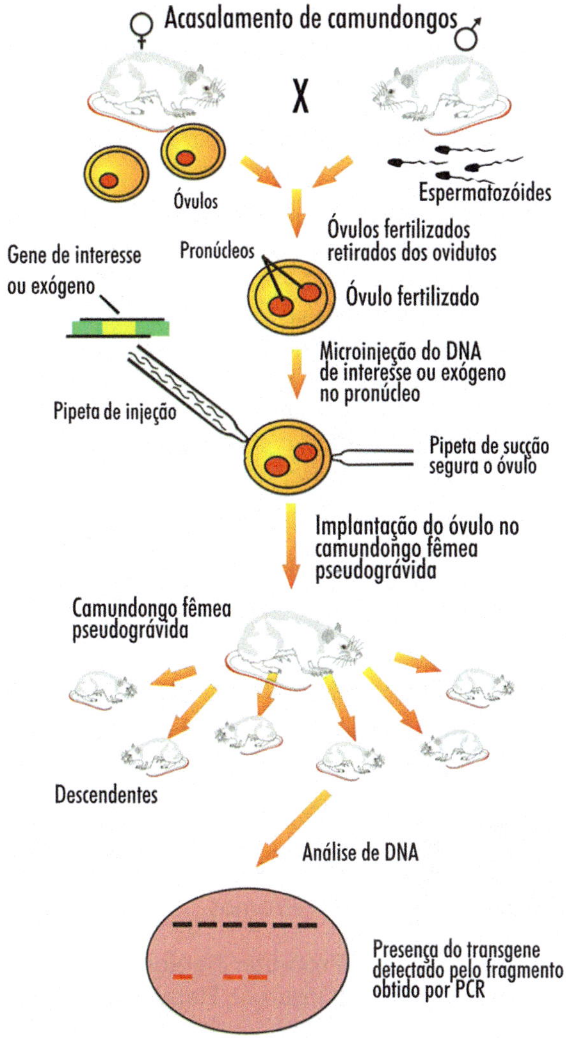

Fig. 11.1 General scheme of transgenic animal generation using the pronuclear microinjection method

yet unknown mechanism involving illegitimate recombination. However, since the integration never happens at the same chromosomal site, the generation of at least two distinct lines of transgenic animals with each construct avoids misinterpretation of the phenotype. The microinjection technology includes the following steps: design and generation of the transgene construct, superovulation of donor animals, isolation of fertilized oocytes, and microinjection of a small volume of solution containing the transgene construct into one pronucleus of the zygote, transfer of injected embryos into a foster mother and identification of transgenic animals in the offspring. The efficiency of transgenic animal production depends on multiple factors; one of the most important is the skill of the investigator. Published values range from 3 % to 5 % transgene-positive offspring (founder) per injected zygote for mice (Brinster et al. 1985).

The manipulated fertilized ovum is transferred into the oviduct of a recipient female or foster mother that has been induced to act as a recipient by mating with a vasectomized male and will give birth to the transgenic animals, which will be genotyped to check the presence in their genome of the foreign DNA or transgene. As such, it is important that the genetic alteration is transmitted to the descendents. The pronuclear microinjection steps are:

- Selection of mice strain to produce a transgenic organism
- Design and purification of the transgene construct for pronucleus microinjection (Fig. 11.2)
- Superovulation of female donors
- Isolation of fertilized oocytes and microinjection of the transgene construct into the zygote (Fig. 11.3)
- Transfer of injected embryos into the oviduct of the foster mother
- Identification of heterozygous transgenic animals
- Homozygous strain acquisition

The detailed manipulation protocol for a gene introduction can be found in some literature reviews and manuals (Pesquero et al. 2002; Pinkert 2002; Nagy et al. 2003; Hogan et al. 1994; Godard and Guénet 1999).

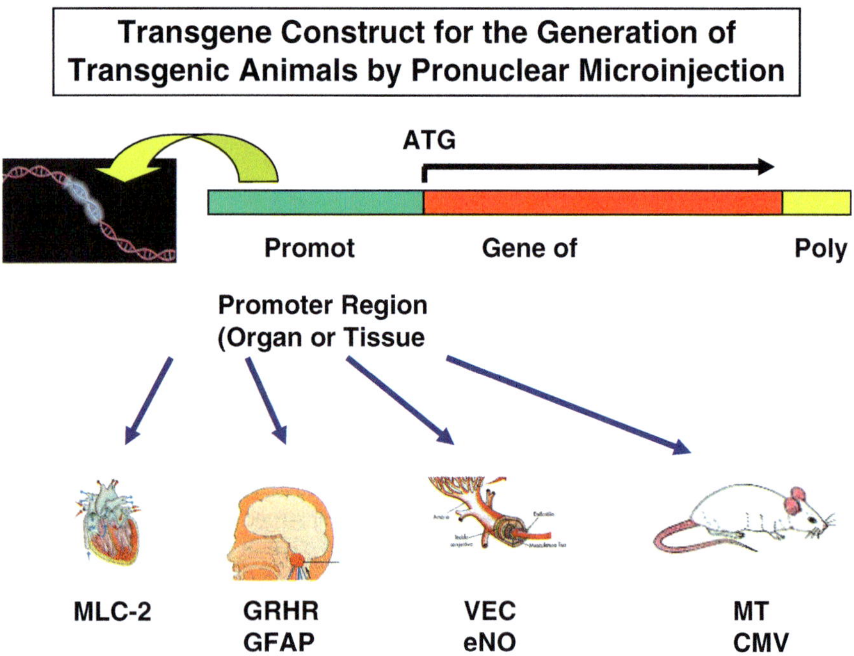

Fig. 11.2 Scheme of the DNA construct for the generation of transgenic animal by pronuclear microinjection containing promoter region, coding region and poly-A. It is also showed some tissue-specific promoters: *MLC-2* myosin light chain, *GHRH* Gonadotropin-releasing hormone receptor, *GFAP* Glial fibrillary acidic protein, *VEC* vascular endothelial cells, *eNOS* endothelial nitric oxide synthase, *MT* metalotionein, *CMV* cytomegalovirus

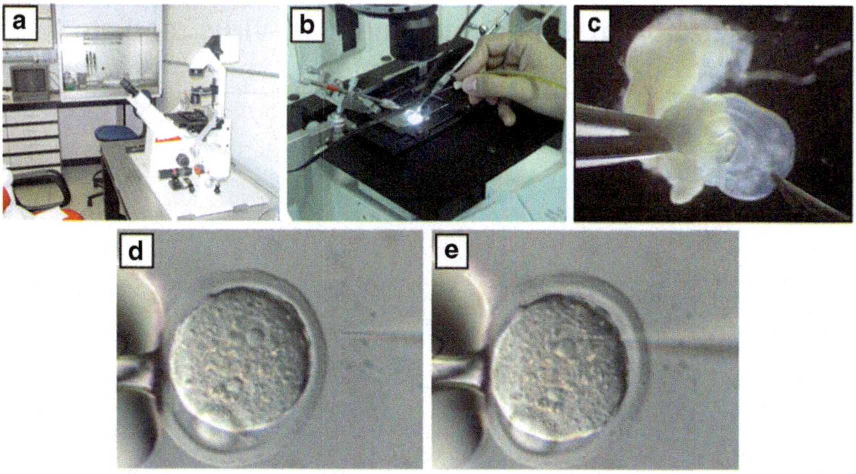

Fig. 11.3 (**a**) Micromanipulator attached to the inverted microscope for embryo pronuclear microinjection. (**b**) Detail of the suction and injection needles part of the micromanipulation system. (**c**) Photograph of the mice oviduct containing granulosa cell mass and embryos (×20 magnificence). (**d**, **e**) Pronuclear DNA microinjection (×400 magnificence)

Sperm Mediated Gene Transfer

In 1971, Bracket et al. demonstrated for the first time that SV40 simian virus could bind to the surface of rabbit sperm, transferring thereby exogenous DNA into the cell. However, this work was not reconsidered until the end of the 1980s when two other independent groups described the production of transgenic mice (Lavitrano et al. 1989) and sea-urchin (Arezzo 1989) after *in vitro* oocyte fertilization with sperm incubated with DNA. Several papers can be found today in the literature demonstrating the capacity of sperm from different species to uptake exogenous DNA (Smith and Spadafora 2005). However, due to the low frequency of integration of the transgene, this technique is still not commonly used. One controversial point concerning the technique is that it enables the possibility of sperm uptaking free genomic sequences (DNA molecules) present in the reproductive tract. If true, this phenomenon could lead to evolutionary chaos.

Despite the barriers linked to the technique, an evolution of the method was recently published by Chang et al. (2002), who were successful in the generation of transgenic pigs and mice expressing human alkaline phosphatase and interferon β using a new technique based on a monoclonal antibody that is able to bind to an antigen expressed in the sperm of several species like ovine, rodents, swine, birds and humans (Fig. 11.4). This antibody is also able to bind to DNA by ionic interactions, due to its basic properties, carrying the DNA to the nucleus after fertilization.

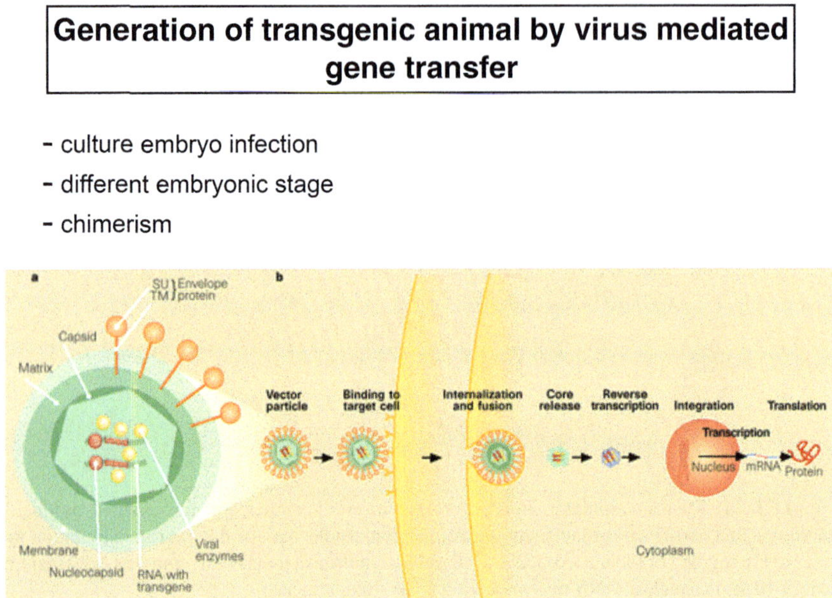

Fig. 11.4 Generation of transgenic animal by sperm mediated gene transfer

Virus-Mediated Gene Transfer

Historically, the first successful trials to generate transgenic animals were performed by infecting mice embryos with retrovirus (Jaenisch and Mintz 1974; Jaenisch 1976). Later, this method was developed by the injection of the virus into the perivitelinic space of fertilized oocytes or by incubation of the virus in embryos without zona pelucida. Initially, one vector is constructed with retroviral sequences linked to the gene to be inserted into the genome. Thus, only one copy of the transgene is specifically added to the genome by the virus integration machinery, which makes it possible to obtain transgenic animals of any species without expensive equipment and experience. In addition, this technique is an interesting option to generate transgenics from species not accessible to pronuclear microinjection, like swine and birds. Recently, the technology using lentivirus, a system based in the human immunodeficiency virus (HIV) was established for cellular culture (Naldini et al. 1996; Miyoshi et al. 1998). The system (Fig. 11.5) was shown to be a powerful tool for generating transgenic rats and mice by infecting differentiated cells (Lois et al. 2002; Pfeifer et al. 2002). An important aspect of this technique is its contribution to the development of gene and cell therapy, where viral vectors are used to insert genes into stem cells from embryos and bone marrow, and are later integrated into the organism to treat different diseases.

Generation of transgenic animal by sperm mediated gene transfer

BMC Biotechnology BioMed Central

Research article

Effective generation of transgenic pigs and mice by linker based sperm-mediated gene transfer.

Keejong Chang†[2,3], Jin Qian†[1], MeiSheng Jiang†[4], Yi-Hsin Liu†[5], Ming-Che Wu[6], Chi-Dar Chen[2], Chao-Kuen Lai[2], Hsin-Lung Lo[3], Chin-Ton Hsiao[2], Lucy Brown[7], James Bolen Jr[7], Hsiao-I Huang[1], Pei-Yu Ho[2], Ping Yao Shih[2], Chen-Wen Yao[8], Wey-Jinq Lin[9], Chung-Hsi Chen[3], Fang-Yi Wu[3], Yi-Jen Lin[3], Jing Xu[1] and Ken Wang*[1]

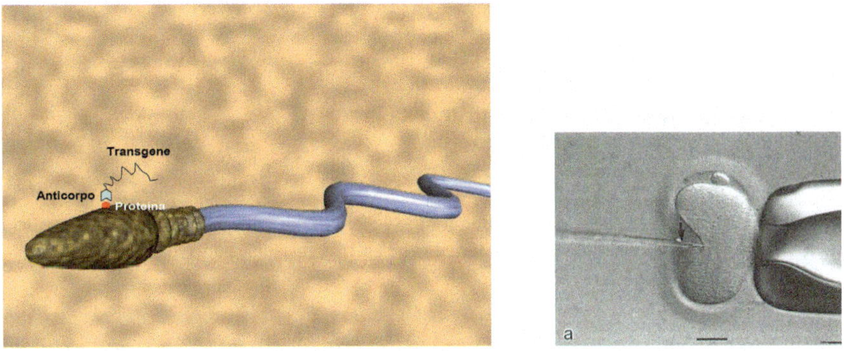

Fig. 11.5 Generation of transgenic animal by virus mediated gene transfer

Knockout Model

At the same time the technique for generating overexpression transgenic models was established, it was also shown to be possible to isolate and culture ES cells, keeping them in an undifferentiated state by the use of specific inhibitors. Therefore, the first *knockout* model was generated. The knockout technique allows the analysis of different aspects of the gene function in vivo, making it possible to generate specific genetic alterations. The principle of the method is to replace a functional gene by a mutated sequence in the chromosome, inactivating the endogenous gene (Fig. 11.6). The insertion of the exogenous DNA sequence takes place by homologous recombination in ES cells in culture. This phenomenon was described for the first time in the work of Thomas and Capecchi in 1987. After selection, mutated ES cells are microinjected into blastocysts and transferred to the uterus of pseudo pregnant females. The resulting animals are called chimeric mice, which contain two lineages of cells: one originating from the injected modified ES cells and the other from the resident ES cells present in the receptor blastocyst. This technique allows the insertion of mutations inside any gene and the study of their function. However, the method can be applied only in mice, because it is the only species, besides humans, for which the conditions for cultivating undifferentiated ES cells have been established (Thomson et al. 1998).

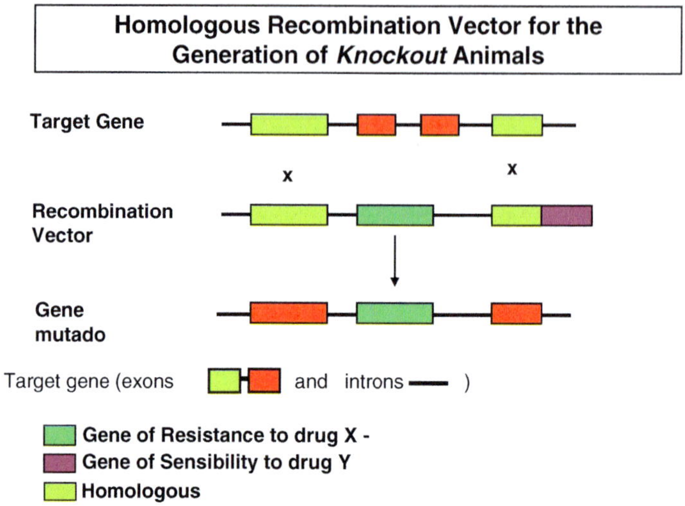

Fig. 11.6 Scheme of the DNA construction for the generation of knockout animals (target vector) including the homologous regions of the target gene and the resistance and sensibility genes

New methods to generate time and site-specific mutations have been described recently. Today, it is possible to specifically inactivate one gene in mice in a certain tissue or at a certain time after birth (Fig. 11.7). The method is based on the Cre recombinase of the P1 bacteriophage, which recognizes a 34 bp sequence called loxP (locus of crossover in P1) and mediates site-specific recombination (Gu et al. 1994).

Applications in Medicine, Agriculture and Industry

One of the most intriguing applications of transgenesis is the use of genetically modified animals as models for the study of the causes, progression, stages and symptoms of cardiovascular, autoimmune or neurological diseases, among others (Silva Jr et al. 2000). Because it allows a detailed analysis of a disease's physiopathology, transgene technology will favor the development of novel treatments and diagnostic tests, more efficient and cheaper therapeutic drugs, as well as the establishment of gene therapy protocols. In 2000, our group demonstrated, using the transgene technology, the physiopathological importance of the kinin B1 receptor, a protein that is present in the cell's membrane and which takes part in the process of pain and inflammation. Further, we recently found that this protein could be an important tool in the treatment of obesity and metabolic disorders.

The transplant of organs and tissues of animals in humans, or xenotransplantation, has been a research focus for over a century. There is a worldwide lack of organs for transplantation and, unfortunately, many patients die waiting.

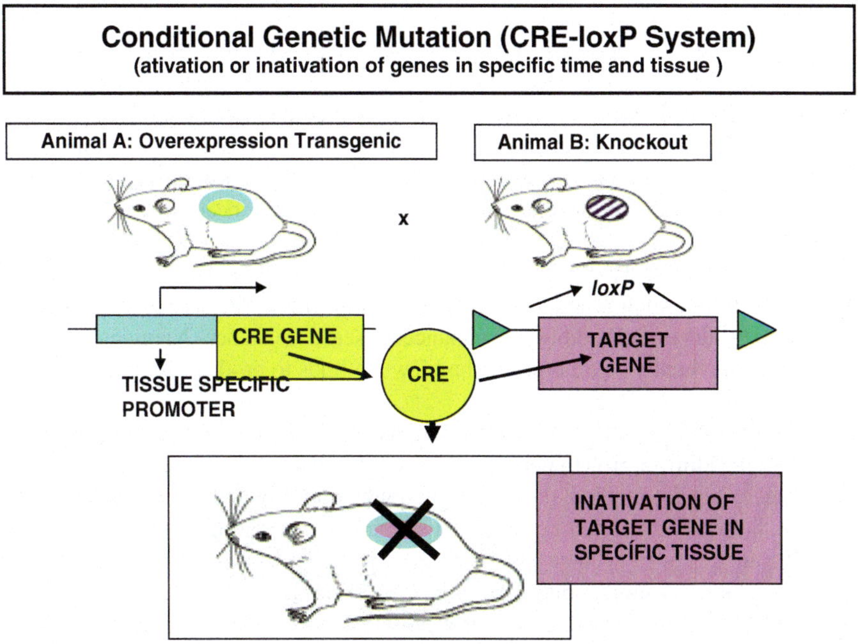

Fig. 11.7 Scheme of the CRE-LoxP system for the generation of conditional transgenic animals

The advantages of the xenotransplant include, therefore, an unlimited number of organs, besides the fact that the organ would not be injured by *post-mortem* effects, such as hemorrhage and metabolic disorders.

Presently, researchers are developing transgenic pigs whose organs can be safely used for human transplant. Safety was achieved through the strategy of introducing into the pig's genome a gene that codes for a human cell surface protein, thereby preventing the human immune system from recognizing and destroying the foreign organ that does not possess specific human markers on its surface.

Another strategy under development, aiming to minimize rejection of organs for xenotransplantation, consists of eliminating from the pig's genome by the *knockout* method the gene that codes for α-1,3-galactosiltransferase, an enzyme that is present on the cell surface and is recognized by the human immune system. Without this enzyme, the first step toward rejection is not triggered.

In degenerative diseases of the nervous system, the transplant of pig fetal neurons represents a potential therapy. To achieve this goal, transgenic pigs expressing a human molecule that inhibits immunosuppressor T cells was generated. The embryonic neurons of such animals, secreting the molecule *in vitro*, were thereafter transplanted in rats, followed by a 50 % reduction in the degenerative response. These data suggest that neurons from transgenic pigs can be used for pre-clinical tests in xenotransplantation.

In spite of the good perspectives about xenotransplantation, its use still needs careful evaluation. There is a possibility that the transplanted tissues could carry microorganism(s), which, once introduced in the receptor organism, would cause infection. The technique also raises ethical concerns that should yet be extensively discussed. However, even in face of such problems, the xenotransplantation technique seems to be rather promising and should lead to the establishment of novel protocols for treatment, bringing hope to those who have been long waiting for a new organ.

With respect to agriculture and cattle breeding, typical applications of the transgene technology include improved carcass composition, lactational performance, and wool production, in addition to enhanced disease resistance (Niemann and Kues 2007). Until recently cattle breeders had few methods to produce animals bearing a desired trait.

To enhance milk production or accelerate growth, the only alternatives were selective breeding or growth hormones. However, selective breeding is an extremely slow and expensive method, the results are not guaranteed, and the use of hormones, in turn, is highly criticized because of its residuals remaining in the product. The new techniques of molecular biology make it possible to introduce desired characteristics in animals more rapidly and precisely. Examples of this are transgenic cows that are able to produce more milk or milk with a lesser amount of lactose or cholesterol; pigs and cows with more meat; and sheep that produce more wool. Thus, after approximately 7,000 years of domestic animal selection based on the random mutations caused by radiation and oxidative injury to the genome, technology is now available to introduce or remove known genes with known functions. Because of the impact and the ethic complexity involved in the use of transgenic animals for human nourishment, the Food and Agriculture Organization (FAO) and World Health Organization (WHO) of the United Nations created the *Codex Alimentarius*, a committee composed by members of many countries who are in charge of the discussion and establishment of rules of international applicability concerning the safety of food produced by means of animal transgenic technology.

Finally, protection of the environment has not been forgotten. One of the attempts to address this important question has been the development of environmentally friendly pigs that carry a bacterial phytase gene under the control of a salivary gland-specific promoter. This modification allows pigs to digest plant phytate, preventing phytate phosphorus to pass undigested into manure, which can pollute the environment (Golovan et al. 2001).

Targeted genetic modifications for the various purposes described tend to become more precise as the complete genomic sequences of farm animals become available. Taken together, all of these approaches will be instrumental in reducing the environmental impact and meeting global challenges in agricultural production in the future.

Another important application of animal transgenesis is the possibility of creating bioreactor animals. Bioreactors are usually domestic animals used to produce human recombinant proteins of great biological or commercial interest like enzymes, hormones and growth factors. This technology entails the production of recombinant biologically active proteins in the mammary glands of transgenic animals, overcoming the limitations of conventional and recombinant production systems for pharmaceutical proteins. The mammary gland is the production site of choice, because of the amount of protein that can be produced in this organ using mammary gland-specific promoter sequences and established methods for extraction and purification of the protein. The technology for using the mammary gland as a bioreactor has been developed to the point that pharmaceuticals derived from the milk of several transgenic farm animal species are currently in the advanced stages of clinical trials (Dove 2000).

As an example, we can cite Rosie, the first transgenic bovine, which was genetically modified to express in its milk the human protein lactalbumin, making its production cheaper and more efficient. This transgenic milk is more nutritious for humans than natural milk, and could be introduced into the diets of children who lack specific nutrients.

Some other experiments are aimed at producing transgenic milk that contains proteins necessary for the treatment of diseases like phenylketonuria, hereditary emphysema and cystic fibrosis. Recently the European Medicines Agency approved the pioneering production of human antithrombin by the American company GTC Biotherapeutics, using transgenic goats as bioreactors. The antithrombin is a protein that is present in the blood of normal individuals, possessing anti-inflammatory and anticlotting activity. The hereditary antithrombin deficiency—a rare genetic disease—renders one vulnerable to thrombosis. In addition to human antithrombin, GTC has also developed other therapeutics that should soon be launched in the market, such as recombinant human alpha 1 antitrypsin, used in the treatment of certain forms of emphysema, and a transgenic vaccine against malaria.

In addition, the Dutch company *Pharming* is engaged in the production of substances in bovine milk, such as fibrinogen, collagen, C1 esterase inhibitor and human lactoferrin, the last two being used in the treatment of hereditary angioedema and inflammation, respectively.

In 2001, two scientists from the Canadian *Nexia Biotechnologies* managed to introduce the gene responsible for the production of spider webs in the mammary gland cells of lactant goats. These animals started to express the silk protein in their milk that, after manufacturing, became a light, flexible and resistant material that can be used in the production of military uniforms, microsuture threads and tennis racket strings.

Rabbits have also been extensively used as bioreactors. Although these animals are not conventional dairy livestock, they are smaller and thus less expensive to maintain than larger animals. Moreover, their reproductive cycle is much shorter, with multiple offspring per litter, stable paternal transmission of the transgene and acceptable milk yield. The mammary glands of rabbits have been employed for the production of recombinant human blood clotting factor VIII (Chrenek et al. 2007), and the recombinant human alpha-glucosidase used in replacement therapy for patients with Pompe disease (Van den Hout et al. 2001), among other proteins. In Brazil, our group, in collaboration with Brazilian Agricultural Research Corporation (Embrapa), the University of Brasília (UnB) and Hospital de Apoio de Brasília, succeeded in producing the human factor IX in mouse milk, a protein whose absence in hemophiliacs hampers the blood clotting process. Our group's next step is to develop cloned transgenic bovines able to express this factor in milk.

Another Brazilian university, Ceará State University, has focused on the use of animals for the production of modified human proteins. The project targets production in transgenic goats expressing granulocyte-colony stimulating factor (G-CSF) in milk, a protein with the ability to stimulate the production of white cells and recruit stem cells from the medulla. Because of such abilities, G-CSF is employed in the treatment of patients with immunodeficiencies.

Conclusions

The genetic modifications developed in animals over the last three decades have led to a breakthrough in the field of biology, enabling the analysis of genetic function in vivo. It is interesting to observe that, by focusing on the genetic and molecular levels of the biological system, one achieves a comprehensive understanding of the pathophysiological mechanisms underlying not only a specific disease, but a whole range of diseases. Biomedical research based on genetic alterations in animal models offers great promise in the search for cures for the many diseases afflicting humankind, and the use of transgenic animals is a pivotal tool therein. Corroborating the importance of transgenesis, 2007 Nobel Laureates in Medicine and Physiology were the creators of gene targeting in mice, which as been referred to as an immensely powerful technology, applied to virtually all areas of biomedicine—from basic research to the development of new therapies.

In spite of the importance of transgenic animals in biomedical research, there are some concerns and misconceptions related to their use. The potential risks of transgenesis for animals, humans and the environment are under strict control of governmental and non-governmental agencies of many countries, responsible for addressing such matters. For example, no transgenic animal is allowed to breed with wild populations, a rule which ensures no long-term change in indigenous populations.

Further, the potential cost of introducing a genetic modification in an animal is routinely assessed, taking into account the proposed ethical procedures, which are weighed against the potential benefits that could come from the experiments. The United Kingdom Royal Society concluded in 2001: "Some would regard genetic modification as unacceptable under all circumstances, because such a procedure is disrespectful to the animal or violates its rights." However, many others would regard the appropriate stance as being to minimize the suffering and maximize the gain to medicine, agriculture and fundamental understanding. This is the position adopted by the Royal Society: "The Royal Society concludes that the development of genetically modified animals has been hugely beneficial in many areas, not least into research on the causes and possible treatments of disease."

Over 95 % of the transgenic animals used in biomedical research are mice, although rats, pigs, rabbits and sheep are also used. It is interesting to note that the use of transgenic mice is helping to reduce the number of other animals needed for experimentation. As an example, some pharmaceutical companies have developed safety tests for vaccines that use transgenic mice rather than monkeys. Most importantly, the predictability of many transgenic phenotypes permits the innovative testing of diagnostic and therapeutic agents, while using a reduced population of experimental animals.

Finally, another question raised in this debate is whether *in vitro* experiments could substitute for experiments with animals. The answer to this question lays in the fact that a cell culture or a computer program cannot fully replicate the conditions that exist when the body's systems work together. The effects of a medicine will ultimately depend on what happens when it interacts with all systems in the body working as a whole. That is the reason why, even after extensive and careful testing in "test tubes" or in cellular systems, a compound may present a dramatically different effect in the body. Therefore, well-designed animal studies will remain essential until our biological knowledge is able to overcome the gap between *in vitro* tests and a living organism.

References

Arezzo F. Sea-urchin sperm as a vector of foreign genetic information. Cell Biol Int Rep. 1989;13:391–404.

Brackett BG, Baranska W, Sawicki W, Koprowski H. Uptake of heterologous genome by mammalian spermatozoa and its transfer to ova through fertilization. Proc Natl Acad Sci U S A. 1971;68:353–7.

Brinster RL, Chen HY, Trumbauer ME, Yagle MK, Palmiter RD. Factors affecting the efficiency of introducing foreign genes into mice by microinjecting eggs. Proc Natl Acad Sci U S A. 1985;82:4438–42.

Chang K, Qian J, Jiang M, Liu YH, Wu MC, Chen CD, Lai CK, et al. Effective generation of transgenic pigs and mice by linker based sperm-mediated gene transfer. BMC Biotechnol. 2002;2:1–13.

Chrenek P, Ryban L, Helga Vetr H, Makarevich AV, Uhrin P, Rekha K, et al. Expression of recombinant human factor VIII in milk of several generations of transgenic rabbits. Transgenic Res. 2007;16:353–61.

Doetschman T, Gregg RG, Maeda N, Hooper ML, Melton DW, Thompson S, Smithies O. Targetted correction of a mutant HPRT gene in mouse embryonic stem cells. Nature. 1987;330:576–8.

Dove A. Milking the genome for profit. Nat Biotechnol. 2000;18:1045–8.

Evans MJ, Kaufman MH. Establishment in culture of pluripotential cells from mouse embryos. Nature. 1981;292:154–6.

Fukamizu A. Transgenic animals in endocrinological investigation. J Endocrinol Invest. 1993; 16:461–73.

Gavrilova TN. The contents of labeled DNA in nerve cells of mice of different ages after introduction of H3-thymidine in the period of embryogenesis. Tsitologiia. 1967;9:68–72.

Godard ALB, Guénet J. Genética de Camundongos. Modelos animais de doenças humanas. Biotecnolog Ciênc Desenvolv. 1999;9:96–100.

Golovan SP, Meidinger RG, Ajakaiye A, Cottrill M, Wiederkehr MZ, Barney DJ, et al. Pigs expressing salivary phytase produce low-phosphorus manure. Nat Biotechnol. 2001;19: 741–5.

Gordon JW, Scangos GA, Plotkin DJ, Barbosa JA, Ruddle FH. Genetic transformation of mouse embryos by microinjection of purified DNA. Proc Natl Acad Sci U S A. 1980;77:7380–4.
Gossler A, Doetschman T, Korn R, Serfling E, Kemler R. Transgenesis by means of blastocyst-derived embryonic stem cell lines. Proc Natl Acad Sci U S A. 1986;83:9065–9.
Gregory SG, et al. A physical map of the mouse genome. Nature. 2002;418:743–50.
Gridley T. Insertional versus targeted mutagenesis in mice. New Biol. 1991;3:1025–34.
Gu H, Marth JD, Orban PC. Deletion of a DNA polymerase beta gene segment in T cells using cell type-specific gene targeting. Science. 1994;265:103–6.
Hammer RE, Pursel VG, Rexroad Jr CE, Wall RJ, Bolt DJ, Ebert KM, et al. Production of transgenic rabbits, sheep and pigs by microinjection. Nature. 1985;315:680–3.
Hammer RE, Maika SD, Richardson JA, Tang JP, Taurog JD. Spontaneous inflammatory disease in transgenic rats expressing HLA-B27 and human beta 2m: an animal model of HLA-B27-associated human disorders. Cell. 1990;63:1099–112.
Hogan B, Constantini F, Lacy E. Manipulating the mouse embryo: a laboratory manual. Cold Spring Harbor: Cold Spring Harbor Laboratory Press; 1986.
Hogan B, Beddington R, Costantini F, Lacy E. Manipulating the mouse embryo: a laboratory manual. Cold Spring Harbor: Cold Spring Harbor Laboratory Press; 1994.
Jaenisch R. Germ line integration and Mendelian transmission of the exogenous Moloney leukemia virus. Proc Natl Acad Sci U S A. 1976;73:1260–4.
Jaenisch R. Germ line integration of moloney leukemia virus: effect of homozygosity at the m-mulV locus. Cell. 1977;12:691–6.
Jaenisch R. Transgenic animals. Science. 1988;240:1468–74.
Jaenisch R, Mintz B. Simian virus 40 DNA sequences in DNA of healthy adult mice derived from preimplantation blastocysts injected with viral DNA. Proc Natl Acad Sci U S A. 1974;71:1250–4.
Justice MJ, Noveroske JK, Weber JS, Zheng B, Bradley A. Mouse ENU mutagenesis. Hum Mol Genet. 1999;8:1955–63.
Lavitrano M, Camaioni A, Fazio VM, Dolci S, Farace MG, Spadafora C. Sperm cells as vectors for introducing foreign DNA into eggs: genetic transformation of mice. Cell. 1989;57: 717–23.
Lois C, Hong EJ, Pease S, et al. Germline transmission and tissue-specific expression of transgenes delivered by lentiviral vectors. Science. 2002;295:868–72.
Loyter A, Zakai N, Kulka RG. Ultramicroinjection of macromolecules or small particles into animal cells. A new technique based on virus-induced cell fusion. J Cell Biol. 1975;66:292–304.
Meisler MH. Insertional mutation of classical and novel genes in transgenic mice. Trends Genet. 1992;8:341–4.
Miyoshi H, Blomer U, Takahashi M, et al. Development of a self-inactivating lentivirus vector. J Virol. 1998;72:8150–7.
Mullins JJ, Peters J, Ganten D. Fulminant hypertension in transgenic rats harbouring the mouse Ren-2 gene. Nature. 1990;344:541–4.
Nagano M, Brinster CJ, Orwig KE, Ryu BY, Avarbock MR, Brinster RL. Transgenic mice produced by retroviral transduction of male germ-line stem cells. Proc Natl Acad Sci U S A. 2001;98:13090–5.
Nagy A, Gertsenstein M, Vintersten K, Behringer R. Manipulating the mouse embryo. Cold Spring Harbor: Cold Spring Harbor Laboratory Press; 2003.
Naldini L, Blomer U, Gallay P, et al. In vivo gene delivery and stable transduction of nondividing cells by a lentiviral vector. Science. 1996;272:263–7.
Niemann H, Kues W. Transgenic farm animals: an update. Reprod Fertil Dev. 2007;19:762–70.
Orban PC, Chui D, Marth JD. Tissue- and site-specific DNA recombination in transgenic mice. Proc Natl Acad Sci U S A. 1992;89:6861–5.
Palmiter RD, Brinster RL, Hammer RE, Trumbauer ME, Rosenfeld MG, Birnberg NC, et al. Dramatic growth of mice that develop from eggs microinjected with metallothionein-growth hormone fusion genes. Biotechnology. 1982;24:429–33.

Pesquero JB, Magalhães LE, Baptista HA, Sabatinni RA. Animais transgênicos. Biotecnolog Ciênc Desenvolv. 2002;27:52–6.

Pfeifer A, Ikawa M, Dayn Y, et al. Transgenesis by lentiviral vectors: lack of gene silencing in mammalian embryonic stem cells and preimplantation embryos. Proc Natl Acad Sci U S A. 2002;99:2140–5.

Pinkert CA. Transgenic animal technology: a laboratory handbook. New York: Academic; 2002.

Pursel VG, Pinkert CA, Miller KF, Bolt DJ, Campbell RG, Palmiter RD, et al. Genetic engineering of livestock. Science. 1989;244:1281–8.

Richa J, Lo CW. Introduction of human DNA into mouse eggs by injection of dissected chromosome fragments. Science. 1989;245:175–7.

Schwartzberg PL, Goff SP, Robertson EJ. Germ-line transmission of a c-abl mutation produced by targeted gene disruption in ES cells. Science. 1989;246:799–803.

Shastry BS. Gene disruption in mice: models of development and disease. Mol Cell Biochem. 1998;181:163–79.

Silva Jr JA, Araujo RC, Baltatu O, Oliveira SM, Tschöpe C, Fink E, et al. Reduced cardiac hypertrophy and altered blood pressure control in transgenic rats with the human tissue kallikrein gene. FASEB J. 2000;14:1858–60.

Smith K, Spadafora C. Sperm-mediated gene transfer: applications and implications. Bioessays. 2005;27:551–62.

Thomas KR, Capecchi MR. Site-directed mutagenesis by gene targeting in mouse embryo-derived stem cells. Cell. 1987;51:503–12.

Thomson JA, Itskovitz-Eldor J, Shapiro SS, Waknitz MA, Swiergiel JJ, Marshall VS, et al. Embryonic stem cell lines derived from human blastocysts. Science. 1998;282:1145–7.

Van den Hout JM, Reuser AJ, de Klerk JB, Arts WF, Smeitink JA, Van der Ploeg AT. Enzyme therapy for Pompe disease with recombinant human alpha-glucosidase from rabbit milk. J Inherit Metab Dis. 2001;24:266–74.

Wall RJ, Pursel VG, Shamay A, Mcknight RA, Pittius CW, Hennighausen L. High-level synthesis of a heterologous milk protein in the mammary glands of transgenic swine. Proc Natl Acad Sci U S A. 1991;88:1696–700.

Wood SA, Pascoe WS, Schmidt C, Kemler R, Evans MJ, Allen ND. Simple and efficient production of embryonic stem cell-embryo chimeras by coculture. Proc Natl Acad Sci U S A. 1993;90:4582–5.

Zhang Y, Proença R, Maffei M, Barone M, Leopold L, Friedman J. Positional cloning of the mouse obese gene and its human homologue. Nature. 1994;372:425–32.

Chapter 12
Measuring Electrocardiogram and Heart Rate in Small Laboratory Animals with Radio-Telemetry

Klaas Kramer and René Remie

Introduction

The last few years, several studies have shown that measurements of physiological variables, such as electrocardiogram (ECG) and heart rate (HR) from freely moving rats and mice by using implantable radio-telemetry were more efficient, reliable and less labour intensive when compared to measurement techniques described in the literature so far. Nowadays, measurement of ECG and HR by radio-telemetry has been described and validated for many laboratory animal species, including rats and mice. The implantable radio-telemetry technique can circumvent many of the problems associated with the conventional method (exteriorised catheters) of ECG and HR monitoring in mice and rats. This chapter describes not only the surgical aspects of the radio-telemetry techniques currently used to monitor and measure ECG and HR in awake animals, but also the ECG waveforms in small laboratory animals obtained with this technique.

For several centuries, scientists have adapted available technologies to study animals in their quests to unravel and understand biological functions and processes. In the most recent of these endeavours, scientists have applied radio-telemetry technology. Radio-telemetry combines miniature sensors and transmitters to detect and broadcast biological signals in animals to remote receiver.

K. Kramer, Ph.D. (✉)
Department of Health, Safety and Environment, Free University Amsterdam, Van der Boechorststraat 1, Amsterdam 1081 BT, The Netherlands
e-mail: k.kramer@vu.nk

R. Remie
René Remie Surgical Skills Center, Madernastraat 21, Almere, Flevoland 1323 HE, The Netherlands
e-mail: r.remie@rrssc.eu

M.L. Andersen, S. Tufik (eds.), *Rodent Model as Tools in Ethical Biomedical Research*, DOI 10.1007/978-3-319-11578-8_12

As with all new technology, scientists should be sceptical and demand validations of the new technology versus current conventional measurement techniques, to the extent possible. However, new technologies make it possible to perform measurements under conditions that have not previously been possible, making direct comparisons with current conventional measurement techniques sometimes impractical.

With the exception of studies of anaesthetic agents and certain other types of experiments involving acutely painful/stressful procedures, it is generally acknowledged that the quality of physiological measurements collected from conscious unstressed animals is superior, since they are collected under conditions that best represent the normal state of the animal, are least influenced by chemical, stress, and psychological factors, and (were appropriate) are most predictive of the results that would be achieved in human beings (Brockway and Hassler 1991; Kramer 2000; Kramer and Kinter 2003). When monitoring physiological parameters in conscious animals, it is easiest to use non-invasive methods such as surface electrodes for monitoring an ECG. However, use of non-invasive techniques can introduce significant artefacts as these techniques often require physical restraints and high levels of technician interaction to ensure that surface connections stay in place and do not accidentally damage, or be damaged by the animal. Temporary use of invasive methods (including wireless radio-telemetry) to implant sensors under the skin and/or within body cavities can eliminate the experimental artefacts associated with non-invasive procedures. Although wireless radio-telemetry technology for monitoring laboratory animals has existed for at least 50 years (Brockway and Hassler 1991; Kramer et al. 2001), it has only been in the last 10 years that affordable, reliable, and easy-to-use commercial products have been really available for monitoring physiological signals in laboratory animals (Kramer et al. 2001). Understandably, availability has resulted in a significant increase in the use of implantable radio-telemetry in biomedical research and several authors have highlighted its advantages:

- Reduction of distress when compared to conventional measurement techniques. Radio-telemetry represents the most humane method for monitoring of physiological parameters in conscious, freely moving laboratory animals (Brockway and Hassler 1991).
- Elimination of use of restraints, which can alleviate a potential source of experimental artefact and inter-animal variability (Schnell and Gerber 1997).
- Reduction of animal use by 60–70 % in single studies (Van Acker et al. 1996), and by more than 90 % in multiple studies (Kinter 1996).
- Refinement in animal use permitted virtually unrestricted continuous data collection (days, weeks, months or more) without the need of any special animal care (Brockway and Hassler 1991).
- Available for use in all laboratory species, from mice to monkeys (Kramer et al. 2001), and most recently fish (Snelderwaard et al. 2006).

And considerations:

- Cost to acquire the requisite equipment (usually completely recovered through reduced animal use).
- Specialised training/certifications to surgically prepare and study telemeterized animals.
- Dedicated space within the animal facility in which to conduct studies.

Although the use of telemetry technology is well established in many and diverse areas of animal research, we have limited the scope of this paper to recent telemetry applications in small laboratory animals.

Surface Electrodes/Lead Wires

It is well documented (Brockway and Hassler 1991; Kramer 2000; Kramer and Kinter 2003) that in pharmacological or toxicological research, measurements from the circulatory system of small laboratory animals, such as blood pressure (BP), the ECG, and HR, can play an important role. The most common conventional measurement techniques include surface electrodes, connecting sensors, lead wires, and catheters to a restrained or tethered awake animal, or taken the desired measurements when the animal is under anaesthesia. One of the problems is that these invasive methods may directly affect the animals' physiological functions, which results in a high variability of the experimental results or even erroneous measurements. The literature extensively documents the effects of restraint on laboratory animals, including increases in body temperature, HR and BP, plasma levels of epinephrine and norepinephrine, changes in responses to pharmaceuticals and decreased food intake and body weight (Brockway and Hassler 1991; Kramer et al. 2001). In the cases of surface electrodes and the lead wires the fundamental nature of this technique is such that measurements can be unreliable or inaccurate if:

- The animal moves resulting in noisy ECG's.
- The animal is subjected to loud noises or other stressors.

Radio-Telemetry

The radio-telemetry technique can circumvent many of the problems associated with conventional methods (surface electrodes/lead wires) of ECG monitoring in small laboratory animals. The implantable devices provide accurate and reliable measurements of an ECG as well as HR from freely moving rats (group housed) and mice (single housed) in their home cages (Brockway and Hassler 1991; Van Acker et al. 1996; Kramer 2000).

Telemetry and Data Acquisition System

The commercially available radio-telemetry system (Fig. 12.1; Data Sciences International (DSI), St. Paul, MN, USA) for small laboratory animals measures an ECG as well as HR, body temperature and locomotor activity (Kramer et al. 1993, 1995; Sgoifo et al. 1994; Dejardins et al. 1996; Van Acker et al. 1996; Gras et al. 1996; Meijer et al. 2006). Components of the radio-telemetry have been described in more detail before (Kramer et al. 1993; Kramer 2000). In brief, the implantable rat- and mouse-transmitter (DSI) provides a direct measure of an ECG. The transmitter passes the ECG signal to a receiver (DSI) located beneath the animal cage via a radio signal. The signals from the receiver are consolidated by the multiplexer (DSI) and are stored and analyzed by an IBM-compatible personal computer with analyzing software (DSI), as shown in Fig. 12.1. ECG traces and HR can be monitored and stored at different time intervals. HR data are extracted from the ECG traces.

ECG Transmitters

The implantable transmitter (DSI) consists of a hermetically sealed thermoplastic housing (for rat and guinea pig, length 32 mm, width 16 mm and height 12 mm; and for hamster and mouse, length 19, width 12 mm and height 5 mm), coated with

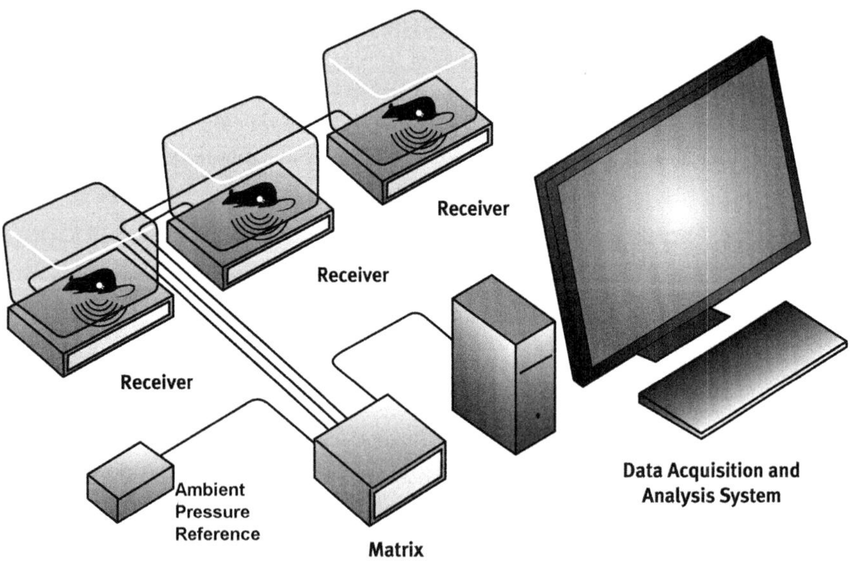

Fig. 12.1 Schematic drawing of a telemetry system for measuring physiological parameters in laboratory animals

silicone elastomer to provide the necessary biocompatibility. Total weight of the rat implant is 7.0 g and of the mouse implant 1.6 g; volume displacement is 3.0 mL and 1.1 mL, respectively.

A transmitter contains an amplifier, a battery (rat transmitter: battery life 9 months; mouse transmitter: battery life 2 months), radio-frequency electronics, a pair of flexible leads with a length of 10 cm and a magnetically activated switch that allows the device to be turned on and off in vivo.

The leads extending from the housing are a pair of 0.94 mm (rat) and 0.64 mm (mouse) diameter flexible helically wound stainless-steel lead wires insulated with silicone tubing (Fig. 12.2). In order to obtain a positive deflection from the analog output of the receiver, the white coloured (–) lead must be positioned subcutaneously at the right shoulder and the red coloured (+) lead subcutaneously towards the lower left chest (Lead II position).

Methodology for Implantation of ECG Transmitters

Acclimatization

Animals should be allowed to acclimatise to their new environment for a period of 7–14 days. This is to make sure that metabolic and hormonal changes, as a result of a stressful transport, are eliminated. It is a good habit to keep a record of body weight, food and water consumption. A severe loss of body weight and a reduced

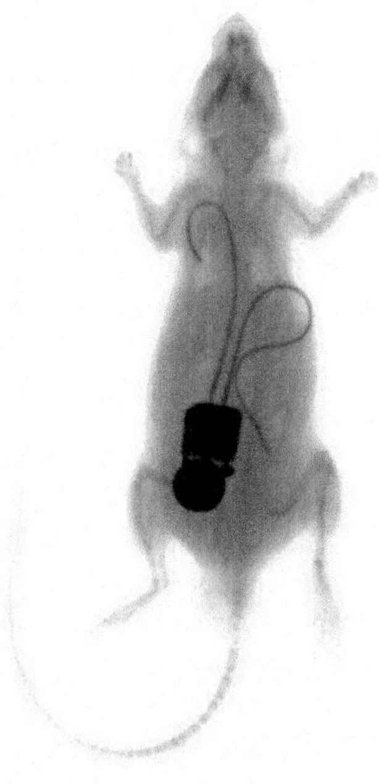

Fig. 12.2 X-ray of a mouse with a radio-telemetry transmitter with the ECG leads of the transmitter sutured subcutaneously in the lead II position and with the body of the transmitter implanted in the peritoneal cavity to measure ECG, HR and BT

intake of food and water is a strong indication that the animal is in pain and adequate measures should be taken. Given the fact that small rodents do not vomit during induction of anaesthesia, it is not necessary to deprive them of food prior to the operation.

Anaesthetic

Inhalation anaesthesia is preferred above injectable anaesthesia, as it allows for a better control of the anaesthetic depth. We routinely use Isoflurane in combination with oxygen and nitrous oxide or air (50–50 %). The concentration of Isoflurane strongly depends on the strain and the concomitant use of nitrous oxide, but is mostly around the 2 %. Following induction it is most convenient to maintain anaesthesia using a face mask. Ensure that waste and excess anaesthetic gases are removed using a gas-scavenging system in order to protect the technician from unknown side effects. Since anaesthetics block the thermo-regulating centre in the brain, animals should be kept at normal body temperature. Hypothermia can easily cause the untimely death of an animal.

Suture Materials

- *Sutures*—As a rule of thumb always use synthetic absorbable suture materials unless something requires fixation. In the operations described in this paper non-absorbable sutures are used only for fixation of the transmitter body. In rats, the maximum suture size is 4-0 for closure of the abdominal cavity, in mice a 5-0 or 6-0 suture will do. Inside the animal only 6-0 and 7-0 sutures should be used. Avoid excessively tight sutures as this allows bacteria to be protected in tissues made ischemic by pressure. Also avoid too many sutures making large ischemic portions, which can result in infection.
- *Surgical Microscope*—Do not use a binocular dissection microscope as it has a very small depth of focus.

Surgical Tools

- Scalpel
- Anatomical forceps (straight and 90° angled)
- Jewellers forceps (Dumont no. 4 and 5)
- Irridectomy scissors
- 14 Gauge needle
- Cotton wool sticks
- Sterile gauze

Solutions

- Saline (0.9 g NaCl in 100 mL water)
- Iodine solution (0.1 % in water)
- Chlorhexidine in alcohol (0.5 % in alcohol 70 %)
- Lidocaine solution (2 % in water)

All solutions used should be sterile and used at body temperature.

Radio Telemetry System

- Implantable rat transmitter (CTA-F40, Data Sciences International, St. Paul, MN, USA) or implantable mouse transmitter (ETA-F20, DSI).
- Receiver (RPC-1, DSI)
- Multiplexer (Data Exchange Matrix™, DSI)
- IBM-compatible personal computer with data collection and analysing software (Dataquest™, A.R.T.™, DSI).

Transmitter Implantation

Preoperative Procedures

- The small telemetric transmitters are implanted under anaesthesia (injected or inhaled) in the peritoneal cavity (Fig. 12.2) of the animals after a midline laparotomy (*see* **Note 1**).
- All animals are prepared for surgery in a room separate from the operating theatre.
- Clip the abdominal area (*see* **Note 2**).
- Disinfect the animal (iodine or chlorhexidine in alcohol).
- Place the animal on a sterile silicon plate and cover the animal with a sterile drape.
- Place the animal under a binocular operation microscope on a thermostatically controlled heating pad that maintains body temperature at 36–37 °C during the whole operation procedure.

Surgery

- Make a 2–3 cm long midline incision immediately caudal to the xiphisternum, large enough to insert the implant body.
- Open the straight abdominal muscle over the white line (linea alba).

- Place the body of the transmitter on top of the intestines in a saggital plane, with the biopotential leads directed cranially.
- Pass a 14 gauge needle through the abdominal wall lateral to the cranial aspect of the incision, going from the outside into the abdominal cavity (*see* **Note 3**).
- Pass the negative lead (white insulation) through lumen of the needle and out of the abdomen.
- Withdraw the needle leaving the lead externalised.
- Shorten the lead material to the appropriate length with a pair of scissors.
- Cut around the silicone tubing at the tip of the lead using a sharp sterile blade and remove 1 cm of the silicone exposing the coiled stainless steel wire.
- Place a pink tip cover onto the end of the exposed lead wire rotating the tip cover counter-clockwise until it is firmly seated, this should leave at least 5 mm of the wire coil exposed (*see* **Notes 4** and **5**).
- Using a pair of mosquito forceps or a suitable trocar, tunnel subcutaneously up to the right pectoral muscle.
- Grasp the terminal end of the lead material with the mosquito forceps and direct the exposed portion of the lead to the right pectoral muscle. Release the lead and withdraw the forceps.
- Ensure that the pink tip cover of the lead is lying flat against the muscle wall (*see* **Note 6**).
- Modify and repeat steps 4–12 for the positive lead (red insulation) by appropriately modifying the positioning for right lower chest.
- Using absorbable suture material (4-0 for rats and a 5-0 for mice) to secure the leads in place by putting two Lambert stitches to the pectoral muscle.
- Position the transmitter body in the peritoneal cavity, and incorporate the suture rib on the transmitter body into the abdominal wall closure with an absorbable suture (*see* **Note 7**). Alternatively, the suture rib on the transmitter body can be sutured to the internal abdominal muscle using non absorbable suture material.
- To complete the operation, close the skin incision again using absorbable suture material (4-0 for rats and 5-0 for mice).

Post-operative Care

- After surgery, place the animals in an incubator (32 °C) for 1 h. Return them together with their non-implanted cage mates to a clean home cage that is partially placed on a heating pad for at least 24 h after surgery.
- To prevent harassment of the implanted animals by their cage mates, gently rub the abdomen of each non-implanted animal with gauze moistened with 70 % ethanol thus ensuring that all animals carry a novel, strange odor for a short period of time.
- Besides normal food and water, provide Solid Drink® (Triple-A Trading, Otterloo, The Netherlands), moistened food pellets and food 'porridge' made with 3 % glucose solution for 4 days.

- After implantation, provide adequate analgesia. Administer the NSAID Rimadyl (carprofen, Pfizer Animal Health) for 2 days after surgery. The dose for carprofen in rodents is 5–10 mg/kg PO or parenterally. A single dose after surgery, or a dose every 3 h during a period of 12 h has a beneficial effects on post-surgical recovery. Administration may be by gavage or in Jello-O.

Surgical Notes

1. Before implanting the transmitter body in the peritoneal cavity make sure that the transmitter is moistened and at least at room temperature or even better at body temperature.
2. Do not use cream to remove the hair from the abdominal area of the anaesthetized animals, since they contain toxic components.
3. Take care not to damage the abdominal organs when passing the 14 gauge needle through the abdominal cavity.
4. The pink lead tip cover will be delivered by the manufacturer together with the telemetry transmitter.
5. Another method for placing a tip cover is as follows:
    a. The leads are placed into the desired position on the animal.
    b. At the area on the lead where the wire should be exposed, the silastic tubing is carefully cut and pulled apart to expose the desired length of wire. The silastic is then secured at this area by a non-absorbable suture just above the exposed wire. Another suture is applied below the exposed wire to prevent body fluids from entering the silastic tubing. This suture should be tied very tightly to prevent the wire from slipping out of the tubing.
    c. The lead can be cut just above this suture.
6. When the lead is lying flat against the wall, this will avoid irritation of the tissue near the lead tip.
7. When closing the abdominal muscular layer, take care not to place the suture rib between the two muscular layers, as this will result in a hernia. Make sure that the muscles are properly approximated without putting too much tension on the suture and the knots.

ECG Waveforms

Electrocardiogram and Heart Rate

Radio-telemetry has been used to record ECG and HR in rats and mice (Kramer 2000; Fig. 12.3). We have recorded ECG and HR in mice in response to surgery, anaesthesia, and handing by using radio-telemetry (Kramer et al. 1993; Van Acker

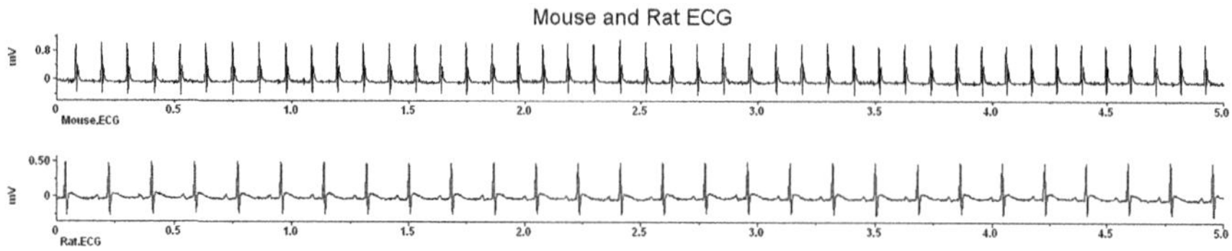

Fig. 12.3 Representative ECG waveforms measured in a freely moving mouse and rat with the ECG leads of the transmitter sutured subcutaneously in the lead II position and with the body of the transmitter positioned in the peritoneal cavity

et al. 1996, 2000; Meijer et al. 2006; Capdevilla et al. 2007). The conscious ECG signal we measured, by using an Einthoven lead II configuration in the mouse, corresponds to the ECG signal detected in anaesthetised or restrained mice by using conventional measurement techniques (Farmer and Levy 1968; Monath et al. 1978; Kishimoto et al. 1984; Postan et al. 1987; Mottram et al. 1987, 1988). The ECG in mice differs from that of man: the T wave immediately follows the QRS complex, thus there is no ST segment (Kramer et al. 1993; Mitchell et al. 1998). It might be that the sensitivity of both radio-telemetry and conventional systems is too low to detect the ST segment at these high heart rates. This hypothesis, however, needs to be explored further.

Mitchell et al. (1998) described a system for long-term monitoring and evaluation of electrocardiographic RR- and QT-intervals in transgenic mice with ion-channel defects. These intervals were measured by using radio-telemetry in conscious as well as anaesthetised mice. These investigators also were unable to find an identifiable ST segment. Surface ECG findings were consistent with QRST morphology observed in the telemetric studies. Ketamine anaesthesia caused a markedly increased duration and variability in RR and QT intervals. This difference led Mitchell et al. (1998) to the conclusion that telemetric studies in the conscious animals allow for more accurate characterisation of the electrocardiographic phenotype in transgenic mice, without the effects of acute anaesthesia and experimental manipulation.

Van Acker et al. (1996) followed the ECG changes in mice by monitoring 2 days per week during doxorubicin (Dox), a widely used anti-tumor agents which clinical use is largely limited by occurrence of cardiotoxicity, which manifests itself in congestive heart failure, treatment for 6 weeks. It was found that the widening of the ST interval became significant during the second week and continued to increase during the treatment period.

In addition, using the radio-telemetry technique, it is possible to collect ECG over a prolonged period of time without disturbing the animals and to obtain high-quality ECG recordings by computer averaging in order to establish small changes in PQ-, QRS-, and QT-time. In literature, data clearly demonstrate that radio-telemetry can be used to collect ECG recordings in small laboratory animals that are at least as accurate as those collected by using conventional techniques (Brockway and Hassler 1991; Kramer 2000).

Lead Placement for Reliable ECG Recordings

The use of radio-telemetry implies that the animals may be moving actively, and this possibility means that proper positioning of ECG electrodes is very important for collecting reliable ECG data. Improper placement of the electrodes will cause the ECG waveform to be noisy and/or distorted by electrical potentials by local muscle activity (Hassler 1994).

The bipolar configuration of the radio-telemetry system is the standard limb-lead system commonly used for Lead I, II and III configurations (Einthoven et al. 1950; Einthoven 1957). However, the two sensing leads of the radio-telemetry system are usually located subcutaneously in Lead II configuration, i.e., the negative electrode at the right shoulder and the positive electrode at the lower left chest (Stöhr 1988; Kramer et al. 1993; Hassler 1994; Kharidia and Eddington 1996; Sgoifo et al. 1996; Chagoya De Sanches et al. 1997; Johansson and Thorén 1997; Stiedl and Spiess 1997; Kayar et al. 1998; Pelissier et al. 1998; Tomiyasu et al. 1998; Wickman et al. 1998). Sgoifo et al. (1994) described placement of the leads in the Lead I configuration, i.e., the two recording leads were pulled along two subcutaneous tunnel towards the left and right shoulder. Mitchell et al. (1998) showed ECG traces from mice obtained with a lead placement in which the negative lead was sutured subcutaneously at the right scapula whereas the positive lead was sutured to the chest wall overlying the apex of the heart.

Although useful ECG recordings have been obtained with the lead placements just described, they are associated with some drawbacks mainly related to muscle activity and/or other artefacts. For this reasons, Sgoifo et al. (1996) performed a study with rats in which the two sensing leads (wire loops) were fixed on the dorsal surface of the xiphoid process and placed into the anterior mediastinum close to the right atrium. According to the authors, this location of the leads ensured a low level of noise and minimised the effects of intense physical activity on the amplitude of the signal. Therefore, several factors should be considered when deciding on electrode placement in small laboratory animals, including:

- The specific needs in terms of the ECG waveform. The information one needs to extract, e.g., whole-arterial contraction (depolarisation: P wave and P-R segment), ventricular contraction (QRS complex) and/or whole-ventricular depolarisation (T wave and ST interval) will help to direct the best location for the electrodes.
- Voltage instability and the superimposed electrical noise of the signal during intense physical activity (Sgoifo et al. 1996).
- Accurate representation of ECG features throughout the cardiac cycle, in baseline and stress conditions (Sgoifo et al. 1996).
- Complete recovery from surgical intervention and persistence in time of high quality signals (Sgoifo et al. 1996).

Placement of one or two electrodes directly on the heart would show a strong ECG signal without any interference of muscle activity. However, such an open-thorax operation in small laboratory animals might prove to be difficult and very aggravating for the animals. First, endotracheal intubation of small laboratory animals (especially with mice), without damaging the vocal cords, is not easy to perform. Second, after being guided subcutaneously from the peritoneal cavity in the direction of the groin area, both leads have to enter the pleural cavity through one or two incisions between the fifth and sixth rib, creating a risk of pneumothorax. However, placing the negative lead subcutaneously at the right shoulder and the positive lead anywhere on the abdomen gives a biphasic QRS complex and considerable disturbance of the ECG signal because of muscle damage or activity (Brockway 1998). The preferred lead placement found in the studies of Brockway (1998) was archived by placing the positive lead near the lower thorax and the negative electrode at the right shoulder. This placement minimises muscle artefacts from the abdominal muscles and offers a strong, positive, monophasic QRS complex. This positioning of leads resembles the Lead II configuration described for several laboratory animals (Kramer 2000).

Mitchell et al. (1998) implanted the transmitter body subcutaneously on the back of the mouse, with the negative lead subcutaneously placed at the right scapula and the positive lead on the chest wall. This modification does not seem to be an improvement in lead placement, because muscle activity interferes with the ECG signals from this lead placement. However, more importantly, implantation of the transmitter body on the back of the animal, which is necessary for this lead placement, causes specific surgical and postsurgical problems in these small animals and is not well tolerated (Kramer et al. 2001).

The lead placement used by Sgoifo et al. (1996) ensured a low level of noise and minimised muscle activity. However, this result could not reproduced by others researchers (Kramer et al. 2001). Brockway (1998) mentioned that the lead placement of Sgoifo et al. (1996) generates a biphasic (or sometimes negatively biased) QRS complex, which is not often handled well by ECG analysis programs. However, Brockway (1998) agrees that the lead placement Sgoifo et al. (1996) minimises muscle artefacts.

Considerable effort has been put into optimising the location of telemetry ECG electrodes in small laboratory animals. It seems that the two sensing leads of the radio-telemetry system yield the best results when positioned in the Lead II configuration, i.e., with the negative electrode at the right shoulder and the positive electrode at the lower left chest. Minor artefacts mainly related to physical activity probably could be overcome by placing one or two electrodes directly on the heart, but this solution would require open thorax surgery.

The conventional measurement techniques that are used in obtaining physiologic data from small laboratory animals can be subdivided into invasive (sensors, and/or electrodes) and non-invasive (surface electrodes) methods. As reported in literature, these conventional measurement techniques have their limitations, are often stressful, and may provide artefact-compromised data (Brockway and Hassler 1991; Kramer 2000).

Studies have shown that baseline values of the physiological parameters are considerably lower in animals implanted with a radio-telemetry transmitter then the values obtained with conventional measurement techniques (Kramer et al. 2001). Nearly all-conventional measurement techniques require acutely anaesthetised or chronically instrumented and trained animals. These methods may directly affect physiological functions, introducing substantial variations into the experimental results and even leading to erroneous conclusions. In comparison, radio-telemetry allows continuous monitoring of the same parameters in conscious, unrestrained small laboratory animals without the artefacts of anaesthesia, restraint, and excessive handling. Although the radio-telemetry technique often is an invasive method, and therefore the weight and volume of the implanted transmitter may cause discomfort (Hawkins et al. 2004; Arras et al. 2007), the animals seem to tolerate the implantation of the transmitters without problems (Moran et al. 1998; Baumans et al. 2001; Chin 2005; Greene et al. 2007). Radio-telemetry can be considered as the technique of choice for measuring many physiological parameters at relatively low levels of stress. Moreover, radio-telemetry is time-efficient and reduces the number of laboratory animals needed in experiments.

Final Considerations

It can be concluded that radio-telemetry measurements of physiologic parameters like ECG and HR have adequately been validated. It has shown that:

- Dynamic physiological data (ECG waveforms) collected by using radio telemetry technology are qualitatively similar to those collected using conventional measurement techniques.
- Absolute values of physiologic parameters collected by using radio telemetry technology are quantitatively similar to those collected by using conventional measurement techniques.

References

Arras M, Rettich A, Cinelli P, et al. Assessment of post-laparotomy pain in laboratory mice by telemetric recording of heart rate and heart rate variability. BMC Vet Res. 2007;3:16.

Baumans V, Bouwknecht JA, Boere H, et al. Intra-abdominal transmitter implantation in mice: effects on behavior and body weight. Anim Welfare. 2001;10:291–302.

Brockway RV. Unpublished data. Data Sciences International, St. Paul, MI, USA; 1998.

Brockway BP, Hassler CR. Application of radio telemetry to cardiovascular measurement and recording of blood pressure, heart rate, and activity in rat via radio telemetry. Clin Exp Hypertens. 1991;3:885–95.

Capdevila S, Giral M, Ruiz de la Torre JL, et al. Acclimatization of rats after ground transport to a new animal facility. Lab Anim. 2007;41:255–62.

Chagoya De Sanches V, Hernandez-Munoz R, Lopez-Barrera F. Sequential changes of energy metabolism and mitochondrial function in myocardial infarction induced by isoproterenol in rats: a long-term and integrative study. Can J Physiol Pharmacol. 1997;75:1300–11.

Chin I. Effects of environmental factors on laboratory mice. PhD thesis, University of Sydney, Australia; 2005.
Dejardins S, Cauchy MJ, Kozliner E, et al. The running cardiomyopathic hamster with continuous ECG: a new heart failure model to evaluate 'symptoms', cause of death and heart rate. Exp Clin Cardiol. 1996;1:29–36.
Einthoven W. The electrocardiogram (trans: Blackburn HW). Am Heart J. 1957;53:602–15.
Einthoven W, Fahr G, De Waart A. On the direction and manifest size of the variations of potential in the human heart and on the influence of the position of the heart on the form of the electrocardiogram (trans: Hoff HE, Sekeli P). Am Heart J. 1950;40:163.
Farmer JB, Levy GP. A simple method for recording the electrocardiogram and heart rate from conscious animals. Br J Pharmacol Chemother. 1968;32:193–200.
Gras J, Llenas J, Palacios JM, et al. The role of ketoconazole in the QTc interval prolonging effects of H1-antihistamines in a guinea pig model of arrhythmogenicity. Br J Pharmacol. 1996;119:187–94.
Greene AN, Clapp SL, Alper RH. Timecourse of recovery after surgical intraperitoneal implantation of radiotelemetry transmitter in rats. J Pharmacol Toxicol Methods. 2007;56:218–22.
Hassler C. ECG lead placement in quadrupeds. Technical note, Data Sciences International, St. Paul, MI, USA; 1994.
Hawkins P, Morton DB, Bevan R, et al. Husbandry refinements for rats, mice, dogs and non-human primates used in telemetry procedures. Seventh report of the BVAAWF/FRAME/RSPCA/UFAW Joint Working Group on Refinement, Part B. Lab Anim. 2004;38:1–10.
Johansson C, Thorén P. The effects of triiodothyromine (T_3) on heart rate, temperature, and ECG measured with telemetry in freely moving mice. Acta Physiol Scand. 1997;160:133–8.
Kayar SR, Parker EC, Aukhert EO. Relationship between T-wave and oxygen pulse in guinea pigs in hyperbaric helium and hydrogen. J Appl Physiol. 1998;85:798–806.
Kharidia J, Eddington N. Effects of desethylaminodarone on the electrocardiogram in conscious freely moving animals: pharmacokinetic and pharmacodynamic modelling using computer-assisted radiotelemetry. Biopharm Drug Dispos. 1996;17:93–106.
Kinter LB. Cardiovascular telemetry and laboratory animals welfare: new reduction and refinement alternatives. Abstract, General Pharmacology/Safety Pharmacology Meeting, Philadelphia, PA. USA; 1996.
Kishimoto C, Matsumori A, Ohmae M, et al. Electorcardiographic findings in experimental myocarditis in DBA/2 mice: complete atrioventricular block in the acute stage, low voltage of the QRS complex in the subacute stage and arrhythmias in the chronic state. J Am Coll Cardiol. 1984;3:1461–8.
Kramer K. Applications and evaluation of radio-telemetry in small laboratory animals. PhD thesis, University of Utrecht, Utrecht, The Netherlands; 2000.
Kramer K, Kinter LB. Evaluation and applications of radio-telemetry in small laboratory animals. Physiol Genomics. 2003;13:197–205.
Kramer K, Van Acker SABE, Voss H-P, et al. Use of telemetry to record electrocardiogram and heart rate in freely moving mice. J Pharmacol Toxicol Methods. 1993;30:209–15.
Kramer K, Grimbergen JA, Van der Gracht L, et al. The use of telemetry to record electrocardiogram and heart rate in freely swimming rats. Methods Find Exp Clin Pharmacol. 1995;17:107–12.
Kramer K, Kinter L, Brockway BP, et al. The use of radio-telemetry in small laboratory animals: recent advances. Contemp Top Lab Anim Sci. 2001;40:8–16.
Meijer MK, Kramer K, Remie R, et al. The effect of routine experimental procedures on physiological parameters in mice kept under different husbandry conditions. Anim Welfare. 2006;15:31–8.
Mitchell GF, Jeron A, Koren G. Measurement of heart rate and Q-T interval in the conscious mouse. Am J Physiol. 1998;274:H747–51.

Monath TP, Kemp GE, Cropp CB, et al. Necrotizing myocarditis in mice infected with western equine encephalitis virus: clinical, electrocardiographic, and histopathologic correlations. J Infect Dis. 1978;138:59–66.

Moran MM, Roy RR, Wade CE, et al. Size constraints of telemeters in rats. J Appl Physiol. 1998;85:1564–71.

Mottram PL, Smith JA, Mason A, et al. Electrocardiographic monitoring of cardiac transplants in mice. Cardiovasc Res. 1987;22:315–21.

Mottram PL, Smith JA, Mason A, Mirisklavos A, Dumble LJ, Clunie GJ. Electrocardiographic monitoring of cardiac transplant mice. Cardiovasc Res. 1988;22:315–21.

Pelissier AL, Gantenbein M, Bruguerolle B. Chronopharmacological effects of nicotine repeated administration on heart rate, body temperature and locomotor activity circadian rhythms in rats. Life Sci. 1998;63:2189–97.

Postan M, Bailey JJ, Dvorak JA, et al. Studies of Trypanosoma cruzi clones in inbred mice. III. Histopathological and electrocardiographical responses to chronic infection. Am J Trop Med Hyg. 1987;37:541–9.

Schnell CR, Gerber P. Training and remote monitoring of cardiovascular parameters in non-human primates. Primate Rep. 1997;49:61–70.

Sgoifo A, Stilli D, Aimi B, et al. Behavioural and electrocardiographic responses to social stress in male rats. Physiol Behav. 1994;55:209–16.

Sgoifo A, Stilli D, Medici D, Gallo P. Electrode positioning for reliable telemetry ECG recordings during social stress in unrestrained rats. Physiol Behav. 1996;60:1397–401.

Snelderwaard PC, Van Ginneken V, Witte F, et al. Surgical procedures for implanting a radio-telemetry transmitter to monitor ECG, heart rate and body temperature in small Carassius auratus and Carassius gibelio auratus under laboratory conditions. Lab Anim. 2006;40:465–8.

Stiedl O, Spiess J. Effect of tone-dependent fear conditioning on heart rate and behavior of C57BL/6N mice. Behav Neurosci. 1997;111:703–11.

Stöhr W. Longterm heart rate telemetry in small mammals: a comprehensive approach as a prerequisite for valid results. Physiol Behav. 1988;43:567–76.

Tomiyasu T, Chishaki A, Nkamura M. Magnesium deficiency in adult rats promotes the induction of ventricular tachycardia by the administration of epinephrine. Heart Vessels. 1998;13:122–31.

Van Acker SABE, Kramer K, Grimbergen JA, et al. Doxorubicin-induced cardiotoxicity monitored by ECG in freely moving mice. A new model to test potential protectors. Cancer Chemother Pharmacol. 1996;38:95–101.

Van Acker AA, Van Acker ABE, Kramer K, et al. 7-Monohydroxy-ethylrutoside protects against chronic doxorubicin-induced cardiotoxicity when administered only one per week. Clin Cancer Res. 2000;6:1337–41.

Wickman K, Nemec J, Gendler SJ. Abnormal heart rate regulation in GIRK knockout mice. Neuron. 1998;20:103–14.

Chapter 13
Gene Expression Studies Using Microarrays

Camila Guindalini and Renata Pellegrino

Introduction

The sequencing of the human genome and other organisms has been accompanied by major methodological and scientific advances in biology and molecular genetics technologies. Currently, in the post-genomic era, it is expected that the data accumulated for over 15 years of projects are finally translated into practical applications. This has generated a growing interest in the scientific community and a series of expectations about future applications of genetics in the understanding and diagnosis of complex diseases like cancer, diabetes, psychiatric and neurological disorders in general.

Among the new emerged technologies, the development of microarrays or DNA chips should be highlighted. This technique allows the investigation of thousands of genes simultaneously and promises to revolutionize predictive medicine, diagnostic and pharmacology, by substantially increasing the analytical capacity of the molecular processes.

Today, the availability of this new research method has allowed scientists to examine global gene expression that occurs in different cell types or a specific tissue, when subjected or exposed to a certain pathological or experimental conditions. Moreover, it is also possible to examine structural variations in DNA sequence that may contribute to increased susceptibility to diseases, in a quick, economical and systematic approach.

Thus, the focus of studies on the pathophysiology of complex diseases tends in the short term, move from the characterization of individual processes and

C. Guindalini (✉) • R. Pellegrino
Department of Psychobiology, Universidade Federal de São Paulo (UNIFESP),
São Paulo, São Paulo, Brazil
e-mail: camila.guindalini@afip.com.br

M.L. Andersen, S. Tufik (eds.), *Rodent Model as Tools in Ethical Biomedical Research*, DOI 10.1007/978-3-319-11578-8_13

mechanisms to the investigation of biological systems as a whole, integrating and generating data that are more realistic and closer to the complexity of an organism. The basic concepts that underlie this technique, as well as the important points to consider in designing an experiment using microarrays, its advantages, prospects and future scientific directions will be discussed.

Gene Expression and Microarrays

The complete genome of a given organism is composed of thousands of genes. Genes are selected regions of the DNA molecules that serve as templates for synthesizing RNA, in a process called transcription (Fig. 13.1). In turn, RNA is, in the majority of the cases, used to guide the synthesis of polypeptides, which subsequently form proteins either directly or by supporting the different stages of gene expression. The RNA molecules which specify a particular polypeptide are known as messenger RNA (mRNA). In this sense, mRNA may be seen as an intermediate product and proteins as the major functional end-points of the DNA template. This, on the other hand, is not the case for non-coding RNA genes, which are genes that encode a functional RNA molecule that is not translated into a protein and include: transfer RNA (tRNA), ribosomal RNA (rRNA), as well as, microRNAs and short interfering RNA (siRNAs), molecules recently described to play a crucial role in gene expression. However, not all genes are active in every cells all of the time. Some are expressed in specific cell types, at particular stages of development, or even in a precise period of the day. In genetics, gene expression is the most basic level at which genotype influences the phenotype.

With the advent of the microarray technology, today scientists have the possibility to analyze the expression of thousands of genes in parallel and use this information to determine gene expression profiles. The analysis of all expressed genes in a target sample is also entitled transcriptome analysis and is increasing being conducted using microarray based approach. In this specific type of experiment, the aim

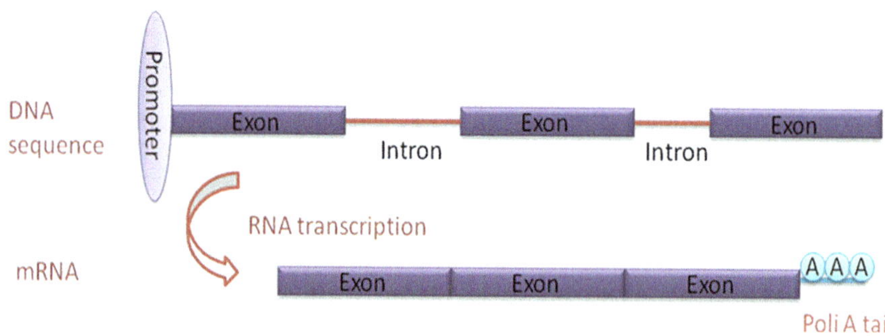

Fig. 13.1 Schematic representation of mRNA transcription process

is to quantify the types and amounts of mRNA molecules present in a particular sample. The number of mRNA molecules derived from a given gene may be seen as an approximate estimate of the level of its expression. The idea is to identify variations in the level of gene expression that may occur as natural biological responses due to the presence of particular disease, or some other experimental or pathological condition, with the assumption that the mRNA levels will reflect protein abundance and help explain the phenotype of interest.

About 20 years ago, the microarray technology was known as *macroarrays* with experiments performed on large membrane sheets made of nitrocellulose spotted with complementary DNA (cDNA), representing around 1–10,000 genes, and were used for comparative hybridization of RNA samples. This technology, although an advancement in comparison to classic methods such as Northern and Southern blotting has moved through to the chip technology, which is available today. Microarrays are small, solid supports onto which the sequences of cDNA or oligonucleotides derived from thousands of different gene sequences, hereafter called *probes*, are immobilized at specific locations in an orderly and fixed manner. The solid supports are typically glass microscope slides, silicon chips or nylon membranes, where the probes are attached to a chemical matrix via surface engineering by a covalent bond. There are a number of different variations on the microarray technology and there are different names for the commercial microarrays, such as DNA/RNA Chips, BioChips or GeneChips. The protocol basically starts with the extraction of total RNA from the specimen and the isolation of the mRNA. The mRNA transcripts are then converted to a form of fluorescent dye labeled nucleotides, normally referred as *targets*, and subsequently, hybridized to the microarray (Fig. 13.2). During the hybridization, the target will bind to the probes on the array by sequence complementarily and the excess sample will be submitted to a washing off procedure. At this point, each probe on the microarray should be bound to a quantity of labeled target that is proportional to the level of expression of the gene represented by that probe. The amount of fluorescent emission on each probe will be used to generate a signal intensity, which will afterwards be processed by bioinformatics tools and provide information on the level of expression of all the corresponding genes.

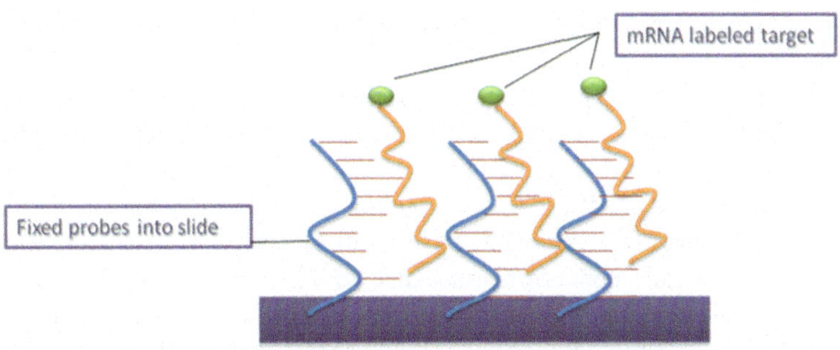

Fig. 13.2 Representation of hybridization of the fluorescently labeled target RNA sample to the synthesized probes immobilized at specific locations on a solid support of the microarray

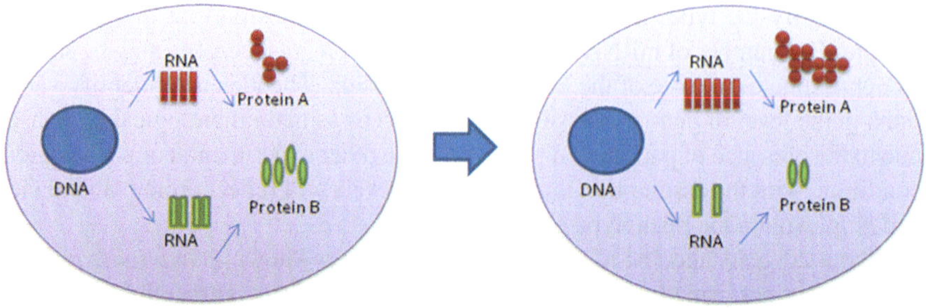

Fig. 13.3 An example of the microarray technology application. The baseline gene expression of two hypothetical genes and their expression levels modified, as a consequence of an experimental manipulation, or altered physiological condition

Microarrays have been applied to many types of biological approaches such as, responses to environmental changes, classifications of tumors, characterization of therapeutic drugs, among others (Fig. 13.3). At the present time, the main large-scale application of microarrays is comparative gene expression analysis. Because of the greater facility in acquiring samples and the nature of the disease itself, the most successful application of microarray technology has been to the study of tumor tissues. In recent years, the technology has been applied to the identification of specific patterns of gene expression that characterize different types of cancer, predict prognosis and responses to specific therapies. However, the efficiency and robustness of microarray analysis have been presented in areas as diverse as: neurological diseases, asthma, psychiatry diseases and cardiovascular diseases, with very interesting and promising results.

Technical Considerations

Experimental Design

A proper experimental design is crucial for obtaining useful conclusions from a project. The choice of the design ideally includes an assessment of the biological variation, the technical variation, the cost and duration of the experiment, as well as the availability of biological material (Fig. 13.4). The experimental plan can also depend on the methods that will be used to analyze the data afterwards. In certain cases, the parameters needed to find the optimal design must be obtained by a pilot experiment. Microarray experiments have multiple sources of variation, including variation from measurement errors associated with the array assays, laboratory process, and biological variation, representing the variability among the subjects under study. Therefore, experimental designs should ensure that effects of interest are not confounded with ancillary effects. For example, it is well-known that even when genetically identical, variability between animals in the same group may be observed. Therefore, it is very important to have a maximum control of experimental

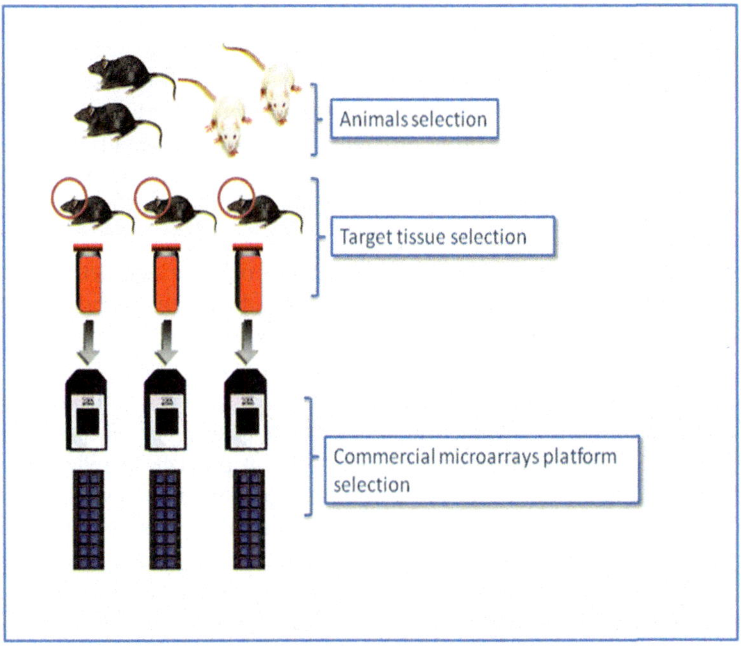

Fig. 13.4 The experimental design overview of a microarray experiment

conditions, establishing uniform procedures for the handling and treatment of the animals. Moreover, the number of animals per cage, diet, gender, age, length of fasting, circadian patterns, stress conditions and the random assignment of the animals to the different treatment groups are important factors that should be carefully established, in an attempt to eliminate potential source of variability. Notably, when separating the tissues or cell lines for the microarray experiment, gloves should be used at all times during the extraction procedure and while handling materials and equipment to prevent contamination. All equipment should be as free as possible from contaminating RNases and should, as often as possible, be treated with diethylpyrocarbonate (DEPC) and autoclaved with baking. The collection of the sample is also crucial and should be performed in a minimum period of time to prevent RNA degradation, since RNA integrity is critical for successful quantitation. In addition, the sample should be immediately snap-frozen in liquid nitrogen or dried ice and kept at –80 °C until the RNA extraction procedure takes place.

The Importance of Replicates

In microarrays studies, there are common strategies to control for technical assays and biological variations. Performing technical and/or biological replicates of the experiment being conducted is one of the classical approaches used by researchers to increase the power of the study. The technical replicate relates to the multiple labeling of the same RNA sample, with the motivation to reduce the variability related to assays and laboratory conditions (array to array difference, reagent lots, dye incorporation, apparatus, and operator, among others). Biological replicates involve the isolation of RNA from different samples independently (multiple cell lines, multiple biopsies, multiples animals, and multiple patients). The main principle of a biological replicate is to control for the biological diversity between samples.

There is no precise rule to define the number of replicates needed per microarrays experiments. However, for statistical instance, the maximum number of samples that can be handled within the biological experiment is important for the accuracy of the experiment. At the present time, it is advocated by experts in the field that, when possible, one should always substitute technical replicates with biological replicates, since individual variability are suppose to be higher then the variability derived from the technical process. A recent published guideline suggest three biological replicates for cell line work, six for animal tissues and at least ten for human samples. Furthermore, when considering two or more groups for analysis, more samples per conditions are required. Conversely, when running a time course experiment, fewer replicates per time point should be sufficient.

Another point to consider when designing an experiment is that the choice of tissue to be analyzed by the microarray technique should be based on its relevance to the physiology of the pathology of interest and/or to the location where a specific process is taking place. In addition, it is important to note that gene expression may not be only tissue-specific, but also cell-specific. Thus, the expression profile of certain population of cells may be modified, if analyzed together with different cell populations. Accordingly, new technologies such as microdissection and laser capture, which allow the extraction of specific individual cells, are already being used by several groups. As a result, the microarray technology should constantly adapt to enable the achievement of highly specific and accurate results from ever smaller amounts of RNA.

The Impact of Pooling

In microarray experiments, sometimes pooling RNA samples before labeling and hybridization may be considered, in cases where there is insufficient RNA from each individual sample, or to reduce the number of arrays for the purpose of saving cost or of simplifying the laboratory procedures. The basic assumption of pooling is that the expression of a particular mRNA molecule in the pool is close to the average expression from individuals that comprise the pool. However, it has been exhaustively discussed that pooling individual samples has a number of disadvantages: (1) the potential risk for pooling bias, e.g. significant differences between the gene expression indentified from the pooled sample and the average signal that would be derived from the individual measurements; (2) the impossibility of detecting and eliminating technical or biological outliers, which would have an effect on the data obtained from the pool; (3) the loss of information about the individual variability, which would eliminate the feasibility of indentifying specific characteristic of a given individual or clustering samples in clinical or pharmacologic subgroups; (4) difficulty in estimating variance between samples, and relying only on the observed fold-change to select genes, since it would not be possible to incorporate any statistical assessment regarding the reliability of the findings.

Nevertheless, if pooling is chosen as the research strategy, one may consider using as many independent pools as possible, so that the sets of pooled samples in each array will represent a biological replication. It has been demonstrated that in certain situations, pooling an increased number of specimens allows the researcher to reduce the number of arrays without losing precision. In addition, being more specific in the biological question under study and considering the results of a pooling experiment only as a screening exercise for future in depth analysis, while recognizing the possibility of detecting false negative and positive findings, may also help researchers to extract reliable information from a pooling experiment.

The Extraction and Quality Control Checking

The RNA quality is the most important factor that will establish the success or failure of any microarray assay. The artifacts caused by nuclease activities, potential cold shock reactions and contaminations can be avoided if the experiment process is strictly controlled and well planned. In this sense, the collection and preprocessing stage are crucial for high-quality RNA isolation. When it is not practical to extract RNA from tissue samples immediately, the samples should be snap frozen in liquid nitrogen or dried ice within 30 min after dissection. As an alternative, RNA stabilizing solutions can be used in an attempt to maintain the integrity of RNA during longer periods of time. Several methods are available to adequately isolate RNA from tissue and cell lines samples. The most common of these is the guanidinium thiocyanate-phenol-chloroform extraction. The method is very useful in providing high-quality concentration of RNA, however technical guidelines suggest that this method should not be used alone. The microarrays assays are very sensitive and since phenol may remain in the RNA solution after extraction, lowering the efficiency of the experimental reactions, the subsequent purification of the sample using a column-based method to remove the phenol residues and keep the purity of RNA is highly recommended. The procedure of RNA extraction is further complicated by the ubiquitous presence of ribonuclease enzymes in cells and tissues, which can rapidly degrade RNA. Therefore, maximum care should be applied during the entire process from tissue collection to RNA purification, in an attempt to maintain the integrity of the samples.

Prior to running a microarray experiment, RNA quality must be adequately checked. There are three characteristics of the isolated RNA that may be measured: quantity, quality and integrity. The most commonly used method to perform the inspection is spectrophotometer analysis though UV absorption measurements. This will allow the determination of the sample concentration and the presence of contaminants, such as proteins and phenol residues. In brief, the absorbance is measured at 260 and 280 nm and the ratio of absorbance at 260 and 280 nm is used to assess the purity of DNA and RNA. A ratio of ~2.0 generally indicates pure RNA. Since RNA has its maximum absorption at 260 nm, if the ratio is appreciably lower, it may indicate the presence of protein, phenol or other contaminants that absorb strongly at or near 280 nm. RNA quality is also usually assessed by electrophoresis on an agarose gel, followed by staining with ethidium bromide (Fig. 13.5). The presence of clear 28S and 18S ribosomal RNA bands are indicative of non-degraded RNA. However, it is important to remember that a number of technical conditions such as saturation of ethidium bromide fluorescence, the amount of sample loaded, agarose quality and concentration may influence the visual evaluation and should always be taken into consideration and standardized as accurately as possible. Moreover, it is also not clear if clear 28S and 18S bands do reflect the characteristics of the underlying mRNA population, which are know to present a more rapid degradation. One excellent alternative to improve the assessment of RNA quality and to standardize the process of RNA integrity interpretation is pro-

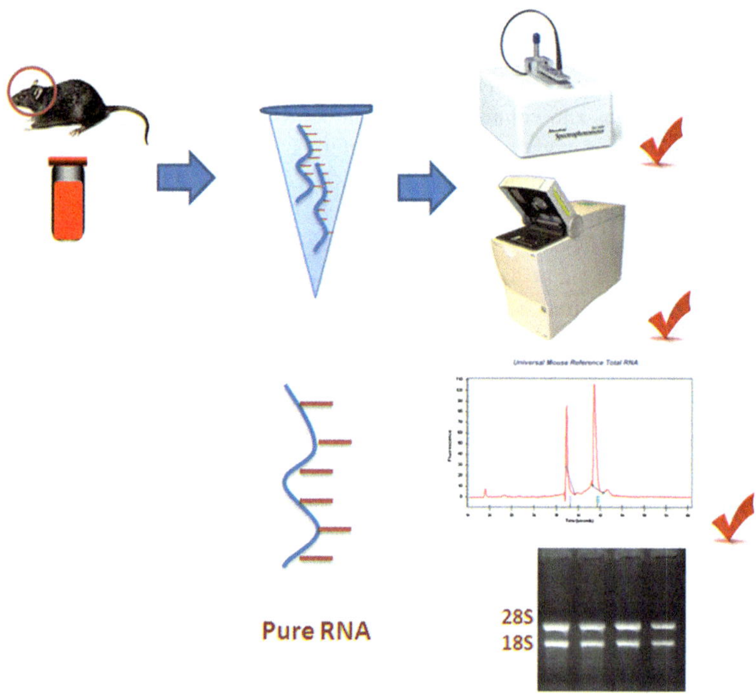

Fig. 13.5 The RNA extraction and quality control checking: sample selection, RNA extraction, spectrophotometer inspection, capillary electrophoresis and ribosomal band visualization on the agarosis gel

vided by the equipment Agilent 2100 bioanalyzer (Agilent Technologies Inc., Palo Alto, CA), a commercially available system that employs chip-based nucleic acid separation technology. RNA samples are separated by capillary electrophoresis on a microchip device (LabChip 7500; Caliper Technologies, Mountain View, CA) and subsequently detected via laser induced fluorescence detection. An electropherogram and gel-like image will be generated by the software, providing the sample concentration, the ratio of the 18S to 28S ribosomal subunits and a more accurate and standardized visualization of the RNA quality and integrity. This new technology introduces an interesting tool for RNA quality assessment, which is called RNA Integrity Number (RIN) and was developed to reduce the subjective interpretation and potentially incorrect determination of RNA quality. The software classifies eukaryotic RNA according to a numbering system that ranges from 1 to 10, with 1 indicating important levels of degradation and 10 representing highly intact and pure RNA. The acceptable number of RIN for microarrays experiments is 6 or higher. The entire process from RNA extraction to quality control samples is represented in Fig. 13.5.

A Typical Experimental Protocol

After sample quality control checking, a typical microarrays protocol may be performed using either total RNA or mRNA. The experiment starts with the target RNA being first reverse transcribed using a T7-Oligo(dT) Promoter primer in the first-strand cDNA synthesis reaction. Subsequently to the second-strand cDNA synthesis mediated by RNase H, the resulting double-stranded cDNA is purified and serves as a template in the following *in vitro* transcription (IVT) reaction. In this step, the complementary RNA (cRNA) is synthesized in the presence of T7 RNA Polymerase and a biotinylated nucleotide analog/ribonucleotide mix. The labelled cRNA products are then cleaned up and submitted to a fragmentation reaction to finally be hybridized to the microarray slide. Immediately following the hybridization, the microarray is submitted to a washing off procedure for the removal of nonspecific bonding sequences. The amount of remaining hybridized target molecules is proportional to the quantity of the originally isolated mRNA. Finally, the microarray slide is scanned while connected to specific software that processes the data and quantifies the intensity of fluorescence at each point. This information will then be used for the relative quantification of differently expressed genes. The detailed assay is shown in Fig. 13.6.

Data Analysis

Microarray data sets are commonly very large, and analytical precision is influenced by a number of variables. Statistical challenges include taking into account effects of background noise and appropriate normalization of the data using

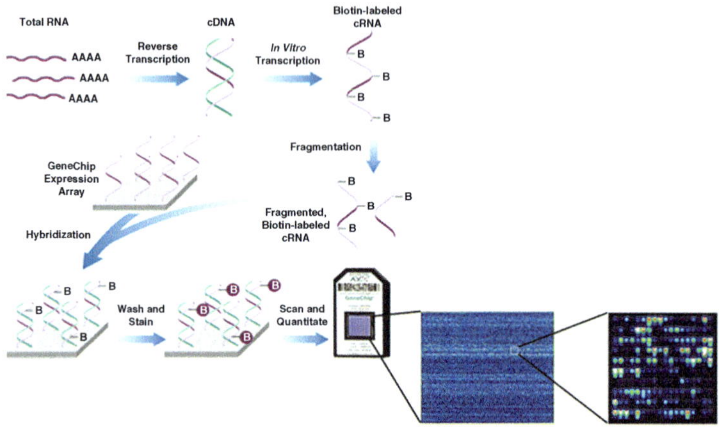

Fig. 13.6 Example of typical microarray experimental protocol. Reproduced with authorization of Affymetrix Inc

algorithms methods. Nowadays, a number of powerful freely available as well as commercial software's packages that incorporate different microarray analysis algorithms have been developed to allow researchers to capture, manage, and analyze effectively data from DNA microarray experiments. After the capture and imaging, the data obtained by a microarray experiment are subjected to a series of analytical processes, which involves the standardization and elimination of experimental noise so that obtained expression estimates reflect the true changes in mRNA abundance, as precisely as possible. After background correction and normalization, the transformation of intensity values into adequate data for statistical analysis is the subsequent step for indentifying differently expressed genes between groups. The most straightforward approach to select potentially regulated genes is ranking the results with respect to fold change. For example, genes demonstrating a twofold or greater change may be regarded as differently expressed and further selected for in-depth examination. However, this approach will have an obvious drawback, in the way that such a selection does not provide the investigator with any measure of the reliability of the observed change, since it does not take inter-experimental variability into account. Moreover, it is possible that a given gene with high fold change may as well as be greatly variable, and therefore its selection will end up providing poor information on its regulation and will be fairly imprecise. In this sense, statistical tests, such as t-tests, Analysis of Variance, Maximum-likelihood analyses, F-statistics, and the non-parametric equivalents, as well as the new generation of modified t-statistics, are alternative methods to identify significant changes between group means. Nevertheless, the choice of method used to identify differently expressed genes can have an important influence on the selected gene list and ultimately, this decision should be based on biological, rather than on statistical considerations. If the research question relies on the identification of absolute changes in gene expression, and not in the variation within the groups, the use of the fold change is recommended. On the other hand, statistical tests or the combination of both methods are more appropriate if one is interested in changes in gene expression relative to the underlying noise for a given gene.

Because of the large number of genes being tested in one experiment, the probability of identifying false positives is substantially enhanced when increasing number of tests are being performed. Therefore corrections for multiple testing methods, such as False Discovery Rate (FDR) and Family-Wise Error Rate (FWER), should be performed before the differentially expressed genes are selected and further analyses are conducted. Once a definitive list of potentially regulated genes is produced, the next step is to biologically interpret the data, using clustering and functional analyses methods for a more detailed understanding of the gene expression profile observed. Clustering is an exploratory data analysis tool that aims to group similar objects to respective categories, according to some measure of similarity. Typically, clustering is used as a strategy to present and summarize the microarray data in the format of dendrogram or heatmaps. Both samples and genes can be clustered, therefore highlighting the overall similarity of samples within a given group, providing discriminative information based on certain selection of genes, indentifying groups of possible related genes, or even providing an illustration of existing gene patterns within the set of microarrays. Functional analysis, also known as functional enrichment, is a method that integrates the gene list indentified by the experiment with the available literature, normally public databases, extracting information on potential biological pathways altered by the experiment (Fig. 13.7). This approach is especially important since analyzing genes as independent entities disregard the fact that genes do not work in isolation but in pathways. Available online programs such as Gominer™, The Database for Annotation, Visualization and Integrated Discovery (DAVID) and Ingenuity Pathway Analysis® provide information on enriched biological themes, gene ontology terms, enriched functional-related gene groups, other functionally related genes not in the list, gene-disease associations, among others.

Validation of Differentially Expressed Genes

Finally, the verification of positive results using a second independent technique is a well-established strategy performed among microarray users to validate their findings. This replication of data is extremely important, since small inconsistencies in protocols may cause subtle changes in expression levels of genes, increasing the chance of indentifying false-positive and false-negative signals. The most common method to confirm microarrays findings is the quantitative real-time polymerase chain reaction (qRT-PCR). The method is a rapid, sensitive and less complex technique for gene expression analyses and offers the opportunity for the investigation of multiple targets in a relative small standardization time. The selection of the gene set for validation analysis depends on many factors such as the original experimental design, relative difference in expression among the samples, biological function and availability of appropriate reagents (primers and antibodies). One important

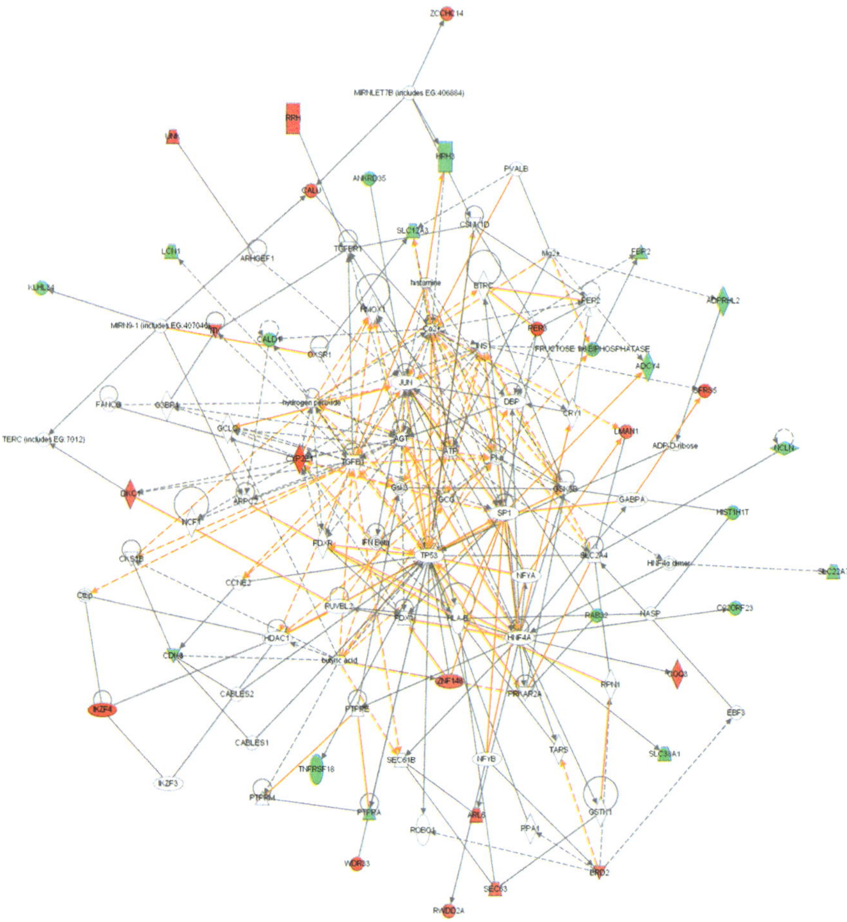

Fig. 13.7 Example of a functional pathway analysis result integrating the list of down (*green*) and up (*red*) regulated genes indentified by the experiment with the available literature and with other functionally related genes (*black*) not indentified

point to consider is that usually, commercial arrays contain a number of different isoforms of the same gene. In this sense, if an inadequate probe selection is performed, the researcher may design primers and perform qRT-PCR of transcripts that were originally not altered on the original experiment. This is one of the main reasons of inconsistencies between arrays and qRT-PCR results. As an additional recommendation, when possible, microarrays findings should be replicated using the original samples (technical replication), as well as using independent new samples (biological replication). Other methods, such as Northern and Western blot analyses, which measures RNA and protein levels, respectively, are also frequently used to validate the microarrays findings. The limitations of those techniques include time for conducting the experiment and the small number of genes that can be interrogated.

Of note, in the last years, the advances in research and the development of more robust platforms have increased the confidence in gene expression data derived from microarray experiments and it is probable that, in the future, the validations procedures may be an optional step to be performed by the researcher.

Future Perspectives

The number of studies involving the use of microarrays for the identification of new genes and molecular mechanisms has grown exponentially. We are moving to a new scientific level, in which the complex pathophysiology mechanisms of a number of diseases are now closer to be understood. The promise for the future is that biomarkers identified by this new technology will help the understanding of a number of conditions, and will eventually be directly incorporated into the diagnosis and treatment of diseases. However, despite robust and conceptually simple, the use of microarrays is still rather limited due its cost, which is considered an important bottleneck for a number of research groups. In addition to a careful experimental design, which involves the acquirement of high quality samples to the choice of the correct and most appropriate platform of analysis, another point to consider is the need for the implementation of bioinformatics tools and statistical analysis capable of managing and interpreting the massive amount of data that is generated, after each experiment. The latter seems to be a crucial factor for the success and reproducibility of the assays. Nevertheless, the next few years await further advances in the development of the technique, making it more accessible to the scientific community, both in financial and analytical terms. If used in an appropriate context and accompanied by appropriate biostatistics methods of analysis, microarray technology can be an important screening tool, which is capable of revealing valuable clues related to the pathophysiology of complex diseases, ultimately offering conditions for the development of new research strategies.

References

Chuaqui RF, et al. Post-analysis follow-up and validation of microarray experiments. Nat Genet. 2002;32:509–14.

Fodor SP, et al. Multiplexed biochemical assays with biological chips. Nature. 1993;364:555–6.

Göhlmann H, Talloen W. Gene expression studies using affymetrix microarrays. 1st ed. Chapman & Hall/CRC: Boca Raton; 2009.

Guindalini CSC, Tufik S. Use of microarrays in the search of gene expression patterns—application to the study of complex phenotypes. Rev Bras Psiquiatr. 2007;29:370–4.

Jafari P, Azuaje F. An assessment of recently published gene expression data analyses: reporting experimental design and statistical factors. BMC Med Inform Decis Mak. 2006;6:27.

Kendziorski CM, et al. The efficiency of pooling mRNA in microarray experiments. Biostatistics. 2003;4:465–77.

Koremberg MJ. Microarrays data analysis: methods and applications, Series methods in molecular biology, vol. 377. New York: Humana Press; 2007.
Morey JS, et al. Microarray validation: factors influencing correlation between oligonucleotide microarrays and real-time PCR. Biol Proced Online. 2006;8:175–93.
Schena M, Shalon D, Davis RW, Brown PO. Quantitative monitoring of gene expression patterns with a complementary DNA microarray. Science. 1995;270:467–70.
Shendure J. The beginning of the end for microarrays? Nat Methods. 2008;5:585–7.
Shi L, et al. Reproducible and reliable microarray results through quality control: good laboratory proficiency and appropriate data analysis practices are essential. Curr Opin Biotechnol. 2008a;19:10–8.
Shi L, et al. The balance of reproducibility, sensitivity, and specificity of lists of differentially expressed genes in microarray studies. BMC Bioinformatics. 2008b;9:S10.
Shih JH, et al. Effects of pooling mRNA in microarray class comparisons. Bioinformatics. 2004;20:3318–25.
Slonim DK, Yanai I. Getting started in gene expression microarray analysis. PLoS Comput Biol. 2009;5:1–4.
Stanislav M, et al. Sources of variation in Affymetrix microarray experiments. BMC Bioinformatics. 2005;6:214.
Zhang SD, Gant TW. A statistical framework for the design of microarray experiments and effective detection of differential gene expression. Bioinformatics. 2004;20:2821–8.

Chapter 14
Brain Microdialysis

Carlos Eduardo Antunes de Macedo, Gabriel Cuadra, Sergé Gobaille, and Luiz Eugênio Araújo de Moraes Mello

Behavioral, cognitive, and emotional phenomena are ultimately encoded by intricate neural networks. During neurotransmission, which involves the movement of several chemicals and ions across the neural membrane, chemical messengers released during brief time intervals into the synaptic gap interact with receptors on the postsynaptic neuron. This can lead to cellular changes, such as an altered membrane potential, gene expression, the release of growth factors, metabolism, and altered responsiveness to further stimuli. It is also well known that numerous neurotransmitters coexist in a given synaptic bouton and that they could be released at different times. In addition, axon terminals of amino acidergic, cholinergic, neuroaminergic, and peptidergic neurons converge in the same brain region. These terminals can become active at different times yet participate in the same function. *In vivo* experiments have yielded many insights into the regulation of neurotransmission in neuronal circuits and the extracellular microenvironment. However, *in vivo* measurements are complicated by several factors, such as the wide variety of brain chemicals (~200 neuroactive compounds identified), the rapid fluctuations in neurotransmitter levels by the reuptake process and enzymatic degradation, the structural heterogeneity of the brain, and the occurrence of other cell types that also may

C.E.A. de Macedo
Department of Psychobiology, Universidade Federal de São Paulo (UNIFESP), São Paulo, São Paulo, Brazil

G. Cuadra
Department of Pharmacology, Universidad Nacional de Córdoba, Córdoba, Argentina

S. Gobaille
Faculté de Médecine, Université Louis Pasteur Strasbourg, Strasbourg, Alsace, France

L.E.A.M. Mello, M.D., Ph.D. (✉)
Department of Physiology, Universidade Federal de São Paulo (UNIFESP),
Rua Pedro de Toledo 669, 3 Floor, São Paulo 04039-032, Brazil
e-mail: Lemello@unifesp.br

M.L. Andersen, S. Tufik (eds.), *Rodent Model as Tools in Ethical Biomedical Research*, DOI 10.1007/978-3-319-11578-8_14

release neurotransmitters. This neurochemical complexity demands sensitivity and versatility of the analytical methods used for chemical determination.

Brain microdialysis is a powerful and well-established technique that provides continuous sampling of endogenous compounds in the neural microenvironment within a specific region of brain (Ungerstedt 1991; Benveniste et al. 1990; Westerink, 1995). This versatile *in vivo* sampling method produces a minimal perturbation to the system and represents an elegant technique based on a simple principle. Dialysis is explained in terms of the diffusion process, whereby concentration gradients drive the movement of molecules through small-diameter pores of a semi-permeable membrane introduced into the tissue. This membrane is perfused with a liquid that equilibrates with the fluid outside the membrane by diffusion in both directions. Ideally, the perfusate imitates the ionic composition of the brain extracellular fluid to prevent undesired changes in the activity of neurochemicals that surround the microdialysis membrane. Microdialysis permits the removal of chemical substances from the extracellular fluid without removing the liquid, and it allows the delivery of substances such as drugs or toxins without the injection of fluid (Fig. 14.1). The delivery technique is actually called retrodialysis or reverse microdialysis. The resulting microcollected solution is referred to as the dialysate, which consists of a

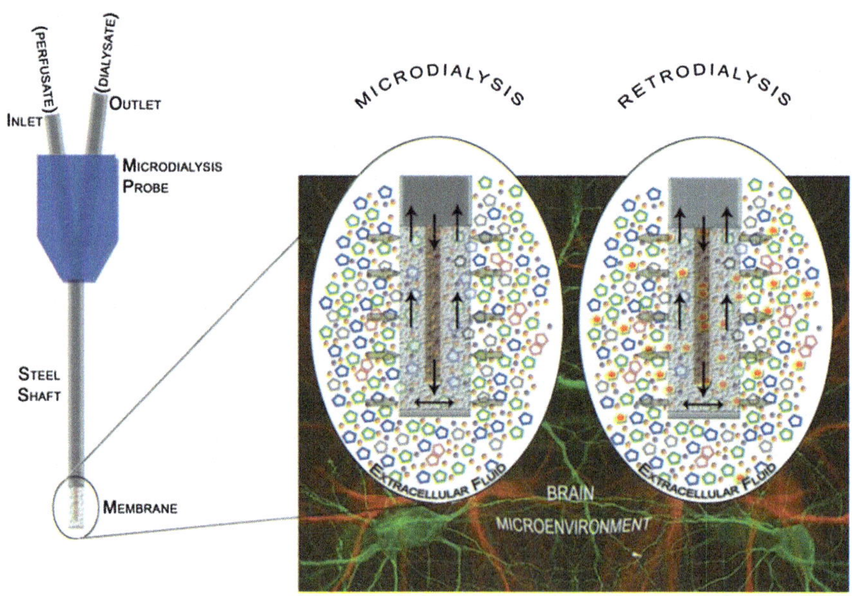

Fig. 14.1 Concentric microdialysis probe with a semipermeable membrane ate the tip. The probe can be inserted into brain tissue by means of a guide cannula. The semipermeable membrane at the probe tip allows exchange of soluble molecules between the probe and the surrounding brain extracellular fluid. When the probe is implanted into tissue microenvironment, molecules continuously diffuse out the interstitial space fluid into the perfusion medium (microdialysis). An endogenous compound can be collected at the same time that an exogenous compound is introduced into the tissue (retrodialysis). Samples are continuously collected and analyzed by standard chemical analytical techniques

blend of substances from both the perfusate and the extracellular fluid. Once they are removed by microdialysis into the flowing stream of liquid inside the probe, the substances may be analyzed with various techniques. The classical method of choice is often liquid chromatography with electrochemical, fluorescence, ultraviolet, or conductance detection.

The introduction of microdialysis can be found in the report by Bito et al. 1966 and Delgado et al. 1972; however, in 1974 Ungerstedt at the Karolinska Institute in Stockholm reported the first *in vivo* microdialysis using perfusion needle-like probes based on a dialysis semi-permeable membrane that acted as an "artificial blood vessel" to sample the extracellular fluid in the brain (Ungerstedt and Pycock 1974). Since 1974, this group in Sweden, as well as other groups around the world, have facilitated the development of the microdialysis method to advance progress in probe design and to promote increases the accessibility.

The *in vivo* microdialysis has been the foremost method for sampling extracellular fluid from discrete brain regions. During the first years of its basic application in neuroscience, various "classical neurotransmitters," such as dopamine, serotonin, noradrenaline, GABA and glutamate were investigated (Abercrombie et al. 1989; Gobaille et al. 1999; Cuadra et al. 2000; Di Chiara 2002; Macedo et al. 2005). Therefore, quantification of these molecules places elevated demands on the detection limits of analytical methods. Without the advance of analytical techniques, microdialysis would have no practical significance. The introduction of highly sensitive analytical techniques for the detection of neuroamines by high performance liquid chromatography with electrochemical detection, and the detection of amino acids with fluorescence accelerated the development and applications of microdialysis in pre-clinical and clinical research.

Since its introduction, more than 12,000 scientific and clinical papers have been published using the microdialysis technique (PubMed, November 2009), with over 9,000 published studies that included the terms "microdialysis" and "brain" (~7,500 performed in rats). The method that originated in brain research is still mostly focused on the central nervous system, but the number of applications in peripheral system is also rapidly growing. Microdialysis probes have been placed in virtually every tissue and organ in the body, including the liver, fat, heart, skin, blood, placenta, stomach, ear, ocular, corpora lutea and muscle. Microdialysis has become a common tool for the monitoring and interpretation of extracellular analyte concentrations in a variety of disciplines, including pharmacology, physiology, neurochemistry, sports medicine, and tumor oncology, in combination with diverse physiological, pathological, and pharmacological stimuli. This basic multidisciplinary potential can be supplemented by a number of monitoring techniques used to collect information on other relevant biochemical or physiological variables. The combination of neurochemistry, pharmacology, and behavioral study has been very constructive for research into the neurochemical control of feeding, addiction, stress, fear, anxiety, depression, learning and memory, and sexual and maternal behaviors.

Microdialysis has been successfully applied in experimental neuropharmacology studies that investigated the mechanisms of drug action. Microdialysis is very suitable for the study of blood–brain transport mechanisms and protein binding, drug development and the pharmacokinetic and pharmacodynamic profiles of therapeutic and toxicological agents; PK–PD models (Pan et al. 2007; De Lange et al. 1997, 2000). For pharmaceutical companies active in the therapeutic area of the central nervous system, microdialysis indubitably provides the fundamental data on the mechanisms of action and functional validity of candidate drugs. It is of great importance that the method is also pertinent to genetically modified mice (De Lange et al. 1998; Kehr et al. 2001). Reverse microdialysis has been mainly used to study the effect of local drug administration on the brain concentration of neurotransmitters and metabolites. Recently, reverse microdialysis has been introduced for the evaluation of pharmacodynamics of therapeutic agents in the clinical setting (Höcht et al. 2007). In the clinical setting, *in vivo* microdialysis is used to monitor the neurochemical changes associated with numerous brain disorders, such as Parkinson disease, epilepsy, and malignant neoplasia (Ronquist et al. 1992). Recently, microdialysis has shown strong promise for clinical use as a neuromonitoring technique in intensive care, ischemia, and metabolic control (Thomas et al. 2005; Ungerstedt and Rostami 2006).

It is also possible to put multiple microdialysis probes in a single animal (Martinez et al. 2008). Therefore, it is feasible to measure blood, brain, and tissue concentrations of drugs or endogenous substances simultaneously. If the animal is awake, these measurements can be correlated with behavior. The simultaneous perfusion of various brain locations has enabled neuroscientists to study the specifics of individual pathways and the role of feedforward and feedback loops. With dual-probe microdialysis, one probe can be inserted into the cell terminal region (e.g., in the nucleus accumbens), while the other is inserted into the cell body region (e.g., in the ventral tegmental area), in the case of the dopaminergic pathway (Westerink et al. 1996). More recently, dual probe approaches have been used to quantitatively evaluate the diffusion rate of drugs through brain tissue from a microdialysis probe. Drug diffusion rate measurement is extremely important for determining an effective concentration of a neuroactive compound. Infusion of drugs in this paradigm allows one to study the effects of the auto-, homo- and heteroreceptors. This paradigm has been used in the assessment of novel therapies for diseases such as basal ganglia-associated neurological disorders, such as schizophrenia, as well as for novel antiepileptics. It has also aided in the study of drug transport mechanisms.

Dual- or three-probe approaches have been used to quantitatively evaluate the diffusion rate of drugs from a microdialysis probe through the brain and others tissues. A simultaneous microdialysis of blood and brain structures offers a very useful tool for monitoring and comparing peripheral and central phenomena. Some investigators have combined microdialysis and electrophysiological recordings in behaving animals. Electroencephalography is the most frequent electrophysiological variable recorded concomitantly with microdialysis, particularly during epilepsy and sleep studies. This method has proven very useful for correlating electrophysiological changes with neurotransmitter effluxes associated, for example, with epileptic seizures (Obrenovitch et al. 1995; Bourne and Fosbraey 2000).

Overview of Brain Microdialysis Technique

The "dialysis", i.e., the diffusion of molecules between the extracellular fluid and the perfusion fluid, takes place while the perfusion fluid passes between the inner tube and the dialysis membrane. The rate of analyte transport by diffusion can be described mathematically by Fick's law of diffusion, which states that: $J = D^*\Delta c/\Delta r$ (where J is the flux or moles of solute transported over the area of diffusion; D is the diffusion coefficient; Δc is the change in concentration of the analyte over the area of diffusion; and Δr is the small length over which diffusion occurs. The assumptions underlying this expression are that the transport of the analyte occurs through an ideal solution that is stable and homogeneous in composition). It is important to realize that there is an exchange of molecules in both directions. The difference in concentration through the membrane governs the direction of the gradient. An endogenous compound can be collected (microdialysis) at the same time that an exogenous compound, such as a drug, is introduced into the tissue (retrodialysis).

Several variants of concentric or linear microdialysis probes of rigid (for intracerebral use) or flexible (intravenous and peripheral organs) design with their respective guide probes are commercially available. Many laboratories have designed their own probes. In addition, some clinical microdialysis catheters have been developed and commercialized for use in clinical research. All these probe and catheter constructions use a polymeric dialysis membrane, which is a tubular hollow fiber with an outside diameter typically in the range of 220–600 mm, depending on the polymer material and the supplier. The tubular membranes are widely used in perfusion cartridges of kidney dialyzers. Typical membranes used for the construction of microdialysis probes include the following materials: polyacrylonitrile AN69, cuprophane (regenerated cellulose), polycarbonate, polyamide, polysulfone, or polyarylethersulfone.

Brain microdialysis is often carried out in rodents, frequently rats (larger animals such as cats and monkeys have also been used). After inducing general anesthesia, a microdialysis probe (or its guide probe) is implanted with small screws and dental cement on the skull of the animal. The neurosurgical procedure uses the same stereotaxic implantation technique of guide-cannulae assemblies into discrete regions of the brain (Fig. 14.2). As microdialysis combined with a guide probe is well tolerated in freely behaving animals; many scientists studied the behavior-linked chemistry with remarkable success. Animals that feed, run, mate or crave for drugs of abuse can be equipped with probes and their neurotransmitter profile recorded.

It should be noted that sampling methods cannot be employed to study the physiology of behavior without giving careful reflection to the changes introduced by the technique itself. Microdialysis is an invasive technique. Probe implantation injures tissue by disrupting cells and blood vessels. During the first hours after probe implantation, several brain physiological processes are affected. This mechanical damage of the tissue has been considered to be short-term, thus allowing microdialysis experiments to take place after a given period. The integrity of the blood-brain barrier is disturbed—there is reduced blood flow and oxygen uptake in the

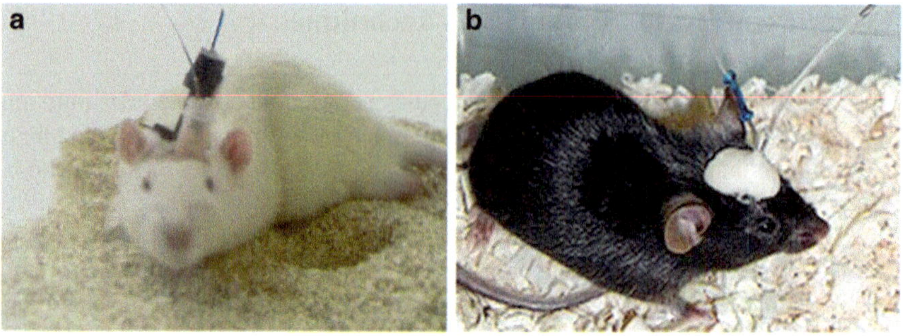

Fig. 14.2 Microdialysis probe implanted with dental cement on the skull of the rat (**a**) and mouse (**b**)

tissue surrounding the microprobe, and an abnormal release of neurotransmitters occurs. However, 2–24 h after implantation of the microprobe these processes appear to recover substantially. In the case of dopamine, it has been shown that during the first hours after implantation a considerable part of the dialysate levels of dopamine are derived from damaged terminals. However, 2–4 h after implantation of the probe virtually all sampled transmitter originates from neuronal activity. A glial barrier around the probe tract has been described 3 days after continuous microdialysis. This glial barrier is probably responsible for the fact that the extracellular levels of neurotransmitter gradually decrease after implantation. Much of this interference could be overcome by the use of a guide probe, in which a removable probe can be re-inserted when the animal is sufficiently recovered. Extracellular dopamine was unchanged after seven daily re-insertions of a removable microprobe across a guide probe. For behavioral studies, it is of course of great importance to know to what extent the anesthesia and neurosurgery will affect the animal's behavior. This implies that when behavioral experiments are carried out in animals equipped with a permanent implanted microdialysis probe, it should be taken into account that certain behavioral and autonomic functions can be disturbed for at least 1 week after implantation. Again, guide probes are recommended to overcome these limitations. We suggest that it is advisable to implant the guide probe at least 5 days before the dialysis experiments are carried out; however, this does not imply that applicable neurochemical or behavioral experiments cannot be carried out during the first days after implantation of a fixed microdialysis probe.

When considering microdialysis for temporally resolved measurements, analyte recovery by the probe is an important issue. Recovery can be defined as absolute or relative. Absolute recovery refers to the mass of analyte collected over a period of time (mol/time unit), whereas relative recovery refers to the concentration of analyte in the dialysate divided by the concentration in the sampled media. Perfusion flow rate strongly controls recovery such that increasing flow rates through the dialysis probe increase the absolute recovery while decreasing the relative recovery. Absolute recovery increases because higher flow rates create a steep concentration gradient between the probe and the extracellular fluid, thus enhancing the flux of molecules to the probe. Relative recovery decreases as the flow rate is increased because there is less time for equilibrium to be reached between the solution flowing through the probe and the extracellular space. Therefore, one must choose a flow rate that allows both adequate relative recovery to meet the instrument's concentration detection limit and adequate absolute recovery to meet the instrument's mass detection limit. Consequently, perfusate flow rates are chosen on the basis of a cost/benefit analysis and typically range between 0.1 and 2 μL/min (recoveries for small molecules are usually between 10 and 40 % at these flow rates, but if a flow rate of 100 nL/min is employed, the relative recovery approaches 100 %).

Lower flow rates are preferred because they minimize tissue disturbance and disequilibrium. In addition to flow rate, other factors also affect recovery. Recovery increases with increasing molecular weight cut-off of the membrane (up to a limit), temperature (increasing the temperature of the perfusate increases the movement of molecules and thus promotes the diffusion of substances across the membrane and elevates recovery), and active membrane length (increasing the length and thus the area of the microdialysis membrane leads to an increase in relative recovery). In general, higher recovery (especially absolute recovery) aids in achieving high temporal resolution. When considering microdialysis for temporally resolved measurements, analyte recovery by the probe is an important variable.

The perfusion itself may also induce non-physiological changes in the surrounding brain tissue. Due to the continuous removal of material by the probe, constant concentration gradients may extend as much as several millimeters radially into the tissue. Important metabolic factors can also be removed from the cellular microenvironment. These extractions can affect the physiology and biochemistry of the microdialysis brain area.

It is essential that the composition, pH, and temperature of the perfusion media mimic brain extracellular fluid. Perfusion media used in microdialysis experiments vary widely in composition and pH (Table 14.1). The ideal composition, ion strength, osmotic value, and pH of the perfusion solution should be as close as possible to those of the extracellular fluid of the dialyzed tissue. In general, the perfusate is an

Table 14.1 Different electrolyte concentrations (mM) compositions of perfusion solutions commonly used when sampling from brain interstitial fluid

	NaCl	KCl	$CaCl_2$	$MgCl_2$	
Artificial CSF	148	2.7	1.2	0.9	
Solution de Ringer	147	4	1.3	0	
Modified Ringers solution	145	2.7	1.2	1.0	+0.2 mM ascorbate
Buffered Ringers solution	147	2.8	3.4	0.6	+0.6 mM K_2HPO_4 and 114 mM ascorbate
Krebs Ringer solution	138	5	1.0	1.0	+11 mM $NaHCO_3$; 1 mM Na_2HPO_4 and 11 mM glucose
Perfusion fluid CNS (CMA microdialysis)	147	2.7	1.2	0.85	
Artificial CSF perfusion fluid (Harvard)	150.0	3.0	1.4	0.8	

aqueous solution of sodium and potassium salts and other ions in a minor proportion, with no or a very small concentration of proteins. The omission of critical substances present in the extracellular fluid (e.g., Calcium) from the perfusate will impair synaptic transmission. One component that is commonly omitted is glucose, which possibly results in the extraction of glucose from the extracellular fluid and alteration of neural functioning. In some studies, the substances are included in the perfusate, but their concentration does not mimic that found in the extracellular fluid. For example, brain extracellular glucose levels are estimated to range from 1 to 2 mM, but some researchers have utilized as high as 8 mM glucose in the perfusate because this high glucose concentration increase extracellular levels of various neurotransmitters. In some cases, proteins should be added to the perfusion medium to prevent the adhesion of drugs to the microdialysis probe and tubing connections. Another important requirement in reverse microdialysis experiments is to ensure that the added substance does not change pH or tonicity of the perfusion medium. Such alterations could affect transmitter release and consequently the results of the study.

The microdialysis technique does not allow 100 % recovery of compounds. Some mathematical models have been proposed for quantifying microdialysis (for review see Parsons and Justice 1994). Some of these models calculate the actual or "true" extracellular concentration *in vivo* by fluctuating the perfusion flow rate or substance concentration. The values are extrapolated back to zero flow by a nonlinear regression model. Other models rely on systemic, local, or endogenous markers as a reference for calculating extracellular concentrations of drugs or endogenous compounds.

Quantitative microdialysis techniques permit the assessment of changes in the extracellular levels of an analyte, as well as in vivo probe recovery or extraction fraction (which in turn provides information about the effects of a particular manipulation on clearance mechanisms). As such, they are powerful techniques for characterizing changes in neurotransmitter dynamics that occur in response to a particular manipulation. However, these studies are time-consuming and require good organizational ability. Varying concentrations of the analyte to be measured are infused via the probe, and care must be taken to ensure that equilibration periods are adequate and that the perfusate is prepared and switched correctly. A single steady-state experiment normally takes a full day.

The most extensively used quantitative methods are the zero-net-flux method and the varied flow rate method, both of which have been questioned regarding their ability to determine the true "unperturbed" basal concentration of extracellular neurotransmitters. This incapacity is attributed to the inevitable disruption of both neurotransmitter release and uptake sites during microdialysis probe implantation. Model simulations show that if microdialysis impairs the neurotransmitter release site more than the corresponding uptake sites, underestimation of the unperturbed basal concentration by these quantitative methods occurs and the trend of the *in vivo* recovery of the neurotransmitter with the uptake inhibition would reverse, compared with that of the microdialysis extraction.

In the no-net-flux method, during calibration, different known concentrations of the analyte of interest are added to the perfusate, and the net loss or gain of the substance in the dialysate is measured and plotted in relation to its concentration in the perfusate. The point of no-net-flux correlates with the tissue concentration. The slope of the regression line provides an estimate of recovery for the substance.

Before the neurotransmitters in dialysate can be considered neuronal in origin, experimental criteria must be satisfied. Two methods that can be used effectively to verify that an alteration in a neurotransmitter is derived from neuronal sources examine the dependency of this elevation on action potentials and calcium-dependent exocytosis. If an increase in extracellular fluid neurotransmitter levels is stimulated by action potentials, then adding tetrodotoxin (TTX, a sodium channel antagonist) to the perfusate should significantly decrease dialysate levels of that neurotransmitter. Similarly, basal levels of neurochemicals that are neuronal in origin should decrease when calcium is omitted from the perfusate. These two criteria have been utilized most clearly for dopamine and acetylcholine. In contrast, dialysate levels of serotonin, norepinephrine, glutamate, gamma-aminobutyric acid (GABA), and glycine, appear to be partially calcium-independent and TTX-insensitive, suggesting that their efflux is at least partially non-neuronal in origin. When interpreting data from microdialysis studies, it is essential to be cognizant that one is sampling extracellular levels of a substance, that alterations can reflect clearance and/or release, and that the source of the variation is not essentially neuronal. However, it is equally imperative to consider that experimentally induced alterations in dialysate levels of a neurochemical are meaningful, regardless of the source of that change.

A membrane with a low molecular weight cut-off (MWCO) filters the samples by excluding large molecules such as proteins (the typical MWCO for several neurotransmitter is 6–20 kD). Recently, a 100 kD MWCO microdialysis membrane was introduced to allow the detection of larger molecules, such as cytokines. The molecular weight of the compounds to be administered is an important factor to be considered with regard to the MWCO microdialysis membrane in a reverse microdialysis experiments.

The choice of size for the microdialysis membrane should take into account the size of the central structure and that most of the compounds have slow infusion rates and do not diffuse more than 1 mm from the membrane into the tissue.

In general, microdialysis is a sampling technique in which a dialysis probe (with a semi-permeable membrane 1–2 mm in length) is stereotaxically implanted into a discrete brain region. A dialysis buffer that is similar to cerebrospinal fluid in ionic strength and composition is perfused through the membrane via tubing attached to a syringe pump (inlet tube). Substances in the brain tissue that are not present in the microdialysis buffer and that are small enough to pass through the dialysis membrane (including all small molecule neurotransmitters and some peptides), will diffuse into the dialysis membrane and will flow through another tube for collection (outlet tube). Samples may be collected using a fraction sample collector or manually (off-line method), or they may be injected into the analytical system (on-line method). In studies of awake and freely moving animals, the experimental animal is housed in a containment system fitted with a liquid swivel. The swivel allows free movement of the animal without entangling or breaking the connective tubing (Fig. 14.3).

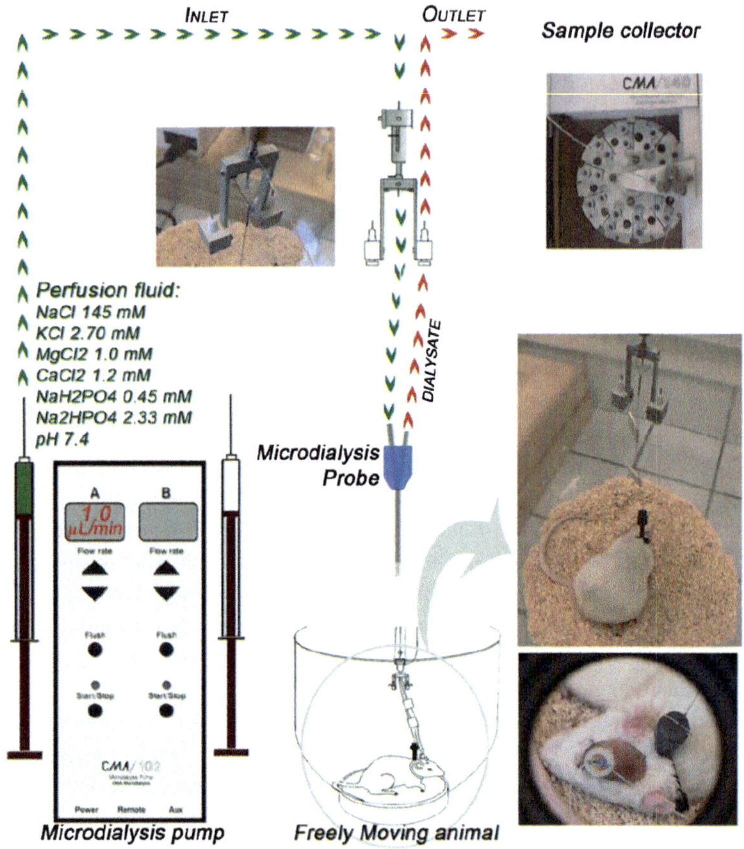

Fig. 14.3 Experimental set-up for the recovery of endogenous extracellular substances in a awake animal, showing microdialysis pump (calibrated to ensure an accurate flow rate with a specific syringe), liquid swivel (this counterbalanced apparatus allows the animal free movement while allowing the collection of perfusate; the animal is usually connected to this apparatus via a collar), the observation cage (allowing easy observation and unrestricted movement), and the microfraction collector (small tubes are loaded into this equipment to collect the perfusate)

The contents of the dialysis samples (dialysates) can then be determined by several methods, usually high-pressure liquid chromatography (HPLC) coupled to fluorescent or electrochemical detectors. Likewise, compounds in the perfusate can be delivered directly to the physiological site of interest (retrodialysis).

Several parameters are always considered during the initial stages of development of an experiment, as they may impose upon a successful study if ignored. The recovery of specific molecules is dependent on a number of variables, including temperature, pH, molecular weight, shape and charge. It is extremely dependent on the surface area and characteristic of the dialysis membrane, the flow of the perfusion liquid, and the speed of diffusion of the substance through the extracellular fluid.

The protocol (Box 14.1) describes the steps necessary to collect samples from awake, freely moving rats. There are obviously many ways to run a microdialysis experiment.

It is expected that prior to applying this procedure the investigator has made numerous decisions regarding the microdialysis parameters suitable for the particular experimental design. The basic protocol describes the conventional microdialysis method in which the dependent measure is dialysate levels of an analyte under study. The protocol describes the application of a quantitative microdialysis method that allows the investigator to obtain estimates of extracellular concentration during steady-state and transient conditions.

Box 14.1: General Protocol for Brain Microdialysis

The essential material and basic procedures in a brain microdialysis technique necessary to collect samples from awake, freely moving rats: microdialysis guide-cannulae and probe, prefusate solution, perfusion pump, glass microsyringes, FEP and connective tubing, liquid swivel system mounted on the testing cage and collection vials.

Procedures

Probe Conditional

The poros of the dialysis membrane is usually filled with glycerol, which prevents it from drying out. Conditioning of all probes, whether they are made in the laboratory or obtained commercially, is necessary prior to use. The protocol is based on three perfuse steeps (distilled water—70 % ethanol—perfusate; 10-3-10 min in flow rate 5 μL/min). After probe conditioning, place microdialysis probe into a vial filled with perfusate solution until ready to implant the probe into the rat.

Pre-implantation of microdialysis probe

Fill microsyringes with perfusate and attach it in microdialysis pump.

Turn on pump and adjust it to the appropriate flow rate (4 μL/min) before attaching the microdialysis probe in the animal's guide probe.

Verify that liquid is coming out through the tip of the syringe and attach FEP tubing from the microsyringe pump to the liquid swivel (inlet).

Attach other FEP tubing from swivel to the microdialysis probe.

Flush the system by running the pump at 4 μL/min (5 min) and observe the probe and the various tubing and connectors. Check for air bubbles inside the membrane.

Attach FEP tubing from the microdialysis probe to the swivel (samples collected using a fraction sample) or to the vial (manual sample collection).

Collect perfusate samples into collection vials.

(Make sure that all air is out of the microdialysis system and collect perfusate in a vial to check the volume; determine the dialysate weight: 1 mg ~ 1 μL. Lower volumes can indicate the presence of leaks or air bubbles in the membrane or tube lines).

(continued)

Box 14.1: (continued)

Implantation of microdialysis probe

Remove the dummy cannula from guide probe implanted in the animal.

Remove the microdialysis probe from vial (see probe condition).

Implant the microdialysis probe through the guide probe while continuously perfusing the probe.

The microdialysis probe is inserted while the animal is awake without any more distress to the animal than the handling (some researchers have used a brief inhalatory anesthesia).

The insertion in itself does not cause any apin.

Attach the tether assembly to the animal (it is important that the rat is accustomed to the cage and tether system). Following insertion of the microdialysis probe, allow an equilibration period of 2–4 h (or ~10 h overnight with a flow rate between 0.1 and 0.2 μL/min, but allow an equilibration of 2–4 h before the microdialysis experiments) for recovery of the tissue and blood vessels from the mechanical damage. The basal concentration of the dialysate samples collected may be both neuronal and non-neuronal in origin. It is essential to demonstrate neuronal release of neurotransmitters in initial studies, for example by infusing tetrodotoxin (TTX) or by varying the composition of the perfusate, such as the Calcium, via the microdialysis probes. Once neuronal release of analyte is determined, experiments can begin. Return the rat on testing cage. Collect samples into vials and store samples for later quantification. Remove samples immediately following their collection to help prevent sample degradation or evaporation. Neuroamines oxidize rapidly and should be frozen. Although the presence of antioxidants in the perfusate or collection vial retards degradation (e.g., ascorbate is often used for quantification of neuroamines), samples should be frozen at –80 °C if more than 6 h will elapse between sampling and quantification. Polyethylene vials are suitable for the storage of neuroamines and amino acids, but other sticky analytes (for example, opioid peptides) can be adhere to the walls of certain plastic or glass vials. In vitro tests should be undertaken for a less-studied analyte. The physiochemical properties of the analyte should be considered when selecting collection vials and storage.

Using the analytical method of choice

Make sure that all air is out of the microdialysis system.

In a conventional dialysis experiment, a series of dialysis samples is first obtained for determination of basal analyte levels. Repeated samples are collected until stable baseline levels are achieved (<2–4 h). Depending upon the experimental protocol, the same probe can then be used for sampling as well as for drug delivery. A liquid switch can be used for the delivery of different perfusate into the region dialyzed. With practice, this can also be done manually.

The size of the site to be dialyzed determines the length of the active membrane. The analyte to be sampled should be considered when selecting a dialysis membrane.

Accurate neurosurgical implantation into the site to be dialyzed is crucial for reproducible microdialysis.

Analytical Methods for Microdialysis: High Temporal Resolution

Reports in the literature regarding the methods of microdialysate analysis are dominated by studies that use reverse-phase liquid chromatography (LC) in the analysis of dialysate samples. LC coupled with fluorescence, ultra-violet, electrochemical, chemiluminescence, or mass spectrometric (MS) detection has been frequently reported, and it is possible to analyze almost every known small neurotransmitter and metabolite using HPLC. However, laser-induced fluorescence (LIF), electrochemical, and MS detection have been favored due to their exceptional low-level detection capabilities.

The time-resolution is dependent on the detection limit of the assay, the sample volume requirements of the analytical system, and the recovery of the transmitter from the extracellular fluid. A sample time of 5–20 min is typically needed for the analysis by HPLC. Although sustained changes in neuronal firing or metabolism (such as those that may occur during a pharmacological intervention) can occur over a longer time scale, whereas oscillations in neurotransmitter levels can also occur on the millisecond time scale. If the objective of microdialysis experiments is to measure the neurochemical changes to correlate them with behavior, drug effect, and other phenomena, there is a risk of erroneous conclusions if the sampling time is too long. In this case, temporal resolution is critically important for *in vivo* brain chemical measurements. For example, in samples collected every 1 min, Bert et al. (2002) showed that the administration of NMDA by retrodialysis to the dorsal striatum lead to an increase of glutamate and dopamine; however, after the perfusion dopamine decreased while glutamate increased again. When the samples were evaluated every 10 min, the dopamine increase and glutamate's bimodal oscillation were completely dampened. Taking another example, axon terminals can release glutamate or GABA for a few milliseconds and nevertheless induce biologically significant alterations in the central excitatory and inhibitory states of other neurons. In this way, insufficient temporal resolution masks the concrete temporal course of synaptic activity.

Capillary electrophoresis (CE) is a high efficiency analytical technique that has had a great impact on biomedical research, clinical and forensic practice, and neuroscience studies (Kennedy et al. 2002; Nandi and Lunte 2009). CE is well suited for achieving high temporal resolution because it has both high mass sensitivity and the potential for high throughput. CE has received a lot of attention as an analytical technique of choice for dialysate analysis. This is primarily due to the high separation efficiency values (1,000,000/m). CE methods have yielded high efficiency, short analysis time, and greater selectivity due to the sharp bands. Electrophoretic techniques are either capillary- or chip-based, and therefore sample volume requirements are in the order of nanoliters as opposed to microlitres (required for HPLC).

Many microdialysis-CE experiments involve fraction collection and off-line analysis with sampling times that range from 2 to 30 min. CE coupled with LIF, MS/ESI, MALDI-TOF, and electrochemical detection has been used to monitor neurotransmitters. Mass detection limits by CE-LIF can be in the yoctomoles range, suggesting the potential for subsecond temporal resolution. Such impressive detection limits are generally achieved only with LIF, but electrochemical detection can achieve attomoles LOD, and ultraviolet detection can achieve femtomoles LODs. Microdialysis combined with capillary electrophoresis and mass spectrometry is still being developed.

CE-LIF has been used to measure the release of amino acids and neuroamines across the sleep-wake cycle in rats (Léna et al. 2005). Microdialysis probes were implanted in the medial prefrontal cortex and the nucleus accumbens. Each sample (2 min) was then sorted by whether it was collected during wakefulness, slow-wave sleep, or REM (paradoxical) sleep. The excitatory amino acids glutamate and aspartate only decreased in the nucleus accumbens as the animal went into slow-wave sleep and REM sleep. This decrease was imitated by norepinephrine in the nucleus accumbens and medial prefrontal cortex. However, dopamine release was higher in both structures during wakefulness and REM sleep compared with slow-wave sleep. These fluctuations in extracellular dopamine levels were hypothesized to be a consequence of the cognitive processes. GABA can also be monitored effectively with this technique. Analysis of GABA levels, in conjunction with electroencephalographic recordings, revealed for the first time that the extracellular concentration of GABA in epileptic rats increases with seizure activity (Parrot et al. 2004).

Figure 14.4 shows the GABA and glutamate electropherogram of one rat with a dual probe implanted into the amygdala and dorsal periaqueductal gray. The temporal resolution was 30 s. In this microdialysis experiment, a high concentration of potassium (K+, 100 mM) was added to the perfusate for general neural stimulation in the microdialysis sites by retrodialysis.

As these examples show, high temporal resolution measurements have been used in a wide variety of applications, including stress responses, learning, sleep, addiction, stroke pathophysiology, and feeding. These studies suggest that the impact of high temporal resolution dialysis methods is only beginning to be realized. Particularly promising is the capability to correlate behaviors, which are typically rated on 5–30 s intervals, with the neurochemical changes that may occur on the same time scale.

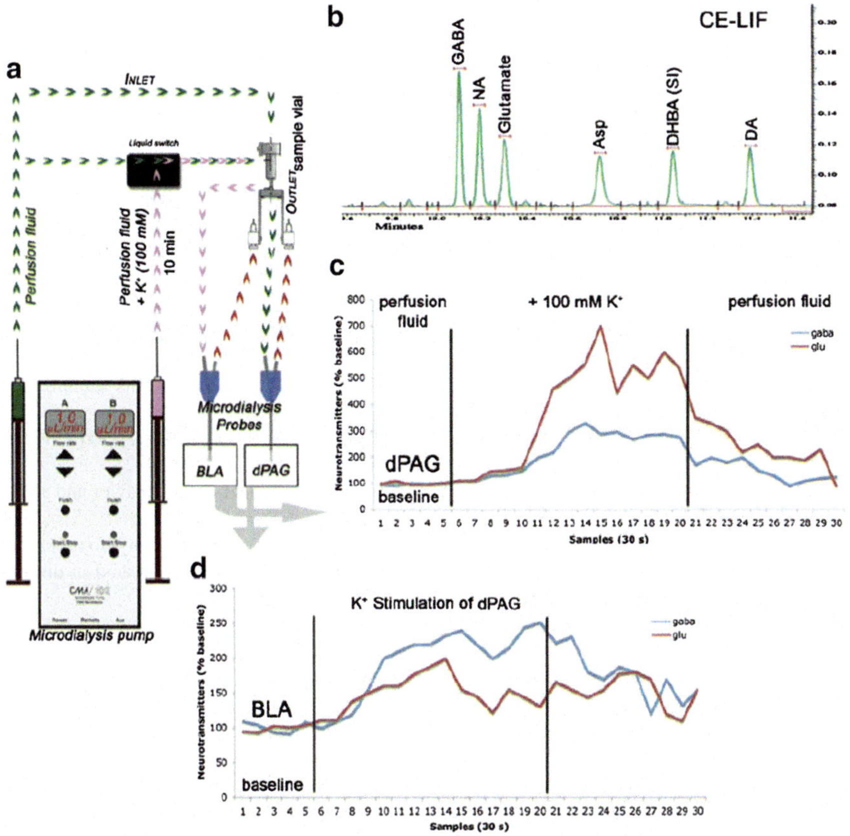

Fig. 14.4 Experimental set-up. (**a**) For dual-probe microdialysis for fast evaluation of glutamate and GABA activity in the basolateral complex of amygdala (BLA) and dorsal periaqueductal gray (dPAG). (**b**) Analyses by capillary electrophoresis with laser-induced fluorescence detection (CE-LIF). Electropherogram of glutamate and GABA (10^{-6} mol/L). (**c**) Potassium stimulation (100 mM) of BLA (6–20 samples) and extracellular glutamate/GABA activity in BLA and dPAG of a freely-moving rat. Baseline (1–5 samples) and post-stimulation (21–30 samples)

References

Abercrombie ED, Keefe KA, DiFrischia DS, Zigmond MJ. Differential effect of stress on in vivo dopamine release in striatum, nucleus accumbens, and medial frontal cortex. J Neurochem. 1989;52:1655–8.

Benveniste H, Christian P, Huttemeier PC. Microdialysis—theory and application. Prog Neurobiol. 1990;35:195–215.

Bert L, Parrot S, Robert F, Desvignes C, Denoroy L, Suaud-Chagny MF, Renaud B. In vivo temporal sequence of rat striatal glutamate, aspartate and dopamine efflux during apomorphine, nomifensine, NMDA and PDC in situ administration. Neuropharmacology. 2002;43:825–35.

Bito L, Davson H, Levin E, Murray M, Snider N. The concentrations of free amino acids and other electrolytes in cerebrospinal fluid, in vivo dialysate of brain, and blood plasma of the dog. J Neurochem. 1966;13:1057–67.
Bourne JA, Fosbraey P. Novel method of monitoring electroencephalography at the site of microdialysis during chemically evoked seizures in a freely moving animal. J Neurosci Methods. 2000;99:85–90.
Cuadra G, Zurita A, Macedo CE, Molina VA, Brandão ML. Electrical stimulation of the midbrain tectum enhances dopamine release in the frontal cortex. Brain Res Bull. 2000;52:413–8.
De Lange EC, Danhof M, de Boer AG, Breimer DD. Methodological considerations of intracerebral microdialysis in pharmacokinetic studies on drug transport across the blood-brain barrier. Brain Res Brain Res Rev. 1997;25:27–49.
De Lange EC, de Bock G, Schinkel AH, de Boer AG, Breimer DD. BBB transport and P-glycoprotein functionality using MDR1A (–/–) and wild-type mice. Total brain versus microdialysis concentration profiles of rhodamine-123. Pharm Res. 1998;15:1657–65.
De Lange EC, de Boer AG, Breimer DD. Methodological issues in microdialysis sampling for pharmacokinetic studies. Adv Drug Deliv Rev. 2000;45:125–48.
Delgado JM, DeFeudis FV, Roth RH, Ryugo DK, Mitruka BM. Dialytrode for long term intracerebral perfusion in awake monkeys. Arch Int Pharmacodyn Ther. 1972;198:9–21.
Di Chiara G. Nucleus accumbens shell and core dopamine: differential role in behavior and addiction. Behav Brain Res. 2002;137:75–114.
Gobaille S, Hechler V, Andriamampandry C, Kemmel V, Maitre M. Gamma-hydroxybutyrate modulates synthesis and extracellular concentration of gamma-aminobutyric acid in discrete rat brain regions in vivo. J Pharmacol Exp Ther. 1999;290:303–9.
Höcht C, Opezzo JA, Taira CA. Applicability of reverse microdialysis in pharmacological and toxicological studies. J Pharmacol Toxicol Methods. 2007;55:3–15.
Justice Jr JB. Quantitative microdialysis of neurotransmitters. J Neurosci Methods. 1993; 48:263–76.
Kehr J. A survey on quantitative microdialysis: theoretical models and practical implications. J Neurosci Methods. 1993;48:251–61.
Kehr J, Yoshitake T, Wang FH, Wynick D, Holmberg K, Lendahl U, et al. Microdialysis in freely moving mice: determination of acetylcholine, serotonin and noradrenaline release in galanin transgenic mice. J Neurosci Methods. 2001;109:71–80.
Kennedy RT, Watson CJ, Haskins WE, Powell DH, Strecker RE. In vivo neurochemical monitoring by microdialysis and capillary separations. Curr Opin Chem Biol. 2002;6:659–65.
Léna I, Parrot S, Deschaux O, Muffat-Joly S, Sauvinet V, Renaud B, et al. Variations in extracellular levels of dopamine, noradrenaline, glutamate, and aspartate across the sleep-wake cycle in the medial prefrontal cortex and nucleus accumbens of freely moving rats. J Neurosci Res. 2005;81:891–9.
Macedo CE, Martinez RCR, De Souza Silva MA, Brandão ML. Increases in extracellular levels of 5-HT and dopamine in the basolateral, but not in the central, nucleus of amygdala induced by aversive stimulation of the inferior colliculus. Eur J Neurosci. 2005;21:1131–8.
Martinez RCR, Oliveira AR, Macedo CE, Molina VA, Brandão ML. Involvement of dopaminergic mechanisms in the nucleus accumbens core and shell subregions in the expression of fear conditioning. Neurosci Lett. 2008;446:112–6.
Nandi P, Lunte SM. Recent trends in microdialysis sampling integrated with conventional and microanalytical systems for monitoring biological events: a review. Anal Chim Acta. 2009; 651:1–14.
Obrenovitch TP, Zilkha E, Urenjak J. Intracerebral microdialysis: electrophysiological evidence of a critical pitfall. J Neurochem. 1995;64:1884–7.
Pan YF, Feng J, Cheng QY, Li FZ. Intracerebral microdialysis technique and its application on brain pharmacokinetic-pharmacodynamic study. Arch Pharm Res. 2007;30:1635–45.
Parrot S, Bert L, Mouly-Badina L, Sauvinet V, Colussi-Mas J, Lambás-Señas L, et al. Microdialysis monitoring of catecholamines and excitatory amino acids in the rat and mouse brain: recent

developments based on capillary electrophoresis with laser-induced fluorescence detection—a mini-review. Cell Mol Neurobiol. 2003;23:793–804.

Parrot S, Sauvinet V, Riban V, Depaulis A, Renaud B, Denoroy L. High temporal resolution for in vivo monitoring of neurotransmitters in awake epileptic rats using brain microdialysis and capillary electrophoresis with laser-induced fluorescence detection. J Neurosci Methods. 2004;140:29–38.

Parsons LH, Justice Jr JB. Quantitative approaches to in vivo brain microdialysis. Crit Rev Neurobiol. 1994;8:189–220.

Ronquist G, Hugosson R, Sjölander U, Ungerstedt U. Treatment of malignant glioma by a new therapeutic principle. Acta Neurochir. 1992;114:8–11.

Thomas PM, Phillips JP, O'Connor WT. Microdialysis of the lateral and medial temporal lobe during temporal lobe epilepsy surgery. Surg Neurol. 2005;63:70–9.

Ungerstedt U. Microdialysis—principles and applications for studies in animals and man. J Intern Med. 1991;230:365–73.

Ungerstedt U, Pycock C. Functional correlates of dopamine neurotransmission. Bull Schweiz Akad Med Wiss. 1974;30:44–55.

Ungerstedt U, Rostami E. Chapter 7.4 microdialysis in the human brain: clinical applications. Handb Behav Neurosci. 2006;16:675–86.

Westerink BHC, Kwint H, DeVries JB. The pharmacology of mesolimbic dopamine neurons: a dual-probe microdialysis study in the ventral tegmental area and nucleus accumbens of the rat brain. J Neurosci. 1996;16:2605–11.

Part IV
Behavioral Models in Rodents

Chapter 15
Sexual Behavior in Rats: An Animal Model for the Study of the Neuroendocrine System

M. Jennifer Rodriguez, Tathiana Aparecida Fernandes Alvarenga, Edith Monroy-López, Armando Ferreira-Nuño, Adriana Morales-Otal, and Javier Velázquez-Moctezuma

Although the influence of gonads on sexual behavior has been empirically known for a number of centuries, it was not until the pioneering studies of Brown-Sequard and Steinach at the beginning of the twentieth century that this relationship was scientifically analyzed. Moreover, the clever experiments conducted by these rescarchers suggested the existence of humoral factors that regulate not only sexual behavior, but also the morphological characteristics that define gender.

A look into the history of behavioral neuroendocrinology reveals that transcendental discoveries regarding brain-hormone interactions have been achieved using various small animal species, including roosters, rabbits, guinea pigs, hamsters and rats, as well as large species, such as pigs, lambs and goats. Currently, the rat is the species that is most often used in behavioral neuroendocrine studies, although other small species are also used, such as hamsters, gerbils, ferrets and mice.

M.J. Rodriguez (✉)
Fisiología y Medicina del Sueño, Anáhuac Cancún,
Mora 65, col Jardínes del sur II, Cancún 77536, Mexico
e-mail: jenniferrodva@gmail.com

T.A.F. Alvarenga
Department of Psychobiology, Universidade Federal de São Paulo (UNIFESP), São Paulo, São Paulo, Brazil

E. Monroy-López
Master in Psychobiology, Universidad Nacional Autonóma, Mexico

A. Ferreira-Nuño • A. Morales-Otal
Departamento de Biología de la Reproducción, Universidad Autónoma Metropolitana, Iztapalapa, Mexico

J. Velázquez-Moctezuma
Departamento de Biologia de la Reproduccion, Universidad Autonoma Metropolitana-Iztapalapa, Mexico City AP 09340, Mexico
e-mail: jvm@xanum.uam.mx

M.L. Andersen, S. Tufik (eds.), *Rodent Model as Tools in Ethical Biomedical Research*, DOI 10.1007/978-3-319-11578-8_15

The use of rats in the study of behavioral neuroendocrinology is due not only to their availability and ease of handling, but also because of their sexual behavior pattern, which is characterized by stereotypical motor patterns that are easily identifiable and readily manipulated by hormones. Both male and female rats have peculiar sexual behavior patterns in which it is relatively easy to analyze the influence of gonadal hormones and their interactions with neurotransmitter systems. In this chapter, we discuss the basics of rat sexual behavior and its use as a tool to understand the neuroendocrine mechanisms underlying this phenomenon.

The rat is a nocturnal mammal, that is, most of their activity occurs during the dark phase of the 24-h cycle. Therefore, rat behavior is commonly recorded in a room with an inverted light-dark cycle. Sexual behavior is generally recorded under a dim red light, which is undetectable to the rat and allows the researcher to clearly see the events. Rats are placed inside a plexiglass cylinder (45 cm diameter and 50 cm high) with sawdust on the floor of the cylinder. Usually, the male rat is first placed into the arena for a 5-min habituation period, and thereafter the female is placed in the arena. Evaluation of male sexual behavior requires the participation of a receptive female, who provides the stimulus. By contrast, assessment of female sexual behavior requires the participation of well-trained and vigorous stud males.

Male Sexual Behavior

Rats are constantly exploring their environment, mainly through the olfactory system and also through the vomeronasal system. Both systems are connected to the nose, but one is specialized in odor perception and the other in the detection of chemical signals known as pheromones. Usually, the presence of a female in the vicinity is detected very efficiently by a male using both systems. Males and females usually approach and explore each other. The male pays particular attention to exploring female genitalia by sniffing and licking. If the female is sexually receptive, and often even if she is not, the male will try to mate with her. Males approach females from the rear, use their front limbs to flank her and perform fast pelvic thrusting during which they tap the perineal region of the female. These movements lead to three different behavioral patterns known as mount, intromission and ejaculation, which are the main components of male sexual behavior in rats.

Mount. As was mentioned above, males perform pelvic thrusts during which they tap the perineal region of the female. It has been suggested that these movements induce the female sexual response and allow the male to locate the vagina with his penis. Therefore, in the absence of an erection, pelvic movements become slower until they stop completely and, at the same time, the male will release the female from their front limbs, with which they hold the flanks of the female, and go back to exploring the environment. The whole behavioral mount pattern lasts for about 3 s. Figure 15.1 shows some photographs depicting typical postures during a mount.

Intromission. This behavioral pattern is quite similar to the mount. The male pursues the female, takes her by the flanks and performs pelvic thrusts. The difference here is that the erect penis protrudes and, after a few pelvic movements,

Fig. 15.1 Typical postures displayed by the males during a mount

the penis penetrates the vagina with a deep pelvic thrust, which is immediately followed by a quick withdrawal of the male, who takes two or three steps backing away from the female. Then, the male starts to intensely explore his own penis by sniffing and licking. It has been suggested that these actions allow the male to become aware of the hormonal status of the female and provide him additional stimulation. The intromission pattern is faster than the mount pattern and lasts for about 1 s. Male rats usually restart sexual activity seconds later, usually within a minute. Figure 15.2 shows some photographs depicting typical postures during an intromission.

Ejaculation. Male rats repeat the pattern of mount and/or intromission and, after approximately ten intromissions, the rat ejaculates. The ejaculation pattern is similar to those of mount and intromission, but instead of the rapid withdrawal observed during intromission, the male performs a few intra-vaginal pelvic movements after he achieves vaginal penetration, followed by a deep pelvic thrust and a marked contraction of his entire body, during which he pulls the female towards him with his front limbs. The male rat subsequently releases the female by opening his front

Fig. 15.2 Typical postures displayed by the males during an intromission

limbs. The male then stands up and remains quite still for about 2 s before slowly withdrawing from the female. He then goes to lie down separately. The entire ejaculatory pattern lasts for about 4 s. Figure 15.3 shows some photographs depicting typical postures during an ejaculation.

A trained male can usually reach the first ejaculation about 7 min after a receptive female is placed in the arena. As has been mentioned, after an ejaculation, the male remains lying down and far away from the female; it is common for him to sleep during this time. After approximately 6 min, the male recovers his interest in the female and again begins to actively seek her and re-initiate copulatory activity with mounts and intromissions.

Figure 15.4 shows a graphic representation of male rat copulatory activity. To assess copulatory activity, several parameters can be used. Mount and intromission latencies are defined as the time that elapses from the introduction of the female until the presentation of the first mount or intromission. When the first event is an intromission, the mount latency is equal to the intromission latency. Ejaculatory latency

Fig. 15.3 Typical postures displayed by males during an ejaculation

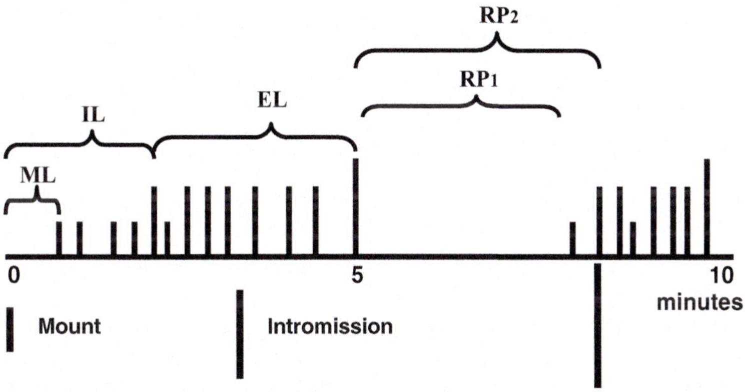

Fig. 15.4 Typical male copulatory activity. *ML* mount latency, *IL* intromission latency, *EL* ejaculatory latency, *RP* post-ejaculatory refractory period

is defined as the time that elapses from the first intromission to the ejaculation. The number of mounts and number of intromissions that precede an ejaculation are also physiologically important parameters. In addition, some physiologically relevant indices can be obtained. The Inter-Intromission Interval (III) is the mean time that elapses between intromissions. The Inter-Copulatory Interval (ICI) is the mean time that elapses between a mount or intromission. The Hit Rate (HR) is obtained by dividing the number of intromissions by the sum of intromissions and mounts.

Furthermore, after an ejaculation, the Refractory Post-Ejaculatory Period is the time that elapses between the ejaculation and the first event of the new copulatory

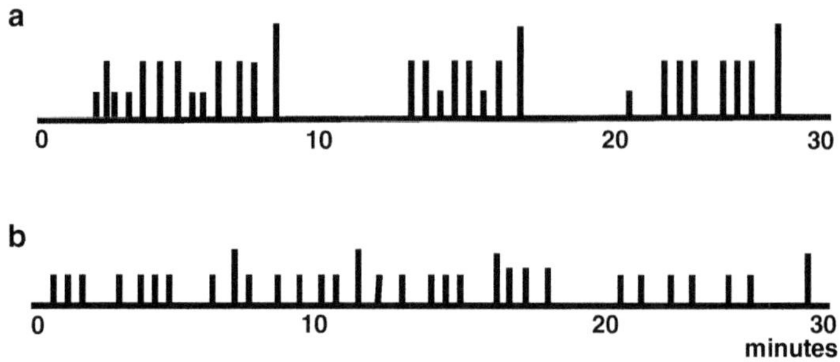

Fig. 15.5 Graphical examples of masculine sexual performance. (**a**) Shows a subject that displays three ejaculations in 30 min. (**b**) Shows a subject with erectile dysfunction. Note the excess of mounts, with few intromissions and no ejaculation

series. If this first event is a mount, it is called Refractory Period 1 (RP1) and the first intromission of the new series is called Refractory Period 2 (RP2). When the first event is an intromission, RP1 = RP2.

Male sexual behavior can be assessed by observing a single ejaculatory series, regardless of its duration. Alternatively, sexual behavior can be observed over a 30-min period by recording all events that take place during that time. In this second case, if the male rat does not display any activity for 15 min, the test ends. In addition, sexual exhaustion can be assessed by allowing the male to freely copulate for several hours. Two behavioral criteria are applied to determine sexual exhaustion: the male rat spends more than 30 min without displaying any sexual activity and the male rat spends more than 90 min without an ejaculation.

Figure 15.5 shows some examples of related pathophysiological conditions. Panel A shows the graph of a 30-min test in which a male rat with erectile dysfunction displays an excessive number of mounts with very few intromissions and no ejaculation. It must be pointed out that the motivational component is intact, which is why the male makes repeated attempts to copulate. Panel B shows an example of a male rat that has been stimulated pharmacologically or by means of direct electrical stimulation in specific brain areas. Note that there are almost no mounts, and only a few intromissions are needed to achieve an ejaculation. Intromission and ejaculatory latencies are shortened and the ejaculatory frequency is increased compared to intact males.

Female Sexual Behavior

Sexual behavior in females largely depends on fluctuations of their hormonal condition. The estrous cycle of the female rat has been classified in four stages: diestrus, metaestrous, proestrous and estrous. The full picture of sexual behavior is displayed mainly during proestrous, which lasts approximately 18 h. The two main components that have been identified for female sexual behavior are proceptivity and receptivity.

Proceptivity includes the display of peculiar behavioral motor patterns known as hopping, darting and ear wiggling. During this period, the female actively seeks sexual intercourse, seeking out the male and calling his attention by constantly approaching him, jumping on him or even biting him. When the male approaches the female, she "freezes" while she waits for minimum contact with the male, after which she displays hopping (fast and repetitive semicircular movements) or darting (fast movements away from the male that are usually no longer than 10 cm) and ear wiggling. It has been suggested that these movements are displayed to inform the male that the female is available to mate.

Receptivity has to do with the lordosis reflex. As was mentioned above, the male mounts the female from the rear and performs pelvic thrusts. In the receptive female, these movements induce the lordosis response in which the female adopts a posture that facilitates penetration of the penis into the vagina. The female arches her back and elevates her head and hind quarter. The most common parameter of female sexual behavior is the Lordosis Quotient, which is obtained by dividing the number of mounts received by the female by the number of displayed lordosis reflexes. Figure 15.6 shows some examples of the typical postures adopted by a receptive female.

It must be pointed out that males occasionally display the female sexual behavior pattern and that females can display the male sexual behavior pattern. The term heterotypical sexual behavior refers to a situation in which a male is mounted by another male and displays the lordosis response; the same is true for a female that displays mounting and pelvic thrusting. These behavioral patterns are rarely spontaneous, but are frequently observed when the normal process of sexual differentiation is altered by hormonal manipulation, mainly in the early stages of development.

Sexual Differentiation

The process by which a chromosomic XX subject becomes an adult female or an XY subject becomes a male is known as sexual differentiation. The phenomenon of sexual differentiation is a process that involves a cascade of events. These events start with the genetic determination of gender at conception, continue during the embryonic stage and last throughout neonatal life and up to puberty. During the early stages of development, steroid hormones play a central role in these processes by exerting what are called organizational actions. When the subject reaches puberty, steroid hormones are active again, but now exert what are called activational effects. When the organizational effects of steroid hormones have been altered and show deficiencies, then the activational effects of steroids cannot be properly expressed and consequently, alterations in the expression of sexual behavior may occur.

At the chromosomic level, subjects are differentiated into XX or XY, although chromosomic aberrations do exist. The first critical period is the differentiation of

Fig. 15.6 Typical postures displayed by a receptive female

gonads into either ovaries or testicles. If the process of gonadal sexual differentiation develops adequately, then the gonads will begin to produce specific steroid hormones, mainly testosterone for males and estrogens and progesterone for females. Therefore, the next critical period for sexual differentiation involves the release of gonadal steroids from the fetal gonads, which act directly on the brain.

The actions of steroid hormones on brain tissue at this developmental stage are critical for the differentiation of the brain into a female or a male brain. The critical period during which fetal brain tissue is sensitive to steroid hormones and can be permanently altered begins in the last trimester of pregnancy and lasts for about 8 days after delivery.

It has been suggested that the brain is female by default. Thus, the presence of testosterone is essential for the development of a male brain. The presence of testosterone requires the existence of a well differentiated gonad, the testicle, and information that induces gonadal differentiation, which is contained in the genes of the short arm of the Y chromosome. These genes encode for a group of proteins that are required for the development of a differentiated testicle in which Leydig cells synthesize and release testosterone.

The metabolism of steroid hormones is complex and most of the intermediary products also have intense effects on their target sites. Testosterone is metabolized into dihydrotestosterone (DHT) by the action of 5 alpha-reductase and to estradiol by the action of an aromatase. It has been reported that the action of testosterone requires these conversions. When a castrated male loses sexual behavior, it can be fully restored by the administration of DHT and estradiol. It has been suggested that DHT is the metabolite responsible for the peripheral effects of testosterone, while estradiol is responsible for the central effects of testosterone.

Brain differentiation into a male brain requires the presence of testosterone and its metabolic products. Thus, both the proper levels of testosterone and the proper action of its metabolizing enzymes are essential for the masculinization of the brain. Hormonal manipulation of the different steps in this process results in dramatic alterations of sexual behavior, which are a reflection of morphological alterations of the brain. Administration of aromatization blockers during the perinatal period results in a dramatic change in sexual behavior of the male offspring during adulthood, including loss of male sexual behavior and an increase in the display of female sexual behavior.

Cerebral differences between a male and a female include the size of some nuclei, neuronal and dendritic density, enzymatic activity, neurotransmitter activity, and expression of specific proteins and mRNAs. Thus, the behavioral expression of masculinity or femininity during development and especially during adulthood will depend on the proper sexual differentiation of each and every one of these components. Deficiencies in the sexual differentiation process will result in a decrease in the expression of behaviors that are characteristic of the gender of the subject, and an increase in the expression of behavioral patterns characteristic of the opposite sex.

Studies that have been performed mainly in rats have indicated that the hypothalamus and adjacent structures show sexual dimorphism. Receptors for androgens (AR), estrogens (ER) and progesterone (PR), which are involved in the regulation of autonomic functions, reproduction, sexual behavior, aggression and learning, display a typical pattern linked to the gender of the subject.

In addition, high levels of aromatase activity have been detected in the preoptic area of the hypothalamus and in the sexual dimorphic nucleus of the preoptic area

(SDN-MPOA). Aromatase activity is also high in other sexual dimorphic nuclei located in the limbic area, which have afferent and efferent connections to the SDN-MPOA. These nuclei include the nucleus of the cortical and medium amygdala, the bed nucleus of the stria terminalis and the ventromedial nucleus of the hypothalamus. Furthermore, after sexual activity, these nuclei show an increase in immunoreactivity to C-fos, which is a marker of neuronal activity.

A few decades ago, Roger Gorski and his colleagues published a series of studies in which they focused their attention on the SDN-MPOA. In these experiments, they showed that this area is seven times larger in male rats. In addition, this group reported that hormonal manipulations are capable of inducing morphological changes in this area, but only during a critical period that extends from the last days of pregnancy to the first 5 days after birth. Aside from the morphological changes in the brain, hormonal treatment during the critical period permanently alters sexual behavior in these subjects.

These pioneering animal studies have encouraged the search for sexual differences in human brains. The biological bases of human sexual orientation have been a matter of controversy for decades. Since the 1990s, several groups have reported morphological evidence of altered brain sexual differentiation in bi-sexual and trans-sexual subjects. Postmortem studies of these subjects' brains have shown deficiencies in sexual differentiation in the sexual dimorphic nucleus of the preoptic area of the hypothalamus (areas known as INAH-1, INAH2 and INAH3); the suprachiasmatic nucleus (SCN) and the bed nucleus of the dorsal and central stria terminalis. These studies have allowed us to establish that sexual orientation in human beings has a biological substratum.

Neuronal Control of Sexual Behavior

In the analysis of the regulatory mechanisms of sexual behavior, several approaches have successfully revealed the complex network that underlies the components of this behavior. Experiments using electrical or pharmacological stimulation, selective neurotoxic or electrolytic lesions and, recently, markers of neuronal activity, have pointed to an undoubtedly critical role of several brain nuclei as determinants of both male and female sexual activity. Although the brain as a whole is involved in this complex behavioral pattern, some structures have shown a more intense participation and will be summarized below.

Male Sexual Behavior

Aside from the assessment of male copulatory parameters, the study of male sexual behavior also implies the motivational component. Researchers working on the participation of cerebral structures in the regulation of male sexual behavior must

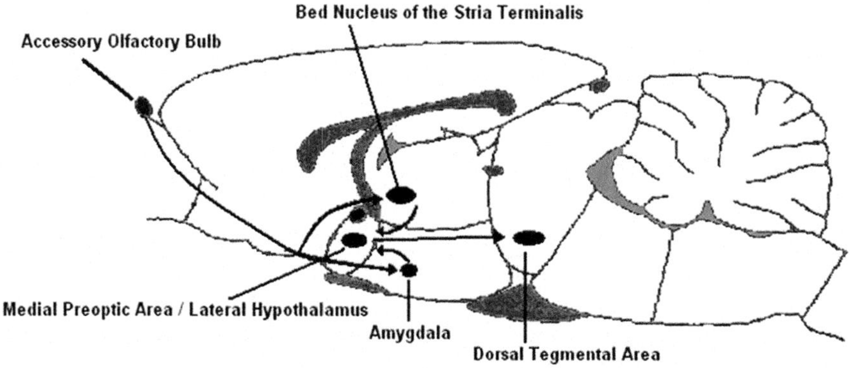

Fig. 15.7 Schematic representation of neural control in male sexual behavior

clearly differentiate the motivational and execution component. Figure 15.7 shows a schematic representation of brain structures involved in the regulation of male sexual behavior.

Olfactory Bulbs and Vomeronasal Organ

The vomeronasal organ (VNO) is a chemoreceptor involved in male sexual behavior, which processes sexually relevant cues. Together with the olfactory bulbs, the VNO has an important role in the neural regulation of copulation. The VNO projects to the accessory olfactory bulb (AOB) via the vomeronasal nerve, which travels to the main olfactory bulb on its way to the accessory olfactory bulb. Removal of the VNO results in an increase in the intromission latency and a decrease in the hit rate. The AOB projects to the medial amygdala (MEA). Specifically, the corticomedial amygdala travels via the stria terminalis, innervating the bed nucleus of the stria terminalis (BNST) as well as the medial preoptic area (MPOA). Removal of the AOB in the male rat reduces the percentage of males that reach ejaculation during copulation.

Bed Nucleus of the Stria Terminalis

In male rats, chemosensory investigation of a female induces neural activation in subregions within the posteromedial bed nucleus of the stria terminalis (BNSTpm).

Amygdala

The medial amygdala may mediate different aspects of male sexual behavior. Lesion of the medial amygdala in the male rat, hamster or gerbil, increases mount and ejaculation latencies and increases the number of intromissions preceding ejaculation. Also, sub-regions within the medial amygdala may regulate different components of sexual behavior. In hamsters, lesion of the anterior medial amygdala results in a greater impairment of sexual behavior, compared to lesion of the caudal medial amygdala. In the rat, lesions of the dorsolateral part of the medial amygdala impair ejaculation.

Ejaculation is under descending inhibitory influence and excitatory influence of supraspinal sites. These sites include the medial preoptic area (MPOA), the paraventricular nucleus of the hypothalamus (PVN) and the nucleus paragigantocellularis (nPGi). These sites form a network and orchestrate the expression of sexual behavior. The nPGi appears to mediate a powerful inhibitory influence over the ejaculatory reflex, projecting serotonergic neurons to the lumbosacral spinal cord. Serotonin appears to mediate a tonic inhibition of the spinal ejaculation generator, since application of serotonin to the spinal cord suppresses spinal ejaculatory reflexes. In contrast, the PVN and MPOA appear to exert an excitatory influence on the spinal ejaculation generator. PVN neurons that project to the lumbosacral spinal cord, release mainly oxytocin, and its levels in the circulation increase following ejaculation in humans, rabbits and rats. In addition, administration of oxytocin (systemic or intracerebroventricular) facilitates copulation by shortening the ejaculation latency, the postejaculatory interval and by reducing the number of intromissions before ejaculation. The PVN however, is not essential for ejaculatory or erectile function, as its lesion does not abolish these responses. It may play a modulatory role in seminal emission and stimulation of the PVN can elicit penile erection and ejaculation.

Medial Preoptic Area (MPOA)

The MPOA is an important site for male sexual behavior. Its stimulation elicits erection and contraction of the pelvic striated muscles. The dopaminergic neurons of the MPOA appear to facilitate ejaculatory reflexes via D2 receptors. The MPOA is a critical structure for male sexual behavior. Extensive lesion of this area abolishes all aspects of sexual behavior in many vertebrates. Cell bodies of neurons located within the MPOA are essential for the display of male sexual behavior, which can be restored by administration of testosterone to this area, but not when it is administered to adjacent brain sites. Large lesions of the MPOA can result in a lack of sexual motivation, a reduced or absent ability to execute copulatory motor patterns and a lack of erection. Unconditioned sexual motivation disappears after preoptic lesions.

Lateral Hypothalamus

Lesions of the lateral hypothalamus severely affect the display of ejaculation, but do not affect the ability to display mount and intromission behaviors. Serotonin in the anterior lateral hypothalamus (ALH) may influence sexual motivation as well as ejaculatory behavior. Administration of a selective serotonin reuptake inhibitor in the ALH increases latencies to mount, intromission and ejaculation.

Female Sexual Behavior

Figure 15.8 shows the schematic representation of the neuroanatomy of female sexual behavior.

Medial Preoptic Area (MPOA)

The MPOA tonically inhibits female sexual behavior, as its destruction facilitates lordosis behavior. This site exerts tonic inhibition on pathways that mediate estrous behavior. In hormone primed ovariectomized females, MPOA lesioned females display a higher lordosis quotient than control MPOA lesioned females receiving vehicle, suggesting that local preoptic neurons inhibit the lordosis reflex. Moreover, males lesioned in the MPOA display facilitation of lordosis, suggesting that this site inhibits the occurrence of lordosis in the male.

Nucleus Accumbens

There is no clear evidence that the nucleus accumbens is involved in the control of female sexual behavior. However, its lesion increases the number of rejection responses to male mount attempts without modifying lordosis in the rat. The increase in rejection was associated with an increase in general hyper-reactivity that reduced the female's tolerance to male intromissions.

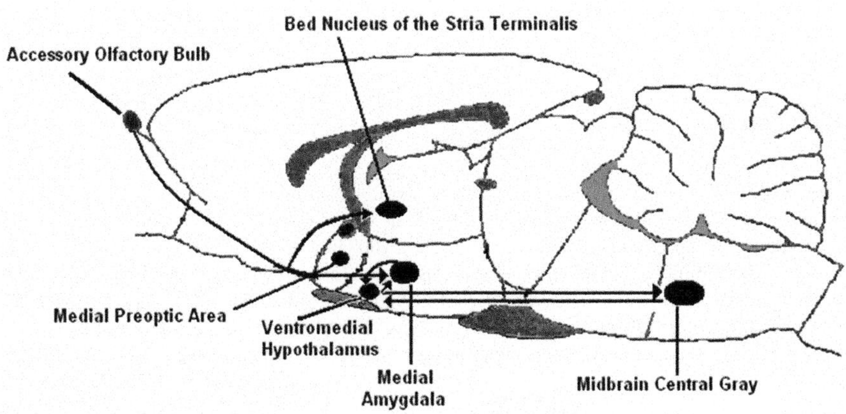

Fig. 15.8 Neuroanatomy involved in the regulation of female sexual behavior

Anterior Hypothalamus

It has been demonstrated that selective lesions of neural cell bodies in the anterior hypothalamus disrupt proceptivity without affecting sexual receptivity; however, in another study, electrolytic lesions in the anterior hypothalamus completely eliminated lordosis.

Ventromedial Hypothalamus

The ventromedial hypothalamus plays a critical role in the control of the lordosis reflex, as its destruction consistently inhibits the lordosis response in many female mammals. The ventromedial nucleus is also involved in both, receptivity and proceptivity.

Concluding Remarks

As can be concluded from the above discussion, both male and female sexual behavior is the result of complex interactions between neurotransmitters, enzymes, hormones, and other proteins, which must interact in a perfectly harmonious sequence to express desire and the ability to mate. The motivational component and execution of one subject must occur in the presence of a partner of the opposite gender that exhibits the same level of motivation and is capable of mating.

The most basic determinants of sexual behavior exist at the genetic level. These determinants also exist at specific developmental stages and in adulthood, with the display of a behavior that allows for preservation of the species. A number of manipulations can alter the proper development of these capabilities. Thus, sexual behavior during adulthood can serve as a reliable and powerful tool for analysis of the neuroendocrine system.

References

Arnold AP. Sex chromosomes and brain gender. Nat Rev Neurosci. 2005;5:1–8.

Blaustein JD. Neuroendocrine regulation of feminine sexual behavior: lessons from rodent model and thoughts about humans. Annu Rev Psychol. 2008a;59:93–118.

Blaustein JD. Feminine reproductive behavior and physiology in rodents: integration of hormonal, behavioral and environmental influences. In: Pfaff DW, Arnold AP, Etgen AM, Fahrbach SE, Rubin RT, editors. Hormones, brain and behavior. 2nd ed. San Diego: Elsevier; 2008b.

Brown RE. An introduction to neuroendocrinology. New York: Cambridge University Press; 1994.

Goldstein I. Male sexual circuitry. Working group for the Study of Central Mechanisms in Erectile Dysfunction. Sci Am. 2000;283(2):70–5.

Hull EM, Rodriguez-Manzo G. Male sexual behavior. In: Pfaff D et al., editors. Hormones, brain and behavior. 2nd ed. San Diego: Elsevier; 2009. p. 5–65.

Hull EM, Dominguez JM. Sexual behavior in male rodents. Horm Behav. 2007a;52:45–55.
Hull EM, Wood RI, McKenna KE. The neurobiology of male sexual behavior. In: Neill J, Pfaff D, editors. The physiology of reproduction. 3rd ed. New York: Elsevier; 2006. p. 1729–824.
Hull EM, Dominguez JM. Sex behavior. In: Nelson RJ, Gallagher M, editors. Comprehensive handbook of psychology. Biological psychology. New York: Wiley; 2003. p. 321–53.
Hull EM, Dominguez JM. Sexual behavior in male rodents. Horm Behav. 2007b;52(1):45–55.
Knobil E, Neill JD, editors. The physiology of reproduction. New York: Raven; 1994.
Le Vay S. The sexual brain. Cambridge: MIT Press; 1993.
Meston CM, Frohlich PF. The neurobiology of sexual function. Arch Gen Psychiatry. 2000;57:1012–30.
Morris JA, Jordan C, Breedlove M. Sexual differentiation of the vertebrate nervous system. Nat Neurosci. 2005;7:1034–9.

Chapter 16
Maternal Behavior

Elizabeth Teodorov, Luciano Freitas Felicio, and Maria Martha Bernardi

According to the current literature, maternal behavior fits into the category of parental behavior, which is defined as any behavior of a member of a species toward a reproductively immature member to ensure that this member will survive to maturity. Research on maternal behavior and physiology strives to catalog all physiological and behavioral changes that occur in the maternal organism, and to understand the function of each change with respect to infant care. The underlying mechanisms of maternal behavior are sensory, hormonal, neural and genetic.

Several books have been written on the sociobiology and evolution of maternal behavior. These behaviors were first investigated in rodents by Sturman-Hulbe (1929), who went on to publish work concerninwg maternal behavior in albino rats. Another example of classic work in this field is that of Weisner and Sheard (1933), who showed that the onset of maternal responsiveness is synchronized with the birth of the young; at parturition, the mother was found to care for her own or alien foster young, whereas normal females and pregnant nulliparous females did not show maternal responses toward foster young. As a result, Weisner and Sheard (1933) suggested that "some internal changes occur in the highly pregnant or parturient organisms which awaken the maternal drives." Another important contribution of

E. Teodorov, Ph.D. (✉)
Universidade Federal do ABC—Centro de Matemática, Computação e Cognição,
Av. dos Estados, 5001, Bloco A, Torre 2, Santo André, São Paulo 03210-000, Brazil
e-mail: Elizabeth.teodorov@ufabc.edu.br

L.F. Felicio, D.V.M., Ph.D.
Department of Pathology, School of Veterinary Medicine, University of São Paulo (USP),
Av. Orlando Marques de Paiva, 87, São Paulo, São Paulo 05508-270, Brazil
e-mail: ifelicio@usp.br

M.M. Bernardi, Ph.D.
Instituto de Ciências da Saúde, Universidade Paulista,
Rua Dr. Bacelar, 1212-Vila Clementino, São Paulo, São Paulo 04026-002, Brazil
e-mail: marthabernardi@gmail.com

M.L. Andersen, S. Tufik (eds.), *Rodent Model as Tools in Ethical Biomedical Research*, DOI 10.1007/978-3-319-11578-8_16

this work was the observation that the physiological state associated with pregnancy and parturition was not necessary for the occurrence of maternal behavior. Lebolnd and Nelson (1937) and Leblond (1938, 1940) showed that young pups could induce maternal behavior in hypophysectomized females, and thus suggested that "a nervous mechanism, which may be stimulated without harmonic influences, seems to be a essential factor of maternal instinct in mice and rats. It may be controlled by hormonal factors in normal animals after parturition."

Research conducted on rats by Rosenblatt and his colleagues (1979) suggested that although the onset of maternal behavior at parturition is hormonally mediated, its maintenance during the postpartum period has a non-hormonal basis. Studies carried out by Lebolnd and Nelson (1937) with hypophysectomized lactating mice (in this technique, the pituitary gland is removed) revealed that although lactation ceased, maternal behavior continued normally. Finally, Beach and Jaynes (1956) indicated that maternal behavior in the rat was under multisensory control; the maternal female reacted to many stimuli from her pups as observed by the act of pup retrieval.

In studies on the neural basis of maternal behavior, Beach (1937) and Stone (1938) investigated the role of the neocortex, and found that large neocortical lesions eliminated maternal responsiveness. Subsequent studies in the early 1970s on the neural basis of maternal behavior revealed that lesions, hormonal implants, or chemical changes produced in specific areas of the brain could provide information about neuronal function in maternal behavior. Thus, if a particular area of the brain is critical for maternal behavior, then it is important to determine the structures to which that area projects. The following chapter discusses some aspects of maternal behavior in rats and how they are modulated and controlled.

The Onset of Maternal Behavior

Non-human mammalian mothers typically display a stereotypical set of behavioral responses to their newborns. They exhibit species-specific methods of transporting, holding, feeding, and grooming their young, and of protecting them from predators and other dangers. These behaviors promote physiological and immunological resilience, physical maturation, and species-typical social and emotional development of the young. The expression of pup-induced maternal behavior has been studied extensively; it is virtually indistinguishable from hormonally-induced maternal behavior, except that parturient females nurse their pups.

Here we present a brief description of the methods used for studying maternal behavior in laboratory rodents. During the first 3 days following parturition, the mother and the litter should not be manipulated. Afterwards, the bedding can be changed. Basic care aspects that apply to any laboratory animal are particularly important for mother rodents. Thus, food and water should be checked daily. Animals should be handled with gloves and the observer should not use any type of fragrance, since odors can be stressor stimuli. Usually, rodents have a well-developed auditory system that is relevant for both social interactions and reproduction in general. Aversive sounds can disturb both mothers and their litters; thus, noise in the animal room should be avoided.

Fig. 16.1 The nest is destroyed before maternal behavior analysis

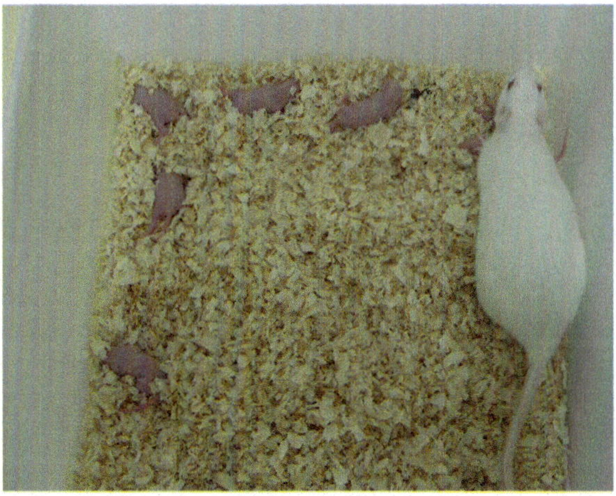

Fig. 16.2 The eight pups are placed back into the cage for maternal behavior analysis

The morning of the testing day, pups and food pellets are removed from the home cage and the nests are destroyed (Fig. 16.1). Sixty minutes later, eight pups are placed back into the cage, and maternal behavior testing begins (Fig. 16.2). Latencies for pup contact, retrieval, grouping, nest building, nursing and full maternal behavior (FMB) responses are assessed in seconds (Fig. 16.3). Animals are scored as fully maternal if they retrieve all eight pups to the nest and nurse them for three consecutive minutes. If animals are not fully maternal after 30 min of continuous observation, they are spot-checked every 15 min until FMB is recorded. Events observed after the first 30 min of continuous observation are recorded at the time they are first observed. For example, if the female is first seen to nurse the young at 60 min, the nursing latency is scored as 60 min.

Most rodent young are helpless, essentially immobile, and incapable of temperature regulation at birth; thus, the young are kept in a nest that the mother builds prior

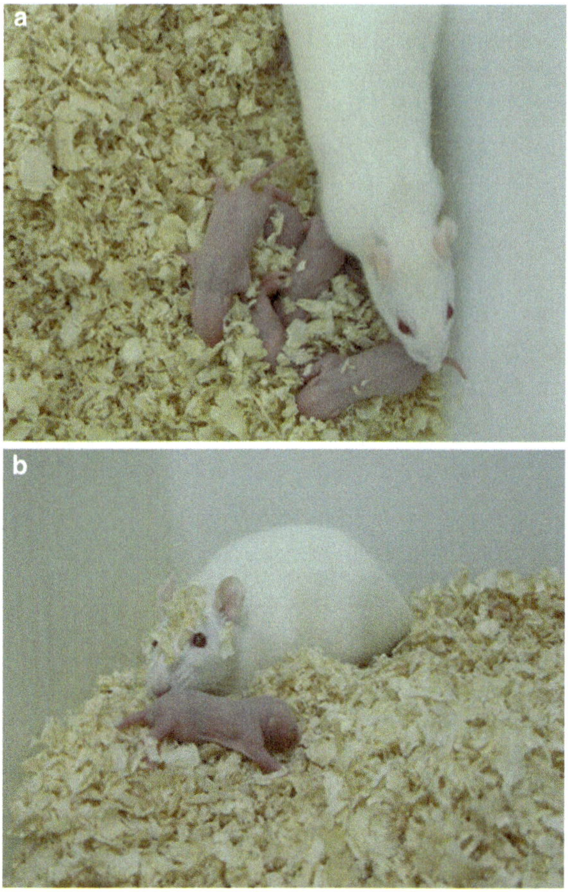

Fig. 16.3 Maternal behavior parameters for pup contact (**a**), retrieval (**b**), grouping (**c**), nursing (**d**) and full maternal behavior (**e**)

to parturition in order to insulate the young during the mother's absence (for example, when she is foraging for food) and to offer protection against intruders. When intruders approach the nest, the mother reacts with maternal aggression.

Maternal stimulation, such as handling and anogenital licking, is directed more to males than females of a litter (Fig. 16.4); it has been suggested that differences in

> This behavior is important for the prevention of infanticide by a non-familiar male and is modulated by the suckling stimulus of the pups during lactation.

maternal stimulation may contribute to the development of differences in adult male sexual behavior. While in the nest, the mother warms the young and exposes her mammary region through nursing behavior. The mother also licks her pups in order

Fig. 16.3 (continued)

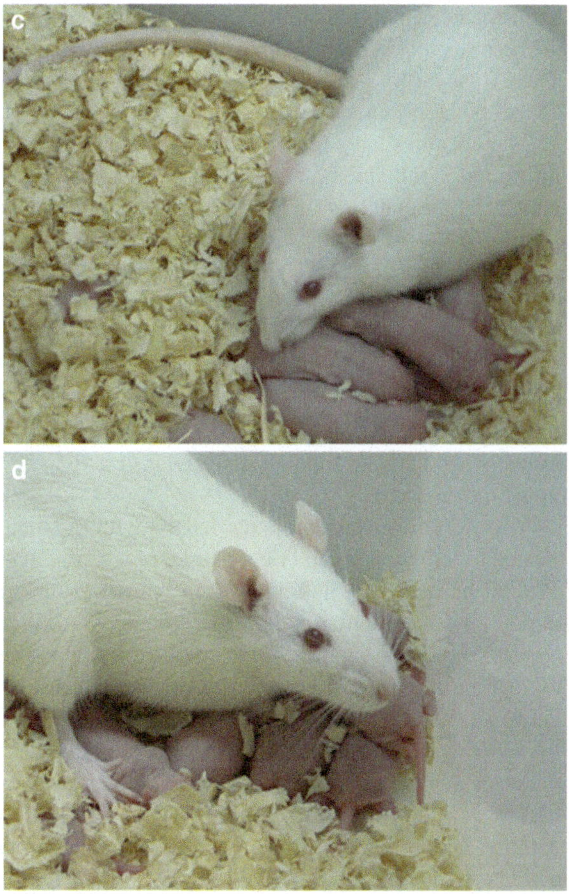

Fig. 16.4 The dam is licking her pups

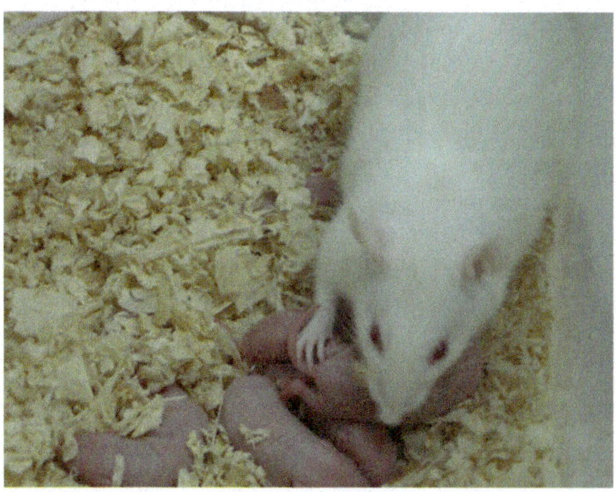

to stimulate urination and defecation. If one of the young becomes displaced from the nest, the rodent mother carries it in her mouth back to the nest (retrieval behavior). Thus, transport of young, protection of young, nursing, and nest-building behavior can all be easily understood as important for infant survival.

Many studies have focused on the role of auditory and olfactory stimulation of the young in influencing maternal responses. In rats, the ultrasonic vocalizations of pups have the function of maintaining high maternal responsiveness. These vocalizations occur when the pups are out of the nest or when their body temperature declines to critical levels. These ultrasonic vocalizations result in maternal attention and are an efficient strategy for maintaining important maternal responses (especially licking of the anogenital region). It has been found that lactating rats are more likely to leave their nests and investigate the environment in the direction of recorded pup ultrasonic vocalizations than to leave their nests in response to other sounds. It was determined that whereas both ultrasonic sounds and olfactory stimuli arouse mothers and activate searching behavior, it is the auditory cues that provide directional information.

There are other behaviors associated with the maternal state that are indirectly related to infant survival. For example, pregnant female rats change their grooming pattern via increased licking of their mammary region; this has been found to be important for mammary gland development. Additionally, lactating females are hyperphagic, show an increased appetite for salt, and engage in placentophagia. The mother has a limited physical capacity for nurturing her pups. Thus, once the number of pups exceeds the physical limit of the mother, she reduces the size of her litter by rejecting the weak or sick via selective infanticide. Infanticide is also related to maternal inexperience, food shortage during pregnancy, and prenatal stress.

The development of the capacity to both express and modify maternal behavior patterns in adulthood may depend on mechanisms that were themselves activated and later tuned by early experiences. The interaction between the newborn and the mother bilaterally alters basic mechanisms of behavioral expression. The response of the mother determines, in part, the neurobiological and behavioral changes in the infant. Changes in the infant may then eventually be expressed in adulthood by the individual response to offspring. It has been shown that adult virgin female rats display maternal behavior when they are exposed to foster young for five to six consecutive days. These nulliparous rats, like lactating dams, will retrieve foster pups to the nest, group the pups together and crouch over them to provide them with a source of warmth and protection.

The degree of pup development at birth influences various patterns of maternal care among mammals in crucial ways (Fig. 16.5). Initially, the mother is in constant contact with her infant, who clings to the mother for transport. As the young advance in age, their increasing independence coincides with decreased maternal responsiveness; both of these processes allow weaning to occur.

Non-behavioral changes in maternal physiology also greatly affect the success and nature of maternal behavior. The most common of these include the endocrinological changes that underlie lactation. The capacity to carry out adequate

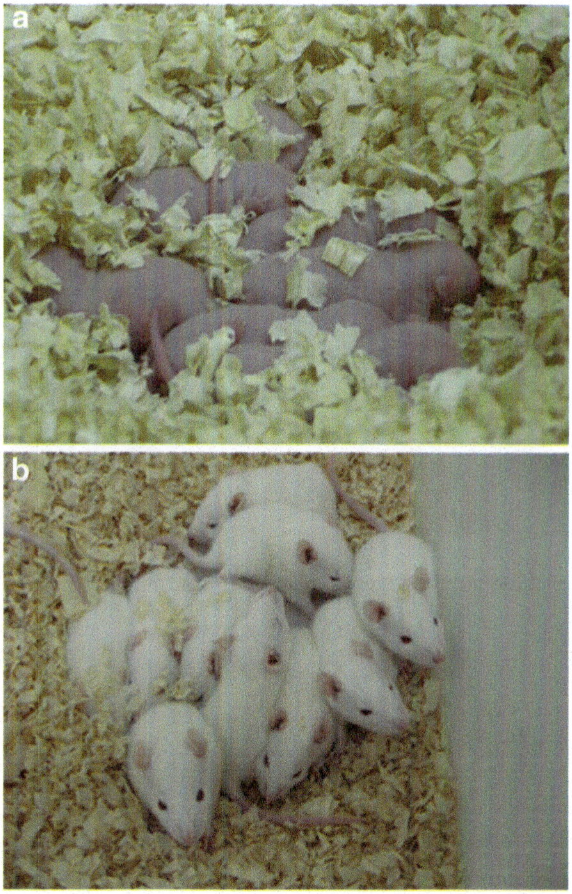

Fig. 16.5 Littermates at 6 days (**a**) and 21 days (**b**) after parturition

maternal behavior is not only conditioned by genetic or purely physiological factors associated with parturition, but also depends on the individual experiences of the subject.

Neurobiological Mechanisms of Maternal Behavior Activation Include Neuroanatomy, Neurotransmission and Hormones

Neuroanatomy

Neurobiological mechanisms are also reported to activate and establish maternal behaviors in mammals. Key components of the neural circuits that mediate maternal behavior include the medial preoptic area (MPOA), the medial (MA) and cortical amygdale (CA), the nucleus accumbens (NAC), and the paraventricular nucleus

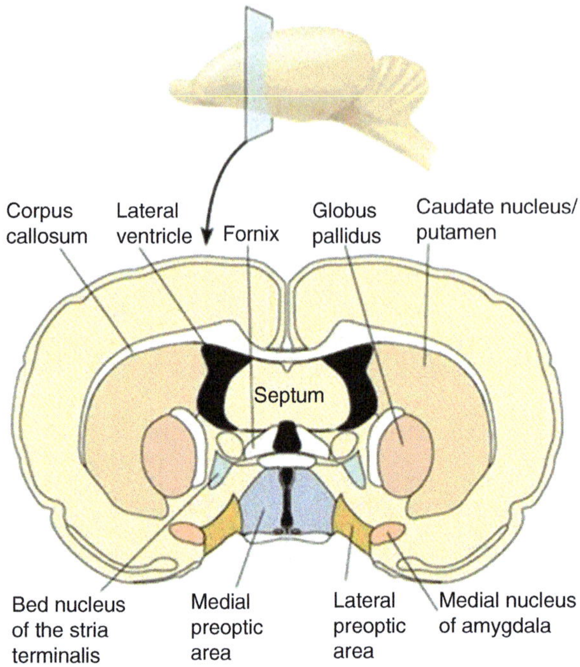

Fig. 16.6 Areas involved in the modulation of maternal behavior (reproduction authorized by Dr. Annabel Ferreira, Montevideo University)

(PVN) in addition to other limbic and hypothalamic structures, as depicted in Fig. 16.6. Most of the understanding of the neural circuits that mediate maternal behavior has been gathered from studies using adult animals. Recent studies have also shown an important role of the lateral periaqueductal gray (PAG) in the modulation of maternal behavior in lactating rats.

The MPOA is necessary for both the expression of and the motivation to perform maternal behavior. Lesions of the MPOA severely and selectively impair behaviors such as the retrieval of pups and nest building, and also reduce the motivation of the dams to crouch over the pups. In adults, both large and small lesions of the MPOA disrupt maternal behavior, while in juvenile or pubertal rats, only large lesions disrupt this behavior. The MPOA in particular is the primary area for estrogen action. In non-maternal females, the MA inhibits the MPOA under the influence of the olfactory system. This effect should be counteracted or even reversed at parturition in order to allow the care to the young. However, unlike the MPOA, which stimulates maternal behavior, areas of the hypothalamus inhibit this display. Lesions of the anterior hypothalamus (AH) as well as of the dorsomedial (DMH) and ventromedial (VMH) nuclei of the hypothalamus decrease the latency to displaying maternal behavior in steroid-primed, ovariectomized virgin rats.

Inhibition of maternal behavior has also been demonstrated in aversive situations, such as the exposure of a virgin female to newborn pups. It has been shown that in this situation, a mobilization of the medial hypothalamic defensive system occurs and that lesions of hypothalamic areas, namely the anterior and ventromedial nuclei, promote the maternal response. Also, both electrical and 6-hydroxydopamine lesions of the ventral tegmental area (VTA) disrupt maternal behavior.

In non-maternal adult females, neuronal activity in the MA and CA inhibits maternal behavior via a mechanism involving the processing of olfactory cues from the initial exposure to pups. Prolonged pup exposure also reduces this basal inhibition, allowing the onset of pup-induced maternal behavior. Lesions of the MA in adult female rats shorten the time of pup exposure needed to induce maternal behavior. In postpartum animals, lesions of the amygdale and MPOA promote deficits in pup retrieval and maternal responses in the home cage. These results indicate that systems associated with the MPOA mediate both stereotypical maternal behaviors and pup reinforcement. In contrast, the expression of home cage maternal behavior but not operant responding, is dependent on the integrity of both the amygdala and NAC. This indicates a dissociation of the mechanisms of species-typical maternal behaviors and pup reinforcement.

The NAC is necessary for the retrieval component of maternal behavior, as well as for the processing of pup-related stimuli likely to be involved in motivation. The paraventricular nucleus (PVN) of the hypothalamus also plays an important role; it is the source of intracerebral oxytocin secreted at parturition and it facilitates the expression of maternal behavior via numerous projections and diffusion of its neurochemicals throughout the cerebral ventricles.

The PAG is known to play an important role in the modulation of nociceptive sensory transmission, regulation of the cardiovascular system and in the expression of a variety of behaviors including defensive, sexual, maternal, and feeding behaviors. Throughout the past decades, however, two themes have largely dominated research on PAG: inhibition of nociception and the integration of behavioral responses to threatening or stressful stimuli. The pioneering studies of Hunsperger (1956) led to the widely accepted view that the PAG is critical for defensive behaviors. Subsequent studies have shown that PAG stimulation can induce a pattern of somatomotor and autonomic responses reminiscent of the behavior of animals facing natural threats. Recent studies have shown that rostrolateral PAG activation with low doses of morphine in previously morphine-sensitized dams can increase the motivational drive for predatory (and perhaps foraging) behaviors, thus hampering the expression of maternal responses. Using this pharmacological paradigm, it has been suggested that the rostrolateral PAG may account for switches in adaptive responses of lactating rats, in this case from maternal to predatory behavior. Under natural circumstances, it is reasonable to believe that a particular behavior might be inhibited in order to favor the expression of a more adaptive response. Thus, in this case, morphine challenge favored the occurrence of predatory activity over maternal behavior.

Taken together, these studies suggest that the neuroanatomy of maternal behavior is complex and involves multiple systems that interconnect with the MPOA and other regions.

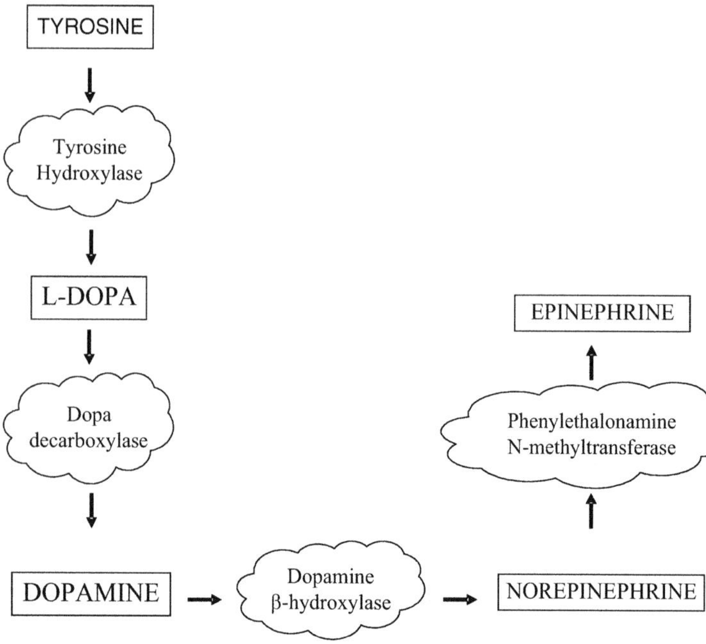

Fig. 16.7 Steps in the enzymatic synthesis of dopamine and other catecholamines

Neurotransmission and Neurotransmitters

Knowledge of the neurotransmitters that mediate maternal behavior is limited to a few molecules and to the adult model. Available literature provides the clearest case for a role of dopamine (DA) in maternal behavior (Fig. 16.7). Extensive lesions of the dopaminergic system disrupt maternal behavior, and dopamine receptor antagonists infused locally into the NAC inhibit the retrieval and licking components of maternal behavior, suggesting that these behaviors require DA. Increased levels of DA and the DA metabolites dihydroxyphenylacetic acid (DOPAC) and homovanillic acid (HVA) in the extracellular space of the ventral striatum after separated mother rats and pups were reunited suggest a role in the motivational components of the behavioral response. It has been demonstrated that DOPAC levels in the MPOA are significantly lower during late pregnancy and on the day of parturition as compared to early pregnancy and during late lactation. Although these results can only correlate monoamine activity with changes in maternal behavior, they suggest that reduced dopaminergic activity in the MPOA around the time of parturition followed by an increase during lactation might be related to the onset and maintenance of maternal responsiveness.

Systemic or widespread central manipulations of dopaminergic activity severely impair many aspects of maternal behavior in rats. This is supported by the importance of elevated DA release in the MPOA for sexual motivation and performance during copulation, which are in some ways similar to maternal motivation and performance. These results suggest that increased DA activity in the NAC might occur throughout the reproductive cycle in order to prepare the dam for her maternal role. Studies that have manipulated the dorsal striatum in lactating rats have focused on medial areas of this region and have found few effects on maternal behavior.

The exploration of roles for other neurotransmitters in maternal behavior has been limited; however there is modest evidence that serotonin (5-HT) could be involved. It has been found that serotonergic lesions in the median raphe nucleus caused short-term disruption of all aspects of maternal behavior except the prolactin response to the suckling stimulus. Microdialysis studies have shown an increase in the concentration of 5-hydroxyindoleacetic (5-HIAA), the metabolite for 5-HT, in the ventral striatum when dams were reunited with pups; thus it is possible that motivational components of maternal behavior are supported by this system.

Infusions of cocaine into the NAC or MPOA disrupt adult maternal behavior, suggesting that a critical level of DA and possibly of 5-HT is important for these behaviors. Indeed, the 5-HT turnover rate in the MPOA was lower for virgin females as compared to early pregnant and parturient/lactating females. Changes in the activity of any of these neurotransmitters in the MPOA during pregnancy or lactation may influence the dam's maternal behavior. However, it appears that MPOA norepinephrine and 5-HT are not critical for this behavior.

Gamma-aminobutyric-acid (GABA) is involved in the development of sexual differences in the brain. Neonatal administration of the $GABA_A$ agonist diazepam induces a feminine neuromorphological pattern with respect to the volume and number of neurons in the male accessory olfactory bulb. Moreover, injection of the $GABA_A$ antagonist picrotoxin disrupts maternal behavior in female rats. Fluctuations of GABA and GABA receptor levels have been observed in the cerebral cortex, hippocampus and midbrain throughout gestation, delivery, and lactation. It is known that in the MPOA, estrogens act to increase GABA release and reuptake as well as $GABA_A$-receptor expression. In addition, it has been reported that steroid manipulation modulates mRNA levels of GAD (glutamic acid decarboxylase); this effects depends on the brain area considered.

Hormones

A number of studies have demonstrated a relevant role for the steroids estrogen and progesterone and the lactogenic hormones prolactin and placental lactogen in the onset of maternal behavior in newly parturient rats.

Briefly, pregnancy is characterized by a significantly longer period of substantially elevated estrogen and progesterone as compared to the estrous cycle. The pattern of hormone secretion that occurs near the end of pregnancy in a variety of species has been analyzed in order to gain some insight into possible hormonal candidates that may arouse maternal responsiveness at the end of pregnancy. The periparturient female begins to display nurturing behaviors within a few minutes of birth.

This behavioral change is thought to critically depend on the ovarian and pituitary hormone fluctuations that occur during pregnancy and parturition. After parturition, once mothering is well established, however, hormones are thought to have little influence on the maintenance of the dam's maternal behavior and instead, sensory cues from pups sustain her maternal responsiveness.

In adulthood, estrogens control the reproductive system, modulating neural circuits of the central nervous system (CNS) and consequently neuroendocrinological, behavioral and cognitive functions. Females are especially sensitive to estrogens during pregnancy and lactation. In particular, the expression of maternal behavior during pregnancy and in the first hours after delivery depends on estradiol, which appears to activate estrogen receptors of neural areas such as the hypothalamus that are critically involved in maternal behavior. About 30 h before parturition, circulating levels of progesterone decrease, levels of estradiol rapidly rise and prolactin levels increase slightly. During the pregnancy/lactation transition, there is an overlap between the effects of estradiol stimulation, which gives rise to maternal behavior, and that of pup stimulation, which maintains this behavior; during lactation, suckling by pups becomes the main stimulation.

For a variety of species, the termination of pregnancy is associated with a decrease in the progesterone-to-estrogen ratio. In the rat, peripheral plasma levels of progesterone are high throughout most of the pregnancy, reaching peak levels on days 14–15 (commonly with 21 days of duration), whereas plasma levels of estrogen continue to rise throughout the terminal stages of pregnancy. Progesterone levels decline sharply beginning on day 19 of pregnancy whereas estradiol levels are elevated at this time. These hormonal changes indicate that a large decrease in progesterone is a necessary precursor to the onset of parturition.

Altered sensitivity of estrogen receptors in relevant brain areas during critical reproductive periods may affect subsequent behavior, specifically maternal behavior. Numerous areas of the brain likely respond to ovarian and pituitary hormones, preparing the dam for maternal behavior. The MPOA and the other regions described above are thought to be particularly important. It has been shown that pregnancy and its associated hormonal fluctuations can alter neurons that regulate non-pup-directed components of maternal behavior (e.g., nest building) or behaviors that support maternal behavior (e.g., foraging, associative memory).

> In female rats, high levels of nuclear estrogen receptors in the MPOA (the main area for mediating estrogen stimulation of maternal behavior) are normally established during the second half of pregnancy and are necessary for estrogen stimulation of maternal behavior.

Under normal conditions, increased estradiol levels in the late phase of pregnancy are necessary for the onset of maternal behavior; this effect is mediated by the oxytocinergic system. The estrogen-oxytocin interaction is responsible for the expression of specific elements of maternal behavior (i.e., licking–grooming and the arched-back posture), even during the first postnatal days (i.e., postnatal day 3), suggesting that aspects of maternal behavior are still hormonally dependent in this phase. Oxytocin is a hormone associated with parturition and is involved in causing the uterine contractions that result in the expulsion of fetuses at the end of pregnancy. This hormone is released into the general circulation by the neural lobe of the pituitary, and its levels increase in the peripheral plasma during parturition. The stimuli that promote the release of oxytocin include decreased plasma progesterone and/or increased estrogen, as well as vaginal, cervical, and uterine distension.

Lactogenic hormones (protein hormones that promote mammary gland development) are released into the blood from the anterior pituitary and placenta. Placental hormones are referred to a placental lactogens, and the anterior pituitary hormone prolactin, which is luteotrophic, promotes progesterone secretion from the corpora lutea throughout pregnancy. Prolactin appears to act centrally for stimulation of the neural lactogenic system. Prolactin can access the brain by crossing the blood–brain barrier via receptor-mediated active transport in the choroid plexus, or it can be produced by the brain itself. Prolactin that originates from either the anterior pituitary gland or other brain areas binds to prolactin receptors that are found in a variety of tissues including the brain, liver, kidney, adrenal gland, ovary, uterus, testis and, especially, the mammary gland. It has been established that prolactin as well as rat (r) placental lactogen I (PL-I) and rPL-II act within the brain to bring about the rapid onset of maternal behavior at parturition in the female rat. Central infusions of these lactogenic hormones into the MPOA shorten the latency to maternal behavior in adult progesterone- and estradiol-primed nulliparous rats.

Throughout most of the second half of pregnancy, prolactin is present at low levels in the blood, but a final peak occurs over the final 2 days of pregnancy. Thus, prolactin seems to be a prime candidate hormone for the facilitation of maternal behavior at parturition. Other candidate hormones include the placental lactogens. Taken together, these results suggest that the central stimulatory actions of prolactin on maternal behavior are dependent upon prior exposure to progesterone and concurrent estrogen exposure.

Maternal behavior also appears to be mediated by oxytocin, endogenous opioids and the gastrointestinal peptide hormone cholecystokinin (CCK). Opioid-induced inhibition of maternal behavior in lactating rats is well-established. This phenomenon can be modulated by factors such as physiological state, endogenous peptides, and non-peptidergic drugs. Opioid effects on maternal behavior vary according to reproductive experience and age; in juvenile rats, maternal behavior is also sensitive to pharmacological opioid manipulations.

It has been shown that maternal behavior is inhibited by opiates. Beta-endorphin, an endogenous opioid, dose-dependently blocks normal maternal behavior when infused into the ventricular system of lactating rats, suggesting that changes in endogenous opiate levels alter maternal responsiveness in rats. Thus, endogenous opiates are directly involved in the regulation of various events surrounding parturition and subsequent lactation.

The onset of maternal behavior has been closely linked to opiate action in the MPOA during lactation. Administration of morphine either systemically or locally into the MPOA disrupts maternal behavior during lactation; this effect can be reversed by concurrent treatment with the opiate antagonist naloxone. Indeed, it has been shown in rats that both endogenous opiate concentrations and MPOA opiate receptor levels are reduced during lactation (Felicio et al. 1991).

CCK has been proposed to act as a neurotransmitter or neuromodulator in the CNS. CCK peptide and CCK mRNA have been found to be widely distributed throughout the mammalian brain. There are two CCK receptor subtypes, CCK_1 and CCK_2 (previously named CCK_A and CCK_B), and these two receptors have been shown to play different roles in maternal behavior.

CCK reduces the action of opioids on maternal behavior. Additionally, CCK can act as an endogenous opioid antagonist for regulation of ongoing maternal behavior. Both lateral ventricle and MPOA infusions of CCK have been shown to block the inhibitory effects of beta-endorphin on the maintenance of maternal behavior in lactating rats. Thus, CCK antagonism of opioid disruption of maternal behavior appears to occur via both CCK_1 and CCK_2 receptor subtypes.

Molecular Biology: Gene Expression and Maternal Behavior

The interaction of both inherited and environmental influences on the genome is a major factor influencing behavior. Gene expression in the brain is the result of the interaction between hereditary and environmental factors. Inherited influences include variations (polymorphisms) in DNA sequences transmitted from generation to generation. DNA polymorphisms can affect protein activity and gene expression in the brain when considerable amounts of each protein are produced. Genomic techniques that allow more sensitive, efficient and comprehensive expression analyses have permitted the investigation of the relationship between gene expression and behavior using animal models. Behaviors are the result of the actions of many genes. Paternally expressed imprinted genes should tend to increase pup physical development, survival and ultimate reproductive success. Expression of some paternal genes is essential for activation of genetically programmed maternal behavior in rodents. In mice, one interesting example is provided by the Mest gene. This gene is imprinted by the father, giving males a role in determining the female's maternal behavior (Lefevre et al. 1998; Keverne 2001).

Some reports indicating a role for gene expression and its consequent effects on maternal behavior in rats remain controversial due to technical issues such as problems with DNA/RNA extraction and amplification of specific segments, as well as the involvement of multiple genes in the control and/or modulation of maternal behavior. Nevertheless, some consistent data have been generated. Increased expression of prolactin receptor mRNA in the MPOA during late gestation, lactation, and in the presence of pups has been demonstrated. In addition, treatment with prolactin, growth hormone, or ovarian steroids increases the expression of the prolactin receptor in female rats, suggesting that the MPOA prolactin receptor plays a key role in the hormonally-induced onset of maternal behavior (Lucas et al. 1998).

The steroid glucocorticoid hormone is an important component of the system that coordinates behavioral responses to stress in vertebrates. Rats with increased expression of the glucocorticoid receptor gene in their brains are more tolerant of stress than individuals with lower expression. These differences could explain variations in maternal care exhibited by different mother rats. Variations in rat maternal care are inherited; pups that receive minimal care from their mothers grow up to provide the same level of care to their own offspring. Pups experiencing indifferent care show profound changes in brain gene activity, including decreased expression of the glucocorticoid receptor gene. These inherited differences in gene expression and behavior occur even in the absence of DNA polymorphisms. In the case of neglected rat pups, epigenetic modification of the glucocorticoid receptor gene via methylation is involved in their altered adult behavior. Hence, environmental influences on behavior can cause epigenetic changes to the genome that are inherited (Brunton et al. 2005; Brunton and Russell 2008).

In the past, some social and behavioral scientists did not consider animal models for behavioral genetics due to the complexity of human behavior. Yet, behaviors such as foraging and care of offspring, each involving molecules known to be present in humans, illustrate complex performances taking place over the course of days, weeks or even a lifetime. These behaviors have learning components in a social context (Miranda-Paiva et al. 2007; Teodorov et al. 2006; Sukikara et al. 2007).

The value of animal models can be further enhanced by applying genomics to the generation of large-scale expression profiles of individuals with different genotypes tested in different environments. In addition, the application of informatics should enable new literature-based comparative analyses of behaviors across different species.

According to Isles et al. (2004), "A subset of mammalian genes is subject to genomic imprinting. These imprinted genes show parent of origin specific monoallelic or parental allele-biased expression, such that for some genes, it is mainly the maternally inherited allele that is expressed, whereas for others, expression occurs mainly from the paternal copy. Evolutionary theory predicts that these genes will have a role in the mother-offspring interaction in mammals, and indeed many imprinted genes have a role in growth and placental function, and consequently influence prenatal development. In addition to the developing fetus, there is increasing evidence to suggest that imprinted genes influence the pre-weaning mother-offspring relationship, and consequently the development of the offspring into adulthood". Nutrition-related genes, such as the gene encoding insulin, seem to play a role in the evolutionary milestone of parental care (Hughes et al. 2008).

The imprinted gene *Peg3/Pw1* produces neonatal mice with deficits in postnatal growth, suckling and thermoregulation. As a consequence, these mice display delayed weaning and onset of puberty. This study suggests that a Peg3 mutation in the mother produced a variety of dysfunctions that were examined independently of the mutation in pups. Mutant females failed to increase their food intake in the early stages of pregnancy. They ate less throughout pregnancy and carried over less body weight reserve into the post-partum period. Milk let-down was also impaired in mutant females during the post-partum period and their pups lost weight on day 1 following birth.

Finally, some genes involved in rat maternal behavior might be pacemakers; thus their identification may lead to molecular pathways that are critical to brain circuitry modulating behavior. The identification and analysis of the promoters and enhancers that regulate these genes should also provide important insights the effects of inherited and environmental factors on brain and behavior.

References

Beach FA. The neural basis of innate behavior. I. Effects of cortical lesions upon the maternal behavior pattern in the rat. J Comp Psychol. 1937;24:393–436.

Beach FA, Conovitz MW, Goldstein AC, Steinberg F. Experimental inhibition and restoration of mating behavior in male rats. J Genet Psychol. 1956;89:165–81.

Bredy TW, Zhang TY, Grant RJ, Diorio J, Meaney MJ. Peripubertal environmental enrichment reverses the effects of maternal care on hippocampal development and glutamate receptor subunit expression. Eur J Neurosci. 2004;20:1355–62.

Bridges RS, Numan M, Ronsheim PM, Mann PE, Lupini CE. Central prolactin infusions stimulate maternal behavior in steroid-treated, nulliparous female rats. Proc Natl Acad Sci U S A. 1990;87:8003–7.

Bridges RS, Robertson MC, Shiu RP, Sturgis JD, Henriquez BM, Mann PE. Central lactogenic regulation of maternal behavior in rats: steroid dependence, hormone specificity, and behavioral potencies of rat prolactin and rat placental lactogen I. Endocrinology. 1997;138:756–63.

Brown RE. Hormones and paternal behavior in vertebrates. Am Zool. 1985;25:895–910.

Brunton PJ, Russell JA. The expectant brain: adapting for motherhood. Nat Rev Neurosci. 2008;9:11–25.

Brunton PJ, Meddle SL, Ma S, Ochedalski T, Douglas AJ, Russell JA. Endogenous opioids and attenuated hypothalamic-pituitary-adrenal axis responses to immune challenge in pregnant rats. J Neurosci. 2005;25:5117–26.

Byrnes EM, Rigero BA, Bridges RS. Induction of maternal behavior in adult female rats following chronic morphine exposure during puberty. Dev Psychobiol. 2003;43:367–72.

Caldji C, Tannenbaum B, Sharma S, Francis D, Plotsky PM, Meaney MJ. Maternal care during infancy regulates the development of neural systems mediating the expression of fearfulness in the rat. Proc Natl Acad Sci U S A. 1998;95:5335–40.

Comoli E, Ribeiro-Barbosa ER, Canteras NS. Predatory hunting and exposure to a live predator induce opposite patterns of Fos immunoreactivity in the PAG. Behav Brain Res. 2003;138: 17–28.

Curley JP, Barton S, Surani A, Keverne EB. Coadaptation in mother and infant regulated by a paternally expressed imprinted gene. Proc Biol Sci. 2004;271:1303–9.

Della Seta D, Minder I, Dessi-Fulgheri F, Farabollini F. Bisphenol-A exposure during pregnancy and lactation affects maternal behavior in rats. Brain Res. 2005;65:255–60.

Felicio LF, Mann PE, Bridges RS. Intracerebroventricular cholecystokinin infusions block beta-endorphin-induced disruption of maternal behavior. Pharmacol Biochem Behav. 1991;39:201–4.

Fries E, Moragues N, Caldji C, Hellhammer DH, Meaney MJ. Preliminary evidence of altered sensitivity to benzodiazepines as a function of maternal care in the rat. Ann N Y Acad Sci. 2004;1032:320–3.

Hughes WO, Oldroyd BP, Beekman M, Ratnieks FL. Ancestral monogamy shows kin selection is key to the evolution of eusociality. Science. 2008;320:1213–6.

Hunsperger RW, Roman D. The integrative role of the intralaminar system of the thalamus in visual orientation and perception in the cat. Exp Brain Res. 1976;25:231–46.

Isles AR, Holland AJ. Imprinted genes and mother-offspring interactions. Early Hum Dev. 2005;81:73–7.

Kendrick KM, Keverne EB, Chapman C, Baldwin BA. Microdialysis measurement of oxytocin, aspartate, gamma-aminobutyric acid and glutamate release from the olfactory bulb of the sheep during vaginocervical stimulation. Brain Res. 1988;442:171–4.

Keverne EB. Genomic imprinting and the maternal brain. Prog Brain Res. 2001;133:279–85.

Kinsley CH, Trainer R, Stafisso-Sandoz G, Quadros P, Marcus LK, Hearon C, Meyer EA, Hester N, Morgan M, Kozub FJ, Lambert KG. Motherhood and the hormones of pregnancy modify concentrations of hippocampal neuronal dendritic spines. Horm Behav. 2006;49:131–42.

Leblond CP. Extra-hormonal factors in maternal behavior. Proc Soc Exp Biol Med. 1938; 38:66–70.

Leblond CP. Nervous and hormonal factors in the maternal behavior of the mouse. J Genet Psychol. 1940;57:327–44.

Lebolnd CP, Nelson WO. Maternal behavior in hypophysectomized male and female mice. Am J Physiol. 1937;120:167–72.

Lee A, Clancy S, Fleming AS. Mother rats bar-press for pups: effects of lesions of the mpoa and limbic sites on maternal behavior and operant responding for pup-reinforcement. Behav Brain Res. 2000;108:215–31.

Lonstein JS, Dominguez JM, Putnam SK, De Vries GK, Hull EM. Intracellular preoptic and striatal monoamines in pregnant and lactating rats: possible role in maternal behavior. Brain Res. 2003;970:149–58.

Lubin DA, Cannon JB, Black MC, Brown LE, Johns JM. Effects of chronic cocaine on monoamine levels in discrete brain structures of lactating rat dams. Pharmacol Biochem Behav. 2003;74: 449–54.

Lucas BK, Ormandy CJ, Binart N, Bridges RS, Kelly PA. Null mutation of the prolactin receptor gene produces a defect in maternal behavior. Endocrinology. 1998;139:4102–7.

Mann PE, Babb JA. Disinhibition of maternal behavior following neurotoxic lesions of the hypothalamus in primigravid rats. Brain Res. 2004;1025:51–8.

Mann PE, Bridges RS. Prolactin receptor gene expression in the forebrain of pregnant and lactating rats. Brain Res Mol Brain Res. 2002;105:136–45.

Mann PE, Felicio LF, Bridges RS. Investigation into the role of cholecystokinin (CCK) in the induction and maintenance of maternal behavior in rats. Horm Behav. 1995;29:392–406.

Marmendal M, Roman E, Eriksson CJ, Nylander I, Fahlke C. Maternal separation alters maternal care, but has minor effects on behavior and brain opioid peptides in adult offspring. Dev Psychobiol. 2004;45:140–52.

Miranda-Paiva CM, Felicio LF. Differential role of cholecystokinin receptor subtypes in opioid modulation of ongoing maternal behavior. Pharmacol Biochem Behav. 1999;64:165–9.

Miranda-Paiva CM, Nasello AG, Yin AJ, Felicio LF. Morphine pretreatment increases opioid inhibitory effects on maternal behavior. Brain Res Bull. 2001;55:501–5.

Miranda-Paiva CM, Nasello AG, Yim AJ, Felicio LF. Puerperal blockade of cholecystokinin (CCK1) receptors disrupts maternal behavior in lactating rats. J Mol Neurosci. 2002;18: 97–104.

Miranda-Paiva CM, Ribeiro-Barbosa ER, Canteras NS, Felicio LF. A role for the periaqueductal grey in opioidergic inhibition of maternal behaviour. Eur J Neurosci. 2003;18:667–74.

Miranda-Paiva CM, Canteras NS, Sukikara MH, Nasello AG, Mackowiak II, Felicio LF. Periaqueductal gray cholecystokinin infusions block morphine-induced disruption of maternal behavior. Peptides. 2007;28:657–62.

Numan M, Numan MJ, Schwarz JM, Neuner CM, Flood TF, Smith CD. Medial preoptic area interactions with the nucleus accumbens-ventral pallidum circuit and maternal behavior in rats. Behav Brain Res. 2005;158:53–68.

Olazábal DE, Abercrombie E, Rosenblatt JS, Morrel JI. The content of dopamine, serotonin, and their metabolites in the neural circuit that mediates maternal behavior in juvenile and adult rats. Brain Res Bull. 2004;63:259–68.
Pedersen CA, Caldwell JD, Walker C, Ayers G, Mason GA. Oxytocin activates the postpartum onset of rat maternal behavior in the ventral tegmental and medial preoptic areas. Behav Neurosci. 1994;108:1163.
Robinson GE. Genomics. Beyond nature and nurture. Science. 2004;304:397–9.
Rodriguez C, Guillamon A, Pinos H, Collado P. Postpartum changes in the GABAergic system in the bed nucleus of the accessory olfactory tract. Neurochem Int. 2004;44:179–83.
Rosenblatt JS, Siegel HI, Mayer AD. Progress in the study of maternal behavior in the rat: hormonal, nonhormonal, sensory, and developmental aspects. In: Rosenblatt JS, Hinde RA, Beer CG, Busnel M-C, editors. Advances in the study of behavior. New York: Academic; 1979.
Sapolsky RM. Mothering style and methylation. Nat Neurosci. 2004;7:791–2.
Serafim AP, Felicio LF. Dopaminergic modulation of grooming behavior in virgin and pregnant rats. Braz J Med Biol Res. 2001;34:1465–70.
Serafim AP, Felicio LF. Reproductive experience influences grooming behavior during pregnancy in rats. Braz J Med Biol Res. 2002;35:391–4.
Silva MR, Bernardi MM, Felicio LF. Effects of dopamine receptor antagonists on ongoing maternal behavior in rats. Pharmacol Biochem Behav. 2001;68:461–8.
Silva MR, Bernardi MM, Cruz-Casallas PE, Felicio LF. Pimozide injections into the Nucleus accumbens disrupt maternal behaviour in lactating rats. Pharmacol Toxicol. 2003;93:42–7.
Stern JM, Lonstein JS. Neural mediation of nursing and related maternal behaviors. Prog Brain Res. 2001;133:263–78.
Stone CP. Preliminary note on the maternal behavior of rats living in parabiosis. Endocrinology. 1925;9:505–12.
Sturman-Hulbe M. Maternal behavior in the albino rat. J Comp Psychol. 1929;9:203–37.
Sukikara MH, Mota-Ortiz SR, Baldo MV, Felício LF, Canteras NS. A role for the periaqueductal gray in switching adaptive behavioral responses. J Neurosci. 2006;26:2583–9.
Sukikara MH, Platero MD, Canteras NS, Felicio LF. Opiate regulation of behavioral selection during lactation. Pharmacol Biochem Behav. 2007;87:315–20.
Teodorov E, Salzgeber SA, Felicio LF, Varolli FM, Bernardi MM. Effects of perinatal picrotoxin and sexual experience on heterosexual and homosexual behavior in male rats. Neurotoxicol Teratol. 2002;24:235–45.
Teodorov E, Modena CC, Sukikara MH, Felicio LF. Preliminary study of the effects of morphine treatment on opioid receptor gene expression in brain structures of the female rat. Neuroscience. 2006;141:1225–31.
Teodorov E, Tomita AT, Banon GP, Gil IG, Bernardi MM, Felício LF. Behavioral effects of acute stimulation of kappa-opioid receptors during lactation. Pharmacol Biochem Behav. 2008;90:534–9.
Weaver IC, Szyf M, Meaney MJ. From maternal care to gene expression: DNA methylation and the maternal programming of stress responses. Endocr Res. 2002;28:699.
Weaver IC, Ceryoni N, Champagne FA, DÁlessio AC, Sharma S, Seckl JR, Dymov S, Szyf M, Meaney MJ. Epigenetic programming by maternal behavior. Nat Neurosci. 2004;7:847–54.
Weisner BP, Sheard NM. Maternal behavior in the rat. Edinburg: Oliver and Boyd; 1933.

Chapter 17
Behavioral Methods to Study Learning and Memory in Rats

Jorge Alberto Quillfeldt

Introduction

The only proof of there being retention is that recall actually takes place

—William James, 1872

This century-old observation is still valid today, despite everything we have learned about the mammal nervous system, especially in the area of neurobiology of learning and memory. After "training" an experimental animal, such as a rat or a mouse, the only way to be sure that a "memory" was formed is by evoking it back, i.e., by recalling it in a "test" session: this "memory" is expressed by a behavior that differs from that one emitted in the training session. Until proof to the contrary, the best explanation for this *new response to the same context* is that some kind of internal modification—a "record"—mediates it inside the animal: this is what we call "memory". Everything else is consequence: if recalling depends upon the established memory trace intensity, it will be a function of the experience intensity during the acquisition, or "training", session, and so on.

"Memory" is quite a slippery concept because we still have not produced a complete, consensual notion about the physical nature of its trace. Notwithstanding the fact that we may be getting closer to this aim, the only sure way to *grab at* such phenomenon is *by measuring behaviors and their modifications*, i.e., *indirectly* quantifying it. Such approach is called "phenomenological", and is opposed to the so-called "mechanistic" vision, that departs from previously existent knowledge about the intrinsic machinery operating behind the phenomenon. We all know that modern science is the business of examining—and refuting, if unfit—the best

J.A. Quillfeldt, Ph.D. (✉)
Department of Biophysics, Universidade Federal do Rio Grande do Sul (UFRGS), Av. Bento Goncalves, 9500, Bldg 43422, Room 208[a], Bairro Agronomia, Porto Alegre 91501-970, Brazil
e-mail: quillfe@ufrgs.br

M.L. Andersen, S. Tufik (eds.), *Rodent Model as Tools in Ethical Biomedical Research*, DOI 10.1007/978-3-319-11578-8_17

mechanistic explanations for natural facts, but Neurobiology of Learning and Memory is one of those frontier areas where complexity forces us to begin with phenomenological description and gradually move into a plausible mechanistic explanation.

This chapter aims to briefly describe the behavioral approach to the study of learning and memory.

A Proposed Taxonomy of Memory Types

What need the bridge much broader than the flood?
William Shakespeare

'Much Ado About Nothing', 1599

Memories can be classified according to many different criteria: function (e.g., Working vs. Reference M.), content (e.g., Declarative/Explicit vs. Procedural/Implicit M.—see Squire & Cohen, 1984), duration (e.g., Immediate or Short-Term vs. Long-Term or Remote M.), nature (Associative vs. Nonassociative M.), or motivation (Appetitive/Reward vs. Aversive M.). In this chapter we will outline a simplified taxonomy of memory types, gathering most of the above categories in one classificatory tree (Fig. 17.1).

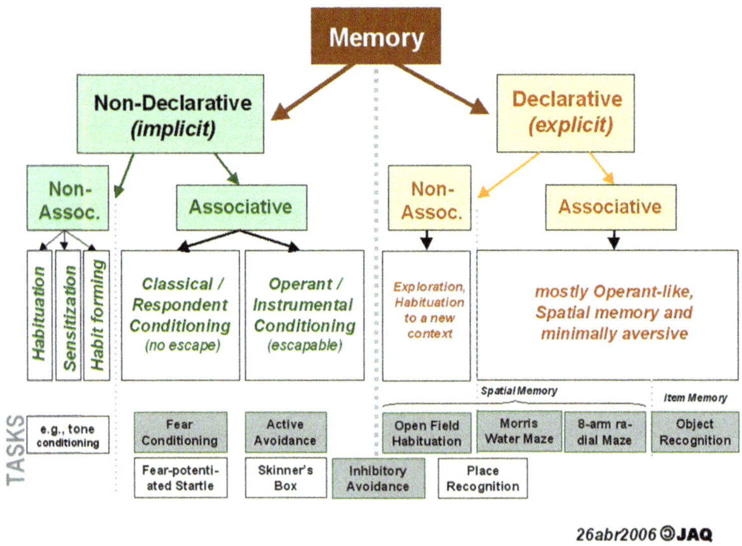

Fig. 17.1 A possible taxonomy for memory types and some behavioral tasks that may have access to them. Please, DO NOT USE THIS TABLE WITHOUT READING THE COMPANION TEXT BELOW, since each mentioned task only serve as example according to the specific protocol employed

Due to the great conceptual diversity found in this area, the very need for such conceptual organization is probably a matter of debate among specialists nowadays. However, we believe there is still a use for a classification, as limited as it can be, at least for educational reasons. Figure 17.1 synthesizes one possible classificatory attempt, mentioning some behavioral tasks as illustrative examples (description below). This taxonomical outline is a mere epistemological *aide-mémoire*, and does not imply the existence of any correspondent fundamental organization behind the natural world (an "ontology") that would be more rigid than the variegated, overlapping experimental approaches employed by different labs around the world.

Notice that Fig. 17.1 does not make reference to time; despite being obviously applicable to *long-term memory* (LTM), it also subserves *short-term memory* (STM) and even working memory (WM). In fact, *Working Memory* differs a lot from STM and LTM due to its essentially "executive" function: it basically stores information for a very short period of time (seconds to minutes) in order to compare it with previous records, and, based upon this, decide which behavior to express. It is also distinguishable from other memory types because it leaves no long-lasting record of any kind, possible relying upon reverberating electrical activity mainly settled in the prefrontal cortex (see Baddeley, 1997 and Goldman-Rakic et al., 1996). To know more about these types of memory, see, e.g., Izquierdo (2002).

Most of the basic behavioral tasks mentioned above can be drawn from three general categories, be it [i] the quantification of a natural response under controlled circumstances (such as fear conditioning), [ii] the suppression of an innate behavior (such as Inhibitory Avoidance), or [iii] the acquisition of a non-natural behavior (such as Skinner's operant conditioning) (see, e.g., Izquierdo 1989).

Figure 17.1 displays a three-levels' logical hierarchy of behavioral types, as follows:

1st Level—Declarative vs. Procedural: Most of these memory classes are derived from human experience, with the obvious difficulties in being extended to animal models.[1] *Declarative* M., for instance, concerns with facts/knowledge ("semantic" M.) or events ("episodic" M.), while *Nondeclarative* M., previously known as *Procedural* M., refers to motor or sensory abilities, or "habits". Human examples of both classes include reminding dates, names, faces or places, as well as autobiographic episodes (Declarative M.), or tasks such as driving a bicycle, swimming butterfly style, reciting multiplication tables or spelling words (Nondeclarative M.). To avoid anthropomorphization problems, it has been suggested to rename these types of memory, respectively, as *Explicit* and *Implicit* M. (Schacter, 1987), a terminology that we will adopt here.

Implicit M., due to its robustness, is a category apart: as soon as it is established, it (a) tends to last for the whole life, (b) is difficult be extinguished, and (c) is less vulnerable to emotional modulation (Izquierdo 2002). Compared to it, Explicit M.

[1] By the other hand, when interpreting behavioral results we must avoid to anthropomorphize them; remember that human memory has at least two very different, "nontranslatable" aspects (to animal models), symbolic language and conscience.

may be (a′) short or long-lasting,[2] and (b′) undergo a complex *consolidation* process, involving different receptors, enzymes, signalling cascades and brain structures (see Izquierdo et al. 1998, 1999, 2002; Izquierdo 2002). Before consolidation, during at least some hours, memory traces are *labile/unstable*,[3] and may suffer weakening or reinforcing disturbances (by head trauma, drugs, different situations, etc.) that may modify the original record in the first hours after acquisition (e.g., McGaugh 1966, 2000; Izquierdo 2002). Short-term M. (STM) lasts a few hours, and usually express itself *inside* the time window necessary for the Long-term M. (LTM) consolidation. Recently the old debate on the relation between STM and LTM was resolved: STM seems not to be the "initial" phase of LTM, and both processes take place in parallel, remaining quite independent one from the other despite sharing the same neuroanatomical substrates, but with different subcellular, neurochemical and/or electrophysiological mechanisms (Izquierdo et al. 1998, 1999, 2002). The loose of explicit M. is usually denominated *amnesia*.

2nd Level—Associative vs. Nonassociative: behavioral tasks that promote associations between stimuli and responses, or between two stimuli, are known as *Associative*. Through them animals learn how to predict future events in order to express a proper, anticipatory behavior. The two main categories of non-declarative (implicit) associative M. are the (a) *Classical* or *Respondent* (*Pavlovian*) *Conditioning* and the (b) *Instrumental* or *Operant Conditioning*. In the first type, contingencies between stimuli and responses are arranged and controlled only by the experimenter (Pavlov 1927), and the associative experience is somewhat unavoidable from the animal's point of view. In the second type, the environment is arranged in order to permit that certain response from the animal is necessary to attain some result, such as avoid a painful stimulus or receive food (Skinner 1953), i.e., the "escape" or avoidance is an option available to an animal that could learn and perform it. Skinner himself coined the expressions *elicited behavior* to describe respondent conditioning, and *emitted behavior* to describe operant conditioning. Thus, in the classical Pavlovian example, a dog was trained to associate an initially neutral stimulus, such as a bell (that will be the *conditioned stimulus*, CS), to a *unconditioned response* (UR), such as salivation (concomitantly provoked by a US—*unconditioned stimulus* such as showing a juicy beef), and obtain, in the end, a *conditioned response* (CR), i.e., "salivate to the bell", a nonnatural response not previously recorded. Both experimental frameworks have a decisive role in the history of behavioral psychobiology, but instrumental conditioning is more flexible, general and spontaneous than respondent conditioning, once this last one is based upon a limited set of innate responses natural to the animal (Sanger & Blackman 1989; Beninger 1989; for a careful characterization of the typical tasks in each of these categories see Hölscher & O'Mara 1997).

3rd Level—Other categories: Subdivisions of the previous level may include the two types of associative conditionings described above, the dichotomy between

---

[2] Some authors mention another type of long-lasting memory, sometimes dubbed *Remote* M. or even *Persistent LTM* (see, e.g., Bekinschtein et al. 2008).

[3] Hence, the term "consolidation", created by Miller and Pilzecker in 1900 (McGaugh, 2000).

aversive (punishing) and appetitive (reward) behaviors, the stronger or weaker spatial nature of the task, or its motivational "drive" (e.g., reactive vs. exploratory vs. decision taking tasks).

This classification may be as inaccurate as any classification built for educational reasons can be, but our aim was fulfilled if a general view about the possible behavioral methodologies was "consolidated": to know which task to select in each situation. As we have previously noticed, active researchers in our research field may not agree with this classificatory attempt, in different levels. Actually, the existence of so many, slightly different (sometimes contradictory) behavioral tasks in the psychobiological literature is due to the fact that every author approaches a limited set of problems and try to adapt the available tools to his aims. If the old tenet that says that "there is no methodology without a theory behind" is true, in the limit there would be no possible classification since every experiment would imply a particular conceptualization that remains essentially nontranslatable to parallel situations.

Two additional difficulties must be mentioned: first, the fact that the tasks described ahead as "typical" instances in each category may be somewhat deceiving, and second, the frustrating fact that the "frontiers" drawn to divide categories may not be as clearcut and solid as expected. The tasks described as examples in each category must not be taken as sole, exclusive instances of that memory type, since slight protocol changes may be enough for it to be adapted to different objectives, eventually displacing it into another category. Some of these protocol modifications will be described for each task, but we may mention two cases: slight adaptations in the 8-*Arm Radial Maze* protocol may turn this explicit/associative/spatial/decision-taking task into an implicit/associative/habit-forming one (see Packard et al. 1989); *Inhibitory Avoidance*, not a pure "spatial" task on itself, may be easily adapted to study some forms of spatial memory (e.g., Cimadella et al., 2000).

Since there are so many possible variations in adaptability for each behavioral task, it is hard to follow all of them as separate threads in the scientific literature. Although there is a positive side on this: the broad margins of manoeuvre for adapting and create new task variations, and any one may advance new suggestions (its methodological validation will just depend upon perspicacity and determination).

The second problem is a consequence of the logical limitations of the suggested classification, unable to tackle the whole richness of possibilities inherent to real-life situations. For instance, *inhibitory avoidance* (according to the protocol described) faces lots of controversy (see, e.g., Xavier 1982) and resists easy classification, as illustrated by its position in the diagram of Fig. 17.1, between the implicit/associative/instrumental and explicit/nonassociative/exploratory classes (this is a really a hybrid task).

To favor intelligibility, Fig. 17.1 omits several important subclasses of memory types and behavioral tasks, as, for instance, *Imprinting*—a type of nondeclarative/nonassociative learning common in birds, and *Priming*—a clue-evoked nondeclarative/associative learning.

With the progressive accumulation of neurobiological knowledge, some of us will feel tempted to substitute much of these inaccurate characterizations for new ones focusing, for instance, the neuroanatomical bases of each behavioral task, or the neurotransmitter(s)/modulator(s) involved, or the electrophysiological mechanism exhibited, etc. Despite possible, this "updating" may be somewhat hasty, and result no more useful than the previously existing classifications.

The Logic of Experimental Design

The purpose of models is not to fit the data but to sharpen the questions

—Samuel Karlin

Memory Phases

The Memory Consolidation Theory, proposed by Müller and Pilzecker in 1900, is a fundamental paradigm in the psychobiology and neuropharmacology of memory (McGaugh, 2000; Dudai 2000). One consequence of the *Consolidation Paradigm* is that we can plan our experimental intervention to take place in two or three different moments around memory *Formation* and *Recall*. Since the formation is not an instantaneous process, we may distinguish two formation phases, *Acquisition*, better known as *learning*, and *Consolidation*, the labile phase during which the memory trace will be physically stored; *Recall*, also known as memory *retrieval*, *elicitation* or *expression*, takes place during the reexposition to the learning context, with or without previously delivered stimuli, and, as we have said before, is the only way to prove that a memory was really formed and stored.

Acquisition can take place after one or several "learning" trials, when the animals are exposed to a context with several controlled variables; immediately after this, and during some hours, the memory trace will undertake a labile phase when the animal is susceptible to disturbances that can modify the original memory record (e.g., McGaugh 1966, 2000; Izquierdo et al. 1999, 2002; Izquierdo 1989; Dudai 2000) (Fig. 17.2).

A Memory Retrieval (Recall) test may be done with different training-test intervals, and LTM is agreed to take place after a minimum period of, say, 6 h; usually it is performed with a 24 h interval, but larger time spans are possible, according

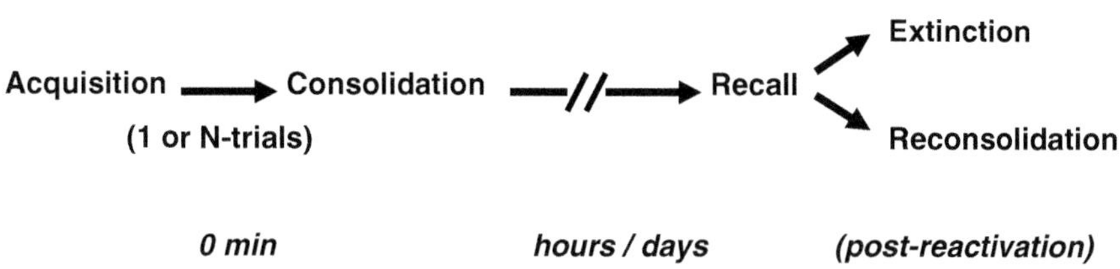

Fig. 17.2 Phases of memory formation (acquisition/consolidation), retrieval (recall), and the two possible post-reactivation phases (extinction or reconsolidation)

to the memory attributes under scrutiny (memory intensity, forgetting, extinction, reconsolidation), going from several days, to months and even years. In rats this interval is limited by its average lifetime of 24–48 months (an 18 month old rat, e.g., may be considered senile and a useful model for memory senescence—Krinke 2000). Studying recall with just a few hours of training-test interval is considered to be dealing with STM (Izquierdo et al. 1998) and if we reduce the interval to some minutes (say, 1–3 min) or less, we may be messing with *Working* M., at least according to some authors (Barros et al. 2002).

A consolidated M. may also suffer new modifications, for instance, be weakened by the repetition of the training context *without* the aversive US (e.g., a shock)—an *extinction*—, or, when this US *is being repeated*, the memory may be strengthened—a *reinforcement* (Izquierdo 2002). Memory *extinction* usually involves the substitution of an old memory for a new one and, on time, has a natural tendency to spontaneously reverse to the original (extinguished) memory trace.

In recent years, however, numerous authors have shown that memories already established can become transiently labile by a recall session—usually using the conditioned stimulus (CS) as a reminder cue of the original learning presented for a limited period of time (Przybyslawski and Sara 1997; Nader, 2003a, b; Debiec and LeDoux 2004; Duvarci and Nader 2004). This phase is followed by a stabilization period, usually defined as *reconsolidation*, which requires *de novo* protein synthesis, at least in the involved brain structures, and *only* after the presentation of this memory reactivation session (Misanin et al. 1968; Przybyslawski and Sara 1997; Przybyslawski et al. 1999; Nader et al. 2000; Eisenberg et al. 2003; Pedreira and Maldonado 2003; Debiec and LeDoux 2004; Duvarci and Nader 2004). However, if re-exposure to the CS extends beyond some critical period, the conditioned response gradually decreases to the well-known process called *extinction*, where the original memory trace is not erased, but transiently replaced by a new active learning: during this acquisition, animals learn that the presentation of the CS no longer predicts the occurrence of the unconditioned stimulus (US) (Bouton et al. 2006; Myers and Davis 2007). Therefore, reconsolidation demands a brief reactivation session, whereas extinction takes place after longer CS presentation, or after repeated presentations of the CS without the US. Consonant with this view, several authors have proposed that the duration of the re-exposure session to the CS is a decisive factor that critically influence which process will predominate: reconsolidation or extinction (Bustos et al. 2006, 2009; Debiec et al. 2002; Pedreira and Maldonado 2003; Boccia et al. 2004, 2007; Suzuki et al. 2004; Tronson and Taylor 2007; de Oliveira Alvares et al., 2008).

According to this sequence of events, any experimental intervention that have consequences upon memory processing (e.g., a pharmacological treatment—the preferred procedure explored in this chapter from now on), depends on *when* it is applied:

Pretraining: any intervention done may affect both acquisition and/or consolidation; if training-test interval is too short, it may affect recall also (this is less probable with a typical 24 h interval, but pay attention to the few, very slowly metabolizing drugs—such as benzodiazepines—that may still be active after this period);

Posttraining: since acquisition already took place, only consolidation may be affected[4] (see above); as said above, recall is hard to be affected if training-test interval is large (e.g., 24 h); different times after-training may give access to different stages of the neurochemical/physiological processes behind consolidation (e.g., Quillfeldt et al. 1996; Izquierdo et al. 1997); since some drugs take some time to act, they may not be suited for posttraining treatment, and pretraining administration, despite its flaws, will remain as the only option if you are looking for an actual "immediate" posttraining action: the "price" to pay is a lot of additional, complementary experiments in the case of finding an affect;

Pretest: intervention only affects recall simply because time only moves forward (e.g., Jerusalinsky et al. 1994; Izquierdo et al. 1997); also, depending on the time between treatment and the test session, some procedures may not work well (some may be done a few minutes before, others should be applied 20–30 min or more before);

Pre/post-reactivation: depending on the need to study post-reactivation phases (extinction or reconsolidation), the "test session" acquires the status of a "reactivation" session, and treatments can be applied before or after it (for results to be observed in the following memory phases).

Controls

Any scientific experiment aims to test for some hypotheses, and in order to make this possible, two conditions must be fulfilled first: (a) there must be *control experimental groups*, and (b) experimental setup must be made *simple and invariant* during multiple sessions/assays of the complete experiment.

The only way to be sure that a drug (or treatment) was *the real cause* of some observed effect (as seen, for instance, in the *treated* group) is by repeating the exact same experimental procedure just *without* the main substance (or intervention) under study: this is the *control* group. In a behavioral pharmacology experiment with a drug, control groups must receive the administration of the exact same amount (volume) of the drug's vehicle (be it a buffered solution or a plain saline solution[5]), under the same protocol (time of infusion, etc.). Some procedures/treatments,

[4]This treatment must be applied in the first few minutes, preferably in less than 1 min (usually termed a "0 min" treatment), in order to still act upon its targets while possible.

[5]In intraparenchymal (intracerebral injections) and intracerebroventricular administrations, special attention must be paid to the vehicle's chemical characteristics in order to assure it is as functionally "neutral" as possible: phosphate buffered saline solution (PBS—a buffered isosmotic 0.9 g % NaCl solution) of pH 7.4 are strongly recommended. If the drug is somewhat lipophylic, trouble may be avoided if the substance is first dissolved in a hydrophobic medium such as ethanol or DMSO, and then suspended in PBS to a reasonable percent: control groups in this case must be exactly like the drug's solvent, just without the drug. Systemic administration (endovenous, intraperitoneal, intramuscular or intradermic) must at least avoid osmotic effects, i.e., plain, distilled water is *never* an acceptable vehicle!

such as surgical procedures, may be controlled by a *sham group* in which the whole procedure is repeated—anaesthesia included—except for the specific, last step under study (e.g., a surgical removal of a brain structure, a blood vessel clamping, etc.). The comparison among the performances of treated and control groups will allow us to decide if there is an effect or not.

Since both control and treated groups involve several animals, *statistical tools* are always necessary in order to analyze these data.[6] In behavioral experiments the N per group should never be smaller than 6–8, and, according to the nature/difficulties of the treatment and/or the task, this may demand up to 20 animals per group.

Simplification of the experimental context is the second necessary condition since behavior is already a complex enough variable to analyze: if one leaves additional variables free to change, how could results be interpreted? For example, it would be really hard to interpret a behavioral experiment made in different times of the day, temperatures, stimuli intensity, etc.: most, if not all basic variables, must be made *as constant as possible* to warrant a nice experimental design, allowing a good control of reproducibility.

Drug administration can be designed to access both *acute* (just one administration) or *chronic effect*, but be aware that some acute treatments may result in chronic changes (e.g., the pilocarpine model of temporal lobe epilepsia or the MPTP model of Parkinsonism). It is always considered elegant to determine the *range of effectiveness* of a drug (or treatment): an assay with several different doses allow the preparation of a *dose-response curve*[7] (some treatments may cause responses arranged in degrees according to its intensity, and an *intensity-response* curve is desirable). In terms of pharmacological studies, choosing the best possible dose[8] is an art to be mastered by practice, but there are no general rules out of the do-it-and-check-for-yourself.[9] If a dose-response curve was previously published (especially if published by your own group), it may not be necessary to repeat it, but it is still recommended to do so in order to support the discussion of the observed (or not observed) effects in terms of drug specificity, selectivity, competitivity, and so on. Finding effective ranges different from those present in the literature does not necessarily mean that your data is wrong, since animals, even from the same strain, may be quite different from lab to lab.

---

[6] On this subject, we recommend two introductory books, Norman and Streiner (1994) and Callegari-Jacques (2003)—this last, in Portuguese; for advanced information, one of the best manuals in biostatistical analysis is Zar (1999). Finally, since nonparametric statistics is frequently necessary to analyze behavioral data, one excellent reference is Siegel and Castellan (1988).

[7] This must be done both for systemic and/or intraparenchymal injections, and every targeted structure into the brain may display its own dose-response curve due to histological particularities.

[8] Even when this (chosen) dose is known, it is recommended to produce a dose-response curve centered in this value.

[9] Intracerebral doses may sometimes be defined taking concentrations effective in *in vitro* experiments and administering a volume containing the substance in a 10–20 times larger concentration: this supposes a reasonable diffusion volume in the quite compact brain parenchyma (but this may vary in different regions, in the presence of nerve fibers, etc.).

Finally, a nice additional check may be performed in the case of pretest treatments. Since most drugs diffuse and/or metabolize after some time, animals can be *retested*, say, 90–120 min after the original test session: if an observed pretest effect disappears, it may have been caused specifically by the drug; if it doesn't, the "effect" may have other cause(s).

Distinguishing Memory from the Rest

Depending on the treatment and on the type of behavioral task employed, the observed behavioral change could be interpreted otherwise. For instance, in the inhibitory avoidance task, a "good" memory is shown by a larger latency to step down from a platform into a previously electrified grid made of bronze or stainless-steel bars, and any drug (or treatment) that changes this latency could be taken as *amnestic* (if decreasing latency) or *facilitatory* (if increasing latency). In order to prove the effect as *genuinely mnemonic*, one must be sure that the drug, in that dose (or that treatment, in that intensity) causes no such behavioral change by itself, for instance, affecting motor performance. This test must be investigate the appropriate behavioral task, according to the behavior one wants to check for, such as, for instance, a free exploration box (an open field would do the job) for gross motor effects, an elevated plus maze for anxiety effects, or a discrimination task for sensory effects. This must be considered a third control "layer" for behavioral experiments, after the fundamental control (vehicle-injected) groups and the simplicity/constancy of the experimental setup, as described in the previous section.

Since the burden of the proof is ours, we must provide results of other behavioral tasks done in parallel with memory tasks themselves in order to eliminate the possibility of a false positive (or false negative). Drugs (or other treatments) under study could always be acting upon unanticipated neural substrates that cause observed behavior that, if unnoticed, may "mask" for a memory effect. Hormonal state, for instance, may affect one or several of these behavioral manifestations, so it must be controlled by additional serum and/or tissue measurements (regular changes resulting from the estrous cycle phase in females are easy to check, for instance).

The list of possible false positive/negative factors is usually finite and not too large, and may include one of those shown in Table 17.1.

If a certain drug (or treatment) have an effect upon the memory task *and also affects* one or more of the non-mnemonic tasks, it is *not* recommended to advance a strong interpretation based on mnemonic mechanisms. A rule of thumb would be "**the 'cleaner' the results, the easier to interpret them**". But never forget that neither mammal nervous systems are simple structures, nor clearcut, straightforward causation can always be shown in behavioral studies. It may be the case that a drug (or treatment) affects both memory neural substrates *and* other, non-mnemonic mechanisms, and we be simply not able to separate them just on behavioral data grounds: in this case, it would also be necessary to collect a broad range of additional data (neurochemical, histological, electrophysiological, etc.), and, as usual, to lay hold of great creativity in order to solve the puzzle. It is never simpler than that.

Table 17.1 Possible factors to be excluded by complementary behavioral tasks

Non-mnemonic factors	Typical behavioral task/measured parameters
Motor performance	Deambulation characteristics in an OF (open field)
	Rotarod test
	Pole or chord climbing
Anxiety	Elevated plus-maze
	Light-dark transition task
	Time in central region of the maze vs. tygmotaxia
Pain sensitivity	Tail flick test
	Hot plate and/or paw pressure test
Sensory perception	Sensory discrimination tasks
Attention	Discrimination/reaction tasks
Arousal level	

Summarizing the experimental controls studied in previous sections, we may speak of three control "layers" to take into account in order to perform a good experimental design:

(1) **Treatment-specific controls**—vehicle, sham, etc. vs. treated groups; retest;
(2) **Reproducibility controls**—simple and invariant experimental setup;
(3) **Behavior-specific controls**—additional non-mnemonic behavioral control tasks.

The Posttraining Paradigm

> (…) Consequently the posttrial injection studies are difficult to interpret in terms of motivational of perceptual effects
>
> —James L. McGaugh (1966)

In the past, researchers in the area of neuropharmacology of learning and memory have been strongly criticized by other specialists based on the above-mentioned difficulties in proving that a supposed mnemonic phenomenon was, in fact, *mnemonic*. Even well-designed experiments with adequate controls, invariant setup, and complementary non-mnemonic behavioral tasks, may not overcome the quite philosophical objection probably inspired by the belief that complex brain functions (such as memory, attention and anxiety) may not be separable at last. This objection, however, was eliminated by a seminal conceptual observation first advanced by James L. McGaugh (1966): since *posttraining* treatments only affect consolidation, being impossible to act upon the acquisition (i.e., learning) that already took place, at least *those* experimental designs may be considered totally *clean* in terms of non-mnemonic "contaminating" parallel effects, and any observed effect can be safely taken as an effect upon memory *consolidation*. This may seem obvious for us now, but we may remember that at that time, 40 years ago, most of the learning and memory studies were performed-by "default"—after pretraining, not-that-*clean* treatments. Science advancement comes usually with simple yet powerful ideas.

The so-called *McGaugh's posttraining paradigm* does not mean that we cannot study memory under pretraining treatments: it only recommends that in the case that an effect is found, additional non-mnemonic behavioral tests will be necessary in order to correctly interpret the results in terms of mnemonic mechanisms.

Finally, recall studies may be considered even "cleaner" than posttraining ones, since they only may affect the memory retrieval process. This completely separates it from any consolidation process, and there are extensive work done showing that these two phases of memory differ in several respects (Izquierdo 2002).

Possible Experimental Interventions

Several *in vivo* treatments/interventions, of different categories, are possible:

(1) Reversible Local Selective Chemical/Pharmacological infusions (same for Cryolesions);
(2) Non-lesioning electrostimulation;
(3) Behavioral manipulation (task, immobilization, stress, fear, defense, etc.);
(4) Mechanical (surgical), Electrical or Chemical Irreversible Lesions;
(5) Transgenic Animal Models;
(6) Concomitant Electrophysiological and/or Imaging recordings.

Interventions 1 and 2, usually delivered in an acute fashion (since being short-lasting, less invasive and even reversible), are suited for application in any of the three *moments* above (pretraining, posttraining and pretest).

Intervention 3 may be suited for all moments if acutely applicable; chronic procedures, however, since they take a long time to be completed (e.g., chronic stress by restraint), should mostly be pretraining (or pretest with a larger training-test interval).

Interventions 4 and 5, since they tend to be irreversible, are more suited for pretraining manipulation protocols, with all its inherent limitations. Slowly diffusing or late acting drugs are also not adequate for the so-called *immediately posttraining*[10] treatment, and must be applied before the training session, fostering the need for the non-mnemonic behavioral assays (mentioned above) in order to assist the interpretation of experimental data.

Classic null-mutated *transgenetic* (knockout) animal models, in particular, demand such large amount of time to be developed that can only be studied under the pretraining protocol, with all the inevitable, additional non-mnemonic behavioral tasks in order to interpret data. These ultrareductionistic models comes with an additional burden since the manipulation takes place *in vitro*, and the animal that

[10]The expression *immediately posttraining* is usually employed for a treatment delivered (such as drug infused) in less than 1 min; despite the resistance of some authors, since this procedure *starts to take place* well between 0 and 59 s (and frequently ends in less than 2–3 min), we may term it as a "0 min" treatment.

develops itself despite (and survive to) the absence of one specific gene most probably adapted in several levels and places, i.e., it is virtually impossible to keep track of most of the possible modified parameters, and these may also impact upon data reproducibility (in this respect see, e.g., Routtenberg 1996). Recently a different approach was developed, the *conditioned knockout* (as is the case, e.g., in Tsien et al. 1996), that may be free from most of the above criticism, but still is less common because it is more complex and frequently hard to reproduce in different labs.

Intervention 6 is mentioned only to emphasize that most *measurements* are always somewhat invasive, and tend to interfere with normal behavior. By the other hand, concomitant recordings may only come in support of *correlational*, not *causational* demonstration.

What Are We Observing?

The effects of the intervention, in any of the above-mentioned *three moments*—pre-training, posttraining or pretest-, may produce only three kinds of results concerned with *memory*[11]:

(1) *amnesia*, i.e., memory reduction or blocking/deficit;
(2) *facilitation*, i.e., memory improvement;
(3) *no measurable effect.*

The stimuli employed may be divided in three categories according to its nature[12]:

(1) *neutral*: a stimulus that originally induces no specific response other than focusing attention; however, its response can be changed through classical or operant conditioning, when it becomes a *conditioned stimulus*.
(2) *aversive*: a negative stimulus, usually painful, distressing or uncomfortable (but not lethal);
(3) *appetitive*: a positive, reinforcing stimulus basically involving any of the instinctive drives needed to maintain organic life such as eating or drinking; sweet food, for instance, may even be a reinforce acting as a *reward*.[13]

Responses (1) and (2) above may admit *degrees of intensity*, usually being more robust/marked when stimuli (both appetitive or aversive) are more intense (*one-trial* training) or repeatedly presented (*multi-trial*, repetitive training).

[11] Supposing we are sure they are specifically mnemonic effects, and not motor, sensory, attentional or emotional *memory-masking* effects (see item 3.3). In the case of pre/postreactivation treatments, we may speak in the blocking or facilitation of extinction or of reconsolidation (it is also accepted to talk about blocking or facilitation of the consolidation of extinction).

[12] The following classification lacks generality due the specific needs of this chapter: a more encompassing classification would employ Skinner's terminology and mention negative reinforces, positive reinforces and punishments (Skinner 1953).

[13] *Rewarding stimuli* comprise a broader category, involving not only appetitive, but other types of pleasant stimuli, such as sexual stimuli.

The observed modifications/effects in responses (1) or (2) may also be detected in different temporal directions in relation to memory acquisition:

(a) *Retrograde* effect: acts upon recently-formed memories;
(b) *Anterograde* effect: acts upon new memories still to be formed, but after the treatment;
(c) *Ambigrade* effect: when it acts in both directions.

Result (3) deserves some additional comments, rarely mentioned in typical course books. First, the absence of an effect is commonly interpreted as a frustrating "no result at all", and is frequently taken as a … "negative result". This, however, is a mistake, since in a well-designed and executed experiment, when it finds no effect at all, *this is also a result*! As a scientific piece of evidence, it may be as—or even more—important than finding an amnestic or facilitatory effect, depending on the hypotheses under scrutiny. It may, anyway, help to prove (or disprove) the hypothesis under scrutiny. Second, even if an expected result is not attained, it still remains valid the old motto that says that "absence of evidence is not proof of absence", and we may still keep looking for new evidence under more refined/modified versions of the same or different experimental designs.

It is quite healthy to remember that nature tends to be much more complex than our limited, reductionist empirical investigation models. When we fail, it is always our move next. That is, in essence, how science works and advances.

A Closer Look into the Real Effects: State Dependency

Difficulties in the interpretation of results may arise not only from possible non-mnemonic effects, but also from the (otherwise) plain fact that in several common experimental protocols, animals are *trained in one state*—say, under the action of a drug, or in a certain hormonal state—and *tested in another state*—for instance, without the drug or not in that hormonal state. The difference in the performance between the training and the test sessions (be it amnestic or facilitatory, it doesn't matter) could be attributed to the simple fact that each session was done with the animals' brain in a different neurochemical/neurohumoral state! Especially *when the same response is observed with the same state* being promoted in the training *and* in the test session: this phenomenon is called *State Dependency*.

State dependency can be promoted by the exogenous administration of drugs, or by the stimulus to generate a certain internal neurochemical/neurohumoral state, a situation known as *endogenous state dependency*. When this type of phenomenon is demonstrated, the *mnemonic* interpretation of the observed effects *may* loose strength, since it may not be caused by the mnemonic mechanisms themselves. Some authors, however, sustain that in the absence of non-mnemonic effects (observed in parallel behavioral tasks—see above), it could still be the case that memory mechanisms *are* being affected because the engram could bear an additional "tag" that records for the concomitant neurochemical/neurohumoral state in

Table 17.2 Experimental design to prove state dependent learning—in this case, an instance of memory facilitation (adapted from Meyer & Quenzer, 2004)

State-dependent learning		Training: No drug	Training: Drug
Test	No drug	**Good recall**	*Less-effective recall*
	Drug	*Less-effective recall*	**Good recall**

which the memory trace was formed (Izquierdo, 1984). Table 17.2 illustrates the classical experimental design necessary to disclose the presence of a *state-dependent learning*.

The Methodological Triangle and Other Limitations

Another fundamental aspect is that there are always limits to what can be derived as *conclusions* from one experiment, notwithstanding the fact that sometimes in history there are some key experiments that really solve questions of great conceptual importance and reach (and who doesn't want to really *do* one of those few!). Of course, most of what we do is plain regular science, something that the philosopher Thomas Kuhn called *ordinary science* in its classic book "The Structure of Scientific revolutions" (1962).

Specifically when doing behavioral neuropharmacology experiments such as those described in this chapter, we must impose *clear limits* to the reaching of any derived conclusions. Although there can be several limiting factors, we cannot ever escape from these three below:

(a) the specific **CNS structure** being targeted (delimited, when accessed through stereotaxically-implanted canullae; or wide-embracing, in the case of an intracerebroventricular or systemic treatment);
(b) the specific **neurochemical/neurohumoral system** being targeted (according to the employed drugs, their selectivity, doses/concentrations, administration pathway, etc.);
(c) the specific **behavioral type of memory** (**neural substrate**) being targeted (according to the specific behavioral task employed); Fig. 17.1 illustrates the diversity of memory types, and could be easily complemented displaying the typical neuroanatomical/neurochemical substrates for each category.

Figure 17.3 illustrates the three inseparable "dimensions" articulated as a "triangle". So, observing the effect of a specific drug applied into certain brain structure in *one* behavioral task does not warrant that the same would take place in *another* behavioral task, or that other neurochemical/neurohumoral target would cause the same effects there, or even that the same would be observed targeting yet another brain structures. For instance, it is common that drugs effective upon aversive tasks (such as, inhibitory avoidance) cause no effect at all upon non-aversive, exploratory tasks (such as open field habituation), and drugs causing one effect

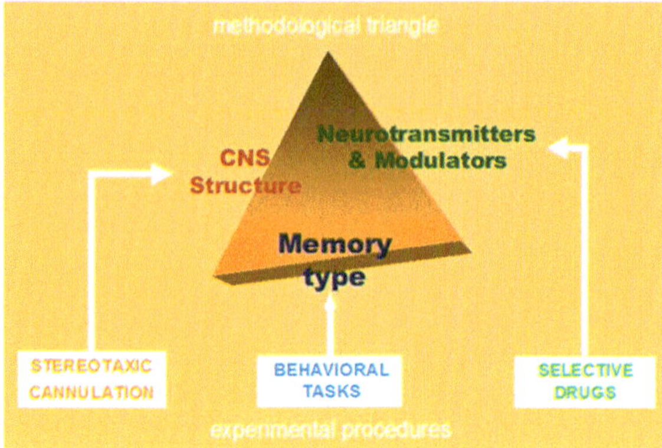

Fig. 17.3 The triangle of methodological limitations of the neuropsychopharmacology of learning and memory: each of the three dimensions demands a specific experimental approach. Conclusions must never extrapolate these limits

when infused into the hippocampus may not necessarily cause the same effect into, say, the amygdala (they may actually differ in a radical way).

By the other hand, more diffused, wide-embracing treatments such as, for instance, a *systemic* drug infusion, may be harder to interpret. Since this treatment may simultaneously target very different CNS structures, it is quite difficult to identify the main neural substrate upon which it would be acting upon, even in the case that *no effect* is detected (brain targets that individually respond in *opposite* fashion may be neutralizing each other's action). This is why systemic treatments, although useful in providing valuable, preliminary information on memory, have limited range in terms of mechanisms' identification. Notice that this comment is valid only in relation to the study of the neural substrates of memory, and may not imply in any criticism to other kinds of systemic treatment.

The situation with systemic treatments may get even more complicated when we use drugs that *cross* the blood-brain barrier (BBB), that may target not only peripheral substrates but also central neural structures. One example of systemic treatment in which these peripheral effects can be simultaneously blocked by another, non-BBB-crossing drug acting upon the same target-receptors, is the pilocarpine epilepsy-inducing treatment, where methyl-scopolamine is previously infused in order to block the undesired peripheral effects (Cavalheiro et al. 1991). Of course, drugs that do not cross the BBB are useful to investigate peripheral neural effects without any central action.

The Biological Continuity Principle and Bioethics

One last consideration concerns how scientific conclusions drawn from experimental animals can be, at last, extrapolated to human clinical cases. Laboratory animals allow us to perform experimental manoeuvres that (a) would never be possible in humans due to obvious ethical reasons, and also (b) allow large scale studies, i.e., large Ns (number of animals used) in order to obtain good statistical reliability. The biological relevance of these results derives from the scientific fact that all of us, humans, monkeys, rats and mice included, are the product of the same *Darwinian Natural Selection Evolutionary Process*, sharing common ancestors in different points of the phylogeny, and, due to this, sharing several fundamental characteristics in terms of metabolism, brain organization and even behavioral strategies. This kinship can be extended much farther, to other vertebrates and even invertebrates, according to the level of complexity of the shared characteristic. The main consequence of this *biological continuity principle* is that results obtained in animal models *can* be at least *similar* (if not *identical*) to those expected/obtained in humans. This is the most fortunate aspect behind the ethical justification to our use of experimental animals in order to understand human characteristics before beginning any real studies directly *in anima nobile*.

All of this does not mean that *everything* is allowed in terms of experimental procedures, and there are ever-growing concern inside and outside scientific community in order to implement better *ethical principles* to guide laboratory animals' use (see specific chapters dealing with this subject in this book).

The biggest human ethical imperative is still to endeavor whatever possible effort to help our fellow conspecifics, but this is being progressively refined and deepened to cover all the species that assist the scientific process. So, there is a strict, solid animal use ethical legislation (and the same should happen with environmental questions). Researchers in the behavioral sciences must be fully committed to the continuous implementation and upgrade of Russell and Burch's (1959) *three Rs principle*—to replace, to reduce, to refine—not only for obvious humane reasons, but because science will still need to use animal models for a long, long time.

Types of Behavioral Tasks

Frustra fit per plura quod fieri potest per pauciora
(It is vain to do with more what can be done with less)

Latin maxim

We will now describe some examples of behavioral tasks that cover most of the fundamental types of memory according to the classification proposed in the previous section (see Fig. 17.1), despite the fact that some of these tasks are quite hard to fit in just one category. To know more about behavioral tasks, including those *not* covered here (operant conditioning, sensitization, etc.), please, refer to the literature suggested in the end of this chapter (in particular the books of Anisman & Bignami 1978 and Boulton et al. 1989; please, see also Izquierdo 2002, and the reviews of Hölscher & O'Mara 1997 and Steckler et al. 1998).

In most (but perhaps not all) of these experimental protocols, it helps a lot if the animal is previously *habituated to the manipulation* in order to avoid behavioral interference of an enormous list of nonspecific aspects of the whole procedure: transporting to and from the experimental room, removing from/returning to home cages, handling, weighing, injection, etc. These procedures should be applied daily, for at least some days. Of course, previous habituation cannot be employed if complete novelty is demanded by the experimental design.

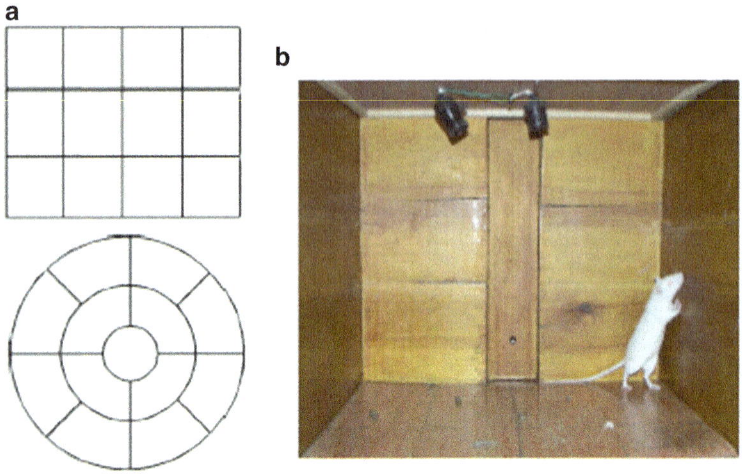

Fig. 17.4 (**a**) Two possible geometrics for open field habituation arenas—rectangular and circular—displaying visible lines that subdivide the floor in regular sectors. (**b**) Typical rearing exploratory posture of a rat

Notice that for practical reasons, the behavioral tasks described are not presented in the same "order" shown in Fig. 17.1.

Open Field Habituation

Open Field Habituation (OF) consists of exposing an animal to an open arena, a new environment without any clearly aversive or appetitive stimuli,[14] and let explore it freely for a fixed amount of time. In this sense, it is the classical non-aversive and non-associative task.

Session duration may range from 2 to 10 min, especially during daytime experimentation; 2–3 min may be the minimum time to assure it *habituates* to the context, and more than 10 min seems to be useless, because the animal will start grooming and/or resting, even sleeping, since there seems to be no novelties/risks around.

The Open Field may have any geometric form, but circular and rectangular arenas are more common. Both types of arenas must display lines subdividing the floor[15] in regular sectors, be it rectangles (in the rectangular arena) or circular sections (in the circular arena)—see Fig. 17.4a. Rectangular arenas may be observed

[14] This may be considered a "neutral" environment, but it is well-known that even the novelty of a new environmental may be stressful for the animal, with intensities that vary according to its intrinsic sensibility.

[15] Linoleum is recommended, because it is easy to clean up: a 70 % alcohol solution is ideal, since it is still somewhat volatile, and yet it does not smells too intensely. Some authors employ different floor textures in order to create subtle context modification.

from above or through a frontal glass wall; circular arenas are usually observed from above. The box may be built in any washable material, such as metal, plastic or plywood, and a typical size (for rats) is 50 cm high, 40 × 60 cm, for the rectangular arena, or a 40–60 cm radius, for the circular one.

During the first exposure—**the "training" session (TR)**—some variables may be quantified in order to measure exploratory behavior. The two most important indexes are (1) the number of *rearings* (REAR) exhibited by the animal, and (2) the number of *crossings* (CROSS) over the lines separating floor sectors. *Rearings* are innate exploratory postures of small rodents, and the newer the environment, the more rearings the animal will exhibit. *Crossings* also express exploration, of course, and have the advantage to measure basal motricity also (some drugs or treatments may affect this function and, in doing so, mask the mnemonic effect if there is one. This task may be automated with a grid of infrared photocells than measures crossings and may record rearings with some confidence (the reliable identification of a rearing posture is somewhat complex).

In order to test for *memory* in this behavioral task, **the "test" session (TT)** is performed after some interval (24 h for long-term memory, less than 6 h for short-term memory, up to 3 min for working memory) putting the same animal in the same arena under the exact same environmental conditions and measuring again the same variables described above.

The rationale of this task is as follows:

- normal memory retention (as should be displayed by control animals) is indicated by a *reduction*, between training and test sessions, of the number of rearings *and/or* crossings;
- this reduction means that the animal has *learned* correctly the task; the difference between training and test sessions, in the number of rearings (or of crossings), is the measure (or score) of retention of the habituation to the open field;
- a behavioral experiment to study memory is valid only if the control group learns adequately the task, otherwise the whole data analysis may not be performed and the experiment should be disconsidered;
- a treatment is *amnestic* when, between training and test sessions, there is no significant difference in both variables (robust amnesia), or in the number of rearings only (partial amnesia[16]);
- a treatment is *facilitatory* when (a) animals *learn* (i.e., their number of rearings *and/or* crossings decreases between training and test sessions), and, moreover, (b) test measurements are *lower* than the control-group test values;
- the observed differences among variable measurements, be it a decrease or an increase, must be *statistically significant* in order to be considered (choose carefully the statistical tests and post hoc tests).

[16] By partial amnesia we understand a situation in which memory was formed, but its trace is less prominent (so, measured memory indexes display lower values).

Results may be classified according to its *degree of robustness*: considering just one of the indexes, we may talk about *partial* or *full* effects; for instance, if the number of rearings does not change between training and test sessions (being not significantly different in statistical terms), we may call it a robust, full amnesia; if the number of rearings in the test session is significantly *higher* than its control-group counterpart, but still significantly different from its training value (i.e., it still learns, so memory is present), we may talk about a partial amnesia.[17]

Important to notice is the attitude of the experimenter: both in the training and in the test session, the animals should be gently placed facing one of the bottom corners and allowed to explore the arena for the specified time. Between training and test sessions, it is recommended to repeat the illumination pattern, temperature, noise level and even the basic odor—remember rodents are hyperosmic small mammals, and this is why it is recommended that the same experimenter be present in both sessions, with similar clothings (odors).

Other variables may be considered, with different meanings: (3) *time to leave the first quadrant*, (TLFQ) usually relates to the anxiety level (despite there being better ways to quantify this—see the chapter on anxiety measurement—a too large leaving time may indicate an abnormal behavior of the animal, possibly *freezing*, and, if this is not related with the experimental design, this situation prompts for the discard of the experimental subject to avoid biased data); (4) *grooming frequency/duration*, a disputable variable indicating both familiarity with the environment and lower levels of anxiety; (5) *defecation boli*, a more controvertible anxiety indicator, for which no agreement is achieved to this point.

The geometry of the arena is mostly a matter of taste: Open Field Habituation may be performed in the circular or the rectangular arenas with similar results. The circular version makes simpler the observation and quantification of *thigmotaxic* vs. *centrophobic* behavior: in the beginning of the exploration session, the animal walks nearer to the walls, guided by tactile/proximal information ("thigmotaxis"), avoiding exposing itself to the open space, probably an evolutionary-selected adaptation; later, after assuring its safety, normal animals explores more the center of the open field (this may change according to the experimental design).

OTHER USES: Since this task permits the measurement of the number of crossings, it may always be used as *a control for the possible motor and general performance effects of the drug* previously administered; in this case, one session would suffice. Another interesting modification is the *Water Licking Task*: the only modification in the protocol described above is the introduction of one small extra detail in the arena, a dropping spout from a water bottle—the animal explores the new environment *and* records the position of the spout; after returning to the home cage, it is water-deprived for 24 h; in the test session, we measure rearings, crossings, and the latency to find the spout and lick the water; this behavior is called *Latent Learning*.

---

[17] An even stronger amnesia is the one observed when these two variables does not change their values and are taken together.

One-Way Step-Down Inhibitory ("Passive") Avoidance

Inhibitory Avoidance (IA) involves learning to inhibit a response in order to avoid an aversive stimulus, and the learning (training) session may be one-trial or multi-trial. Since there is punishment to the natural exploratory drive of a rodent with a non-lethal, pulsating electric footshock, this is clearly an *aversive* task. IA is hard to classify according to criteria discussed (see Fig. 17.1), because it involves both an explicit, associative component (to the context), and an operant-like conditioning[18] component (to the shock), this last being considered a type of implicit memory, especially in the one-trial version of IA.

There are two different approaches to the IA behavior, the step-down IA, here described in more detail, and the step-through IA (see, e.g., Bermudez-Rattoni et al. 1997). The typical step-down IA apparatus is an automatically operated, brightly illuminated box with dimensions around 40.0×25.0×25.0 cm[19] and a frontal glass wall; the floor consists of a grid of parallel 0.1 cm caliber bronze (or steel) bars spaced 1.0 cm apart; the left extremity of the grid is covered by a 7.0–10.0 cm[20] wide, 5.0 cm high formica-covered non-conductive platform. There may be a sliding door separating two halves of the box (as in the step-through IA) and each side may also be painted in different color (e.g., one black, the other, white).

In the one-trial, step-down IA task **training session**, animals are gently held by the body and lowered onto the platform with their noses pointing to the bottom corner, and a chronometer is started. Immediately upon stepping down with their four paws on the grid (when the chronometer is stopped), animals receive a 3.0 s scrambled footshock of 0.2–1.0 mA, according to the experimental design: the stronger the shock, the better the memory retention (and its duration).[21]

The pulsating shock may be delivered during some seconds (3–5 s), and, after that, the animal is removed from the apparatus. Some authors preconize holding the shock until the animal climbs back onto the platform, but this is very stressful if kept for more than 10 s, and could lead the animal to a freezing reaction, and no climbing up at all.

Since the training session is a quick procedure, this task is not simple to be automated.

---

[18] The operant factors are described as follows: "In one-trial inhibitory avoidance (IA), a fear-motivated learning task (Gold 1986), rats associate a conditioned stimulus (CS; an elevated platform present in a given context) with an unconditioned stimulus (US; a shock given to the foot when they step down from that platform)" (Cammarota et al. 2003).

[19] Small variations around these values are possible.

[20] The variation may be larger than this: usually the platform should cover ¼ (or 1/3) of the grid-floor, but depending on the experimental design, it can be larger (e.g., to measure animal activity, Netto and Izquierdo 1985, have used a platform that covered ½ of the floor).

[21] Visible signs of reaction to the shock may include piloerection, back-arching, eyeball-protrusion and even jumping and squeaking, according to the intensity of the shock and/or the sensibility of the animal; extreme reactions such as freezing should be avoided.

In the **test session**[22] the animal is put in the same apparatus, under the exact same environmental conditions, except that no footshock is delivered. A ceiling of 180 s up to 300 s is imposed to the step-down latency, i.e., latencies larger than, say, 300 s, will be counted as 300 s. Notice that if more than one test is made with the same animal, the first test session may involve some degree of memory extinction that would modify the performance in the next one.

The rationale of this task is as follows:

- normal memory retention (as should be displayed by control animals) is indicated by an *increase*, between training and test sessions, of the latency to step-down from the platform;
- this increase means that the animal has *learned* correctly the task; both the difference between training and test session latencies, or the test session latency can be used as retention scores;
- a behavioral experiment to study memory is valid only if the control group learns adequately the task, otherwise the whole data analysis may not be performed and the experiment should be disconsidered;
- a treatment is *amnestic* when, between training and test sessions, there is no significant difference at all in the latency to step-down from the platform between sessions (robust amnesia), or this latency is significantly smaller than the controls one, but still higher that its training value (partial amnesia);
- a treatment is *facilitatory* when (a) animals *learn* (i.e., their latency to step-down from the platform increases between training and test sessions), and, moreover, (b) test measurements are *higher* than the control-group test values; facilitatory drugs/treatments are easier to be detected with lower shock intensities, since higher shocks tend to promote ceiling values in the test latency, and no further increase would be observable;
- the observed differences among variable measurements, be it a decrease or an increase, must be *statistically significant* in order to be considered (choose carefully the statistical tests and post hoc tests).

The same basic attitude aforementioned may be observed by the experimenter, both in the training and the test sessions: between training and test sessions, it is recommended to repeat the illumination pattern, temperature, noise level and the basic odors in the room.

Despite the fact that this task is called *inhibitory avoidance* by some (Izquierdo and Dias 1983) and *passive avoidance* by others (e.g., Anisman 1978), both terms do not have the same meaning: "passive" suggests inactivity, and inhibition refers to a more active restraint. Since it was shown that the retrieval of this task involves a fair amount of activity that was not related to retrieval scores under different shock intensities (Netto and Izquierdo 1985), we understand it is not adequate to call it "passive" (Fig. 17.5).

[22] As described for the OF habituation, test session may be done after an interval of 24 h (for long-term memory), less than 6 h (for short-term memory), or up to 3 min (for working memory).

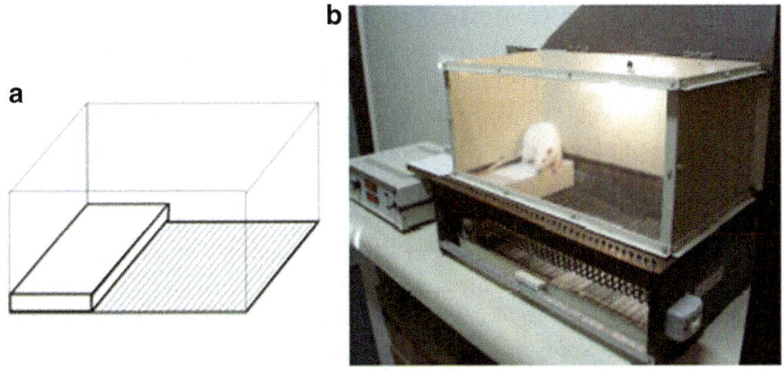

Fig. 17.5 (**a**) Step-down inhibitory avoidance apparatus showing the elevated platform and electrified grid floor. (**b**) In the test session, the animal recalls the aversive experience of the training session—having received a footshock after stepping-down into the grid: the better retention, the larger the latency to descend from the platform

Step-through IA (e.g., Bermudez-Rattoni et al. 1997) employs a trough-shaped alley divided into two compartments separated by a guillotine door that retracts into the floor: a "safe" compartment is illuminated by a fluorescent lamp from above, and is separated from a darkened compartment where animals received the shock (possible dimensions: 90 cm long, 20 cm wide at the top, 6.5 cm wide at the floor, 15 cm deep; safe X dark compartment length proportion is 1:2, i.e., safe compartment would be of 30 cm). Animals are placed in the dark compartment facing away from the door leading to the illuminated compartment and when they turn toward the door, it is opened and a timer is started; a footshock (variable intensities from 0.2 to 1.0 mA) is administered until the animal escapes into the illuminated compartment. The rat is then, retained in the illuminated compartment with the door closed for 60 s. After this, according to the experimental protocol, the animal returns to its home cage (one-trial protocol) or is removed from the lighted compartment and placed back into the dark compartment where the same procedure is followed for the remaining trials (multi-trial protocol). In the test session, the animal is placed in the lighted compartment and the latency to step-through is measured.

Contextual Fear Conditioning

In Fear Conditioning (FC) the animal learns that certain environmental stimuli predict aversive events. Since there is no possibility to escape from the aversive stimulus, this task is an example of a classical, Pavlovian (respondent) conditioning, and represents a defensive behavior selected by evolution in all animals (Maren 2001). The recent interest in this model derives from the fact that this task provides an interface between memory and emotion (LeDoux 2000).

Here we describe the **Contextual Fear Conditioning** protocol, arguably the simplest version of FC once there are only the context (the CS) and the aversive stimulus (the shock, or the US) to be paired. Refer to the literature if you are interested in the more Pavlovian-like auditory FC (Wilensky et al., 2000) or the amygdala-dependent fear-potentiated startle memory (Walker and Davis 2000).

Two conditioning boxes are necessary: they should be placed in different, acoustically isolated separate rooms, and maintained at constant temperature (e.g., 25 °C). The first one (the paired context) has a floor that consists of an electrified grid of bronze (or steel) bars. Despite being similar to the one employed in the IA task, its dimensions are smaller (say, around 20×25×20 cm).[23] Internal illumination, provided by a 2.5 W white light bulb, and background noise (ventilation fans, air conditioning, etc.) should be kept constant in both sessions. The second box (the unpaired context) differ from the first in its size, color, illumination, floor texture, and wall properties to warrant a context as different as possible from the original one (used in training) in order to maximize the possibility of different levels of memory expression. Both chambers (paired and unpaired) should be cleaned (with, e.g., 70 % aqueous ethanol solution) before and after each session.

The measured variable is the time the animal spent *freezing*, taken as an index of fear in rats (Blanchard and Blanchard 1969; Bolles and Collier 1976): an animal is considered to be freezing when crouching, without any visible body movement of the body and head, except for breathing.

The contextual FC training session is made as follows: on the day of conditioning, animals are transported from the housing room and individually placed in the paired context. A 3 min (preshock) habituation (acclimation) period is followed by at least three unassigned scrambled footshocks; to assure a strong aversive learning, a shock of 0.7–1.0 mA is recommended (3 s of duration each and 30 s intershock interval—for this intense protocol randomness among shocks is not necessary). Animals remain in the chamber for another 2 min (postshock period), and then removed back to their home cages (and housing room).

Testing for contextual fear conditioning is assessed 24 h after training: animals are randomly assigned to two subgroups, half of them being reintroduced in the paired context for a 5–10 min period (without shocks), and the other half, exposed for the same period of time to the unpaired context. Freezing is observed (and/or video recorded) during the exposure period, minute by minute (or in consecutive 5 min periods), both in the paired and the unpaired contexts (Fig. 17.6). Total time spent freezing in each period, in each context, is quantified in seconds with a stopwatch.[24]

The rationale of this task is as follows:

- normal memory retention (as should be displayed by control animals) is indicated by a larger percent time in freezing in the test session, compared to the training, or no freezing at all in the training session;

[23] Actually, the same IA apparatus may be used for this task, provided only that a dividing wall is positioned in order use only one side of the box (without a platform).

[24] It is highly recommended to measure freezing behavior at the end of the experiment, preferably by a person who was blind in relation to the treatment applied to each animal (and videotaping is important).

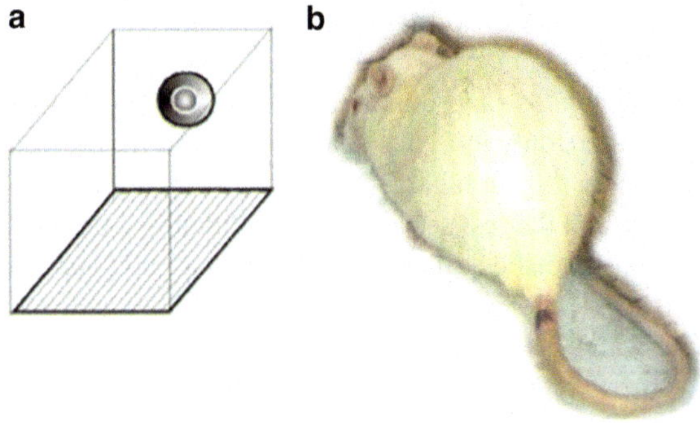

Fig. 17.6 Fear conditioning may be performance in the inhibitory avoidance apparatus (using just half of the grid) or in (**a**) a dedicated box (the loudspeaker is used in the auditory FC protocol, not in the contextual FC); after some footshocks, animals exhibit (**b**) a freezing reaction, a long-lasting absence of movements—except for the respiration (also observable is piloerection, back-arching, and, sometimes, eyeball protrusion)

- this increase means that the animal has *learned* correctly the task; memory is expressed as the percentage of time the animals spent in this defensive behavior; it can be used as retention scores; the better the memory, the more the animal spends in freezing behavior;
- a behavioral experiment to study memory is valid only if the control group learns adequately the task, otherwise the whole data analysis may not be performed and the experiment should be disconsidered;
- a treatment is *amnestic* when, between training and test sessions, there is no significant difference at all in the percent time spent in freezing (robust amnesia, expressed by no freezing at all), or this percent time is significantly smaller than the controls one, but still higher that its training value (partial amnesia);
- a treatment is *facilitatory* when (a) animals *learn* (i.e., their percent time spent in freezing increases between training and test sessions), and, moreover, (b) test values are *higher* than the control-group test values;
- the observed differences among variable measurements, be it a decrease or an increase, must be *statistically significant* in order to be considered (choose carefully the statistical tests and post hoc tests).

Two-Way Active (or Shuttle) Avoidance

In the **Two-Way Active Avoidance** (AA) the animal learns that a random stimulus (a tone, the CS) is a reliable predictor for a coming aversive experience (a shock, the US), and can prompt an evasive action in order to avoid it, i.e., it moves to the other side of the shuttle box (the CR) when the stimuli predict aversive events. Since there is the possibility to learn how to escape, this task may be classified as an operant (or instrumental) conditioning, i.e., the animal must learn the relation between CS (sound) and US (shock) in order to anticipate US with a CR (escape) and avoid it.

This task is also called Shuttle Avoidance, in a reference to the strategy the animal must learn and perform.

The shuttle box apparatus (approximate dimensions 60 × 20 × 30 cm), is similar to the IA box (see above), only it (a) has no platform, and (b) the floor grid is visibly divided at the middle by a 1-cm high acrylic (or similar) hurdle.[25]

Both, the **training** and the **test sessions** have an identical protocol that consist of a fixed number of tone-footshock pairing trials (30 is a good number), in which the CS is a 5 s, 70 dB, 1 kHz tone[26] emitted by a loudspeaker attached in the midline position of the rear wall of the shuttle-box. As soon as the 5 s tone ceases, a 0.5 mA[27] footshock (US) is delivered until the animal crosses the midline; if the animal crosses to the other side of the box[28] *during* the tone (avoidance CR), the shock is interrupted: this must act as a reward.

The shuttle-box should be a fully automated apparatus where it does not matter in which side of the grid the animal is, when the tone comes up, it must move to the other side, without any preferred direction[29]—so, the "two-way" characteristic.

Each session starts with a 3–5 min[30] free exploration of the environment, without any stimuli, and the intertrial interval must vary at random between 10 and 50 s: intertrial interval randomness and the two-way protocol are essential to assure that the only established association is done between the tone and the shock, without other predictive elements such as tone-delivery regularity and/or side of the grid. The shuttle-box apparatus should be placed in a soundproof, dim lighted room.

The automated box should record the total number of crossings, the number of escapes (crossing during the tone) and the number of mistakes; sometimes is useful to record the time receiving shock in each mistaken trial. Usually, the animal makes more mistakes in the beginning, and then starts to perform better; the learning will be expressed in the test session with a lower number of mistakes. Larger training sessions (with more trials) should improve performance, but the stress involved may be a problem due to fatigue or even freezing response (that is why 30 is a good number of trials) (Fig. 17.7).

---

[25] In some cases, the delimitation is made by a wall with an opening (say, 7 × 10 cm) situated on the grid-floor level, and each side is independently illuminated by a 5 W lamp inside the compartment.

[26] Some variation may be introduced in these values without problems, according to the experimental design.

[27] Higher footshock values may be used, up to 1.0 mA.

[28] Delimited by the hurdle. In the automated shuttle-box IR photocells constantly monitor the side the animal is.

[29] The alternative would be the (less employed) One-Way Active Avoidance in which the tone-shock pairing is done in just one and the same side (and the task would be much easier to learn).

[30] Some authors use an elaborate protocol in order to habituate the animals to the situational cues of the apparatus: the pre-habituation may last up to 10 min and be repeated for two consecutive days (Savonenko et al. 2003); in this case, the training session takes place in the following day and the tone-footshock pairings start sooner, after 20 s. Another modification consists of limiting the shock duration up to 30 s.

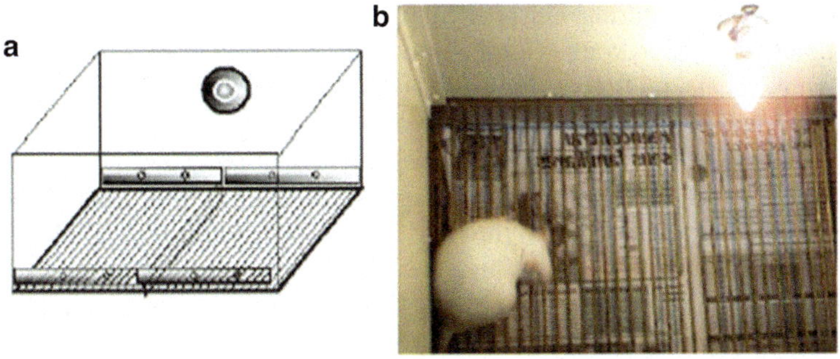

Fig. 17.7 (**a**) The active avoidance apparatus, also known at the shuttle-box, resemble an IA box; only without the platform; each half of the grid is separated by a plastic hurdle and is independently electrified in order to deliver (**b**) the aversive stimulus—a footshock—according to the tone and the side in which the stands (two-way AA protocol)

The rationale of this task is as follows:

- normal memory retention (as should be displayed by control animals) is indicated by higher avoidance responses (or lower number of mistakes) in the test session;
- this increase means that the animal has *learned* correctly the task (and the better the memory, the less mistakes the animal makes); the difference in the avoidance responses between the test and the training sessions is a measure of the memory retention;
- a behavioral experiment to study memory is valid only if the control group learns adequately the task, otherwise the whole data analysis may not be performed and the experiment should be disconsidered;
- a treatment is *amnestic* when, between training and test sessions, there is no significant difference at all in the number of avoidance responses (robust amnesia), or this number is significantly smaller than the controls one, but still higher that its training value (partial amnesia);
- a treatment is *facilitatory* when (a) animals *learn* (i.e., their number of avoidance responses increases between training and test sessions), and, moreover, (b) test measurements are *higher* than the control-group test values;
- the observed differences among variable measurements, be it a decrease or an increase, must be *statistically significant* in order to be considered (choose carefully the statistical tests and post hoc tests).

Additional variables may be recorded during sessions with the assistance of a video camera (or additional shuttle-box gadgets): so, besides the number of avoidance responses, it is possible to measure, e.g., the number of reactions on CS (rearing, turning, freezing, flinching, moving across compartment, vocalization etc.), the latencies of reaction on CS or US,[31] and the intertrial crossings (ITC);

[31] Escape latency in the shuttle box may be affected by the modality of the CS, be it a tone or the illumination level.

the overt behavior during US presentation may be visually discriminated into directional (needs a wall with opening) or nondirectional escape response (Savonenko et al. 2003). The type and number of reactions to CS, besides visually controlled may be divided into three groups according to the definition of these authors: (1) freezing reactions, defined as the lack of any movement except that related to respiration; (2) preparatory responses during CS presentation, i.e., turning of the body and orienting of the head toward the opening during CS presentation, excluding the cases when preparatory response is followed by avoidance reaction, and (3) attention reaction to the CS, i.e., any change in ongoing behavior observed during the first seconds of CS presentation, such as initiation of preparatory response, dissipation of freezing, or the interruption of any previous activity. It is recommended to measure these visually-classified behaviors at the end of the experiment, preferably by a person who was blind in relation to the treatment applied to each animal.

Morris Water Maze

In the beginning of an OF, walking rats navigate mostly based on proximal information obtained with their vibrissae—hence, the "thigmotaxic" behavior (from the Greek *thigma*, to touch). Small rodents are also noticeable for their spatial learning abilities, supposedly dependent upon visual information. Water Maze and Radial Maze have been widely accepted as major spatial learning paradigms (Hölscher and O'Mara 1997). Several variants of each of these tasks may be used in order to obtain abundant behavioral indexes of contextual/spatial habituation, cue-driven navigation, operant-like navigation responses learning, and/or decision-taking.

In the **Morris Water Maze** (MWM or simply WM) task, the animal learns to swim in a water tank, guided by external cues, and find (and climb up to) a submerged platform (Morris 1984). Based upon spatial information, this animal learns how to escape to a platform, so this task may be classified as explicit, associative memory with operant-like spatial learning (see Fig. 17.1).[32]

The water maze is a black-painted[33] circular pool of 120–200 cm diameter, 50 cm high, filled with water to a depth of 25–30 cm. Water temperature is a critical factor

[32] A lot of practical information about water maze techniques can be find at http://www.hvsimage.com/documents/watermaze_tips.pdf

[33] If you use albino or white strains, a black pool maximizes visual contrast for video recording; for dark-haired animals, use white-painted pool. In this last case, some authors, including Morris himself whitens the water with skim powder milk or titanium dioxide in order to increase animal-background contrast and prevent animal from seeing thru the water. However, this may be an overcare, since shuttling the animal from home cage to pool goes by swiftly, and, when on water, the animal cannot easily see anything under waterline; in our experience, it suffices to have a transparent platform.

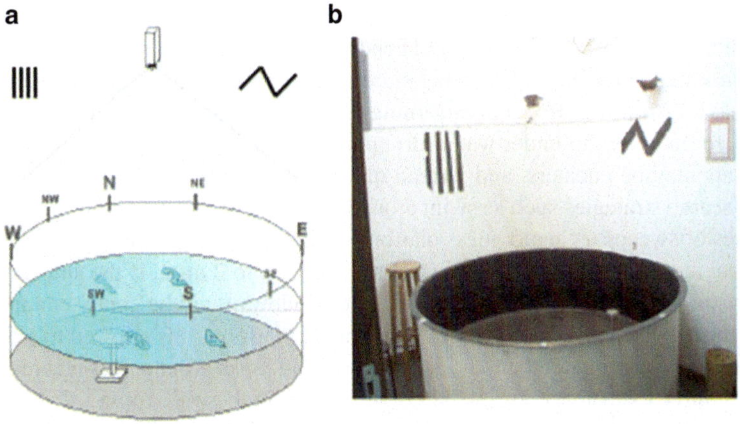

Fig. 17.8 (**a**) The Morris water maze setup, showing the external cues in the walls, quadrants of the tank (with submersed platform), and eight different starting positions. (**b**) A black-painted tank is best for white animals, but room illumination must be adjusted to avoid reflections that disturb videotaping or automatic data acquisition

(optimum is 26 ± 2 °C[34]) as much as the room "decoration": it must be rich in consistently positioned spatial cues, such as the room's door, furniture, noticeable posters (in one or more walls); even the position of the experimenter must be kept constant. The only escape from the water is a platform, with minimum diameter of 10 cm and submerged 0.5–1 cm below the surface. This platform must be invisible to the animal (from its point of view), located in the middle of one of the quadrants (equidistant from the wall and the pool center), and kept in the same quadrant on every trial during the training session.

Rats and mice are natural swimmers, but in this task they just want to get out of the water; swimming for short periods of time does not distress them.[35] Two advantages of Morris water maze over others mazes are that (a) it is a self-driven task (rats want to get out, so it actively searches), and (b) water environment is devoid of local cues, such as scent trails (except for the tank walls).

Training sessions consist of repeating a number of trials, several days in a row (4–8 trials a day, for 2–5 days—or more, when training to a criterion). In each trial, the animal is released from one different starting position randomly selected from eight possible "geographic" points around the perimeter of the pool (Fig. 17.8a). It is important that, during each learning trial, the experimenter is not visible to the animal.[36] The only relevant variable measured in the training trials is escape latency.

---

[34] Although colder water would encourage activity, it may induce hypothermia, known to impair learning; warmer water would favor animal relaxation and decrease exploration.

[35] Swimming for more than 12–15 min without finding any escape is, otherwise, stressful; actually, this is a classic stress model called "forced swimming".

[36] To avoid recue expectation from the animal.

A trial begins by placing the animal in the water, usually facing the pool side (to minimize bias), and timing the latency to reach platform and climb it (escape response). Care must be taken when putting the animal in the water to avoid stress that has a disruptive effect upon learning: gently place it with the tail-end lower, so the head does not dip under water (dropping them in head-first is stressful). Notice that rats may be cheaters, and instead of learning where the platform is, they can learn search strategies such as swim around at some distance from the side, or make a series of sweeps trying to guess platform position.

Trial duration is usually of 60 s (but may take up to 2 min). If the animal fails to climb the platform (escape) within this time window, it will be gently conducted to it by the experimenter. In any case, when on the platform, it is allowed to stay there for 10–30 s to orientate: then, the animal rears and looks around. It is recommended at least three extra-maze visual cues (e.g., posters in the wall, room's door or some furniture). After some swim trials, animals will go directly to the platform.

After each trial, animals are gently lift off, dried,[37] and returned to their home cages until the next trial; at the end of the day (or the training session), home cages are returned to the housing room. For a good retention, an inter-trial interval (ITI) of between a minimum of 10 min and a maximum of 20 min is recommended. All movements (including the path course) are recorded from a camera attached to the ceiling, and either videotaped, or digitally stored in a computer (depending on the automatic setup you have at your disposal[38]) for posterior analysis.

Test session (*probe trial*) takes place 24 h after the last training session: the platform is *not present* and the animal swims only one trial of 60 s (sometimes of 120 s, but data may still be recorded minute by minute), and then the animal is rescued.[39] Measured variables include (a) *latency* (sec) *to reach* the original position of the platform for the first time, (b) number of *crossings* in that exact place, and (c) the *time* (sec) *spent in the target quadrant* (TQT) compared to the opposite one (OQT).

Additionally, other variables can be measured: (d) *path length* (in cm), (e) *mean swimming speed* (cm/s), (f) *departure angle* in relation to the target position, and (f) *time spent in the peripheral ring* vs. central spot (thigmotaxis). Manual record of these variables is pretty hard, but most automated systems (such as HVS) provide them without difficulties. If an altered swim speed is detected, any measured difference between group latencies cannot be clearly interpreted; this is when path length measurements can be helpful: if this variable differs between groups, animals may be experiencing both motor *and* learning impairments; if they are approximately equal across groups, then these latency differences, if detected, may be due to motor, not learning impairments (Hölscher and O'Mara 1997):

---

[37] Dr. Morris' tips (note 23 above) suggest that it is much better to put the animal in a litter of tissues, so it can dry itself.

[38] For instance, the HVS tracking system for water maze, from Dr. Morris' lab (www.hvsimage.com).

[39] Since rodents are good at spatial learning, do this only on trained animals, and not too often: when done at the start, it tests for spatial bias; when done after trainings, it tests for spatial learning.

The rationale of this task is as follows:

- normal memory retention (as should be displayed by control animals) is indicated by (a) small test session *escape latency* (less than 10 s), and (b) significantly higher *mean time spent in the target quadrant* compared to the opposite one; it is also desirable that it crosses the exact place where the platform was more than once;
- the small escape latency (and higher TQT vs. OQT) means that the animal has *learned* correctly the task; the difference in the test session escape latency compared to the first training trial latency is a measure of the memory retention; higher TQT than OQT is another measure, although less robust if taken alone;
- a behavioral experiment to study memory is valid only if the control group learns adequately the task, otherwise the whole data analysis may not be performed and the experiment should be disconsidered;
- a treatment is *amnestic* when, between the first training and the test sessions, there is no significant difference at all in the escape latency (robust amnesia), or when this number is significantly higher than the control one, but still smaller that its training value (partial amnesia);
- a treatment is *facilitatory* when (a) animals *learn* (i.e., their escape latency decreases between training and test sessions, and/or TQT > OQT), and, moreover, (b) test latency is even smaller (or TQT higher) than the control-group test value.
- the observed differences among variable measurements, be it a decrease or an increase, must be *statistically significant* in order to be considered (choose carefully the statistical tests and post hoc tests).

Interesting alternative protocols are possible, such as the Reversal task and the Transfer task (see Hölscher and O'Mara 1997). In the *Reversal task*, platform is moved from one quadrant to the next from trial to trial, and the task consists of learning the new location "overriding" the knowledge about the previous location; this learning is a sensitive way to test animals that have difficulties in learning to "switch" from one procedure to a new one, what is known as *perseverative behavior* (a hippocampus-dependent trait).

Nonspatial WM tasks, such as the Visible Platform and the Visual discrimination tests, not only provide interesting behavioral information, but may also be powerful control tasks. In the *Visible Platform test*, the platform is visible above the water; spatial elements may be removed from the task by randomly moving the platform around in each trial, or simply by wrapping up the WM tank with a curtain that obliterates distal visual cues. This nonspatial visual discrimination task tests if the animals can see the target and move normally to it, but is may also involve a striatum-dependent type of procedural memory, different from the spatial task, that relies on the hippocampus[40] (Packard and Teather 1997): if there is any sensory, motivational or motor impairment, there will be a difference in the latency/distance

[40] These two variants of the WM task have demonstrated a double dissociation of the mnemonic functions of the hippocampus (with the Spatial WM task) and the dorsal striatum (with the Cued

between groups. In the *Visual discrimination tasks*, animals are trained to recognize and discriminate between visual cues: one protocol used two different visible platforms, one stable and the other, floating (but anchored). Since the last one does not support the animal's weight, it must learn to choose which platform to mount in order to escape water; this task controls for general motor skills, visual discrimination, and learning ability (though it does not depend on hippocampus—see, e.g., Bannerman et al. 1994).

8-ARMS Radial Maze

The **8-Ams Radial Maze** (8ARM) apparatus must be elevated from the ground (a minimum of 60 cm is recommended) and may be made of wood or plastic. Dark surface colors are interesting, since rodents are somewhat photophobic, and it will be useful in the case of videotaping and/or automatic tracking due to contrast (at least for white animals).[41] Arms have dimensions of 60 cm long × 10 cm (up to 20 cm) wide, and may or may not have elevated walls—height goes from 2 to 30 cm. Guillotine doors are useful to set different contexts. Central platform size may vary from the regular octagon defined by each of the eight arms width to a somewhat larger arena, i.e., its diameter goes from 25 to 45 cm. Good illumination may be provided from above the maze (in the case of open arms) or independently inside each arm (in the case of walled arms), e.g., with 6 W light bulbs controlled by a switchboard.

Food cups (wells) are drilled into the floor at the end of each arm, to place the baits (food pellets[42]) in order to avoid visibility from the central platform; rebating process may be performed both manually or automatically, depending on the available setup and the experimental design. In order to obtain the "drive" to execute this task, animals are first reduced to 85 % of their ad lib feeding weights; after this, **training sessions** may begin, be it once or more times a day, in consecutive or every other day, for a limited number of days or for an unpredictable number of days (e.g., when training to a criterion).

Radial maze is one of the most versatile and adaptable behavioral tasks, thus, to summarize all the possible experimental designs, is actually impossible. In here we pick two or three simple versions of 8ARM task. Also, if this apparatus is built with detachable arms it will be of great advantage, since (a) it will be appropriate for storage, and (b) may easily be reconfigured into, say, a T-maze, or Y-maze, or a 4-arm "plus" maze.

WM task), a phenomenon also observed with the win-shift and win-stay radial-maze tasks (Packard et al. 1989), and, to some extent, with the allocentric vs. egocentric maze tasks (Kesner et al. 1993).

[41] Tracking programs may assist in measuring animal's running speed, useful to evaluate nonspecific effects upon motor performance.

[42] These may consist of palatable pellets such as peanuts, *Froot loops* (Kellogg's sweet pellets of wheat and corn starch and sucrose) or even special brands, such as Noyes Formula A pellets.

It is important to begin allowing the animal to freely explore the maze (habituation) in the first 1–2 days, with no food available, for 5–10 min each trial. Only in their home cages they will be introduced to the reward (baits), in a limited number (e.g., no more than 8–10 small pellets).

Food trials begin on the following day, and the simplest **training** procedure to study spatial memory consists of baiting only one arm and training the animal to find it, for several trials (this task can be called *spatial delayed matching*). The experimental room setup is similar to the one employed for the WM task (above), with extra-maze visual cues in the walls (posters, door, etc.) and, preferably, without the presence of the experimenter during trial. An arm entry takes place when four paws cross into it. The entry of the animal into an arm previously entered in the same session may also be considered an error (clearly, a working memory deficit), but some protocols tolerate at least two entries during training.

Animals, then, return to their home cages for a delay interval that may vary from 5 s to 24 h or more (depending on which kind of memory is under study). After this, they are put back into the maze for the **retention test**, in which all eight arms are open and only those arms that had not been blocked before the delay contain food. Animals are removed from the maze after all baited arms have been chosen. The entered arms and the order of entry are recorded, including errors.

Considering that healthy rodents explore actively new environments and they naturally tend to (a) alternate between arms and (b) not to visit the same arms twice, a more complicate version of the task may be implemented involving, say, 2, 3 or 4 baited arms—be it in a simple spatial arrangement (e.g., arms 1, 3, 5[43]), be it in a more complex pattern (such as 1, 4, 5). Different authors use different strategies to reinforce learning such as turning lights on and off in baited arms, use floors with distinct textures, or close guillotine doors when animal enters to add an internal delay period (and check for working memory effects). As usual, testing for long-term memory depends only on the duration of the training-test interval (e.g., more than 6 h).

Between trials, and specially, between different subjects, the maze must be cleared of feces and urine vestiges (wipe out and clean with a low % alcohol solution to remove scents).

In another version of the 8ARM, the *spatial delayed nonmatching* or *delayed spatial win-shift task* (Melo et al. 2005), animals already trained make a last (pre-delay) visit to the learned baited arms—nonbaited arms are kept closed, and, then, are removed to home cage for the delay interval (see above).

Thus, the animal is returned to the center of the radial maze, with all eight arms open: the difference is that now only the previously closed/unbaited arms contains the reward, and animals are allowed to complete all choices, i.e., find all the baits (the post-delay recall test). They can be trained until reaching a stable response (e.g., no more than one error in at least three consecutive sessions) and then submitted to a new "challenge" again. This procedure has some similarity with memory

[43] Regular spacing favours search strategies that do not involve long-lasting forms of memory (only, of course, working memory), such as, "entering every other arm to the left".

extinction, and tests for the ability of changing memory referentials (resistant animals are said to be *perseverant*).

Finally, a nonspatial visual discrimination version of this task (the so called *win–stay* protocol) is possible by wrapping up the maze with opaque curtains, and dimly illuminating it from above; rebaiting must be made by an unobtrusive overhead tubing system, and the animal is observed through a slanted overhead mirror from outside the curtain. Maze and pellet habituation sessions (2 days) are similar to those described above. On each food trial (training session), four randomly selected maze arms are illuminated *and* baited with a single food pellet in the food cup after eating it and returning to the center platform, the arm remains lit and is rebaited; after the consumption of the second pellet in that arm, light is turned off and no rebaiting takes place in that arm. Each daily training session ends when animals eat eight pellets (within a trial) or 10 min have passed. Entered arms are recorded and visits to the unlit/unbaited arms are scored as errors; food trials are run once a day for 6 days. This task may involve habit formation and some degree of associative memory.

The spatial delayed matching and the win-shift versions of the 8ARM (see above) are good examples of explicit, associative memory tasks, only with an appetitively-motivated (rewarded) spatial learning component (see Fig. 17.1). A comparison between the MWM and the 8ARM is shown in Table 17.3 (Fig. 17.9).

Object Recognition Task

Recognition is defined as the process by which a subject is aware that a stimulus has been previously experienced. It is strongly dependent on memory, since it requires a series of cognitive operations (such as perception, discrimination, identification and comparisons) that rely on previously known information in order to "match" the observed event against a memory of the previously experienced ones.

Recognition tasks may also be classified in two major paradigms, *object recognition* (item memory) and *place recognition* (spatial memory). Since this last one has obvious relations both with exploratory behavior and some elements discussed for the OF habituation task (see Fig. 17.1, for instance), and may involve even spatial elements, it will not be discussed here. But this does not mean it is less important, its just a matter of personal choice. A very important type of place recognition task, for instance, is the *Spontaneous Alternation Behavior* (SAB), that employs the classical "T" (ot "Y") Maze: about this task, please, refer to the excellent review of Hughes (2004).

Since recognition memory is based in the general principle of "matching", experiments consist of three clearly defined phases: a sample phase, a delay phase (or retention interval) and a choice (or comparison) phase. These phases are analogous to the scheme we have been using till now, with a training session, a training-test interval, and a test session. In this framework, two rules of response are possible to be learned, *matching* and *non-matching*, so the great variety of experimental designs, such as DMS, DNMS, etc. (a complete discussion is in Steckler et al. 1998).

Table 17.3 Advantages and disadvantages of the two major paradigms of spatial learning (adapted with permission from Hölscher and O'Mara 1997)

	Water maze	8-Arm radial maze
Advantages	• No motivational problems—problem cases (e.g. aged or uncooperative animals) are easily tested	• Distinguishes between motor impairment and spatial learning—if the animal has motor problems it can still choose an arm
	• Olfactory cues are not present	• Different memory types can be tested: working memory, long-term (reference) memory, motor (egocentric) memory
	• Animals learn readily and quickly	• A tracking program is not essential
		• Cheap, simple set-up, which can be dismantled and easily stored after use
Disadvantages	• Differences in platform finding latency are the product of learning *and* motor performance; it is not possible to subtract one from another; if the test group displays motor impairments it is impossible to separately assess spatial learning impairments	• Animals must be well-handled and motivated to perform the tasks
	• Working memory cannot be tested independently	• Appetitive task: the animals must be mildly food-deprived to be motivated to look for baited arms; may be a problem in some cases (e.g. in aged rats which need full food supplies to stay healthy)
	• A videotracking system is essential for evaluating the swim tracks of animals	Training the animals usually takes much more time than MWM
	• A fairly large pool has to be installed in a dedicated room	

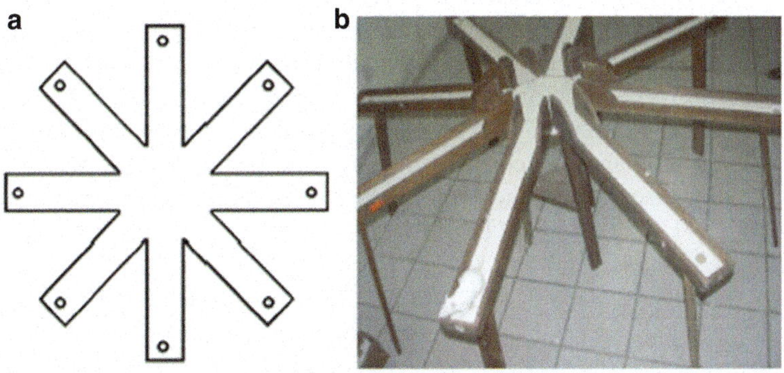

Fig. 17.9 (**a**) 8-arm radial maze schematics showing each aisle (arm) with food cups carved in its extremity. (**b**) The apparatus must be elevated from the floor and external cues, similar to those used in the water maze, must be showing in the walls

Object Recognition (OR) may be performed in any simple box, with or without a transparent wall (if it is the case, the animal is to be observed from above). A typical apparatus has a 50 cm high, 40×60 cm box made of wood (or plastic) with a frontal glass wall, the inside of which is painted with clear colors. Usually the recognition objects are made of plastic or metal to allow easy cleaning between sessions with different animals.

Before starting the trainings, all animals have at least two free exploration sessions for contextual habituation, with no objects inside the box. **Training sessions** (sample phase) consist of allowing the animal to explore two different objects during a certain fixed amount of time (e.g., 5 min). It is important that (a) the objects have a "neutral" shape[44] in terms of its significance to the animal, (b) be devoid of any marked characteristics, such as odor and movement, for instance, and (c) both must be positioned more centrally in the box, at least 10 cm from the side wall, to avoid accidental touching during the initial, thigmotaxic exploration.

The total time spent exploring the two objects is recorded by the experimenter with the assistance of two stop watches: "object exploration" is defined as directing the nose and vibrissae to the object at a distance of less than 2 cm, as if "smelling" it with caution; bumping, turning around or sitting upon the object are not considered exploratory behaviors. At the end of the training trial, the animal is removed from the box and returned to its home cage (Fig. 17.10).

After an interval (the retention delay) that may be of 15 min, the animal is reintroduced into the box for another trial, the **test session** (choice phase), now with a different set of objects—one familiar (identical to, but not the same one previously explored object), and the other, a new/unexplored object—both placed in the same position as the sample stimuli. Due to its particular characteristics, object recognition is usually more employed to investigate short-term memory.[45]

According to the phase analysed, different indexes can be used: a *discrimination index*, the difference in time spent exploring each of the two objects in the *choice phase* (i.e., time with B minus time with A), and the *discrimination ratio*, which is the difference in exploration time, expressed as the ratio of total exploration time with *both* objects in the *choice phase* (this ratio allows to adjust for individual or group differences in the total amount of exploration time).

Recognition memory raises some complex conceptual questions: (1) pure *recall* may sometimes mask an effective *recognition* procedure; (2) at least two types of very different recognition memories can be distinguished, *familiarity/novelty* and *recency*; and (3) different cognitive strategies, from the simplest stimulus-response association to more intricate concept formation, may be involved to a varying extent (for a detailed discussion of these topics, see Steckler et al. 1998).

---

[44] As far as it could be ascertained, the objects should have no natural significance for the rats, to avoid being associated with a reinforcer (or an aversive stimulus).

[45] By the other hand, place recognition tasks are more versatile and may allow the investigation of long-term memory processes (with "delays" of more than 6 h).

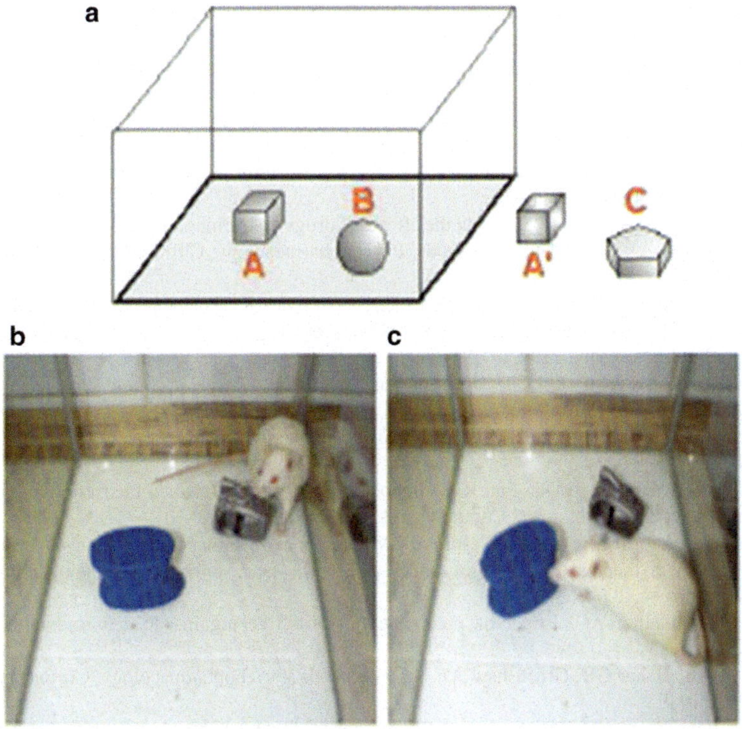

Fig. 17.10 (**A**) Object recognition task setup: two different plastic (washable) objects, **a** and **b**, are explored by the animals in the sample phase (training session); in the choice phase (test session) they are substituted by **a′** (identical to **a**) and **c** (totally unfamiliar). (**B**) Object exploration is defined as directing the nose and vibrissae to the object at a distance of less than 2 cm

Acknowledgements Science involves a long apprenticeship and, at last, always remains a collective endeavor. Anyway, some people deserve to be mentioned with special care: I would like to thank my former tutor, Dr. Ivan Izquierdo (PUCRS, Brazil), for the privilege of his friendship. I would also like to acknowledge Dr. James McGaugh (UCI, USA) for all the precious lessons over the years. Finally, I would like to thank my good friends Dr. Diana Jerusalinsky (UBA, Argentina) and Victor Molina (UNC, Argentina). I also thank each one of them for kindly reading this manuscript, but I, alone, am responsible for any errors found here.

References

Anisman H. Aversively motivated behavior as a tool in psychopharmacologic analysis. In: Anisman H, Bignami G, editors. Psychopharmacology of aversively motivated behavior. New York: Plenum Press; 1978. p. 1–62.

Anisman H, Bignami G. Psychopharmacology of aversively motivated behavior. New York: Plenum Press; 1978.

Bannerman DM, Chapman PF, Kelly PA, Butcher SP, Morris RG. Inhibition of nitric oxide synthase does not impair spatial learning. J Neurosci. 1994;14:7404–14.
Barros DM, Pereira P, Medina JH, Izquierdo I. Modulation of working memory and of long- but not short-term memory by cholinergic mechanisms in the basolateral amygdala. Behav Pharmacol. 2002;13:163–7.
Bekinschtein P, Cammarota M, Katche C, Slipczuk L, Rossato JI, Goldin A, Izquierdo I, Medina JH. BDNF is essential to promote persistence of long-term memory storage. Proc Natl Acad Sci U S A. 2008;105:2711–6.
Beninger RJ. Methods for determining the effects of drugs on learning. In: Boulton AB, Baker GB, Greenshaw AJ, editors. Neuromethods. Psychopharmacology. Clifton: Humana Press; 1989. p. 623–85.
Bermudez-Rattoni F, Introini-Collison I, Coleman-Mesches K, McGaugh JL. Insular cortex and amygdala lesions induced after aversive training impair retention: effects of degree of training. Neurobiol Learn Mem. 1997;67:57–63.
Blanchard RJ, Blanchard DC. Crouching as an index of fear. J Comp Physiol Psychol. 1969;67:370–5.
Blanchard RJ, Blanchard DC. Dual mechanisms in passive avoidance I & II. Psychon Sci. 1970;19:1–4.
Blanchard RJ, Blanchard DC, Fial RA. Hippocampal lesions in rats and their effect on activity, avoidance, and aggression. J Comp Physiol Psychol. 1970;71:92–101.
Boccia MM, Acosta GB, Blake MG, Baratti CM. Memory consolidation and reconsolidation of an inhibitory avoidance response in mice: effects of i.c.v. injections of hemicholinium-3. Neuroscience. 2004;124:735–41.
Bolles RC, Collier AC. Effect of predictive cues on freezing in rats. Anim Learn Behav. 1976;4:6–8.
Boulton AB, Baker GB, Greenshaw AJ. Neuromethods. Psychopharmacology. Clifton: Humana Press; 1989.
Bouton ME, Westbrook FR, Corcoran KA, Maren S. Contextual and temporal modulation of extinction: behavioral and biological mechanisms. Biol Psychiatry. 2006;60:352–60.
Brillaud E, Morillion D, de Seze R. Modest environmental enrichment: effect on a radial maze validation and well being of rats. Brain Res. 2005;1054:174–82.
Bustos SG, Maldonado H, Molina VA. Midazolam disrupts fear memory reconsolidation. Neuroscience. 2006;139:831–42.
Bustos SG, Maldonado H, Molina VA. The disruptive effect of midazolam on fear memory reconsolidation: decisive influence of reactivation time span and memory age. Neuropsychopharmacology. 2009;34:446–57.
Callegari-Jacques SM. Bioestatística—princípios e aplicações. Porto Alegre, Brasil: Artmed; 2003.
Cammarota M, Bevilaqua LR, Kerr D, Medina JH, Izquierdo I. Inhibition of mRNA and protein synthesis in the CA1 region of the dorsal hippocampus blocks reinstallment of an extinguished conditioned fear response. J Neurosci. 2003;23:737–41.
Carobrez AP, Bertoglio LJ. Ethological and temporal analyses of anxiety-like behavior: the elevated plus-maze model 20 years on. Neurosci Biobehav Rev. 2005;29:1193–205.
Cavalheiro EA, Leite JP, Bortolotto ZA, Turski WA, Ikonomidou C, Turski L. Long-term effects of pilocarpine in rats: structural damage of the brain triggers kindling and spontaneous recurrent seizures. Epilepsia. 1991;32:778–82.
Debiec J, LeDoux JE. Disruption of reconsolidation but not consolidation of auditory fear conditioning by noradrenergic blockade in the amygdala. Neuroscience. 2004;129:267–72.
Debiec J, LeDoux JE, Nader K. Cellular and systems reconsolidation in the hippocampus. Neuron. 2002;36:527–38.
Dudai Y. The neurobiology of consolidations, or, how stable is the engram? Annu Rev Psychol. 2000;55:51–86.
Duvarci S, Nader K. Characterization of fear memory reconsolidation. J Neurosci. 2004;24:9269–75.

Eisenberg M, Kobilo T, Berman DE, Dudai Y. Stability of retrieved memory: inverse correlation with trace dominance. Science. 2003;301:1102–4.

File SE, Gonzalez LE, Gallant R. Role of the basolateral nucleus of the amygdala in the formation of a phobia. Neuropsychopharmacology. 1998;19:397–405.

Flecknell P. Laboratory animal anesthesia. 2nd ed. London: Academic; 1996.

Frenkel L, Maldonado H, Delorenzi A. Memory strengthening by a real-life episode during reconsolidation: an outcome of water deprivation via brain angiotensin II. Eur J Neurosci. 2005;22:1757–66.

Gold PE. The use of avoidance training in studies of modulation of memory storage. Behav Neural Biol. 1986;46:87–98.

Hölscher C, O'Mara SM. Model learning and memory systems in neurobiological research: conditioning and associative learning procedures and spatial learning paradigms. In: Lynch MA, O'Mara SM, editors. Neuroscience Labfax. London: Academic; 1997.

Hughes RN. The value of spontaneous alternation behavior (SAB) as a test of retention in pharmacological investigations of memory. Neurosci Biobehav Rev. 2004;28:497–505.

Izquierdo I. Different forms of post-training memory processing. Behav Neural Biol. 1989; 51:171–202.

Izquierdo I. Memória. Porto Alegre: Artmed; 2002.

Izquierdo I, Dias RD. Effect of ACTH, epinephrine, beta-endorphin, naloxone, and of the combination of naloxone or beta-endorphin with ACTH or epinephrine on memory consolidation. Psychoneuroendocrinology. 1983;8:81–7.

Izquierdo I, Quillfeldt JA, Zanatta MS, Quevedo J, Schaeffer E, Schmitz PK, Medina JH. Sequential role of hippocampus and amygdala, entorhinal cortex and parietal cortex in formation and retrieval of memory for inhibitory avoidance in rats. Eur J Neurosci. 1997;9:786–93.

Izquierdo I, Barros DM, Mello e Souza T, de Souza MM, Izquierdo LA, Medina JH. Mechanisms for memory types differ. Nature. 1998;393:635–6.

Izquierdo I, Medina JH, Vianna MR, Izquierdo LA, Barros DM. Separate mechanisms for short- and long-term memory. Behav Brain Res. 1999;103:1–11.

Izquierdo LA, Barros DM, Vianna MR, Coitinho A, de David e Silva T, Choi H, et al. Molecular pharmacological dissection of short- and long-term memory. Cell Mol Neurobiol. 2002;22:269–87.

Jerusalinsky D, Quillfeldt JA, Walz R, Da Silva RC, Bueno e Silva M, Bianchin M, et al. Effect of the infusion of the GABA-A receptor agonist, muscimol, on the role of the entorhinal cortex, amygdala, and hippocampus in memory processes. Behav Neural Biol. 1994;61:132–8.

Kelley AE, Cador M, Stinus L. Exploration and its measurement. A psychopharmacological perspective. In: Boulton AB, Baker GB, Greenshaw AJ, editors. Neuromethods. Psychopharmacology. Clifton: Humana Press; 1989. p. 95–144.

Kesner RP, Bolland BL, Dakis M. Memory for spatial locations, motor responses, and objects: triple dissociation among the hippocampus, caudate nucleus, and extrastriate visual cortex. Exp Brain Res. 1993;93:462–70.

Krinke GJ. The laboratory rat. San Diego: Academic; 2000.

Kuhn TS. The structure of scientific revolutions. 1st ed. Chicago: University of Chicago Press; 1962.

LeDoux JE. Emotion circuits in the brain. Annu Rev Neurosci. 2000;23:155–84.

Maren S. Neurobiology of pavlovian fear conditioning. Annu Rev Neurosci. 2001;24:897–931.

McGaugh JL. Time-dependent processes in memory storage. Science. 1966;153:1351–8.

Melo LCS, Cruz AP, Valentim Jr SJR, Marinho AR, Mendonça JB, Nakamura-Palacios EM. Δ9-THC administered into the medial prefrontal cortex disrupts the spatial working memory. Psychopharmacology (Berl). 2005;183:54–64.

Misanin JR, Miller RR, Lewis DJ. Retrograde amnesia produced by electroconvulsive shock after reactivation of a consolidated memory trace. Science. 1968;160:554–5.

Morris R. Developments of a water-maze procedure for studying spatial learning in the rat. J Neurosci Methods. 1984;11:47–60.

Myers KM, Davis M. Mechanisms of fear extinction. Mol Psychiatry. 2007;12:120–50.

Nader K. Memory traces unbound. Trends Neurosci. 2003a;26:65–72.
Nader K. Neuroscience: re-recording human memories. Nature. 2003b;425:571–2.
Nader K, Schafe GE, Le Doux JE. Fear memories require protein synthesis in the amygdala for reconsolidation after retrieval. Nature. 2000;406:722–6.
Nahas TR. A aprendizagem da esquiva. In: Xavier GF, editor. Técnicas para o estudo do sistema nervoso. São Paulo: Plêiade; 1999a. p. 221–41.
Nahas TR. O teste do campo aberto. In: Xavier GF, editor. Técnicas para o estudo do sistema nervoso. São Paulo: Plêiade; 1999b. p. 203–20.
Netto CA, Izquierdo I. On how passive is inhibitory avoidance. Behav Neural Biol. 1985;43: 327–30.
Norman GR, Streiner DI. Biostatistics: the bare essentials. St. Louis: Mosby; 1994.
Packard MG, Teather LA. Double dissociation of hippocampal and dorsal-striatal memory systems by posttraining intracerebral injections of 2-amino-5-phosphonopentanoic acid. Behav Neurosci. 1997;111:543–51.
Packard MG, Hirsh R, White NM. Differential effects of fornix and caudate nucleus lesions on two radial maze tasks: evidence for multiple memory systems. J Neurosci. 1989;9:1465–72.
Pavlov IP. Conditioned reflexes: an investigation of the physiological activity of the cerebral cortex. London: Routledge Kegan Paul; 1927.
Paxinos G, Watson C. The rat brain in stereotaxic coordinates—the new coronal set. 5th ed. New York: Academic; 2004. p. 209.
Pedreira ME, Maldonado H. Protein synthesis subserves reconsolidation or extinction depending on reminder duration. Neuron. 2003;38:863–9.
Przybyslawski J, Sara SJ. Reconsolidation of memory after its reactivation. Behav Brain Res. 1997;84:241–6.
Przybyslawski J, Roullet P, Sara SJ. Attenuation of emotional and nonemotional memories after their reactivation: role of beta adrenergic receptors. J Neurosci. 1999;19:6623–8.
Quillfeldt JA, Zanatta MS, Schmitz PK, Quevedo J, Schaeffer E, Lima JB, Medina JH, Izquierdo I. Different brain areas are involved in memory expression at different times from training. Neurobiol Learn Mem. 1996;66:97–101.
Routtenberg A. Reverse piedpiperase: is the knockout mouse leading neuroscientists to a watery end? Trends Neurosci. 1996;19:471–2.
Russell WMS, Burch RL. The principles of humane experimental technique. London: Methuen; 1959 [reprinted by UFAW, 1992: 8 Hamilton Close, South Mimms, Potters Bar, Herts EN6 3QD England].
Sanger DJ, Blackman DE. Operant behavior and the effects of centrally acting drugs. In: Boulton AB, Baker GB, Greenshaw AJ, editors. Neuromethods. Psychopharmacology. Clifton: Humana Press; 1989. p. 299–348.
Savonenko A, Werka T, Nikolaev E, Zielinski K, Kaczmarek L. Complex effects of NMDA receptor antagonist APV in the basolateral amygdala on acquisition of two-way avoidance reaction and long-term fear memory. Learn Mem. 2003;10:293–303.
Siegel S, Castelan NJ. Nonparametric statistics. 2nd ed. Boston: McGraw-Hill; 1988.
Skinner BF. Science and human behavior. New York: Macmillan; 1953.
Squire LR. Memory and brain. New York: Oxford University Press; 1987.
Squire LR, Kandel ER. Memory: from mind to molecules. New York: WH Freeman; 1999.
Steckler T, Drinkenburg WH, Sahgal A, Aggleton JP. Recognition memory in rats—I. Concepts and classification. Prog Neurobiol. 1998;54:289–311.
Suzuki A, Josselyn SA, Frankland PW, Masushige S, Silva AJ, Kida S. Memory reconsolidation and extinction have distinct temporal and biochemical signatures. J Neurosci. 2004;24:4787–95.
Swanson LW. Brain maps: structure of the rat brain. 2nd ed. Amsterdam: Elsevier; 1998.
Swerdlow NR, Gilbert D, Koob GF. Conditioned drug effects on spatial preference: critical evaluation. In: Boulton AB, Baker GB, Greenshaw AJ, editors. Neuromethods. Psychopharmacology. Clifton: Humana Press; 1989. p. 399–446.
Tronson NC, Taylor JR. Molecular mechanisms of memory reconsolidation. Nat Rev Neurosci. 2007;8:262–75.

Tsien JZ, Huerta PT, Tonegawa S. The essential role of hippocampal CA1 NMDA receptor-dependent synaptic plasticity in spatial memory. Cell. 1996;87:1147–8.

Walker DL, Davis M. Involvement of NMDA receptors within the amygdala in short- versus long-term memory for fear conditioning as assessed with fear-potentiated startle. Behav Neurosci. 2000;114:1019–33.

Xavier GF. A aprendizagem da esquiva ii—a esquiva passiva. Ciência e Cultura. 1982;34: 1587–600.

Xavier GF, Bueno OF. On delay-of-punishment and preexposure time: effects on passive avoidance behavior in rats. Braz J Med Biol Res. 1984;17:55–64.

Zar JH. Biostatistical analysis. 4th ed. Englewood Cliffs: Prentice Hall; 1999. p. 663.

Chapter 18
Animal Tests for Anxiety

Leandro José Bertoglio and Antônio de Pádua Carobrez

In behavioral neurosciences, animal models enable investigation of brain-behavior relations, with the aim of gaining insight into human behavior and its underlying processes. Beyond doubt, the most significant information derives from the study of humans, but this is not always possible due to ethical, methodological, and/or economical constraints. Alternatively, a comparative approach that relies on animal models could be used to reach these goals. This approach is based on the evolutionary theory proposing that fundamental aspects of the behavior of humans are shared with other animals (Nesse 1999; Panksepp et al. 2002). The ground for this proposal was established by Charles Darwin, by means of his book entitled *The Expression of Emotions in Man and Animals* (Darwin 1872).

According to van der Staay (2006), an animal model is a living organism used to study brain-behavior relations under controlled conditions, with the final goal to gain insight into, and to enable predictions about, these relations in humans. An additional implicit purpose is that animal models can provide a simplification of complex phenomena.

Various experimental preparations have been developed to assess behavioral parameters indicating anxiety, especially in rodents such as rats and mice. These protocols may be useful in investigating the fundamental neuronal mechanisms underlying anxiety and may contribute to the development of new medications. The evaluation of anxiety-related behavior in animals submitted to biochemical or gene targeting manipulations is also conducted in this fashion (Carobrez and Bertoglio 2005).

L.J. Bertoglio
Departamento de Farmacologia, Universidade Federal de Santa Catarina, Florianópolis, Santa Catarina, Brazil

A.P. Carobrez, Ph.D. (✉)
Departamento de Farmacologia/CCB, Campus Universitario-Trinidade, Universidade Federal de Santa Catarina, Florianópolis 88040-900, Brazil
e-mail: Padua.carobrez@ufsc.br

M.L. Andersen, S. Tufik (eds.), *Rodent Model as Tools in Ethical Biomedical Research*, DOI 10.1007/978-3-319-11578-8_18

Before describing how to test anxiety in animals, it is important to establish some assumptions and limitations of the animal model. Firstly, the use of animal models stands on the assumption that anxiety in animals is comparable to anxiety in humans. Although it cannot be proven that an animal experiences anxiety in the same way as a human being, it is well accepted that some behavioral patterns in rodents indicate anxiety-like behavior akin to that of humans (Ohl 2003). Secondly, in both humans and laboratory animals this emotional state is not a unitary phenomenon, as it includes innate (trait) anxiety, which is considered to be an enduring feature of an individual, and situation-evoked (state) or experience-related anxiety (van der Staay 2006). Finally, modeling anxiety in animals is critically dependent on the test systems used. As the behavior of a species has been shaped during evolution, *a valid test for anxiety should respect and maximize the animal natural defensive behavior*. Although this may sound trivial, the ethological relevance of behavioral tasks has long been ignored in basic research. In view of these facts, one should always consider that data derived from animal models are of value only to the extent that the models are valid, and that the severity of the disorder evoked in animals may not be equivalent to the level of human disorder being modeled (Kalueff and Tuohimaa 2004). One of the current challenges is therefore to optimize existing neurobiological approaches to anxiety. The present chapter considers important methodological and conceptual issues which may improve the validity, and thus provide a more rational use, of animal tests for anxiety.

General Concepts in the Experimental Modeling

The behavioral repertoire of animals has long been employed to detect effects on and impact of anxiety (Rodgers 1997; Blanchard et al. 2003; Ohl 2003). A number of models based on animal emotional reactivity have been designed and proven to be bidirectionally sensitive to stressful manipulations (Table 18.1). Many of these tests involve simple, rapid, and inexpensive ways of evaluating an animal's condition (Table 18.2).

The classification of experimental tests for anxiety can be based on the nature and the type of stressors employed. They can either induce (by drugs, targeted gene mutations, or brain lesions/stimulations or external stressful factors) or measure anxiety, in terms of behavioral and physiological reactions (Overall 2000; Belzung and Griebel 2001; Uys et al. 2003; van der Staay 2006). With regard to the eliciting stimulus, experimental preparations can be categorized as being "natural" or "artificial." Whereas the former is outlined to maximize the naturally occurring defensive behavior in response to aversive stimuli with ecological meaning to the species (e.g., predator odor and open spaces), the latter employs strong and often painful stressors (e.g., shock) to elicit behaviors not normally seen in natural conditions (Kalueff 2003). Natural animal models allow for a reliable evaluation of a number of external factors including pharmacological agents (Overall 2000). Such ethologically based paradigms are more sensitive to stress when compared to artificial ones (Rodgers 1997).

Table 18.1 Some commonly used animal tests for anxiety[a]

Name	Reference for reviewing details
Elevated plus-maze	Bertoglio and Carobrez (2005)
Mouse defense test battery	Blanchard et al. (2003)
Exposure to predator odors (e.g., cat)	Apfelbach et al. (2005)
Light/dark exploration	Bourin and Hascoet (2003)
Elevated T-maze	Graeff and Zangrossi (2002)
Social interaction	File and Seth (2003)
Open-field (arena)	Prut and Belzung (2003)
Ultrasonic (or stress-induced) vocalization	Sanchez (2003)
Defensive burying	De Boer and Koolhaas (2003)
Vogel conflict test	Millan and Brocco (2003)
Discriminative avoidance task	Silva and Frussa-Filho (2000)

[a]Based on ethopharmacological evidence, it has been proposed that anxiety and fear are categorically distinct entities (Gray and McNaughton 2000). Further details about defensive behaviors thought to reflect fear (e.g., freezing and escape) and relevant experimental tests to study them in animals can be found elsewhere (e.g., Graeff and Zangrossi 2002; Blanchard et al. 2003; Bittencourt et al. 2005)

Table 18.2 Principal behavioral profiles in experimental tests for anxiety[a]

Avoidance	↑
Exploration	↓
Risk assessment	↑ or ↓ (depending on the model)
Self-grooming	↑ or ↓ (frequency; depending on model)
Immobility	↑ (freezing)
Defecation, urination	↑
Aggression	↑
Others (e.g., defensive burying)	↑

Legend: ↑ = increase; ↓ = decrease
[a]Adapted from Kalueff and Tuohimaa (2004)

Clearly, the stressfulness of the test has to be taken into account when analyzing the behavior, as it may significantly affect behavioral performance. The present chapter will focus on the first group of anxiety tests.

Behavioral Dimensions Related to Anxiety

The expressions of species-specific behaviors have been shown to be related to anxiety in rodents. This section concisely describes some of these behaviors and discusses their use in increasing the reliability and sensitivity of tests for anxiety.

Avoidance

It is well known that rodents tend to avoid the unprotected area of a novel environment when first entering it (Belzung and Le Pape 1994). In an experimental setup, usually represented by a defined area, rodents will typically start to explore the environment along the walls while avoiding the open (i.e., unprotected) area. This aversive characteristic can be modulated by illumination levels, by the height of the apparatus, and by enabling the animal to see the edge. Notably, the expression of avoidance behavior depends on the sensorial (visual, olfactory, touch, hearing, taste) capabilities of the animal and can further be influenced by its locomotor activity, motivational factors, and by its exploration strategy (Ohl 2003). Nevertheless, a large body of literature reports avoidance behavior in rodents to be sensitive to compounds, such as benzodiazepines, that show anxiolytic-like activity in humans (Graeff and Zangrossi 2002; Apfelbach et al. 2005; Carobrez and Bertoglio 2005).

Risk Assessment

When confronted with a threatening stimulus, rodents display species-specific patterns of response, such as stretched-attend posture,[1] which is categorized as risk assessment behavior (Blanchard and Blanchard 1989; Rodgers 1997). The biological function of these behaviors is to gather information about the potential threat by cautiously approaching the threatening stimulus or by scanning the surrounding area. Risk assessment behavior is thought to be an active defense pattern (Blanchard et al. 2003), thus being closely related to anxiety. Usually, rodents will still display increased risk assessment behavior even when no longer avoiding an unprotected area. Therefore, risk assessment, representing thus the most enduring behavioral expression of anxiety, may even be more sensitive to anxiety modulating drugs than avoidance behavior (Rodgers and Cole 1994; Roy and Chapillon 2004).

Exploration

When confronted with novelty, the resultant behavior in rodents is determined by the conflict between the drive to explore the unknown area/object and the motivation to avoid potential danger. The exploratory behavior in rodents summarizes a broad spectrum of behavioral patterns such as risk assessment behaviors, walking, rearing, climbing, sniffing, and manipulating objects (Barnett 1975). Exploration is gradually inhibited by anxiety, thereby representing an indirect measurement of

[1] It occurs when the animal stretches forward (Fig. 18.1c) and then retracts to its original position.

anxiety (Bertoglio and Carobrez 2000). Notably, primary alterations in exploratory motivation may confound measures of anxiety (Belzung 1999), which must be considered when behaviorally phenotyping rodents.

Cognition

It has been argued that anxiety-related behavior and cognitive processes may interact in a fundamental manner. In this respect, it has been shown that, in a rat model of innate high or low emotionality, the degree of anxiety is differentially associated with enhanced performance of distinct informational processes (Ohl et al. 2002). It was hypothesized that the increased anxiety-related behavior may be due to these differences in cognitive processing. Consequently, cognitive processes should be carefully controlled, especially when characterizing animal models for anxiety or anxiety-modulating effects of drugs.

Some Methodological Issues

Are animal anxiety models a good approximation to human disorders? Which tests could be classified as a "good model," and which particular subtypes of anxiety do they model? These questions are rarely asked, but are fundamentally important. The use of nearly all animal tests has been extensively criticized in the literature for several reasons. Firstly, many clinically important, especially cognitive-based, symptoms of anxiety cannot be directly modeled in animals. Secondly, behavioral measures are often confounded and reflect changes in general activity, exploration, and anxiety levels (Belzung and Griebel 2001; File 2001). Thirdly, there is often a poor correlation between different behavioral measures taken in the same test, or between the same measures taken in several different tests (Kalueff 2003). For example, even the simplest task, distinguishing between horizontal exploration and locomotion in the open-field, still requires further elaboration (Choleris et al. 2001). Thus, since it is difficult to interpret a subjective anxiety level based on a single behavioral measure, proper understanding of the animal state is only possible through assessment of interaction between behavioral and physiological variables (Calatayud and Belzung 2001). Given that various forms of psychopathologies in animals and humans can be characterized as context-regulation disorders, subjects may sometimes produce "normal" behavior in inappropriate contexts. Thus, special analysis of behavioral contexts may be needed in the field of animal anxiety. Finally, it should be noted that animal emotional behavior is not just "plus" or "minus," but has several dimensions including anxiety, exploration, locomotion, risk assessment, general arousal, and coping (Salome et al. 2002). These dimensions interact with each other as well as with cognitive functions, giving a complex, mosaic picture of behavior. Therefore, the traditional quantitative behavioral methods (i.e., latency, frequency, and duration parameters, and their spatial, temporal, or sequential patterns) to study animal stress are now combined with sophisticated analysis of "not just the presence or absence of these behaviors, but also whether or not the acts, postures and gestures are fully developed in intensity, latency and patterning" (Barrett and Miczek 2000).

Validity and Reliability Aspects

Validity can be defined as the agreement between a test score and the quality that it measures (Kaplan and Saccuzzo 1997). It represents the process by which the reliability and the relevance of a method are established for specific purposes (Kalueff and Tuohimaa 2004). Reliability is characterized by the reproducibility of a test within and between laboratories and over time. Since numerous differences exist between laboratories, good reproducibility at least within the same laboratory needs to be established (Salome et al. 2002). Relevance reflects the meaningfulness and usefulness of results obtained with a particular animal test (van der Staay 2006).

It has been stated that (anxiety) animal models should possess face validity, be predictive, and construct validity (Sarter and Bruno 2002). This order reflects the hierarchy of the categories of validity,[2] where the construct validity is the highest one (Kaplan and Saccuzzo 1997). To be a good model, it should fulfill all these three criteria at the same time. This situation, however, is not often seen in animal modeling (Clement et al. 2002).

An animal model possesses face validity if there are resemblances between the test and the situation to be modeled. Several authors, however, consider face validity to be of limited value (Kaplan and Saccuzzo 1997; Sarter and Bruno 2002). One of the main reasons is that animals have their own species-specific behavioral repertoire to survive in their habitat, and thus, there may be little resemblance between their behavior and that of humans even though similar underlying processes may guide these behaviors (Sarter and Bruno 2002). In relation to the predictive validity, it takes place when the result obtained in the test has some value to the process to be modeled, i.e., if it allows extrapolation of the effect of a particular experimental manipulation from one condition to others (e.g., laboratory to real world). In psychopharmacology, predictive validity usually refers to the ability of a drug screening or an animal model to correctly identify the efficacy of a putative therapeutic approach (van der Staay 2006). Finally, it is assumed that a certain animal model possesses construct validity if its procedures provide a sound theoretical framework (McGuire and Troisi 1998). In this respect, Darwin (1872) advocated

[2] In addition, the genetical validation based on behavioral phenotyping approach is becoming increasingly important (Flint 2003).

that behavioral characteristics are acquired as a result of selective pressure exerted by evolution. As several environmental constraints are similar, many adaptations are general to species, among which stand basic emotions such as anxiety. This view therefore justifies the use of animals other than humans. Finally, one can also say that face validity reflects the isomorphic aspect, predictive validity the correlational aspect, and construct validity the homologous aspect of a model (Kalueff and Tuohimaa 2004).

Panorama of Current Animal Tests in Use: Differences from the Ideal

The preceding sections reviewed some essential concepts related to animal models, the animal behavioral expressions related to anxiety, and important issues to evaluate their reliability, validity, and relevance. Using the elevated plus-maze (EPM) test as an example, this section further stresses the importance of (1) identifying and controlling the major sources of variability; (2) introducing the measurement of more subtle defensive behaviors such as risk assessment; and (3) adopting other methodological refinements such as min-by-min scoring and use of a test/retest protocol.

The EPM (Fig. 18.1) stands as one of the most popular *in vivo* animal tests currently in use. Its popularity, with around 2,700 published papers so far (Web of Science, 2006), is likely due to its obvious and numerous advantages, namely: economy, rapidity, simplicity of design, and bidirectional drug sensitivity, coupled with the fact that it does not require lengthy training procedures, the use of food/water deprivation, or electric shock (Rodgers et al. 1997). This suggests that the popularity of the EPM owes more to practical than theoretical considerations. This assumption is also perhaps valid to some extent for other experimental tests for anxiety (Carobrez and Bertoglio 2005).

The main sources of inter-laboratory variability in the use of the EPM are summarized in Table 18.3. Based on the fact that behavioral responses and pharmacological effects observed in the EPM are under the influence of these variables, it would be imperative that laboratories using or planning to use this test dedicate time and effort in order to define the optimal experimental conditions before starting their respective studies. Moreover, this variability has led to calls for what is known as the ethological approach to data collection (Rodgers et al. 1997; Roy and Chapillon 2004). From such a perspective, the next paragraph focuses on the advantages of applying both ethological measures (instead of conventional only) and detailed temporal analysis (min-by-min scoring) of the rodents' behavior in the EPM. It also deals with the importance of the EPM test/retest protocol.

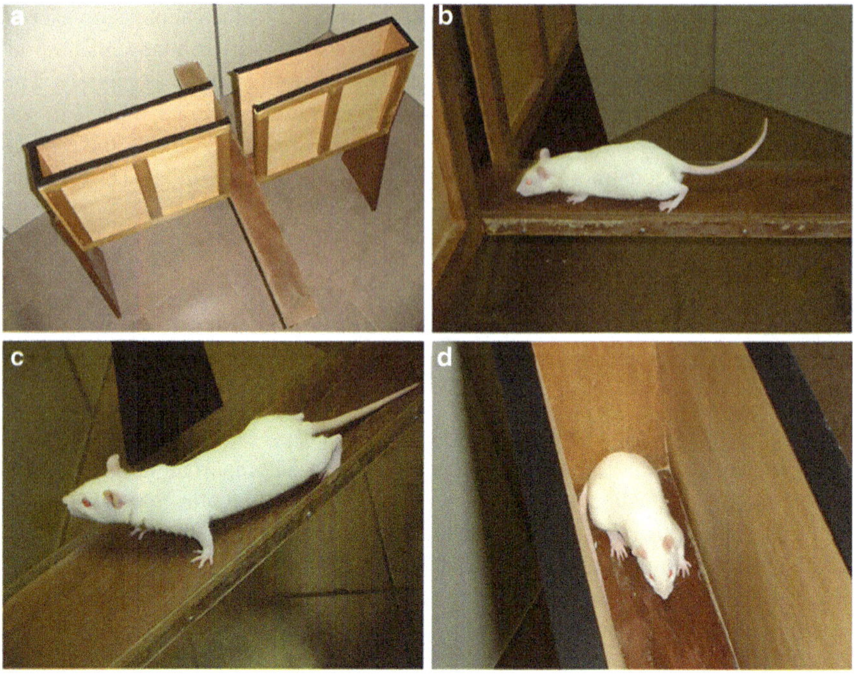

Fig. 18.1 The (rat) elevated plus-maze (EPM) is consisted of two opposite open-arms (surrounded by a small ledge), and two enclosed-arms, about 50 cm above the ground (**a**). Using the EPM to measure anxiety is relatively simple: one may score the number of entries and the time spent on the open-arms (**b**). In addition to these spatiotemporal measures, there are more refined postures associated with anxiety such as the stretched-attend (**c**). They are collectively referred to as risk assessment behaviors. An "anxious" animal is one that displays risk assessment behavior very often and rarely ventures out on the open-arms. In general, whereas anxiolytics (e.g., diazepam) increase open-arms exploration and reduce stretched-attend postures, anxiogenic drugs (e.g., pentylenetetrazole) produce the opposite effect. One possible complication, that an animal might not come out because it is inherently inactive rather than anxious, can be dealt with by scoring the number of enclosed-arms entries, an index of general exploratory activity in this test (panel **d**)

Table 18.3 Variables proven to influence both behavioral and pharmacological responses in the EPM[a]

Type	Examples
Organismic	Species; strain; gender (estrous cycle/lactation); age
Procedural	Housing condition; circadian rhythm/light cycle; prior handling and injection experience; prior stress; apparatus construction (e.g., floor surface, walls/arms color, arm width, open-arm ledges); illumination level; prior test experience; behavioral measures scored: conventional, ethological; definition/validation of measures (e.g., arm entry); method of scoring (e.g., live, manual/automated, videotape)

[a]Adapted from Carobrez and Bertoglio (2005)

Refining the Information Gathered

The primary indexes of EPM anxiety comprise spatiotemporal measures of open-arm avoidance (percentage of entries and of time spent in). Risk assessment is a significant behavioral dimension closely related to anxiety that has been almost entirely ignored not only in the EPM but also in other animal tests (Rodgers et al. 1997). As previously mentioned, the biological function of these acts and postures is to inform behavioral strategies in potentially dangerous situations, therefore strengthening the idea of the risk assessment analysis as a valuable tool to measure more precisely emotional reactivity in rodents (Roy and Chapillon 2004). For instance, while doses of 2.5–5.0 mg/kg of the serotonin 1A receptor partial agonist buspirone induces an anxiolytic-like effect on spatiotemporal measures in the EPM, coupled with a profound suppression of most active behaviors, at lower doses significant anxiolysis was observed in several ethological measures, including reductions in risk assessment behaviors such as stretched-attend postures (Griebel et al. 1997). Moreover, the utility of risk assessment measures in discriminating between anxiogenesis and sedation has also been demonstrated (Weiss et al. 1998). The anxiogenic drug mCPP reduced the time spent in the open areas but increased the occurrence of stretched-attend postures, supporting the conclusion that these effects of mCPP are due to anxiogenesis rather than sedation or locomotor impairment (Shephard et al. 1994).

An interesting feature of the EPM concerns the effect of prior test experience, which usually lasts 5 min, on subsequent behavioral responses. There is growing evidence that it increases open-arm avoidance during Trial 2. This pattern of response appears to be acquired early on Trial 1, as revealed by min-by-min scoring in both mice and rats (Holmes and Rodgers 1998; Carobrez and Bertoglio 2005). More specifically, up to the second min of Trial 1, roughly equal open and enclosed-arm exploration is observed, suggesting novelty/curiosity as the main motivational stimulus during this phase. As the session continues, however, rodents display a clear enclosed-arm preference, which remains in Trial 2 as well. Both risk assessment behavior and general exploratory activity, represented by stretched-attend postures and enclosed-arms entries, respectively, remained stable throughout Trial 1, suggesting that the pattern of change in open-arm exploration cannot be attributed to a general behavioral suppression. In Trial 2, the first minutes are characterized by higher levels of these risk assessment and general exploratory activity measures when compared to the end of Trial 1. This profile seems to reflect an initial and rapid re-familiarization with the EPM apparatus before resorting to the typical open-arm avoidance (Carobrez and Bertoglio 2005).

Besides risk assessment analysis and min-by-min scoring of the rodents' behavior in the EPM task, several laboratories worldwide have also employed a test/retest protocol. This approach has proven that prior EPM experience not only produces enduring changes in behavioral responses, but also strongly affects future drug responsiveness in this test (Table 18.4).

Table 18.4 Experimental evidence showing that prior test experience interferes with both behavioral responses and pharmacological effects of drugs in the elevated plus-maze[a]

	Test session in the EPM (5 min duration)	
Maze type used in the pre-test (5 min duration)	Open-arm exploration	Effect of midazolam
1. None	Normal range (20–30 %)	Anxiolytic
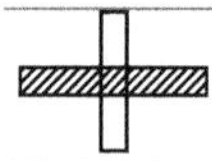 2. Elevated plus-maze	Decreased	None
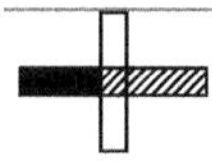 3. Elevated T-maze	Decreased	Untested
 4. Elevated L-maze	Decreased	Untested
 5. Open-arm confinement	Unchanged	Anxiolytic
 6. Enclosed-arm confinement	Unchanged	Anxiolytic
 7. Open-arm + central platform confinement	Unchanged	Untested

Legend: ☐ = Open-arm; ▨ = Enclosed-arm; ■ = Blocked-arm

As observed in the test session, open-arm exploration and midazolam anxiolytic-like effect depend on the maze type used during pre-test

[a]Adapted from Bertoglio and Carobrez (2000, 2002)

The usefulness of the test-retest protocol also rests on the fact that a non-selective and/or false-positive anxiolytic-like effect of a given drug can be detected in the EPM. For example, considering that there was an increase in open-arm exploration after the systemic administration of midazolam and scopolamine in Trial 1, one could advocate that both drugs have anxiolytic-like properties. Nevertheless, while midazolam undoubtedly possesses an anxiolytic-like effect (Bertoglio and Carobrez 2002), scopolamine-induced learning acquisition deficit has been systematically reported using several animal models of learning and memory (Zanotti et al. 1986; De-Mello and Carobrez 2002). Evidence that scopolamine given pre-Trial 1 disrupted open-arm avoidance, the usual behavioral strategy adopted throughout Trial 2 performance, is given by the min-by-min analysis. Whereas the group treated with midazolam prior to Trial 1 displayed a percentage open-arms time score similar to controls in Trial 2, the group receiving scopolamine prior to Trial 1 performed differently (Bertoglio and Carobrez 2004).

Concluding Remarks

As noted by Rodgers (1997), many of the inconsistencies in this area might be resolved if the emphasis were directed not only to pharmacological validation but also to behavioral validation of animal models. Additionally, since not all robust behavioral changes seen in the experiments represent meaningful parameters for evaluation of animal anxiety, there is a need for clear-cut measures, resistant to experimental conditions or apparatus design of particular laboratories, showing reliable and predictable changes following experimental manipulations affecting anxiety (Wall and Messier 2001; Kalueff and Tuohimaa 2004).

It is believed that anxiety disorders represent an inappropriate activation or exaggeration of normally adaptive defensive responses, and their form, function, and mechanisms are highly conserved in evolution (Nesse 1999). Thus, an improvement in test validity could be achieved when the full defensive repertoire of animals is considered (Rodgers 1997). An important caveat is the fact that the defensive repertoire of animals will seldom match a specific anxiety disorder in humans. For example, when analyzing the behavioral symptoms of generalized-anxiety disorder, avoidance and risk assessment appear to be predominant, but flight reactions may also occur. Conversely, risk assessment and avoidance behaviors might arise as a consequence of panic disorder as well. Further, considering that symptoms of psychiatric disorders are often being revised and their pathogenesis revisited (Borsini et al. 2002; Sarter and Bruno 2002), additional caution is needed before claiming or using an animal model of anxiety. In view of this, one should always keep in mind that generating the perfect animal model does not represent a separate goal of research. Rather the model and its constant evolution represent an integral part of neuropsychobiology (McKinney 2001).

In its normal form, anxiety can be divided into two categories: state anxiety, a measure of the immediate or acute level of anxiety, and trait anxiety, which reflects the long-term tendency of an individual to show an increased anxiety response. In its pathological form, anxiety can severely interfere with normal life. This latter has been split into six disorders in humans, namely: generalized anxiety disorder, social phobia, simple phobia, panic disorder, posttraumatic stress disorder, and obsessive-compulsive disorder (American Psychiatric Association 2000). Thus, when assessing anxiety-like behavior in rodents, it is necessary to increase the range of behavioral tests used, including animal models of state and trait anxiety. Many mice with targeted mutations (i.e., transgenic and knockout animals) have been proposed as animal models of pathological anxiety, since they display high levels of anxiety-related behaviors in classical tests (Finn et al. 2003; Flint 2003). However, it is important to emphasize that such mice are animal models of a single gene dysfunction, rather than models of anxiety per se (Belzung and Griebel 2001). It is still a matter of investigation (and debate) whether or not the use of inbred "anxious" strains (van Gaalen and Steckler 2000) and/or selected lines (Liebsch et al. 1998) of rodents to provide models of anxiety that have greater validity than state (see Table 18.1) or single-gene deletion models of anxiety.

Finally, what to say of comorbidities? Although depression and anxiety are considered to be separate entities according to current diagnostic classifications, in clinical practice these two conditions often co-exist (Pollack 2005). Thus, "ideal" modeling of anxiety or depression in animals presumes that, in order to achieve better results, one models these pathologies separately. However, the important problem now is whether animals may possibly have comorbidity of depression and anxiety. Theoretically, there are no reasons to rule out this possibility, and modeling comorbidity may represent certain interest for the researchers. Relatively few such studies have been conducted, and there is a great need in developing specialized models which will allow assessing comorbidity in animals. Thus, measuring comorbidity of anxiety and depression, or their comorbidity with other pathologies such as addiction and alcoholism, may present an important direction for future studies.

Acknowledgements Leandro José Bertoglio and Antônio de Pádua Carobrez are supported by fellowships from FAPESP and CNPq, respectively.

References

American Psychiatric Association. Diagnostic and statistical manual of mental disorders: DSM-IV-TR. Washington, DC: American Psychiatric Association; 2000.

Apfelbach R, Blanchard CD, Blanchard RJ, Hayes RA, McGregor IS. The effects of predator odors in mammalian prey species: a review of field and laboratory studies. Neurosci Biobehav Rev. 2005;29:1123–44.

Barnett SA. The rat: a study in behavior. Chicago: University of Chicago Press; 1975.

Barrett JE, Miczek KA. Behavioral techniques in preclinical neuropsychopharmacology research. In: Bloom FE, Kupfer DJ, editors. Psychopharmacology, the fourth generation of the progress. New York: Raven; 2000. p. 103–23.

Belzung C. Measuring rodent exploratory behavior. In: Cruzio WE, Gerlai TT, editors. Handbook of molecular genetics for brain and behavior research. New York: Elsevier; 1999. p. 77–99.

Belzung C, Griebel G. Measuring normal and pathological anxiety-like behaviour in mice: a review. Behav Brain Res. 2001;125:141–9.

Belzung C, Le Pape G. Comparison of different behavioral test situations used in psychopharmacology for measurements of anxiety. Physiol Behav. 1994;56:623–8.

Bertoglio LJ, Carobrez AP. Previous maze experience required to increase open arm avoidance in rats submitted to the elevated plus-maze model of anxiety. Behav Brain Res. 2000;108: 197–203.

Bertoglio LJ, Carobrez AP. Prior maze experience required to alter midazolam effects in rats submitted to the elevated plus-maze. Pharmacol Biochem Behav. 2002;72:449–55.

Bertoglio LJ, Carobrez AP. Scopolamine given pre-Trial 1 prevents the one-trial tolerance phenomenon in the elevated plus-maze Trial 2. Behav Pharmacol. 2004;15:45–54.

Bittencourt AS, Nakamura-Palacios EM, Mauad H, Tufik S, Schenberg LC. Organization of electrically and chemically evoked defensive behaviors within the deeper collicular layers as compared to the periaqueductal gray matter of the rat. Neuroscience. 2005;133:873–92.

Blanchard RJ, Blanchard DC. Anti-predator defence behaviors in a visible burrow system. J Comp Psychol. 1989;103:70–82.

Blanchard DC, Griebel G, Blanchard RJ. The mouse defense test battery: pharmacological and behavioral assays for anxiety and panic. Eur J Pharmacol. 2003;463:97–116.

Borsini F, Podhorna J, Marazziti D. Do animal models of anxiety predict anxiolytic-like effects of antidepressants? Psychopharmacology (Berl). 2002;163:121–31.
Bourin M, Hascoet M. The mouse light/dark box test. Eur J Pharmacol. 2003;463:55–65.
Calatayud F, Belzung C. Emotional reactivity in mice, a case of nongenetic heredity? Physiol Behav. 2001;74:355–62.
Carobrez AP, Bertoglio LJ. Ethological and temporal analyses of anxiety-like behavior: the elevated plus-maze model 20 years on. Neurosci Biobehav Rev. 2005;29:1193–205.
Choleris E, Thomas AW, Kavaliers M, Prato FS. A detailed ethological analysis of the mouse open field test: effects of diazepam, chlordiazepoxide and an extremely low frequently pulsed magnetic field. Neurosci Biobehav Rev. 2001;25:235–60.
Clement EY, Calatayd F, Belzung C. Genetic basis of anxiety-like behaviour: a critical review. Brain Res Bull. 2002;57:57–71.
Darwin C. The expression of emotions in man and animals. New York: Philosophical Library; 1872 (Reprint 1985).
De Boer SF, Koolhaas JM. Defensive burying in rodents: ethology, neurobiology and psychopharmacology. Eur J Pharmacol. 2003;463:145–61.
De-Mello N, Carobrez AP. Elevated T-maze as an animal model of memory: effects of scopolamine. Behav Pharmacol. 2002;13:139–48.
File SE. Factors controlling measures of anxiety and responses to novelty in the mouse. Behav Brain Res. 2001;125:151–7.
File SE, Seth P. A review of 25 years of the social interaction test. Eur J Pharmacol. 2003;463:35–53.
Finn DA, Rutledge-Gorman MT, Crabbe JC. Genetic animal models of anxiety. Neurogenetics. 2003;4:109–35.
Flint J. Animal models of anxiety and their molecular dissection. Semin Cell Dev Biol. 2003;14:37–42.
Graeff FG, Zangrossi Jr H. Animal models of anxiety disorders. In: D'haenen H, Den Boer JA, Westenberg H, Willner P, editors. Textbook of biological psychiatry. London: Wiley; 2002. p. 879–93.
Gray AJ, McNaughton N. The neuropsychology of anxiety. 2nd ed. New York: Oxford University Press; 2000. p. 72–82.
Griebel G, Rodgers RJ, Perrault G, Sanger DJ. Risk assessment behaviour: evaluation of utility in the study of 5-HT-related drugs in the rat elevated plus-maze test. Pharmacol Biochem Behav. 1997;57:817–27.
Holmes A, Rodgers RJ. Responses of Swiss-Webster mice to repeated plus-maze experience: further evidence for qualitative shift in emotional state? Pharmacol Biochem Behav. 1998;60:473–88.
Kalueff AV. Today and tomorrow of anxiety research. Stress Behav. 2003;8:145–7.
Kalueff AV, Tuohimaa P. Experimental modeling of anxiety and depression. Acta Neurobiol Exp. 2004;64:439–48.
Kaplan RM, Saccuzzo DP. Psychological testing. Principles, applications, and issues. Pacific Grove: Brooks/Cole; 1997.
Liebsch G, Linthorst AC, Neumann ID, Reul JM, Holsboer F, Landgraf R. Behavioral, physiological, and neuroendocrine stress responses and differential sensitivity to diazepam in two Wistar rat lines selectively bred for high- and low-anxiety-related behavior. Neuropsychopharmacology. 1998;19:381–96.
McGuire M, Troisi A. Darwinian psychiatry. Oxford: Oxford University Press; 1998.
McKinney WT. Overview of the past contributions in animal models and their changing place in psychiatry. Semin Clin Neuropsychiatry. 2001;6:68–78.
Millan MJ, Brocco M. The Vogel conflict test: procedural aspects, gamma-aminobutyric acid, glutamate and monoamines. Eur J Pharmacol. 2003;463:67–96.
Nesse RM. Proximate and evolutionary studies of anxiety, stress and depression: synergy at the interface. Neurosci Biobehav Rev. 1999;23:895–903.

Ohl F. Testing for anxiety. Clin Neurosci Res. 2003;3:233–8.
Ohl F, Roedel A, Storch C, Holsboer F, Landgraf R. Cognitive performance in rats differing in their inborn anxiety. Behav Neurosci. 2002;116:464–71.
Overall KL. Natural animal models of human psychiatry conditions: assessment of mechanisms and validity. Prog Neuropsychopharmacol Biol Psychiatry. 2000;24:727–76.
Panksepp J, Moskal JR, Panksepp JB, Kroes RA. Comparative approaches in evolutionary psychology: molecular neuroscience meets the mind. Neuro Endocrinol Lett. 2002;23:105–15.
Pollack MH. Comorbid anxiety and depression. J Clin Psychiatry. 2005;66:22–9.
Prut L, Belzung C. The open field as a paradigm to measure the effects of drugs on anxiety-like behaviors: a review. Eur J Pharmacol. 2003;463:3–33.
Rodgers RJ. Animal models of 'anxiety': where next? Behav Pharmacol. 1997;8:477–96.
Rodgers RJ, Cole JC. The elevated plus maze: pharmacology, methodology and ethology. In: Cooper SJ, Hendrie CA, editors. Ethological pharmacology. New York: Wiley; 1994. p. 56–67.
Rodgers RJ, Cao BJ, Dalvi A, Holmes A. Animal models of anxiety: an ethological perspective. Braz J Med Biol Res. 1997;30:289–304.
Roy V, Chapillon P. Further evidences that risk assessment and object exploration behaviours are useful to evaluate emotional reactivity in rodents. Behav Brain Res. 2004;154:439–48.
Salome N, Viltart O, Darnaudery M. Reliability of high and low anxiety-related behaviour: influence of laboratory environment and multifactorial analysis. Behav Brain Res. 2002;136: 227–37.
Sanchez C. Stress-induced vocalisation in adult animals. A valid model of anxiety? Eur J Pharmacol. 2003;463:133–43.
Sarter M, Bruno JP. Animal models in biological psychiatry. In: D'haenen H, Den Boer JA, Willner P, editors. Biological psychiatry. New York: Wiley; 2002. p. 47–79.
Shephard JK, Grewal SS, Fletcher A, Bill DJ, Dourish CT. Behavioral and pharmacological characterization of the elevated 'zero-maze' as an animal model of anxiety. J Psychopharmacol. 1994;116:56–64.
Silva RH, Frussa-Filho R. The plus-maze discriminative avoidance task: a new model to study memory-anxiety interactions. Effects of chlordiazepoxide and caffeine. J Neurosci Methods. 2000;102:117–25.
Uys JD, Stein DJ, Daniels WM, Harvey BH. Animal models of anxiety disorders. Curr Psychiatry Rep. 2003;5:274–81.
van der Staay FJ. Animal models of behavioral dysfunctions: basic concepts and classifications, and an evaluation strategy. Brain Res Brain Res Rev. 2006;52:131–59.
van Gaalen MM, Steckler T. Behavioural analysis of four mouse strains in an anxiety test battery. Behav Brain Res. 2000;115:95–106.
Wall PM, Messier C. Methodological and conceptual issues in the use of the elevated plus-maze as a psychological measurement instrument of animal. Neurosci Biobehav Rev. 2001;25: 275–86.
Web of Science. According to the search performed in May 2009, using the entering expression "elevated plus-maze." 2009. http://isi9.isiknowledge.com
Weiss SM, Wadsworth G, Fletcher A, Dourish CT. Utility of ethological analysis to overcome locomotor confounds in elevated maze models of anxiety. Neurosci Biobehav Rev. 1998;23:265–71.
Zanotti A, Valzelli L, Toffano G. Reversal of scopolamine induced amnesia by phosphatidylserine in rats. Psychopharmacology (Berl). 1986;90:274–5.

Chapter 19
The Plus-Maze Discriminative Avoidance Task: An Ethical Rodent Model for Concomitant Evaluation of Learning, Memory, Anxiety, Motor Activity and Their Interactions

Roberto Frussa-Filho, Camilla de Lima Patti, Daniela Fukue Fukushiro, Luciana Takahashi Carvalho Ribeiro, Sonia Regina Kameda, and Rita de Cassia Carvalho

The Plus-Maze Discriminative Avoidance Task

The elevated plus-maze consists of two open, elevated arms running along a north-south axis and two arms enclosed by walls running east-west. It combines elements of unfamiliarity, openness and elevation (see Chap. 19). Since its introduction by Handley and Mithani (1984) and experimental validation for rats (Pellow et al. 1985) and mice (Lister 1987), it has been most extensively used to asses anxiolytic and anxiogenic effects of drugs in several hundred studies. Standard anxiolytic and anxiogenic drugs increase and decrease, respectively, both the percentage of time that rats (or mice) spend in the open arms and the percentage of entries that they make into the open arms. However, due to its widespread appeal, the elevated plus-maze has also been used to measure levels of spontaneous anxiety behavior (Goto et al. 1993), to investigate anxiety-induced antinociception (Lee and Rodgers 1990; Frussa-Filho

R. Frussa-Filho (In Memoriam) • S.R. Kameda
Department of Pharmacology, Universidade Federal de São Paulo (UNIFESP), São Paulo, São Paulo, Brazil

C.L. Patti, Ph.D.
Instituto de Genética e Erros Inatos do Metabolismo, Universidade Federal de São Paulo (UNIFESP), São Paulo, São Paulo, Brazil

D.F. Fukushiro, Ph.D. (✉)
Department of Pharmacology, Universidade Federal de São Paulo (UNIFESP), Rua Botucatu, 862, Ed. Leal Prado, 1° andar, São Paulo, São Paulo 04023062, Brazil
e-mail: danifukushiro@gmail.com

L.T.C. Ribeiro, Ph.D. • R.C. Carvalho
Department of Pharmacology, Universidade Federal de São Paulo (UNIFESP), São Paulo, São Paulo, Brazil

M.L. Andersen, S. Tufik (eds.), *Rodent Model as Tools in Ethical Biomedical Research*, DOI 10.1007/978-3-319-11578-8_19

et al. 1992; Conceição et al. 1992; Rodgers et al. 1992) and to evaluate memory. Regarding this latter use, Itoh et al. (1990) proposed that the retesting-induced decrease in the time taken by the animal to move from an open arm to an enclosed arm in the elevated plus-maze could be a measure of memory. This suggestion has been corroborated and extended by others (Frussa-Filho et al. 1991; Graeff et al. 1993; Conceição et al. 1994). For example, Graeff et al. (1993) showed that the retesting-induced increase in the time taken by the animal to move from an enclosed arm to an open arm could be used as an inhibitory avoidance task to measure memory.

On the basis of our previous experience with multifaceted uses of the elevated plus-maze (see above) and considering that (1) alterations in anxiety levels can modify learning and memory processes and (2) modifications of motor activity can lead to important misinterpretations in most animal models of learning and memory, we set out to develop and validate the plus-maze discriminative avoidance task (PM-DAT).[1] This animal model allows a concomitant evaluation of learning, memory, anxiety and motor activity in rodents. Most important, it has proven to be especially useful for experimental investigation of the interactions between these different processes.

Methodology

The apparatus employed is a modified elevated plus-maze made of wood that contains two enclosed arms with sidewalls and no top opposite to two open arms. A 100 W lamp is put exactly over the middle of one of the enclosed arms (aversive enclosed arm, 660 lx at floor level). In order to prevent the lamp from illuminating both enclosed arms, each enclosed arm faces an open arm (see Fig. 19.1). In the training session, each animal is placed in the center of the apparatus facing the intersection of the two open arms. Over a period of 10 min, every time the animal enters the enclosed arm containing the lamp, an aversive situation is produced until the animal leaves the arm. The aversive stimuli consist of a combination of a 100 W light and an 80 dB noise[2] (in the case of experiments with mice) or a combination of a 100 W light and a frontal hot air current produced by a 700 W hair drier placed above the end of the aversive enclosed arm (in the case of experiments with rats). When rats are employed, this aversive enclosed arm is covered with black rubber as an extra spatial cue. On each side of the plus-maze discriminative avoidance apparatus, there are different extra-maze visual cues (door, window, and cupboard) that animals can use to distinguish the location of the different arms of the maze.

In the test session, animals are again placed in the center of the apparatus and observed for 3 min without receiving any stimulation [i.e., lamp (mice and rats) and hair drier (rats) are turned off]. In all experiments, the animals are observed in random order and in a blind manner, and the apparatus is cleaned with a 5 % alcohol solution after each behavioral session. The animal is considered to be in a certain arm when all four paws have passed over the entrance.

[1] The authors would like to acknowledge the fundamental contribution of Dr. Regina Helena da Silva to the development and validation of the PM-DAT.

[2] The noise is produced by the horizontal displacement of a piece of metal against a piece of wood.

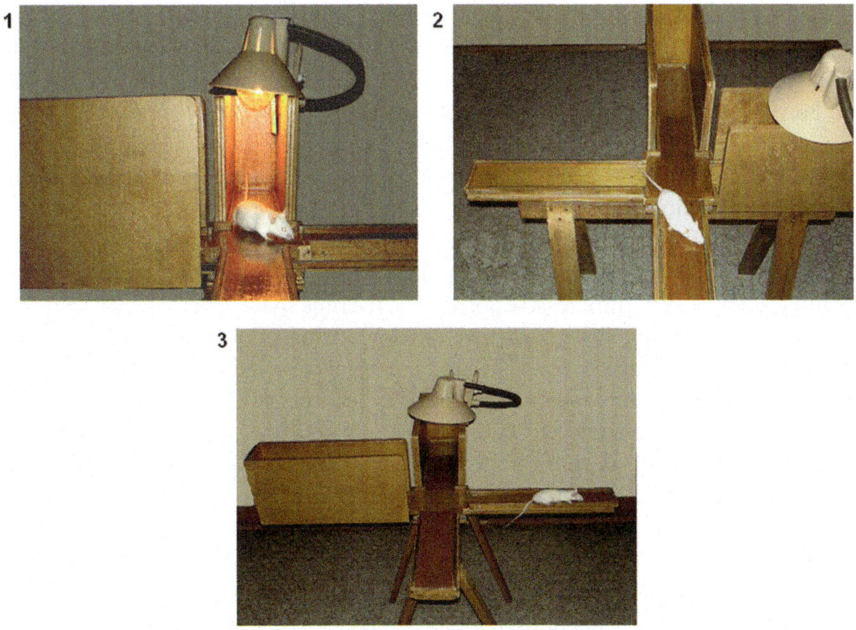

Fig. 19.1 Plus-maze discriminative avoidance task. Mouse submitted to training session and escaping from aversive stimuli in the aversive enclosed arm. Note that the lamp is turned off immediately after the animal leaves the enclosed aversive arm

In the case of experimental protocols using mice, the two enclosed arms with sidewalls measure $28.5 \times 7 \times 18.5$ cm (3 lx at floor level), and the two open arms measure 28.5×7 cm (9 lx at floor level). Conversely, when rats are employed, the two enclosed arms with sidewalls measure $50 \times 12 \times 40$ cm (3 lx at floor level) and the two open arms measure 50×12 cm (9 lx at floor level).

In the PM-DAT, learning is verified by a progressive decrement in the exploration of the aversive enclosed arm throughout the training session (Silva et al. 2004a; Carvalho et al. 2006; Patti et al. 2006), while avoidance of the aversive enclosed arm in the test session (evaluated by the time spent in this arms in comparison to time in the non-aversive enclosed arm) is considered to measure retention. Moreover, this animal model allows a concomitant and independent evaluation of anxiety (by the avoidance of the open arms) and motor activity (by the total number of entries in each arm of the apparatus).

Application: Evaluation of Learning

Memories can be initially formed by engaging an object or performing an action, leading to the recreation of that object or action within the brain (Walker and Stickgold 2004). This process is called learning or acquisition.

In the plus-maze discriminative avoidance task, learning is evaluated by the progressive decrease in the exploration of the aversive enclosed arm where light and sound (mice) or light and noisy hot air (rats) are presented. Since learning has to be verified by a modification of performance in a specified period of time, the learned avoidance to this arm is measured by the percent time spent in the arm (time spent in the aversive arm / time spent in both enclosed arms × 100) throughout the training session.

Regarding the pharmacological validation of this parameter as a learning index, we (Patti et al. 2006) have demonstrated that acute pre-training morphine administration potentiated the progressive decrease in the time spent in the aversive enclosed arm of the apparatus throughout the training session. Thus, morphine enhanced learning levels. This increased learning by morphine-treated mice is in accord with clinical data using choice reaction time (Hanks et al. 1995). On the other hand, acute haloperidol administration (0.3–1.0 mg/kg) impaired the acquisition of the task (Correia-Pinto et al., unpublished data), which also corroborates data found in clinical situations (Goldstone et al. 1979; Levin et al. 1996; Stip 2006).

Our group also verified that partial sleep restriction (21 h of sleep deprivation for 15 days—see a detailed description of this experiment at the end of this chapter) and total and repeated sleep deprivation (72 h of sleep deprivation followed by 72 h of recovery for 15 days—unpublished data) could induce learning deficits in mice, once more corroborating clinical data (see Dinges et al. 1997). Moreover, in a less complex protocol of sleep deprivation (total sleep deprivation for 96 h), Alvarenga et al. (2009) have demonstrated that, similarly to mice, rats present learning deficits in the PM-DAT.

Application: Evaluation of Memory

The retention of learned avoidance behavior can be detected by re-exposure of the animal to the apparatus. In a second trial (test session), the aversive stimuli are not applied. Nevertheless, the memory traces these stimuli previously produced as well as the association between these traces and the enclosed arm that had been aversive because the animal to spend less time in this arm in comparison to the non-aversive enclosed arm. It should be noted that the identification of the arm that had been aversive is given by the presence of the lamp (turned off in this session) and the environmental cues in the experimental room. Using this technique, our group (Silva et al. 1997) demonstrated in the study that originally introduced this model to the literature that conditioned mice preferred to spend more time in the non-aversive enclosed arm when compared to the aversive arm during the test session. Importantly, this preference was not observed in a group of animals that had been exposed to the apparatus without pairing the aversive stimuli with a specific arm.

It should be noted that the test session is shorter (3 min) than the training (10 min) session to prevent the animal from learning that the aversive enclosed arm is no longer aversive.

Detection of Memory-Enhancement Procedures

It has been demonstrated that bovine phosphatidylserine (a phospholipid present in biological membranes) can increase the performance of old rats in different animal models of memory, including passive avoidance (Zanotti et al. 1986), active avoidance (Nunzi et al. 1990), spontaneous alternation (Aporti et al. 1986), and Morris water-maze (Zanotti et al. 1986) tasks. Clinically, phosphatidylserine has been reported to be effective for attenuating memory deficits in Alzheimer's disease patients (Crook et al. 1992) as well as age-related cognitive decline (Schreiber et al. 2000). In order to determine whether the PM-DAT could detect the effects of phosphatidylserine administration on scopolamine-induced amnesia, we administered 50 mg/kg phosphatidylserine to male mice for 5 days (Claro et al. 1999). Twenty-four hours after the last injection, the animals received 1 mg/kg scopolamine. Twenty minutes later, they were submitted to the training session of the model. We found that phosphatidylserine abolished the memory impairment induced by scopolamine. Taken together, our findings suggest that the PM-DAT is not only effective for detecting memory impairment induced by scopolamine, a classic amnestic anticholinergic drug, but also for detecting the effect of a drug that reverses scopolamine-induced amnesia.

To confirm the effectiveness of the PM-DAT in detecting memory enhancement effects, the effects (Silva et al. 1999) of GM1 ganglioside (a glycosphingolipid found at higher concentrations within the nervous system than in any other body tissue; Rapport 1981) were investigated. Numerous investigations in both animals and humans have indicated the involvement of gangliosides in neurobiological events associated with adaptive functions, neuronal plasticity and even formation and improvement of memory (Agnati et al. 1983; Tettamanti et al. 1985; Rahmann 1986; Toffano et al. 1986; Seifert et al. 1987; Miceli et al. 1997). Thus, Wistar male rats and Swiss male mice were treated with 50 mg/kg GM1 ganglioside for 7 or 14 days, respectively. Twenty-four hours after the last injection, animals received 1 mg/ kg scopolamine. Twenty minutes later, the animals were submitted to step-through passive avoidance or to PM-DAT. The results showed that both the passive avoidance task and the PM-DAT could detect scopolamine-induced amnesia in both rats and mice. The administration of GM1 ganglioside abolished scopolamine-induced memory impairment in both animal models.

Detection of Memory-Impairing Procedures

As previously mentioned, rodents treated with scopolamine and submitted to the PM-DAT could not distinguish the aversive enclosed arm from the non-aversive arm in the test session; this was interpreted as memory impairment (Silva et al. 1999; Claro et al. 1999).

Since the PM-DAT is effective for detecting memory-enhancement effects induced by drugs and also memory impairment induced by scopolamine, we aimed to investigate the effects of other drugs on memory. Thus, the effects (Silva and Frussa-Filho 2000) of pre-training chlordiazepoxide administration on memory were investigated. Despite the anxiolytic effects of this drug on learning/memory processes, acute administration of chlordiazepoxide increased the time spent in the aversive enclosed arm in the test session, demonstrating a memory deficit. This effect is well-known in humans (see Beracochea 2006).

Our group has also reported that acute or chronic administration of amphetamine (Silva et al. 2002a) and acute administration of ethanol (Kameda et al. 2007) induce memory impairment. These findings are also important for the pharmacological validation of the memory quantification component of the model, since amphetamine (Rapeli et al. 2005; Tipper et al. 2005) and ethanol (Hernández et al. 2007), have shown amnestic effects in humans.

The effects of age on memory were investigated to evaluate the role of physiological processes in cognition. The well-documented age-induced memory impairment was also detected by the PM-DAT (Ricardo et al., unpublished data). Similarly, we have also studied the interaction between sleep and learning/memory processes. It was found that 72 h of sleep deprivation induced retention deficits (Silva et al. 2004a, b). Moreover, we have found that partial sleep restriction also induces memory impairment in mice submitted to the PM-DAT (Patti et al., unpublished data; a partial presentation and discussion of these data is provided elsewhere in this chapter). The inhibitory effect of sleep deprivation on memory evaluated by the PM-DAT is also an important physiological validation of the model, as considerable data have shown memory impairment by sleep deprivation in both clinical studies (Cochran et al. 1994; Karni et al. 1994; Fluck et al. 1998; Mednick et al. 2002; Isomae et al. 2003) and animal models of memory (Bueno et al. 1994; Smith and Kelly 1998; Silva et al. 2004a, b).

Detection of State-Dependent Memory

State dependence is a condition in which the retrieval of learned information requires that the animal be in a state similar to that in which the memory was acquired (Izquierdo et al. 1981; Bruins Slot and Colpaert 1999). State dependence has been demonstrated after the administration of several psychoactive drugs (Colpaert 1990; Jackson et al. 1992; Bruins Slot et al. 1999; Colpaert et al. 2001). Importantly, state dependence has been implicated not only in retrieval but also in the acquisition and retention phases of the learning/memory process (Colpaert et al. 2001).

Since morphine state-dependent learning has been proposed as a mechanism for opiate dependence (Colpaert 1990, 1996; Spanagel 1995), we (Patti et al. 2006) set out to determine whether the PM-DAT is an effective model for detecting state dependence. We found that morphine given before the training session induced retrieval deficits in mice tested in the PM-DAT. Such deficits were shown to be related to morphine-induced state-dependent learning, since pre-test morphine administration restored retention performance to control levels.

Collectively, the bulk of experimental evidence discussed thus far suggests the PM-DAT is effective not only for bidirectional detection of the effects of drugs or procedures on memory but also for detecting behavioral phenomena related to concomitant alterations of different phases of the learning/memory process.

Application: Evaluation of Anxiety

One of the great advantages of the PM-DAT is the possibility of concomitantly evaluating learning/memory and anxiety levels in the same animal. Considering a possible intrinsic interaction between these processes, we (Silva and Frussa-Filho 2000) investigated the effects of two drugs with well-known opposing effects on anxiety [caffeine (anxiogenic) and chlordiazepoxide (anxiolytic)] on memory/anxiety interactions in the PM-DAT. Acute administration of caffeine decreased the exploration of the open arms of the apparatus in the training session. Conversely, acute administration of chlordiazepoxide increased the exploration of the open arms of the apparatus in the training session. In the test session, both treatments induced memory impairment (determined by an increased amount of time spent in the aversive enclosed arm), despite the previously observed effects in open arm exploration.

Two different possibilities might explain the retention impairment produced by pre-training administration of caffeine or chlordiazepoxide. First, each drug might have produced memory impairment by its own direct amnestic neuronal mechanism. Second, the memory impairment produced by each drug might have been the consequence of a bidirectional alteration of the ***optimum*** level of anxiety necessary for memory retention. In order to investigate the latter possibility, we co-administrated caffeine and chlordiazepoxide before training. As expected, the percent time spent in the open arms of the apparatus was no longer different from that observed for the control group (i.e., while chlordiazepoxide abolished the anxiogenic effect of caffeine, caffeine abolished the anxiolytic effect of chlordiazepoxide). Most importantly, under these conditions, the animals showed no memory impairment. These results suggest that, in addition to being effective for measuring the effects on anxiety of classical anxiogenic and anxiolytic drugs, the plus-maze discriminative avoidance model appears to be uniquely suited to investigate the critical and complex relationship between learning/memory and anxiety. This conclusion was supported by other experimental evidence obtained by our research group.

The effectiveness of the PM-DAT for appropriately measuring anxiety was also demonstrated in a series of experiments in which we verified that it was as effective as the conventional elevated plus-maze and other traditional anxiety animal models for detecting alterations in anxiety. Within this context, the "anxiolytic behavior" of the spontaneously hypertensive rat (SHR) strain was detected in the PM-DAT (Calzavara et al. 2004). This was important for validation of the effectiveness of the

PM-DAT in evaluating rodent anxiety because SHR rats present reduced anxiety behavior in a series of traditional models, including the conventional elevated plus-maze, the open-field test, the social interaction test (Goto et al. 1993) and the light/dark test (Ramos et al. 2002). Similarly, we reported that the potentiating effect of naltrexone on the anxiolytic activity of chlordiazepoxide, which was previously demonstrated in different traditional anxiety models, such as the elevated plus-maze and social interaction tests (Frussa-Filho et al. 1999), was also effectively detected in the PM-DAT (Silva and Frussa-Filho 2002). Interestingly, in that study, we found that naltrexone also potentiated the amnestic effects of chlordiazepoxide, which is in line with the notion that the amnestic effect of this benzodiazepine is directly related to its anxiolytic action.

We also recently performed a study aimed at dissecting the interactions between ethanol-induced modifications of anxiety and memory in mice tested in the PM-DAT (Kameda et al. 2007). All ethanol doses employed in the study (0.3, 0.6, 1.2, 1.8, 2.4 and 3.0 g/kg) impaired retention of the discriminative task. However, only doses higher than 1.2 g/kg induced anxiolytic effects. Thus, the PM-DAT is effective for detection of drug-induced memory modification due to alteration of primary anxiety (as described above); it can also demonstrate that some drug-induced memory alterations are not related to changes in anxiety levels.

Application: Evaluation of Motor Function

The vast majority of animal models of memory involve motor function. Some alterations in this behavioral parameter can modify performance during acquisition or test trials. For instance, sedative doses of ethanol can prevent animals from entering the dark chamber of a passive avoidance apparatus during the acquisition trial, suggesting a false improvement in performance (Kameda et al. 2007). Considering the marked influence of motor behavior on performance in animal models of memory, the total number of entries in all arms and/or the number of entries in only the enclosed arms of the PM-DAT apparatus have been used to evaluate motor function.

The well-known hypolocomotion induced by the catecholamine-depleting agent reserpine (Carvalho et al. 2003; Castro et al. 2006) was effectively detected by the PM-DAT model (Silva et al. 2002b; Carvalho et al. 2006). In this respect, we have recently determined that, although acute administration of reserpine produced both memory and motor function impairment, motor inhibition occurred at higher doses (Carvalho et al. 2006). Importantly, given that reserpine administration to rodents is considered an animal model of Parkinson's disease (Colpaert 1987), the induction of memory but not motor impairment by lower doses of reserpine is in accord with the observation that cognitive deficits precede severe motor disturbance in Parkinson's disease patients (Cooper et al. 1991; Owen et al. 1992).

Regarding the detection of enhanced motor function by the total entries and/or the number of entries in the enclosed arms of the PM-DAT apparatus, we have already demonstrated that hyperlocomotion induced by both drug and non-drug procedures can be quantified by the PM-DAT model. Indeed, this model was sensitive enough to detect the motor stimulant effects of both amphetamine (Silva et al. 2002a) and continuous exposure to light (Castro et al. 2005) under similar experimental conditions to which such hyperlocomotion effects were also detected in the traditional open-field model (Bellot et al. 1996; Abílio et al. 1999).

Kameda et al. (2007) have also recently observed that this animal model can detect the well-known (Cohen et al. 1997) biphasic pattern of ethanol-derived effects on motor activity. While 1.2 and 1.8 g/kg ethanol generated a stimulant effect, 3.0 g/kg ethanol decreased motor behavior in the PM-DAT.

Application: Evaluation of the Interactions Between Memory, Anxiety and Motor Function

We have already described experimental situations in which the PM-DAT was a uniquely appropriate rodent model for evaluating interactions between memory and anxiety as well as interactions between memory and motor activity. We are going now to describe a set of experiments that demonstrate the effectiveness of the PM-DAT for evaluating the highly complex behavioral interactions that can occur among memory, anxiety and motor function. We investigated the effects of acute or chronic administration of amphetamine on the PM-DAT behavior of mice (Silva et al. 2002a). Acute pre-training injection of amphetamine produced both an anxiogenic effect (decreased percent of time spent in the open arms of the apparatus in the training session) and an impairment of memory (increased percent of time spent in the aversive enclosed arm in the test session), but it did not significantly affect motor activity (no change in the total number of entries in the training session). After repeated administration, a challenge pre-training injection with the same dose of amphetamine no longer produced the anxiogenic effect, but it did continue to promote memory deficits and now elicited a previously undetected enhancement of motor function. These results allowed a series of conclusions to be made that would likely not have been possible if using any other model. First, the amnestic effect produced by acute amphetamine injection was not due to modifications of motor function. Second, the amnestic effect of chronic amphetamine was not related to alterations in anxiety levels. Third, repeated administration of amphetamine does not result in tolerance to its amnestic effects. Fourth, chronic administration of amphetamine induces tolerance to its anxiogenic effect. Fifth, the motor stimulant effect of amphetamine is sensitized after repeated administration. Regarding the last two conclusions, the anxiogenic action of amphetamine plus the presence of aversive stimuli during the training session could have caused motor inhibition, which may have prevented the observation of the stimulating effect of the drug. In support of this hypothesis, the anxiogenic effect was no longer present when amphetamine was chronically administered, and its motor stimulatory effect could be observed. This finding may have implications in the phenomenon of behavioral sensitization.

Repeated exposure to psychostimulants such as amphetamine and cocaine produces behavioral sensitization, which is characterized by augmented locomotor response to a psychostimulant challenge injection. Such sensitization has been suggested to be associated with mechanisms that underlie both pharmacological psychosis and compulsive drug intake (Pierce and Kalivas 1997). The PM-DAT data discussed here not only demonstrate that behavioral sensitization to amphetamine can be evaluated by this model, but they also indicate that this phenomenon might be correlated with tolerance to the anxiogenic effect of this drug.

A Brief Experimental Contemplation

For the sake of practicality and with the intention of illustrating the operational approach to specific scientific questions through the PM-DAT, we will describe here an original experiment specifically designed for this chapter. The aim of this experiment was to investigate the effects of partial sleep restriction on male mouse performance. In comparison to the acute total sleep deprivation protocol (i.e., sleep deprivation for a consecutive 72 h; Silva et al. 2004a, b), the partial sleep restriction protocol appears to more closely reflect common human sleep restriction. Partial sleep restriction was achieved using the multiple platform method (Suchecki et al. 2000). Thus, mice were submitted to home-cage control (CTRL; two home cage changes per day), stress control (SC; 21 h of large platforms in water tanks) and sleep restriction (SR; 21 h of small platforms in water tanks) for 15 days. The animals in the SC and SR groups were allowed to sleep from 10 a.m. to 1 p.m. in their home cages. Immediately after the end of the chronic sleep deprivation protocol, animals were submitted to the training session of the PM-DAT. The test session occurred 24 h after training (Table 19.1).

Although the animals could distinguish both enclosed arms (Fig. 19.2), sleep restriction induced learning deficits, as a progressive decrement in the percent time spent in the aversive enclosed arm was not observed throughout the training session (Fig. 19.3a, b). These results are in agreement with those of Bonnet and Arand (1995, 2003), who showed that sleep restriction could lead to sleepiness, fatigue and lack of concentration. Still, an increased exploration of the open arms was presented

Table 19.1 Scheme of the different experimental conditions to which the animals were submitted

	Day 1–15	Training (17th day)	Test (18th day)	Retest (31st day)
Control (CTRL)	Home-cage—two changes (10 a.m. and 1 p.m.)	**Plus-maze Discriminative Avoidance Task (PM-DAT)**		
Stress control (SC)	21 h-large platform exposure and 3 h-home-cage (from 10 a.m. to 1 p.m.)			
Partial sleep restriction (SR)	21 h-sleep deprivation and 3 h-home-cage (from 10 a.m. to 1 p.m.)			

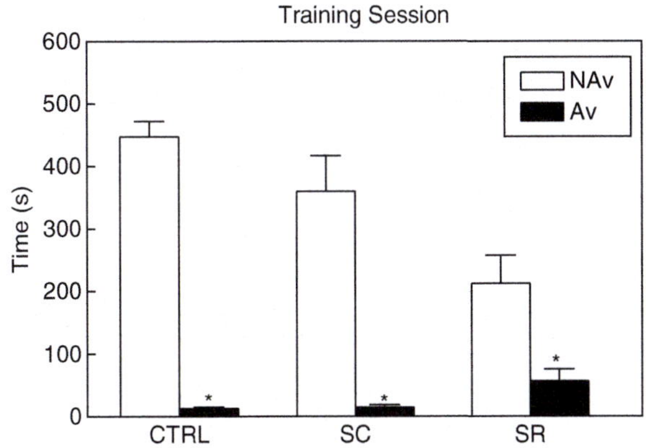

Fig. 19.2 Effects of pre-training sleep restriction on the training performance of mice tested in the plus-maze discriminative avoidance task. Mice were partially sleep-restricted (SR), or exposed to large platforms (stress control—SC) for 21 h per day for 15 days. Animals of the control group (CTRL) were maintained in their home cages. Results are presented as mean ± SE of time (seconds) spent in the non-aversive (NAv) and in the aversive (Av) enclosed arms. *$p<0.05$ compared to the time spent in the non-aversive enclosed arm (Two-way ANOVA and Duncan's test)

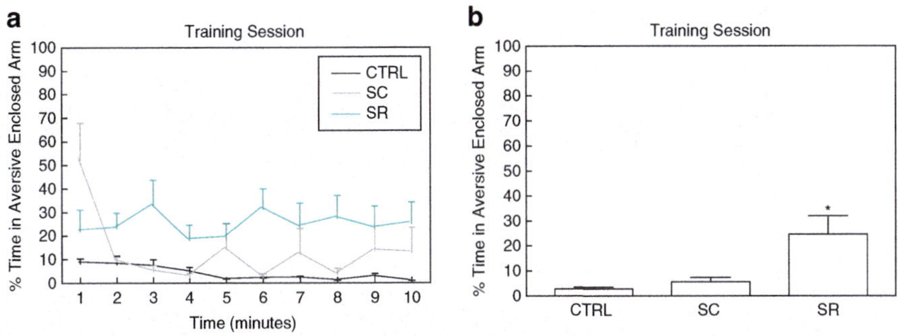

Fig. 19.3 Effects of pre-training sleep restriction on the training performance of mice tested in the plus-maze discriminative avoidance task. Mice were partially sleep-restricted (SR), or exposed to large platforms (stress control—SC) for 21 h per day for 15 days. Animals of the control group (CTRL) were maintained in their home cages. (**a**) Percent time spent in the aversive enclosed arm throughout the training session. ANOVA for the percent time spent in the aversive arm with experimental condition as a between-subject factor and time (minutes of observation) as a repeated-measure revealed significant effects of experimental condition, and an interaction between time/experimental condition and time. (**b**) Percent time spent in the aversive enclosed arm, considering the whole session. Results are presented as mean ± SE. *$p<0.05$ compared to control group (ANOVA and Duncan's test)

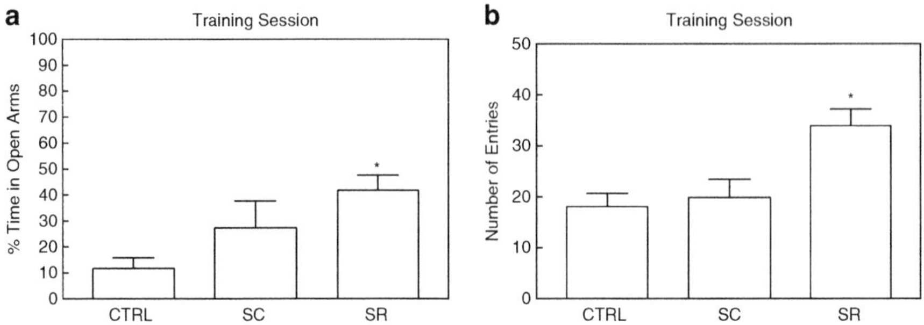

Fig. 19.4 Effects of pre-training sleep restriction on the training performance of mice tested in the plus-maze discriminative avoidance task. Mice were partially sleep-restricted (SR), or exposed to large platforms (stress control—SC) for 21 h per day for 15 days. Animals of the control group (CTRL) were maintained in their home cages. (**a**) Percent time spent in the open arms and (**b**) number of total entries for animals in the training session. Results are presented as mean ± SE. *$p < 0.05$ compared to control group (ANOVA and Duncan's test)

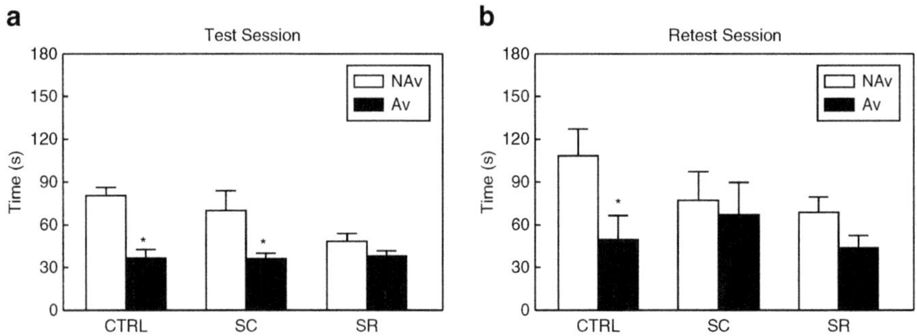

Fig. 19.5 Effects of pre-training sleep restriction on test and retest performance of mice tested in the plus-maze discriminative avoidance task. Mice were partially sleep-restricted (SR), or exposed to large platforms (stress control—SC) for 21 h per day for 15 days. Animals of the control group (CTRL) were maintained in their home cages. Time (seconds) spent in the non-aversive (NAv) and in the aversive (Av) enclosed arms in the test (**a**) or in the retest session (**b**). Results are presented as mean ± SE. *$p < 0.05$ compared to the time spent in the non-aversive enclosed arm (Two-way ANOVA and Duncan's test)

by sleep-restricted mice (Fig. 19.4a); thus, sleep restriction enhanced motor activity (Fig. 19.4b).

Our results demonstrate that sleep restriction induced memory impairment, since the animals presented similar explorations of both enclosed arms in the test session (Fig. 19.5a). In this context, memories persisting for longer than a few minutes are typically parsed into two categories: intermediate-term memories that last a few hours and long-term memories that can last days, months or years (Rosenzweig et al. 1993). In addition to differences in their temporal dynamics, these two categories differ mechanistically; intermediate-term memory requires *de novo* protein synthesis (Mizumori et al. 1987; Sutton et al. 2001) while long-term memory requires both *de novo* protein synthesis and gene transcription (Mizumori et al. 1987). As expected, our results suggest that sleep restriction impaired intermediate-term memory. These results support the previously described sleep deprivation (72 h continuous)-induced memory deficits in mice tested in the PM-DAT (Silva et al. 2004a, b).

To investigate the effects of this sleep deprivation protocol on long-term memory, we performed a retest 30 days after the training session. In this session, only the control group spent more time in the non-aversive enclosed arm than in the arm that had been aversive (Fig. 19.5b), demonstrating retention of the task. As expected, since sleep restriction impaired intermediate-term memory, it also impaired long-term memory. Interestingly, mice that had been submitted to a stress control protocol presented memory deficits. These results suggest a deleterious effect of stress on long-term memory. Indeed, a myriad of previous studies suggested that stress is involved in sleep-deprivation-induced memory impairments (see Youngblood et al. 1997; Siegel 2001) and is able to decrease paradoxical sleep itself (Bouyer et al. 1997).

Together, these results demonstrate the effectiveness of the PM-DAT for concomitantly detecting the effects of chronic partial sleep deprivation on learning, memory, anxiety and motor function of mice.

Ethical and Final Comments

The plus-maze discriminative avoidance task, besides presenting doubtless ethical advantages when compared to other memory models (aversive stimuli are less stressful than electrical footshock, for instance), has the methodological advantage of simultaneously evaluating potentially interacting processes such as learning, memory, anxiety and motor function.

Upon comparing of this model with two other widely employed models in the study of learning/memory (the passive avoidance task and the Morris water maze), it can be concluded that: (1) in contrast to the passive avoidance task, the aversive stimuli employed in the PM-DAT do not involve painful stimulation. Thus, the effects on memory and nociception of drugs such as morphine can be analyzed separately; (2) learning/memory, anxiety and motor activity can be measured in the same animal, in contrast to the two other models. This interactive analysis has important ethical implications since the number of animals used in the experimental protocols is dramatically reduced (positive control experiments are not necessary); (3) in contrast to the other two models, the PM-DAT not only ***evaluates*** motor activity but is also ***less influenced*** by possible motor alterations presented by the animals. Indeed, in the PM-DAT, memory is evaluated by the ***preference*** for the non-aversive enclosed arm, which might be less influenced (in comparison with the two other models—see Chap. 26) by increased or decreased motor function;

Table 19.2 A comparative analysis of rodent models used to evaluate memory

		Behavioural Parameters Evaluated					Duration of the Session	
	Aversive stimuli	Dependence on Motor Function	Learning	Memory	Anxiety	Motor Activity	Training	Test
Passive Avoidance Task	footshock	+	–	+	–	–	Up to 5 min	Up to 5 min
Morris Water Maze	water	+	+	+	–	+	Multiple trials	Multiple trials
Plus-maze Discriminative Avoidance Task	sound and light	+	+	+	+	+	10 min	3 min

(4) the plus-maze discriminative avoidance task effectively evaluates learning since it is measured in an intra-session fashion, in contrast to the Morris water maze. In the latter, learning is evaluated by the improvement of performance in an inter-session fashion and can be influenced by reactivation and extinction processes; (5) finally, when comparing the time spent employing the three animal models, the passive avoidance task seems to be the most practical and time effective, followed (closely) by the PM-DAT and the Morris water maze (Table 19.2).

References

Abílio VC, Freitas FM, Dolnikoff MS, Castrucci AM, Frussa-Filho R. Effects of continuous exposure to light on behavioral dopaminergic supersensitivity. Biol Psychiatry. 1999;45:1622–9.

Agnati LF, Fuxe K, Benfenati F, Batistini N, Zini I, Toffano G. Chronic ganglioside treatment counteracts the biochemical signs of dopamine receptor supersensitivity induced by chronic haloperidol treatment. Neurosci Lett. 1983;40:293–7.

Aporti F, Borsato R, Calderini G, Rubini R, Toffano G, Zanotti A, et al. Age-dependent spontaneous EEG bursts in rats: effects of brain phosphatidylserine. Neurobiol Aging. 1986;7:115–20.

Bellot RG, Camarini R, Vital MA, Palermo-Neto J, Leyton V, Frussa-Filho R. Monosialoganglioside attenuates the excitatory and behavioural sensitization effects of ethanol. Eur J Pharmacol. 1996;313(3):175–9.

Beracochea D. Anterograde and retrograde effects of benzodiazepines on memory. Sci World J. 2006;6:1460–5.

Bonnet MH, Arand DL. We are chronically sleep deprived. Sleep. 1995;18:908–11.

Bonnet MH, Arand DL. Clinical effects of sleep fragmentation versus sleep deprivation. Sleep Med Rev. 2003;7:297–310.

Bouyer JJ, Deminiere W, Mayo W, Le Moal M. Inter-individual differences in the effects of acute stress on the sleep-wakefulness cycle in the rat. Neurosci Lett. 1997;225:193–6.

Bruins Slot LA, Colpaert FC. Opiate states of memory: receptor mechanisms. J Neurosci. 1999;19:10520–9.

Bruins Slot LA, Koek W, Colpaert FC. Ethanol state dependence involving a lever press response requirement in rats. Behav Pharmacol. 1999;10:229–33.

Bueno OF, Lobo LL, Oliveira MG, Gugliano EB, Pomarico AC, Tufik S. Dissociated paradoxical sleep deprivation effects on inhibitory avoidance and conditioned fear. Physiol Behav. 1994; 56:775–9.

Calzavara MB, Lopez GB, Abílio VC, Silva RH, Frussa-Filho R. Role of anxiety levels in memory performance of spontaneously hypertensive rats. Behav Pharmacol. 2004;15:545–53.
Carvalho RC, Silva RH, Abílio VC, Barbosa PN, Frussa-Filho R. Antidyskinetic effects of risperidone on animal models of tardive dyskinesia in mice. Brain Res Bull. 2003; 60:115–24.
Carvalho RC, Patti CL, Takatsu-Coleman AL, Kameda SR, Souza CF, Garcez-do-Carmo L, et al. Effects of reserpine on the plus-maze discriminative avoidance task: dissociation between memory and motor impairments. Brain Res. 2006;1122:179–83.
Castro JP, Frussa-Filho R, Fukushiro DF, Chinen CC, Abílio VC, Silva RH. Effects of long-term continuous exposure to light on memory and anxiety in mice. Physiol Behav. 2005;86:218–23.
Castro JP, Frussa-Filho R, Fukushiro DF, Silva RH, Medrano WA, Ribeiro RA, et al. Effects of baclofen on reserpine-induced vacuous chewing movements in mice. Brain Res Bull. 2006;68:436–41.
Claro FT, Silva RH, Frussa-Filho R. Bovine brain phosphatidylserine attenuates scopolamine-induced amnesia. Physiol Behav. 1999;67:551–4.
Cochran JC, Thorne DR, Penetar DM, Newhouse PA. Decoupling motor memory strategies: effects of sleep deprivation and amphetamine. Int J Neurosci. 1994;74:45–54.
Cohen C, Perrault G, Sanger DJ. Evidence for the involvement of dopamine receptors in ethanol-induced hyperactivity in mice. Neuropharmacology. 1997;36:1099–108.
Colpaert FC. Pharmacological characteristics of tremor, rigidity and hypokinesia induced by reserpine in rats. Neuropharmacology. 1987;26:1431–40.
Colpaert FC. Amnestic trace locked into the benzodiazepine state of memory. Psychopharmacology (Berl). 1990;102:28–36.
Colpaert FC. System theory of pain and of opiate analgesia: no tolerance to opiates. Pharmacol Rev. 1996;47:605–29.
Colpaert FC, Koek W, Bruins Slot LA. Evidence that amnesic state govern normal and disordered memory. Behav Pharmacol. 2001;12:575–89.
Conceição IM, Maiolini Jr M, Mattia MA, Vital MA, Santos BR, Smaili S, et al. Anxiety-induced antinociception in the mouse. Braz J Med Biol Res. 1992;25:831–4.
Conceição IM, Maioline Jr M, Mattia NF, Chang YH, Smaili S, Frussa-Filho R. Effect of withdrawal from long-term nifedipine administration on open-field habituation in the rat. Braz J Med Biol Res. 1994;27:1363–7.
Cooper JA, Sagar HJ, Jordan N, Harvey NS, Sullivan E. Cognitive impairment in early, untreated Parkinson's disease and its relationship to motor disability. Brain. 1991;14:2095–122.
Crook T, Petrie W, Wells C, Massari DC. Effects of phosphatidylserine in Alzheimer's disease. Psychopharmacol Bull. 1992;28:61–6.
Dinges DF, Pack F, Williams K, Gillen KA, Powell JW, Ott GE, et al. Cumulative sleepiness, mood disturbance, and psychomotor vigilance performance decrements during a week of sleep restricted to 4–5 hours per night. Sleep. 1997;20:267–77.
Fluck E, File SE, Springett J, Kopelman MD, Rees J, Orgill J. Does the sedation resulting from sleep deprivation and lorazepam cause similar cognitive deficits? Pharmacol Biochem Behav. 1998;59:909–15.
Frussa-Filho R, Otoboni JR, Uema FT, Sá-Rocha LC. Evaluation of memory and anxiety in rats observed in the elevated plus-maze: effects of age and isolation. Braz J Med Biol Res. 1991;24:725–8.
Frussa-Filho R, Otoboni JR, Gianotti AD, Amaral ACS, Conceição IM. Effects of age on antinociceptive effects of elevated plus-maze exposure. Braz J Med Biol Res. 1992;25:827–9.
Frussa-Filho R, Barbosa-Júnior H, Silva RH, Da Cunha C, Mello CF. Naltrexone potentiates the anxiolytic effects of chlordiazepoxide in rats exposed to novel environments. Psychopharmacology (Berl). 1999;147:168–73.
Goldstone S, Nurnberg HG, Lhamon WT. Effects of trifluoperazine, chlorpromazine, and haloperidol upon temporal information processing by schizophrenic patients. Psychopharmacology (Berl). 1979;65:119–24.

Goto SH, Conceição IM, Ribeiro RA, Frussa-Filho R. Comparison of anxiety measured in the elevated plus-maze, open-field and social interaction tests between spontaneously hypertensive rats and Wistar EPM 1 rats. Braz J Med Biol Res. 1993;26:965–9.
Graeff FG, Vianna MB, Tomaz C. The elevated T-maze, a new experimental model of anxiety and memory: effect of diazepam. Braz J Med Biol Res. 1993;26:67–70.
Handley SL, Mitani S. Effects of alfa-adrenoceptor agonists and antagonists in a maze exploration model of 'fear'-motivated behavior. Naunyn Schmiedebergs Arch Pharmacol. 1984;327:1–5.
Hanks GW, O'Neill WM, Simpson P, Wesnes K. The cognitive and psychomotor effects of opioid analgesics. II. A randomized controlled trial of single doses of morphine, lorazepam and placebo in healthy subjects. Eur J Clin Pharmacol. 1995;48:455–60.
Hernández OH, Vogel-Sprott M, Ke-Aznar VI. Alcohol impairs the cognitive component of reaction time to an omitted stimulus: a replication and an extension. J Stud Alcohol Drugs. 2007;68:276–81.
Isomae K, Morimoto S, Hasegawa H, Morita K, Kamei J. Effects of T-82, a novel acetylcholinesterase inhibitor, on impaired learning and memory in passive avoidance task in rats. Eur J Pharmacol. 2003;465:97–103.
Itoh J, Nabeshima T, Kameyama T. Utility of an elevated plus-maze for the evaluation of memory in mice: effects of nootropics, scopolamine and electroconvulsive shock. Psychopharmacology (Berl). 1990;101:27–33.
Izquierdo I, Perry ML, Dias RD, Souza DO, Elizabetsky E, Carrasco MA, et al. Endogenous opioids memory modulation and state dependency. In: Martinez JL, Jensen RA, Messing RB, Rigter H, McGaugh JL, editors. Endogenous peptides and learning and memory process. New York: Academic; 1981. p. 269–90.
Jackson A, Koek W, Colpaert FC. NMDA antagonists make learning and recall state-dependent. Behav Pharmacol. 1992;3:415–21.
Kameda SR, Frussa-Filho R, Carvalho RC, Takatsu-Coleman AL, Ricardo VP, Patti CL, et al. Dissociation of the effects of ethanol on memory, anxiety, and motor behavior in mice tested in the plus-maze discriminative avoidance task. Psychopharmacology (Berl). 2007;192:39–48.
Karni A, Tanne D, Rubenstein BS, Askenasy JJ, Sagi D. Dependence on REM sleep of overnight improvement of a perceptual skill. Science. 1994;265:679–82.
Lee C, Rodgers RJ. Antinociceptive effects of elevated plus-maze exposure: influence of opiate receptor manipulations. Psychopharmacology (Berl). 1990;102:507–13.
Levin ED, Wilson W, Rose JE, McEvoy J. Nicotine-haloperidol interactions and cognitive performance in schizophrenics. Neuropsychopharmacology. 1996;15:429–36.
Lister RG. The use of a plus-maze to measure anxiety in the mouse. Psychopharmacology (Berl). 1987;92:180–5.
Martinez Jr JL, Schulteis G, Janak PH, Weinberger SB. Behavioral assessment of forgetting in aged rodents and its relationship to peripheral sympathetic function. Neurobiol Aging. 1988;9:697–708.
Mednick SC, Nakayama K, Cantero JL, Atienza M, Levin AA, Pathak N, Stickgold R. The restorative effect of naps on perceptual deterioration. Nat Neurosci. 2002;5:677–81.
Miceli G, Caltagirone C, Gainotti G. Gangliosides in the treatment of mental deterioration. A double-blind comparison with placebo. Acta Psychiatr Scand. 1997;55:102–10.
Mizumori SJ, Sakai DH, Rosenzweig MR, Bennett EL, Wittreich P. Investigations into the neuropharmacological basis of temporal stages of memory formation in mice trained in an active avoidance task. Behav Brain Res. 1987;23:239–50.
Nunzi MG, Milan F, Guidolin D, Zanotti A, Toffano G. Therapeutic properties of phosphatidylserine in the aging brain. In: Hanin I, Pepeu G, editors. Phospholipids. New York: Plenum Press; 1990. p. 213–8.
Owen AM, James M, Leigh PN, Summers BA, Marsden CD, Quinn NP, et al. Fronto-striatal cognitive deficits at different stages of Parkinson's disease. Brain. 1992;115:1727–51.
Patti CL, Kameda SR, Carvalho RC, Takatsu-Coleman AL, Lopez GB, Niigaki ST, et al. Effects of morphine on the plus-maze discriminative avoidance task: role of state-dependent learning. Psychopharmacology (Berl). 2006;184:1–12.

Pellow S, Chopin P, File SE, Briley M. Validation of open: closed arm entries in an elevated plus-maze as a measure of anxiety in the rat. J Neurosci Methods. 1985;14:149–67.

Pierce RC, Kalivas PW. A circuitry model of the expression of behavioral sensitization to amphetamine-like psychostimulants. Brain Res Brain Res Rev. 1997;25:192–216.

Rahmann H. Brain gangliosides: neuromodulators for synaptic transmission and memory formation. In: Matthies H, editor. Advances in the biosciences. Oxford: Pergamon Press; 1986. p. 235–45.

Ramos A, Kangerski AL, Basso PF, Da Silva Santos JE, Assreuy J, Vendruscolo LF, et al. Evaluation of Lewis and SHR rat strains as a genetic model for the study of anxiety and pain. Behav Brain Res. 2002;129:113–23.

Rapeli P, Kivisaari R, Kähkönen S, Puuskari V, Autti T, Kalska H. Do individuals with former amphetamine dependence have cognitive deficits? Nord J Psychiatry. 2005;59:293–7.

Rapport MM. Introduction to the biochemistry of gangliosides. In: Rapport MM, Gorio A, editors. Gangliosides in neurobiological and neuromuscular function, development and repair. New York: Raven; 1981. p. 15–9.

Rodgers RJ, Lee C, Shepherd JK. Effects of diazepam on behavioural and antinociceptive responses to elevated plus-maze in male mice depend upon treatment regimen and prior maze experience. Psychopharmacology (Berl). 1992;106:102–10.

Rosenzweig MR, Bennett EL, Colombo PJ, Lee DW, Serrano PA. Short-term, intermediate-term, and long-term memories. Behav Brain Res. 1993;5:193–8.

Schreiber S, Kampf-Sherf O, Gorfine M, Kelly D, Oppenheim Y, Lerer B. An open trial of plant-source derived phosphatydilserine for treatment of age-related cognitive decline. Isr J Psychiatry Relat Sci. 2000;37:302–7.

Seifert W, Wierasko A, Terlau H, Hollmann M. Gangliosides and neuronal plasticity in the hippocampus. NATO ASI Ser. 1987;7:523–8.

Siegel JM. The REM, sleep-memory consolidation hypothesis. Science. 2001;294:1058–63.

Silva RH, Frussa-Filho R. The plus-maze discriminative avoidance task: a new model to study memory-anxiety interactions. Effects of chlordiazepoxide and caffeine. J Neurosci Methods. 2000;102:117–25.

Silva RH, Frussa-Filho R. Naltrexone potentiates both amnestic and anxiolytic effects of chlordiazepoxide in mice. Life Sci. 2002;72:721–30.

Silva RH, Bellot RG, Vital MABF, Frussa-Filho R. Effects of long-term ganglioside GM1 administration on a new discriminative avoidance test in normal adult mice. Psychopharmacology (Berl). 1997;129:322–8.

Silva RH, Felicio LF, Frussa-Filho R. Ganglioside GM1 attenuates scopolamine-induced amnesia in rats and mice. Psychopharmacology (Berl). 1999;141:111–7.

Silva RH, Abílio VC, Torres-Leite D, Bergamo M, Chinen CC, Claro FT, et al. Concomitant development of oral dyskinesia and memory deficits in reserpine-treated male and female mice. Behav Brain Res. 2002a;132:171–7.

Silva RH, Kameda SR, Carvalho RC, Rigo GS, Costa KL, Taricano ID, et al. Effects of amphetamine on the plus-maze discriminative avoidance task in mice. Psychopharmacology (Berl). 2002b;160:9–18.

Silva RH, Abílio VC, Takatsu AL, Kameda SR, Grassl C, Chehin AB, et al. Role of hippocampal oxidative stress in memory deficits induced by sleep deprivation in mice. Neuropharmacology. 2004a;46:895–903.

Silva RH, Chehin AB, Kameda SR, Takatsu-Coleman AL, Abílio VC, Tufik S, et al. Effects of pre- or post-training paradoxical sleep deprivation on two animal models of learning and memory in mice. Neurobiol Learn Mem. 2004b;82:90–8.

Smith C, Kelly G. Paradoxical sleep deprivation applied two days after end of training retards learning. Physiol Behav. 1998;43:213–6.

Spanagel R. Modulation of drug-induced sensitization processes by endogenous opioid systems. Behav Brain Res. 1995;70:37–49.

Stip E. Cognition, schizophrenia and the effect of antipsychotics. Encéphale. 2006;32:341–50.

Suchecki D, Duarte-Palma B, Tufik S. Sleep rebound in animals deprived of paradoxical sleep by the modified multiple platform method. Brain Res. 2000;875:14–22.

Sutton MA, Masters SE, Bagnall MW, Carew TJ. Molecular mechanisms underlying a unique intermediate phase of memory in aplysia. Neuron. 2001;31:143–54.

Tettamanti G, Sonnino S, Ghiondi R, Masserini M, Venerando B. Chemical and functional properties of gangliosides. Their possible implication in the membrane mediated transfer of information. In: Degiorgio V, Corti M, editors. Physics and amphiphiles: micelles, vesicles and microemulsions. Amsterdam: North Holland Physics; 1985. p. 607–36. XC Corso Soc Ital Fisica Bologna.

Tipper CM, Cairo TA, Woodward TS, Phillips AG, Liddle PF, Ngan ET. Processing efficiency of a verbal working memory system is modulated by amphetamine: an fMRI investigation. Psychopharmacology (Berl). 2005;180:634–43.

Toffano G, Agnati LF, Fuxe KG. The effect of the ganglioside GM1 on neuronal plasticity. Int J Dev Neurosci. 1986;4:97–100.

Walker MP, Stickgold R. Sleep-dependent learning and memory consolidation. Neuron. 2004; 44:121–33.

Youngblood BD, Zhou J, Smagin GN, Ryan DH, Harris RB. Sleep deprivation by the "flower pot" technique and spatial reference memory. Physiol Behav. 1997;61:249–56.

Zanotti A, Valzelli L, Toffano G. Reversal of scopolamine-induced amnesia by phosphatidylserine in rats. Psychopharmacology (Berl). 1986;90:274–5.

Chapter 20
Rodent Models in Psychiatric Research

Norma Lilia Anaya Vázquez, Rosaely Casalegno, Edith Monroy-López, and Javier Velázquez-Moctezuma

Since psychiatric illnesses are commonly perceived as disorders of thoughts and feelings, for a number of years it was widely accepted that it was impossible to study psychiatric diseases in animal models. In recent decades, this notion has drastically changed, and the number of psychiatric disease studies using laboratory animals has been rapidly increasing.

Psychiatry is a branch of medicine that has evolved slowly, mainly because of the lack of experimental activity to elucidate the physiopathological mechanisms underlying these psychiatric diseases. Besides the outstanding descriptions of psychiatric illnesses reported by the pioneers of this discipline, no major advances in the understanding of their etiology were reached. During the early years of the discipline, the range of psychiatric diseases went from major psychotic states to behaviors that were socially and politically unacceptable. The history of the first psychiatric hospitals records the presence of "patients" suffering from psychosis and depression but also suffering from epilepsy, homosexuality, and premenstrual syndrome. The definitions of the field of psychiatry and the characteristics of psychiatric illness have themselves evolved slowly and currently, and the most recent agreement among specialists is reported in the Diagnostic and Statistical Manual of Mental Illness (DSMIV).

Faced with this lack of experimental support, the major advances in therapeutic alternatives in psychiatry were serendipitous. The first antipsychotic drug available

N.L.A. Vázquez • J. Velázquez-Moctezuma (✉)
Departamento de Biologia de la Reproduccion, Universidad Autonoma Metropolitana-Iztapalapa, Mexico City AP 09340, Mexico
e-mail: jvm@xanum.uam.mx

R. Casalegno
Helmholtz Center for Infection Research HZI, Department of Vaccinology and Applied Microbiology, Braunschweig, Germany

E. Monroy-López
Master in Psychobiology, Universidad Nacional Autónoma, Mexico

M.L. Andersen, S. Tufik (eds.), *Rodent Model as Tools in Ethical Biomedical Research*, DOI 10.1007/978-3-319-11578-8_20

was chlorpromazine, which was originally used as a co-treatment in surgical anesthesia. Subjects receiving chlorpromazine became calm and relaxed. Therefore, it was administered to psychotic, agitated patients with outstanding results. In a convergent line of studies, reserpine, which is an alkaloid obtained from a plant (*Rauwolfia Serpentina*) and has been successfully used in cardiovascular diseases, also showed efficiency as an antipsychotic. The discovery of the antipsychotic properties of these drugs heralded the birth of psychopharmacology.

Once psychiatrists had identified two efficient antipsychotic drugs, the search for new antipsychotic drugs began. The use of animals in the search for new antipsychotic drugs was seriously considered for the first time. The strategy was quite simple. The first step was to determine the effects of reserpine or chlorpromazine in animals. The second step was to administer a great number of drugs, even those chemically unrelated to reserpine or chlorpromazine, to animals. Drugs eliciting similar effects in animals to those elicited by reserpine and chlorpromazine were immediately considered as candidates to be tested in humans.

In addition, advances in basic neuroscience were a critical source of information that enriched psychiatry in its efforts to understand psychiatric illnesses and discover new therapeutic tools. For a number of years, psychopharmacology has been geared toward the relationships between psychiatric illnesses and neurotransmitter systems. Today, almost all drugs used in psychiatry exert their effect at synaptic level by interacting with a specific type of receptor.

In general terms, the use of animals in psychiatric research is oriented toward two main objectives. One objective is to elucidate the neural mechanisms involved in a specific mental pathology. The second objective is the possibility of performing a screening test in which new therapeutic alternatives, including pharmacological as well as non-pharmacological manipulations, can be tested. For these purposes, three general strategies can be distinguished: screening tests, behavioral bioassays, and simulations.

Screening tests are experimental set-ups in which a large number of compounds can be tested in a fast, reliable, and inexpensive way. Most of the screening tests follow one of the two main strategies used for the development of new drugs. One strategy is to look for compounds that have a similar action on specific psychiatric illnesses (e.g., antidepressant, anxiolytic, or antipsychotic actions). The procedure implies comparing the actions of the novel compounds with those elicited by drugs that are currently used in clinical practice. An additional strategy is oriented toward the identification of biochemical similarities with drugs already known to be clinically effective. This strategy looks for similarities in the actions of the new drugs with the neurotransmitter system in which the known drugs interact. For instance, the search for an anxiolytic drug implies the search for interactions with the GABA/benzodiacepine receptor. The search for antidepressants might imply analysis of the reuptake system in the serotonergic synapse. Needless to say, both strategies have major disadvantages, mainly concerning novel compounds that unexpectedly show high clinical efficiency by acting on a different neurotransmitter system.

When the whole animal is used as a measuring device to assess the general effect of a drug or a manipulation, the strategy is referred to as a behavioral bioassay. In these assays, it is possible to analyze the mechanisms responsible for changes in brain function after acute or chronic drug administration, brain lesion, brain stimulation, or any other manipulation. The typical example is stereotypic behavior and the increase in locomotor activity observed after the administration of dopaminergic psychomotor stimulants. This strategy has important advantages compared to the preceding strategies. Behavioral methods are non-destructive, and behavioral changes undoubtedly indicate that the drug has reached the brain. In addition, behavioral measures are indicative of a functional state that integrates the activity of the whole brain.

Simulations imply focusing on human behavior and attempting to replicate certain normal or abnormal aspects of it. Normal human behaviors are commonly studied in animals; some examples include experiments of classical conditioning, the control of operant behavior by reinforcement, and experiments on memory and learning. Abnormal behavior simulations involve choosing a specific illness and replicating its main features. In psychiatry, simulation models are linked to specific illnesses such as anxiety, depression, or schizophrenia with the goal of replicating a conspicuous symptom characteristic of the disorder, a group of symptoms, or rarely the complete syndrome. To achieve this goal, researchers have used brain lesions, chemical stimulation, selection of animals for selective breeding, and the application of certain conditions such as stress, social isolation, and aging that allegedly induce mental disorders in human beings. The challenge for each model is to provide information on the etiology, treatment, pathophysiology, and physiological mechanisms involved when treatment is successful. The importance of each proposed animal model depends on its likelihood to achieve goals in each of these venues.

Validity

Since animal models are essentially tools that help us to understand a given illness, their value resides in the extent to which the model can provide us with more accurate information. In the mid-1980s, Paul Willner, the author with the greatest influence on this topic, defined the validity of animal models in psychiatry. Basically, the validity of each model can be defined in three different categories: predictive validity, face validity, and construct validity.

Predictive validity refers primarily to the degree to which an animal model can detect drugs, either known or novel compounds, having clear therapeutic actions. In this regard, an animal model with high predictive validity exhibits some similarities to screening tests. Nevertheless, there are some critical differences with the simple screening tests. The tests are designed to detect drugs or manipulations that improve clinical features of an illness, whereas an animal model should also be able to detect manipulations or drugs that worsen the clinical condition. Moreover, the tests are only good for certain classes of drugs, whereas animal models respond to all classes of drugs that potentially alleviate the symptoms. Finally, animal models show a positive correlation between the potency of a drug and its potency in clinical use. Obviously, animal models should be free of false negatives as well as false positives.

Face validity is regarded as the extent to which a given animal model replicates one or more symptom of the syndrome. In addition, it has been suggested that animal models should resemble the illness in four respects: etiology, symptomatology, treatment, and physiological basis. Furthermore, aside from the phenomenological similarities of the animal model with the disorder that is being simulated, it is expected that the model have no major dissimilarities.

Animal Models of Depression

As was mentioned above, animal models are oriented toward the screening of new treatments or toward the replication of the whole picture of human depression in animals. Fortunately, depression's clinical symptoms are clearly depicted in the DSM-IV. For the diagnosis of depression, two major symptoms have been defined: dysphoric mood and anhedonia. In addition, there are eight secondary signs and symptoms: alterations in food intake, sleep, and motor activity, decrease in aggression, sexual behavior, and self-esteem, the presence of suicidal ideation, and excessive guilt. The clinical symptoms should be present for at least 2 weeks, and patients should exhibit either both of the major symptoms or one major symptom accompanied by at least four secondary symptoms. Anhedonia and some secondary data can be reproduced in laboratory animals. Concerning the search for new treatments, some experimental tests have been shown to be reliable in detecting the antidepressant activity of drugs or manipulations.

Forced Swim Test (FST)

This test was developed during the 1970s and is currently the most widely used test for assessing pharmacological antidepressant activity. Originally, the test consisted of placing a rat in a cylinder containing a column of water from which escape was impossible. Initially, the rat displays agitated motor activity that includes swimming, climbing, jumping, and diving. After a few minutes, the rat adopts a typical posture in which only its head protrudes above the surface of the water. The rat remains immobile, moving only to correct its posture. This situation lasts for 15 min; 24 h later, the same rat is again submitted to the same situation for 5 min. The amount of immobility in this second test is assessed, and the effects of drugs or manipulations are commonly observed in this 5-min test. Figure 20.1 shows typical postures observed during the FST. Panel (a) shows the immobilization posture adopted after a period of struggling, during which the rat is climbing (Panel (b)), swimming (Panel (c)), or diving (Panel (d)).

The relationship between this test and depression arose with the observation that the amount of immobility is modified only after the administration of drugs with well-known antidepressant activity. The forced swim test can detect drugs with

Fig. 20.1 Forced swim test. Panel (**a**) shows the typical posture of an immobile rat. Panel (**b**) shows the rat climbing with the forepaws against the walls. Panel (**c**) exhibits a rat swimming and moving the four extremities. Panel (**d**) presents a diving rat

antidepressant effects—even those that are chemically related, including tricyclic antidepressants, inhibitors of monoamino oxidase, or atypical antidepressants. In addition, some manipulations that have shown the same clinical antidepressant effect, such as REM sleep deprivation, electroconvulsive shock, exercise, and transmagnetic stimulation, also influenced the immobility time.

Thus, the forced swim test has perhaps the highest predictive validity that can be found in the search for new and novel antidepressant treatments. The test has proven to be reliable because it has given only a few false negatives and a few false positives after screening a great number of drugs. False negatives include drugs that induce a psychomotor effect and consequently impact the time of immobility. For this reason, the FST should be accompanied by tests that assess motor activity.

On the other hand, the face validity of the FST has been a matter of controversy for a number of years. According to some authors, the FST should be considered only as a screening test for antidepressants. Other authors, however, argue that the test reveals a depressive behavior. This concept is mainly supported by the fact that the test requires a pretest swim. This manipulation allegedly induces learned helplessness, a condition in which the subject "learns" the uselessness of any activity to avoid a stressful situation. Actually, rats submitted to models of depression that are geared to induce depressive behaviors show an increase of immobility in the FST, suggesting that the test is also capable of detecting a depressive state in the rat. Thus, the FST is also suitable for assessing the effectiveness of procedures that have been proposed for inducing a depressive state in rats.

Since the test has been widely used for a number of years, new peculiarities have arisen. As mentioned above, more immobility is equated with more depression, and a decrease of immobility reflects an antidepressant effect. However, a decrease of immobility indicates an increase of mobility due to either climbing or swimming. It has been suggested that the type of mobility elicited by any given antidepressant reflects the activation of a neurotransmitter system. The increase of climbing activity suggests the activation of the noradrenergic system, whereas the increase in swimming suggests the activation of the serotonergic system.

Intracranial Self-Stimulation (ISS)

In the early 1950s, James Olds discovered the possibility of inducing in the rat a behavior that leads to the delivery of a mild electric shock into the rat's own brain. Similar to many other important scientific findings, the discovery of intracranial self-stimulation (ISS) was serendipitous. While the researchers were looking for the activation-related effects of electrical stimulation in the reticular formation, they misplaced the electrode and found that the rat seemed to be very interested in the stimulus. The experiment evolved to conditions in which the rat was able to decide when and how frequently to apply the ISS. After a number of studies and intense analysis of the results, the authors suggested that ISS was linked to pleasure. After mapping the brain, they proposed the existence of aversive, neutral, and pleasure-inducing areas. After the rat gained experience in ISS, it frenetically activated the lever to obtain the stimulus, forgetting about other rewarding stimuli such as sex or food intake. The animal is also capable of withstanding a high level of pain in order to reach the stimulus lever. Figure 20.2 shows a diagram of the self-stimulation and the self-administration set-up.

Surprisingly, ISS is not considered to be an animal model of depression. It is mentioned here because it offers an approach to the central mechanisms involved in the experience of pleasure. It can be reasonably assumed that the cerebral mechanisms of joy and pleasure are impaired in depressed subjects. ISS allows us to explore the reward system, although the precise relationship between the reward system and depression is still a matter of controversy.

Stress and Depression

A huge volume of literature has been published regarding the relationship between stress and depression. Stress, especially chronic low grade stress, is considered to be a powerful predisposing factor for depression. A stressful event increases by five- or

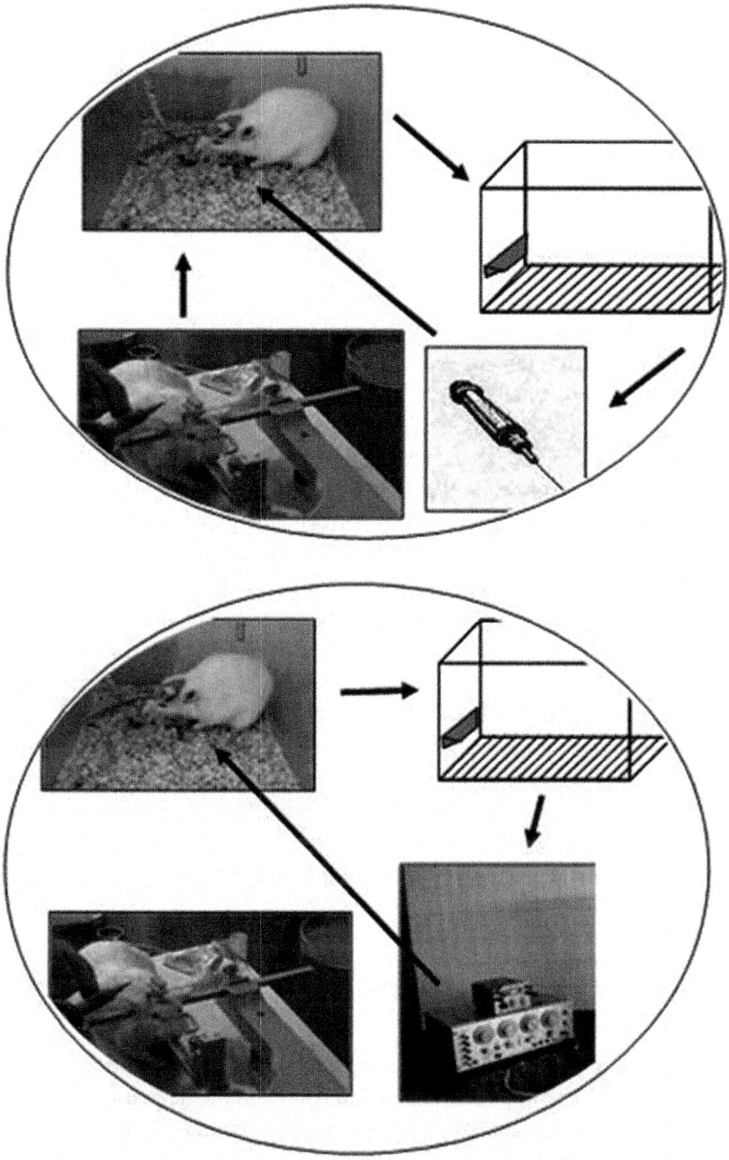

Fig. 20.2 Schematic representation of the set-up for self-stimulation (*upper panel*) and self administration (*lower panel*). The principle is similar. The rat is chronically implanted with either an injection cannula or an electrode. The rat activates a lever that releases a specific amount of a drug (administration) or delivers an electric shock in a rewarding zone (stimulation)

sixfold the likelihood of suffering an episode of depression within the following 6 months. However, the response to stress is characterized by a process of habituation. After repeated exposure to the same stressor, the response of the system progressively declines. This characteristic of the stress response makes it extremely difficult to study the effect of chronic stress in animal models. To elicit a chronic response, the stressor needs to change constantly. The following models are the most conspicuous and successful attempts at modeling the effects of acute and chronic stress on mood.

Learned Helplessness

In the mid-1970s, Seligman and his group described a phenomenon that has been widely used in research laboratories around the world and remains a matter of controversy. In summary, when a dog is placed in a cage with no chance to escape and receives an electric foot shock, it seems that the animal "learns" that any effort to avoid the shock is useless. Then, when the same animal is placed in a situation in which it can escape from the electric shock, the animal is unable to perform any escape action. In addition, some other performance deficits are displayed by the animal in subsequent learning tasks. Other behavioral effects include a decrease in motor activity and aggression, a loss of appetite, and a marked deficit in performance of pleasure-seeking behaviors (including intracranial self-stimulation). Most of the effects disappear within 3 days, although it has been reported that some effects may last for even 7 weeks under special conditions. Animals submitted to exactly the same conditions but with the possibility to avoid the shock do not display learning deficits. Learned helplessness responds positively to treatment with antidepressants. Although there are some reports to the contrary, a number of studies have indicated that acute and chronic antidepressant treatment reverses the behavioral deficits observed in learned helplessness. Moreover, administration of neuroleptics, stimulants, sedatives, or anxiolytics is ineffective.

Chronic Mild Unpredictable Stress

As mentioned above, in addition to a typical response triggered by any stressor, there is also a mechanism of adaptation that results in the decrease of this response. This phenomenon was a challenge for those who attempted to study the effects of chronic stress. To avoid adaptation, animals were chronically exposed to a number of stressors in a random presentation sequence. In the early stage, the method included electric foot shocks, immersion in cold water, immobilization, bright and constant light, and loud noise, among others. After 3 weeks of manipulation, the animals displayed deficits in motor activity that were ameliorated by antidepressant treatment. Nevertheless, the method was widely criticized because of the high level of suffering that the animals endured and because the stressors were not common stressors that an animal like the rat would experience during natural life. Therefore, the method evolved to include more naturalistic conditions, including mild stressors

	Food and water deprivation	Water deprivation	Soiled cage	Stroboscopic lighting	Empty water bottles	Continuous lighting	Cage tilt	Mouse cage
A.M. **Mon** P.M.				08:30 hr			16:00 hr	
A.M. **Tues** P.M.	16:00 hr	10:00 hr 16:00 hr					10:00 hr	
A.M. **Mier** P.M.	10:00 hr		16:00 hr			16:00 hr		10:00 hr
A.M. **Wed** P.M.			10:00 hr		10:00 hr 16:00 hr	10:00 hr	16:00 hr	
A.M. **Thurs** P.M.	16:00 hr						10:00 hr	
A.M. **Fri** P.M.	10:00 hr	10:00 hr 16:00 hr		16:00 hr				16:00 hr
A.M. **Sat** P.M.								10:00 hr
A.M. Sun P.M.					10:00 hr 10:00 hr			

Fig. 20.3 Example of the sequence and duration of some stressors used to induce a decrease in sucrose consumption (anhedonia)

that the animal might experience in its natural life (e.g., short periods of food and water deprivation, small changes in temperature, water in the saw dust bed, tilting of the cage, changes in the light-dark cycle, and similar manipulations). The sequence and duration in which the stressors were applied were varied constantly so that the animal was unable to predict the onset of any stressor. Figure 20.3 shows an example of the sequence and duration of the stressors. Another important improvement in the method is the fact that monitoring of the animal's mood is done by a simple test, the sucrose consumption test. In pre-experimental conditions, rats consume more sucrose solution than simple water due to the hedonistic value of the sucrose solution. After weeks of exposure to chronic mild stress, the animals diminish their consumption of sucrose solution. It has been assumed that animals at this moment are in a state of anhedonia, which is one of the two major descriptors of depression described above. Aside from anhedonia, the animals also display other behavioral deficits that are consistent with the picture of depression. Finally, although relatively few studies have been performed, antidepressant treatment reverses the behavioral deficits even when the stressors remain present.

In addition to the models of depression mentioned, other models have also been studied to a considerable degree. In brief, the separation model has been analyzed mainly in infant monkeys deprived of their mothers. These young monkeys display features of depression that are reversed by antidepressant treatment. Similarly, isolated 1-week-old chicks display a typical distress-calling behavior, the characteristics of which can be modified by antidepressant treatment. In addition, lesioning of brain structures can also induce behavioral changes congruent with depression. Olfactory bulbectomy induces dramatic behavioral changes, such as irritability, hyperactivity, and deficits in passive avoidance learning; it concomitantly produces an increase of circulating corticosteroids, which is also observed in human depression. These changes are reversed by antidepressant treatment.

Animal Models of Anxiety

Anxiety can be briefly defined as the subjective perception of being in a risky situation (i.e., in danger). This is a common experience in the daily life of human beings; when it becomes extreme and negatively influences the life of a subject, however, it is considered an illness. Currently, anxiety has been subdivided into four separate sets of clinical symptoms: obsessive-compulsive disorder, generalized anxiety disorder, panic disorder, and phobic disorders in which the trigger can be a wide variety of stimuli. There is no one effective pharmacological treatment for all of the disorders characterized as anxiety disorders.

Anxiety is commonly used as a synonym of fear in both humans and animals. Fear is, without a doubt, a powerful motivating force that is present along the entire phylogenetic scale. Regularly, fear anticipates danger and evidently has a major role in the survival of the species. Thus, any procedure that induces fear immediately becomes a possible animal model of anxiety. Common procedures in psychopharmacology that involve changes in environmental conditions, deprivation of food, water, or sleep, or any annoying manipulation induce a mild or severe fear and as a consequence can be used as animal models of anxiety.

Besides psychotherapeutic interventions, anxiety is treated mainly with benzodiazepines. Other drugs, especially alcohol and nicotine, have an anxiolytic effect as well. benzodiazepines act in the brain on specific receptors that are linked via a supramolecular complex with the GABA receptor. It has been suggested that benzodiazepines exert their clinical effect through the facilitation of GABAergic neurotransmission, which in general terms induces an inhibition of the activity of other neurotransmitter systems. benzodiazepines receptors are widely present across the phylogenetic scale. Thus, animal models of anxiety are based mainly on fear and the therapeutic effects of benzodiazepines.

The most widely used animal models of anxiety will be briefly presented in the following paragraphs:

Open Field Test

This is perhaps the most widely used animal model of anxiety, primarily because of its simplicity. Rats or mice are placed in a novel and relatively large arena, and the experimenter records parameters of exploration such as motor activity, sniffing, and rearing. In addition, some parameters of fearfulness, such as defecation, urination,

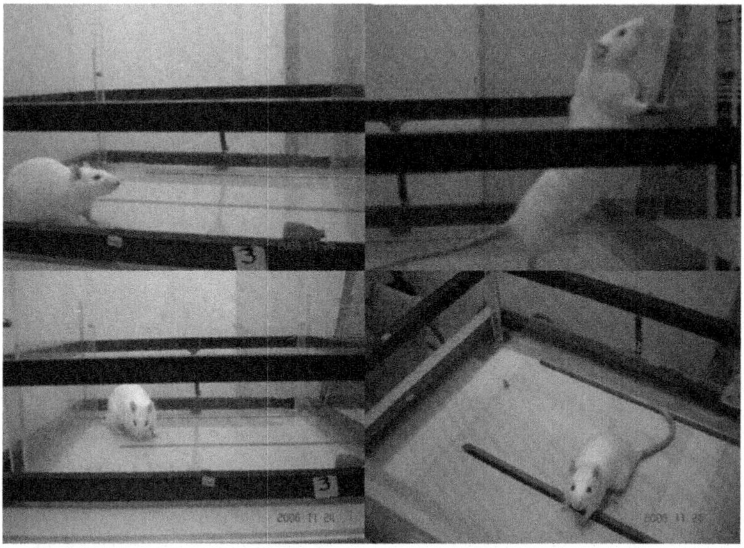

Fig. 20.4 Automatic recording of motor activity. The frames contain infrared beams that send a signal to the computer each time the rat crosses the beam

freezing, and grooming, are also recorded. The effects of well-known anxiolytics are an increase of motor activity and a decrease of rearing and all parameters of fearfulness. In Fig. 20.4, an automatic device for motor activity assessment is shown. Infrared beams are located in the frames; when the rat interrupts the beam, a signal is recorded in a computer. In the picture, the rat is located in a "safe" position near the walls (A and C), is rearing (B), and is exploring the "unsafe" zone in the center. Unfortunately, and despite the wide use of the model, it has been shown that several drugs not related to anxiety also modify the mentioned parameters. Stimulants such as amphetamine, for instance, increase motor activity. In addition, any drug that interacts with motor regulation has an effect on the open field test. Nevertheless, this simple test is a common component of behavioral batteries that attempt to assess the general state of the animal.

Elevated Plus Maze

This is also a widely used test. It is shaped like a T or an X, with two opposed open arms and one or two enclosed arms with walls but no roof. The X or the T usually is elevated 50 cm above the floor. The size of the arms is 50 cm × 10 cm, forming a square (10 cm × 10 cm) in the center (Fig. 20.5). Usually, the tests are done in a separate observation room and under bright illumination. The rat is placed in the center square facing one of the open arms, and the following parameters are recorded for 5 min: open arm latency (the time from placing the rat in the center until it completely (four paws) enters the open arms), the number of entries, and time spent

Fig. 20.5 Elevated T maze

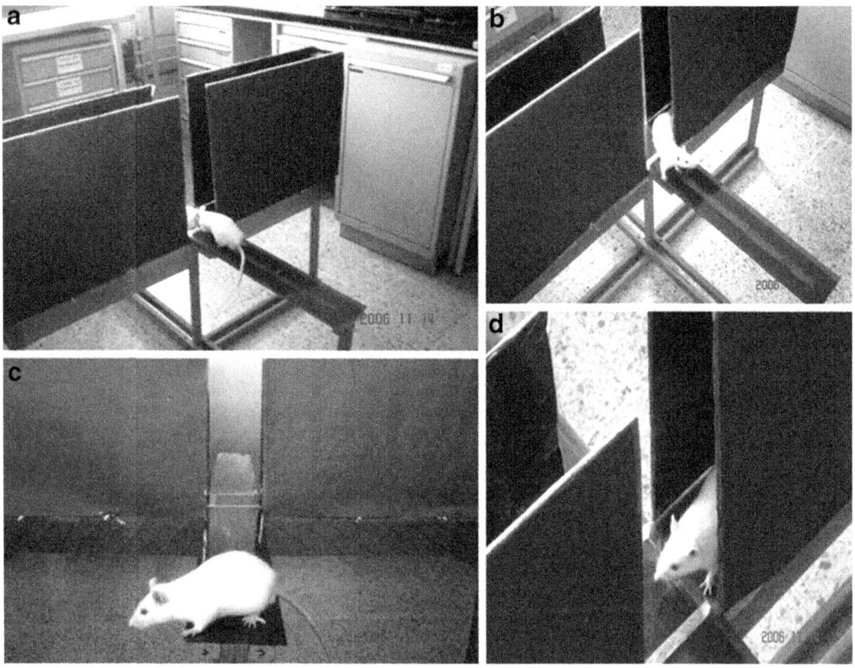

Fig. 20.6 Different moments of rat behavior in a typical T maze test. In (**a**) and (**b**), the rat is near the center. In (**c**), the rat is exploring one of the open arms. In (**d**), the rat is located in one of the "safe" arms

in both the open and the enclosed arms, rearing, grooming, urination, and defecation, and risk assessment. Figure 20.6 shows different views of a rat in a typical test. The test is supported by the widely accepted notion that the open arms will provoke higher anxiety, whereas the enclosed arms will represent a relatively safe situation. Rats that are forced to spend time in the open arms show an increase in blood corticosterone levels and display behaviors such as freezing and defecation more frequently. Rats that have not been treated spend more time inside the enclosed arms. The administration of well-known anxiolytic drugs produces an increase in both the number of entries and the time spent in the open arms as well as a decrease in other parameters of fear. Stimulants, such as amphetamine and even antidepressants, do not modify these parameters. Despite the promising results concerning the detection of anxiolytic drugs, significant variability has been reported among different laboratories and even within the same lab group.

Defensive Burying

Burying behavior is commonly observed in rodents. It seems that this is a species-specific behavior that becomes particularly prominent when rodents are faced with noxious stimuli or threatening situations. It has been suggested that burying is a reaction against something that induces fear. Understanding this response has given rise to several tests related to anxiety. In the most common experimental situation, a rat is placed in a cage in which one of the walls has a metallic tube. When the rat touches the tube, an electric shock is delivered. Subsequently, the rat buries the metallic tube with the saw dust of the cage floor. Figure 20.7 shows some

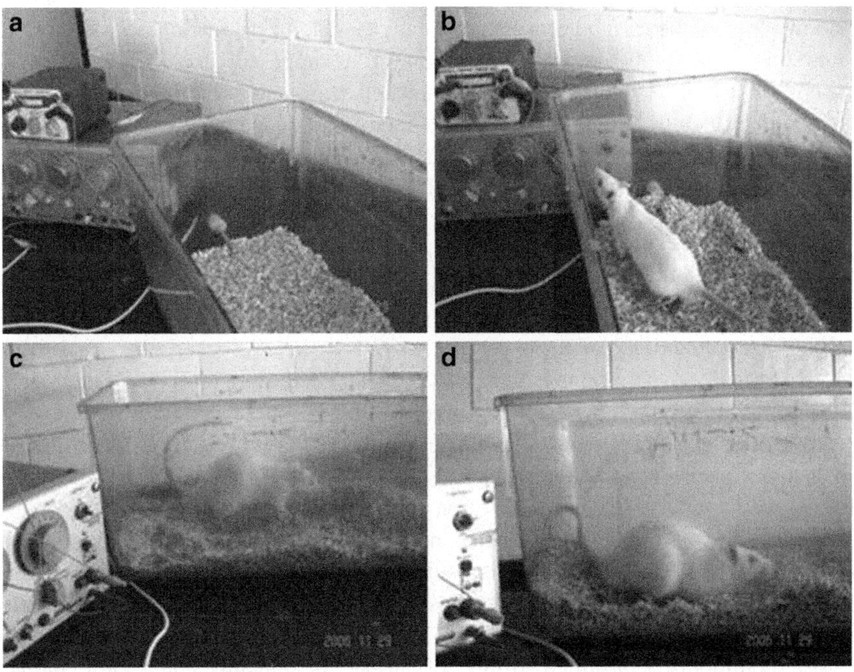

Fig. 20.7 Defensive burying set-up. In (**a**), the stimulator delivers an electric shock through the probe indicated by the arrow. In (**b**), the rat explores the cage and eventually touches the probe. In (**c**) and (**d**), the rat covers the probe with saw dust after touching it

moments of this typical behavior. The administration of anxiolytic drugs decreases defensive burying in correlation with the drugs' clinical potency. Amphetamine and morphine are ineffective, although it must be mentioned that chlorpromazine suppresses defensive burying. An interesting variation to this procedure is the replacement of the electric shock tube with glass marbles (four marbles with a 3.5 cm diameter). To increase the aversive component of the marbles, they are frequently covered with a drop of Tabasco sauce. This procedure also induces a high level of defensive burying.

These are only a few examples of the wide variety of experimental procedures that have been proposed as reliable tools in the search of anxiolytic drugs. Other tests include 1) the light/dark crossing test, which is based on the aversive property of bright light and the safety of a dark and enclosed place, and 2) the social interaction test, which explores behaviors that are normally displayed by rats living in groups and that decrease when rats are faced with an aversive situation such as an unfamiliar chamber or bright light. It has been suggested that this decrease is due to the anxiety elicited by the environmental situation. For a review of these models, see Willner (1991).

Animal Models of Drug Abuse

As mentioned above, laboratory animals are capable of displaying operant behaviors that lead to the self-administration of a supposedly rewarding stimulus (Fig. 20.2). In the technique of intracranial self-stimulation, the rat presses a lever to obtain an electric shock directly in the brain. Similarly, animals can learn to press a lever to obtain a small dose of a drug that we assume has a rewarding effect. In this sense, the self-administration technique fulfills the defining criterion for drug abuse that refers to the expending of extensive efforts and long periods of time in behaviors related to drug-taking. Thus, the self-administration technique can be used to obtain more information about the neurochemical mechanisms involved in drug abuse and may also serve as a screening test to detect drugs that have the potential risk of becoming drugs of abuse.

Historically, experiments in monkeys showed that the animals could easily learn different schedules of cocaine self-administration. A great volume of literature has dealt with this phenomenon, and the characteristics of the effects of different schedules of reinforcement on behavior are already well known. Among the drugs that readily induce self-administration are morphine, heroin, codeine, cocaine, ethanol, some barbiturates, and some anesthetics. Another group of drugs, including nicotine, caffeine, and some benzodiazepines, triggers self-administration, but only to a limited extent. In general terms, antidepressants, neuroleptics, and anxiolytics do not induce self-administration. Surprisingly, cannabis, which is one of the most commonly consumed drugs of abuse in humans, does not induce self-administration.

Presently, the rat is the most widely used model in drug abuse studies. Thus, the reinforcing characteristics of any drug can be determined in self-administration studies. In addition, it is also possible to study relationships between simultaneous addictions to more than one drug. The experimenter can modify the amount of drug delivered and the schedule in which it is delivered and assess the efforts of the animal to obtain the drug. Once the animal has become addicted, several behavioral similarities to the addicted human can be observed.

Although the researcher can modify the availability of the drug in studies of self-administration, the animal itself ultimately determines the amount of drug that it receives. The animal will work more or less, depending on the drug delivery schedule set by the researcher; however, the final amount of drug that the animal receives is kept at a steady level. Therefore, other approaches should be used to study the effects of chronic administration of several doses of drugs. The easiest method allows the researcher to decide when and how much drug the animal will receive as well as when the administration will end, which will trigger the withdrawal syndrome.

Another widely used method involves studying the effects of drug abuse in animals forced to consume the drug. Studies concerning the effects of alcohol, for instance, replace water with solutions of different alcohol concentrations so that the animal has no choice but to drink the required amount of alcohol. In other approaches, however, the animal is allowed to choose between water and an alcohol solution and thereby show a significant preference for alcohol consumption.

Because rats with a spontaneously high preference for alcohol consumption have been found, these rats have been selected and selectively bred. Therefore, special strains of rats with a high preference for some drug (particularly alcohol) are now available.

Needless to say, the detection of abuse-inducing properties in the development of a novel compound that is effective for any given action will lead to its exclusion from human use.

Concluding Remarks

The above paragraphs represent only a brief and limited survey of the great scientific value that the use of laboratory animals has provided regarding mental disease. The notion that psychiatric illness cannot be reproduced in animals has been surrounded by an enormous amount of experimental evidence that has shed light on the mechanisms underlying these pathologies. In addition to the models described above, there are also other experimental tests for schizophrenia, mania, dementia (particularly Alzheimer's disease), and eating disorders. Animal models have not only provided deep knowledge regarding the brain mechanisms involved in psychiatric disease but also shown great potential to generate new insights into mental disease and, more importantly, new therapeutic alternatives for the well-being of psychiatric patients.

References

Klein DF. Endogenomorphic depression. A conceptual and terminological revision. Arch Gen Psychiatry. 1974;31:447–54.

Kłodzińska A, Tatarczyńska E, Stachowicz K, Chojnacka-Wójcik E. The anxiolytic-like activity of AIDA (1-aminoindan-1,5-dicarboxylic acid), an mGLu 1 receptor antagonist. J Physiol Pharmacol. 2004;55:113–26.

Lucki I. The forced swimming test as a model for core and component behavioral effects of antidepressant drugs. Behav Pharmacol. 1997;8:523–32.

Mirmiran M, van de Poll NE, Corner MA, van Oyen HG, Bour HL. Suppression of active sleep by chronic treatment with chlorimipramine during early postnatal development: effects upon adult sleep and behavior in the rat. Brain Res. 1981;204:129–46.

Muscat R, Willner P. Suppression of sucrose drinking by chronic mild unpredictable stress: a methodological analysis. Neurosci Biobehav Rev. 1992;16:507–17.

Olds J. Pleasure centers on the brain. Sci Am. 1956;195:105–16.

Overstreet DH. Behavioral characteristics of rat lines selected for differential hypothermic responses to cholinergic or serotonergic agonists. Behav Genet. 2002;32:335–48.

Porsolt RD, Le Pichon M, Jalfre M. Depression: a new animal model sensitive to antidepressant treatments. Nature. 1977;266:730–2.

Porsolt RD, Anton G, Blavet N, Jalfre M. Behavioural despair in rats: a new model sensitive to antidepressant treatments. Eur J Pharmacol. 1978;47:379–91.

Vogel G, Neill D, Hagler M, Kors D. A new animal model of endogenous depression: a summary of present findings. Neurosci Biobehav Rev. 1990;14:85–91.

Willner P. Animal models of depression: validity and applications. Adv Biochem Psychopharmacol. 1995;49:19–41.

Part V
Other Uses of Rodent Models

Chapter 21
Male Reproductive Toxicology

Renata Mázaro e Costa and Marcos Fernando Oliveira e Costa

Reproduction has garnered attention since time immemorial. Ancient records bear mute testimony to man's attempts to attain control, in one way or another, over reproduction. In the twentieth century, the investigation into the theme intensified, yielding more effective results. In recent years male toxicology has drawn special attention from physicians because of the large number of couples whose infertility involves the male reproductive system. Some studies maintain that this increase in male infertility is due to occupational exposures.

Reproductive toxicology is one of many terms that have arisen as a result of the systematic investigation of reproduction, prompted by two milestone events with tragic consequences. In the 1960s, the teratogenic effects of Thalidomide, prescribed for use during the first quarter of pregnancy with the purpose of curbing nausea and inducing sleep became evident in infants of that generation. Later, mercury contamination in Minamata, Japan, reinforced the perception for the need of establishing strict control, procedures and guidelines for the use and experimentation of substances. Regulation would allow for the thorough scrutiny of drugs for possible effects upon reproduction; these tragic events prompted legislators to outline the necessary protocols for reproductive toxicology.

Experimental protocols adopted in the investigation of male and female reproduction are essential as each protocol is designed to address the physiological, immunological, pharmacological and toxicological processes involved in reproduction.

In this chapter, we examine aspects of male reproduction in rats (*Rattus norvegicus*) as well as contemplate results from the laboratory tests that are most often used to investigate reproductive issues among the male gender. We provide a

R.M. e Costa
Department of Pharmacology, Universidade Federal de Goiás, Goiânia, Goiás, Brazil

M.F.O. e Costa, D.Sc. (✉)
Laboratório de Reprodução Animal, Embrapa Arroz e Feijão, Rodovia GO-462, Km 12, C.P.179, Zona Rural, CEP 75375-000, Santo Antônio de Goiás, Goiás, Brazil
e-mail: mfocosta@hotmail.com

M.L. Andersen, S. Tufik (eds.), *Rodent Model as Tools in Ethical Biomedical Research*, DOI 10.1007/978-3-319-11578-8_21

comprehensive view of the rat male reproductive system and examine some of the methods that are frequently adopted in experimental investigation.

The first reports involving reproductive toxicology date back to 1775 when British physician Percival Pott noted a high occurrence of scrotum cancer among chimney cleaners. In 1977, upon examining workers exposed to dibromochloropropane (DBCP), Whorton described the chemical mechanism by which that substance adversely affected fertility. In the years subsequent to those discoveries, countless other drugs have been added to the list of substances that elicit some sort of alteration in male reproductive function.

As in any other investigation, the adoption of a given experimental protocol is directly dependent upon the specific objectives of the study. The researcher needs control over several experimental parameters, such as the environment, the administration pathway, the type of exposure, and the subjects (animals). Thus, by eliminating the number of variables, the researcher renders observations more reliable. The primary concern is to define the species that will serve as subject (the most ubiquitous is the rat) and to exert control over temperature, water, food and housing conditions. Researchers also need a reliable description of chemical, physical, and behavioral factors to which the animal will be exposed. The researcher needs to choose the form of drug administration: gavage; in the chow; in the water; in another liquid (oil, Tween, etc.); by inhalation or transdermal methods. It is necessary to record the doses, period and mode of administration, as well to define the ontogenetic phase over which the experimental protocol shall be executed (puberty, sexual maturity, etc.). All the factors that could potentially influence the experiment should be described or discussed in detail at some point of the investigation.

In order to understand the experimental protocols described herein, one must become acquainted with the anatomy and physiology of the male reproductive system of the rat.

The Male Reproductive System of the Rat

The male reproductive system of the rat is comprised of a pair of testes, seminal ducts (epididymides and vas deferens), the penis (through which spermatozoids, along with seminal fluid, are ejaculated), and accessory glands (ampulla, seminal vesicles, prostate, bulbourethral glands, coagulating gland and preputial gland). The location of the testis together with the epididymis in the scrotum results in the maintenance of testicular, as well as epididymal, temperature several degrees below that of core body temperature, a factor that is essential for spermatogenesis.

Spermatogenesis and germ cell production occur in the testis, which is particularly sensitive to the action of liposoluble drugs. The testes are encapsulated ovoid organs consisting of seminiferous tubules separated by interstitial tissue containing blood and lymphatic vessels, nerves, a considerable number of macrophages and mast cells and Leydig cells (also called interstitial cells, these are responsible for the synthesis and secretion of testosterone). This hormone level is several times higher within the testis and in the interstitial fluid than in peripheral venous blood. Estradiol is also an important steroid involved in testis dynamics. Estradiol may participate in Leydig cell regulation and modulate the seminiferous epithelium as well as the epididymis.

Spermatogenesis is prompted by steroid action from sexual hormones that trigger events involving paracrine and autocrine signaling mechanisms. Meanwhile, the blood-testis barrier guarantees isolation during germ cell production, providing favorable conditions for spermatogenesis. This barrier is compounded by the Sertoli cell junctional complexes, yielding two compartments within the seminiferous tubule (one basal and another adluminal). The basal compartment between Sertoli cells and peritubular tissue is filled with spermatogonia and preleptotene spermatocytes. The adluminal compartment formed by adjacent Sertoli cells contains rounded, immature and also mature spermatids. Sertoli cells synthesize and secrete *androgen-binding protein* (*ABP*), which carries testosterone into the seminiferous tubule and epididymis and regulates Leydig cell secreting activities by inhibitory or stimulating factors.

Cellular interaction of the Sertoli cells also affects Leydig cells, either directly or at the link between germ cells and myoid cells. Sertoli cells secrete a product that can stimulate Leydig cell steroid production. Paracrine factors rather than cell-cell contacts are believed to be an important means of communication between Leydig cells and the seminiferous tubule.

During puberty (up to 60 days in the rat), the transport of ABP into the testis progressively intensifies and ABP + testosterone controls the development of germ cells. The rate of ABP production and spermatic fertilization capacity seem to be closely associated; ABP can be measured for the purpose of assessing Sertoli cell activity.

In the rat, spermatogenesis occurs according to the spermatogenic cycle, a 14-stage process that ultimately produces the mature spermatid. These stages are distinct and well defined by the arrangement of the germ cells (spermatogonia, primary and secondary spermatocytes, and newly formed and mature spermatids) within the spermatogenic cycle. Those cells that leave the testis are called mature spermatids, but do not possess the capacity for locomotion and fertilization. These properties are acquired during the spermatic maturation process that takes place when the mature spermatids travel along the epididymis. The daily sperm production in an adult rat is on the order of 30 million spermatozoids.

The epididymal duct is a single highly convoluted duct, comprising an initial segment where the ductuli efferentes empties, as well as caput, corpus and cauda segments. The epididymal epithelium is pseudo-stratified and contains different cells whose morphology and abundance vary from one region to another.

Spermatic maturation occurs during sperm epididymal transit, a process that commences at the caput segment and continues through the corpus to the initial portion of the cauda. As it travels, the sperm progressively gains motility and thus increases its chances of survival and success in fertilization. The epididymal fluid to which the sperm are exposed during transit is constantly modified along the duct.

Substantial alterations in composition, osmolarity, ionic proportion, energetic reserves, and protein occur. Estradiol is an important steroid that promotes spermatic maturation in the epididymis, mainly in the initial segment and ductuli efferentes. Many sperm membrane alterations are mediated not only by proteins secreted from specific regions within the duct (some of which are androgen-dependent) but also by subsequent associations through which these proteins engage with the spermatozoid. Spermatic transit time through the entire epididymis is about 8 days: 3–4 days through the caput and corpus segments, and 5–6 days to migrate to the cauda.

The Hypothalamus-Pituitary-Testis Axis: Hormonal Control

The reproductive system exhibits complex neurohumoral regulation that involves central nervous system (CNS) structures like the limbic system, the hypothalamus and pituitary, as well as peripheral structures like the autonomic sympathetic and parasympathetic systems in addition to the reproductive organs and male accessory glands.

After puberty, the hypothalamus secretes, in a pulsating mode, gonadotropin releasing hormone (GnRH) that is synthesized and secreted by the mid-pre optical area and mid supra-optical nucleus. This hormone protein reaches the porta tuberal tract and from there it contacts the adenohypophysis, stimulating gonadotrophs to synthesize and secrete gonadotropins (luteinizing hormone [LH] and follicle stimulating hormone [FSH]). These hormones then reach the bloodstream and arrive at the gonads: in this case, the testes.

Endocrine control over spermatogenesis is exerted mainly by gonadotropic hormones (LH and FSH) and by testosterone. In fact, testosterone effects are mediated by LH. FSH (and perhaps testosterone) is necessary for the initiation of spermatogenesis at puberty; testosterone can maintain this process throughout maturity as well as restore spermatogenesis when a suppression event occurs.

In the rat, FSH seems to be the main regulating agent of Sertoli cell multiplication during fetal and neonatal life and also provides support for the formation of inter-Sertoli-cell tight junctions. During the peripubertal phase, FSH stimulates androgenic receptor production in Sertoli cells and LH receptor production in Leydig cells. At puberty, FSH becomes essential for the commencement of spermatogenesis, initiating Sertoli cell maturation and thereby stimulating the proliferation and differentiation of spermatogonia, giving rise to the beginning of meiotic division. As puberty approaches, the responsiveness of Sertoli cells to FSH decreases but there is a surge of testosterone and a progressive switch from being mainly FSH-modulated to being mainly testosterone-modulated. These changes may result partly from an increase in the levels of androgen receptor in the Sertoli cells, which may be induced by FSH but also appear to be induced by the presence of particular types of germ cells.

The coordinating effect of FSH operates via the Sertoli cell. FSH also appears to play an important role in the regulation and development of the appropriate number of Leydig cells so that high levels of testosterone can be produced when required. FSH induces an increase in the number of LH receptors and in the capacity of the testis to secrete testosterone *in vivo* or *in vitro* in response to LH stimulation.

Testosterone, or dihydrotestosterone, acts via the androgen receptor, which is most confined to Leydig cells, peritubular cells and Sertoli cells. It affects few stages of the spermatogenic cycle directly, but could interfere in the processes mediated by other testicular cells.

Many factors may interfere with spermatogenesis. These may range from occupational exposures to accidental exposures to chemicals (xenobiotic), including the consequences of daily situations with either chronic or acute episodes, like stress.

The sympathetic stimulus also seems to contribute to hormonal control of the hypothalamus-pituitary-testis axis, especially when it comes to stressful situations, when norepinephrine simulates GnRH secretion via α-adrenergic receptors in the hypothalamus.

Adrenergic stimuli seem to play a role in androgen secretion during puberty, most likely through the activation of 17 α-hydroxylase and 17,20-desmolase, which is known to be involved in the conversion of progesterone to androstenedione and in the effects mediated by testosterone throughout the process of steroidogenesis. It is also known that adrenergic innervation surrounds Leydig cells, and that catecholamine may directly stimulate the secretion of testosterone through β-adrenergic receptors.

In order to determine endocrine reproductive function, those hormones found in the hypothalamus-pituitary-testis axis can be measured in plasma by radioimmunoassay analysis; the most frequently examined are FSH, LH, and testosterone. In the testis, there is synthesis and release of testosterone and dihydro-testosterone (predominant male androgen) through steroidogenesis in Leydig cells, using cholesterol as a steroid precursor. Testicular concentrations of testosterone may reach values tenfold greater than found in plasma. Thus, the assessment of steroidogenesis is an important laboratorial tool, by means of which testosterone may be dosed by radioimmunoassay or by chemiluminescence tests using homogenized testicular tissue. At present, the estradiol dosage is under scrutiny, because testosterone is converted into estradiol within the reproductive system; ERβ receptors control spermatic maturation in the epididymis.

Studies of Reproductive Toxicology: Experimental Protocols

Male reproductive toxicology involves the assessment of fertility and male sexual behavior as well as evaluation of the F1 generation. Such examinations are conducted under the criteria of experimental protocols, and these are classified according to the time that the animals are exposed to a certain condition:

Chronic Toxicity Tests

The chronic toxicity test produces data that allow for the examination of toxic effects during the animal's ontogenetic development. Exposure commences after weaning and is terminated 12 or 24 months thereafter. Animals may be submitted to euthanasia after exposure for 6 months. More than merely assessing male reproduction capability and the hormonal profile, this test generates information on the toxicity progression in relation to the animal's development. Diligent execution of this test requires maintaining weekly records of water and food intake, as well as the animal's body weight.

Sub-chronic Reproductive Toxicity Tests

In this test, exposure may commence either in the pre-pubertal phase (35 days) or in puberty (45–60 days) and continue until sexual maturity (70–90 days). The test can also be conducted in adult rats provided they are exposed to xenobiotics for 90 days. That is the time required to expose the animal to two seminiferous cycles (in rat, one cycle takes 48–53 days).

Yet another protocol for male reproductive toxicology is exposing the animal for 13 weeks (52 days) as a way of creating a reproductive screening, often deemed a pre-chronic condition.

Short-Term Toxicity Tests (Less Than One Spermatogenic Cycle)

C1) 4 Days

This protocol was specifically designed to identify those toxic substances whose actions are detrimental to the epididymis, testis and the hypothalamus-pituitary-gonad axis. Exposure occurs for 4 days and euthanasia on the subsequent day (5th) with removal of the biological mass that is to be examined. The 4-day-period is justified because it corresponds to the time it takes for spermatozoids to travel through the epididymis caput, corpus and to the cauda, an androgen-dependent region.

One method that aids in the investigation of a given drug as if affects epididymal function is ligature of the efferent ducts and the administration of testosterone supplements. By doing so, one avoids confounding with testicular events and examines only the epididymis.

C2) 15 Days

This test is used whenever one wishes to assess the spermatoxicity potential from drug investigations based upon structure-activity relationships. Depending on the period of exposure and the date of necropsy, the protocol period may exceed 2.5 weeks (Ex.: exposure for 5 days and necropsy performed on day 14). This short-term test is efficient in readily detecting those drugs that produce moderate to severe alterations in spermatozoids. Multiple assessments complement the interpretation of results, elucidating cell targets and providing insights as to the mechanisms underlying toxicity.

C3) 21 Days

This test is applied to determine those drugs of high testicular toxicity potential. It is not, however, efficient in detecting toxic effects in the initial stages of spermatogenesis; other organs (accessory sex glands or in the CNS) cannot be assessed by this methodology.

In this protocol, the subjects may be exposed, from 2 to 20 days, to three different concentrations of the drug; the resulting groups are tested along with an additional control group. Subjects mate during exposure for 5 days and subsequently the male reproductive system and male fertility are examined.

Assessment of the Male Reproductive System

The descriptions that follow refer to consolidated protocols widely adopted by laboratories currently performing reproduction studies.

The Development of the Sexual Organs

The weight of the male reproductive organs is a measure of reproductive function and is an indicator of alterations in steroidogenesis because the epididymis, prostate and seminal vesicle (this latter may be weighed with or without secretion) are androgen-dependent organs. A reliable weight-reading requires maintaining a score of absolute weight (in some types of investigations, this parameter is denominated wet weight) and relative weight (determined in relation to the weight of the animal or brain weight).

Testicular Histopathologic Assessment

One of the most ubiquitous methods in the investigation of male reproduction is the testicular histopathologic exam, in which the researcher becomes acquainted with the dynamics of spermatogenesis as well as with Sertoli cell counts.

Reliable histopathologic exams require that the material under analysis be handled with extreme care. The testis must be manipulated ever so delicately during extraction from, and subsequent fixation to, the body of the animal. Such diligence in care should prevent the appearance of histological artifacts like spermatids within the lumen of the seminiferous tubules. One staining method that facilitates the identification of various spermatogenic stages of the seminiferous cycle utilizes Periodic acid–Schiff's reaction (PAS) with counterstaining by Harris hematoxylin. This method shows the glycogen encountered in the developing acrosomal cap during various stages of spermatogenesis.

An examination of spermatogenesis as revealed by transversal sections of the seminiferous tubules can be performed with the methodology proposed by Hess (1990), whose criteria identify 14 stages for development of the acrosomal complex.

Mature testicular spermatid counts in homogenized testis/epididymis are a widespread assessment method for determining the absolute and relative amounts of germ cells; this methodology was described by Robb et al. (1978). Such counts also measure daily sperm production, which can also be performed through histological assessments.

Histometric determination (for example, the height of the seminiferous epithelium) is yet another method adopted to assess the male reproductive system. It is therefore necessary to use a computer system with specific software to capture images and perform calculations from histological images obtained with optical microscopy.

Sertoli cells multiply until puberty, do not divide within the adult testes and are unevenly distributed along the seminiferous tubules. Thus they may be used as a reference point or a correction factor of germ cell counts. The rate of germ cell production for each Sertoli cell gives the researcher an estimation of the number of cells originating from the seminiferous epithelium over the course of several germ cell generations. This allows for comparisons between different groups of animals.

Assessment of the Epididymis

A toxic substance may affect maturation, function, or even spermatic survival by either direct action upon spermatozoids or by adversely affecting epididymal function.

Histological epididymal cauda descriptions of the rat are subdivided into 6 and 8 zones. In the cauda epithelium are clear cells, so-called due to their clear, foamy cytoplasm. The epididymal duct is lined with connective tissue containing fibroblasts, collagen, elastic fibers, blood and lymphatic vessels, nerve fibers (cholinergic and adrenergic), macrophages, leukocytes, and concentric layers of smooth muscle.

Sperm transport along the epididymal duct is a consequence of the joint action of intraluminal hydrostatic pressure, peristaltic activity along the duct (promoted by sympathetic and cholinergic innervations and by peptides such as AVP and prostaglandins), and cilia movement in the lining of the duct epithelium that seems to be indirectly dependent upon androgens. The main factor in propulsion of the luminal content is the sympathetic innervation (mediated by norepinephrine), whose density increases along the epididymis, reaching a peak in the cauda where it contributes to the ejaculatory response.

An alteration in transit time, for instance, may directly reflect quantity and quality of the spermatozoids. An accelerated period allows little time for the sperm maturation process because the maturation depends on spermatozoid interaction with the epithelium and luminal fluids.

Sperm counts from homogenized epididymis are a common practice in male reproduction research centers, using the method described by Robb et al. (1978). Results collected from these counts may determine the transit time required for that sperm to cross the epididymis: successful sperm maturation is a time-dependent process.

The main cells, which make up 65 % of the cauda epithelium, are responsible in the caput and corpus, for the secretion of several proteins and glycoproteins like immobilin and inhibin, which interact with the sperm. This interaction results in biochemical and morphological transformations that allow spermatids to progressively acquire motility and the capability to fertilize the oocyte. The epididymal epithelium also becomes capable of participating in endocytosis of material within the lumen. The clear cells, mainly in the cauda, have endocytosis absorption properties, and play an active role in re-uptake of cytoplasmic droplets released by sperm. Clear cells are also responsible for the endocytosis of deceased sperm fragments and giant uni- and multi-nucleated cells found in the lumen of the epididymis.

Sperm Morphology and Motility

The investigation of sperm morphology has been of great value in the determination of toxic sperm effects.

The rat sperm morphology is quite different from that of humans, as the head in the rat spermatozoid is sickle-shaped. The most common alterations are encountered in either the head or in the spermatic flagellum. Head alterations may be characterized as pinhead shaped or banana-shaped, and the flagellum might be curled up, severed, or folded. Sperm may be collected in the cauda epididymal or vas deferens, washed with a formal-saline solution. The obtained solution from either manner of collection is used to perform histological swaps or heat-fixed for staining with hematoxylin-eosin. Researchers can also analyze the position of the spermatic droplet, present in the sperm flagellum.

Sperm motility can be evaluated in fresh samples under phase contrast microscopy, or by sperm images recorded and stored in video or digital format analyzed later, either manually or using computer-aided semen analysis.

Vas Deferens

The vas deferens is the structure through which sperm are conducted from the cauda epididymal to the prostatic urethra. The duct features a narrow passageway and thick walls, comprised of three sub-layers of smooth muscle tissue: the internal and external layers are longitudinal and thinner than those encountered in the middle layer, comprised of circular fibers. The mucosa fold allows for expansion of the duct during sperm ejaculation, and is made of ciliated pseudo-stratified epithelium.

Male Accessory Glands

Prostate

The prostate is an organ composed of a collection of ramified tubuloalveolar glands that empty into the prostatic urethra. The prostate is encapsulated in fibroelastic smooth muscle tissue that sends a septum to the interior of the gland. The central portions of the prostate contain the highest concentration of conducting ducts, where the epithelium is stratified and composed of columnar cells. Glands with higher secretion activity are encountered in the peripheral portions, containing internal pseudostratified epithelium.

Seminal Vesicle

The seminal vesicle is compact glandular tissue arranged in multiple lobes containing a system of ramified secretory ducts. The epithelium is pseudostratified and composed of columnar cells. The smooth muscle tissue that lines the duct is composed of two layers: the inner one comprises circular fibers while the external layer is composed of longitudinal fibers.

The concentration of fructose within the prostate and seminal vesicle is an indirect measure of accessory gland activity and can be gauged by the colorimetric method originally developed by Mann (1948) for the determination of fructose content in semen.

Biochemical studies can be used to evaluate the activity of reproductive tissues (testes, cauda epididymides, seminal vesicle and ventral prostate) to assay for protein, sialic acid, glycogen and cholesterol.

Conclusion

The current approach towards the investigation of male reproduction focuses on assessing xenobiotic toxicity, on biologic targets that control fertility (Ex: male contraception), or on improving male sexual performance. Despite the ever-growing body of literature on the subject, the need for standard parameters remains. Standard testing allows for replication of results at other research facilities and this aspect is of utmost importance in view of the many variables that come into play during the execution of an experimental protocol. Standard parameters are needed particularly when there are numerous variations among the many different strains of species used in the investigation of male reproduction.

References

Bardin CW, et al. The sertoli cell. In: Knobil E, Neill JD, editors. The physiology of reproduction. 2nd ed. New York: Raven; 1994. p. 1291–362.

Eddy EM. Duct system and accessory glands of the male reproductive tract. In: Lamb IV JC, Foster MD, editors. Physiology and toxicology of male reproduction. San Diego: Academic; 1988. p. 46–9.

Filler R. Methods for evaluation of rat epididymal sperm morphology. In: Chapin RE, Heindel JJ, editors. Methods in toxicology: male reproductive toxicology. New York: Academic; 1993. p. 334–43.

Foster PMD. Testicular organization and biochemical function. In: Lamb IV JC, Foster MD, editors. Physiology and toxicology of male reproduction. San Diego: Academic; 1988. p. 11–23.

Griswold MD. Interactions between germ cells and sertoli cells in the testis. Biol Reprod. 1995;52:211–6.

Klinefelter GR, Hess RA. Toxicology of the male excurrent ducts and accessory sex glands. In: Korach KS, editor. Reproductive and development toxicology. New York: Marcel Dekker; 1998. p. 553–91.

Krester DM, Kerr JB. The cytology of the testis. In: Knobil E, Neill JD, editors. The physiology of reproduction. 2nd ed. New York: Raven; 1994. p. 1177–290.

Leblond CP, Clermont Y. Definition of the stages of the cycle of the seminiferous epithelium in the rat. Ann N Y Acad Sci. 1952;55:548–73.

Mann T. Fructose and fructolysis in semen in relation to fertility. Lancet. 1948;1:446–8.

Mazaro R, Di Stasi LC, Filho SA, Kempinas WG. Decrease in sperm number after treatment of rats with *Austroplenckia populnea*. Contraception. 2000;62:45–50.

Mazaro R, Di Stasi LC, Kempinas WG. Effects of the hydromethanolic extract of *Austroplenckia populnea* (Celastraceae) on reproductive parameters of male rats. Contraception. 2002; 66:205–9.

O'Donnell L, Stanton PG, Bartles JR, Robertson DM. Sertoli cell ectoplasmic specializations in the seminiferous epithelium of the testosterone-suppressed adult rat. Biol Reprod. 2000;63:99–108.

Hess RA. Quantitative and qualitative characteristics of the stages and transitions in the cycle of the rat seminiferous epithelium: light microscopic observations of perfusion-fixed and plastic-embedded testes. Biol Reprod. 1990;43:525–42.

Robb W, Aman RP, Killian GJ. Daily sperm production and epididymal sperm reserves of pubertal and adult rats. J Reprod Fertil. 1978;54:103–7.

Sharpe RM. Endocrinology and paracrinology of the testis. In: Lamb IV JC, Foster MD, editors. Physiology and toxicology of male reproduction. San Diego: Academic; 1988. p. 80.

Sharpe RM. Regulation of spermatogenesis. In: Knobil E, Neill JD, editors. The physiology of reproduction. 2nd ed. New York: Raven; 1994. p. 1363–434.

Sharpe RM, McKinnel C, Kivlin C, Fisher S. Proliferative and functional maturation of sertoli cells, and their relevance to disorders of testis function in adulthood. Reproduction. 2003;125:769–84.

Zenick H, Clegg ED, Perreault SD, Klinefelter GR, Gray LE. Assessment of male reproductive toxicity; a risk assessment approach. In: Hayes W, editor. Principles and methods of toxicology. 3rd ed. New York: Raven; 1994. p. 93.

Chapter 22
Sleep Patterns in Rats

Katsumasa Hoshino, Monica Levy Andersen, Ligia Assumpção Papale, and Tathiana Aparecida Fernandes Alvarenga

The rat, a mammal of the *Rodentia* order that encompasses several species with the most ubiquitous being the brown rat (*Rattus norvegicus*) and the black rat (*Rattus rattus*), is present in all latitudes of the globe. Thanks to their tremendous ability to withstand and survive adverse conditions, the rat has been used in biological experimentation for a 100 years. In the first half of the twentieth century, several breeds and strains were created through successive breeding between males and females that possessed specific characteristics. These have been utilized in distinct fields of research because of their known genetic properties. Today, the most widely used experimental animal is the albino rat, selected at the Wistar Institute of Philadelphia, whose fur is entirely white (hence the name *albino rat* or the *Wistar rat*). This strain, called *Rattus norvegicus albinus*, is characterized by the complete absence of melanin in the fur and in the iris. The word albino derives from the Latin adjective *albus*, meaning white.

The rat was introduced in sleep research during the 1960's. The sleep-wake cycle of the species has been thoroughly studied and described since then by Brazilian

K. Hoshino (✉)
Universidade Estadual Paulista, Av Luiz Edmundo Carrijo Coube s/n, Bauru, São Paulo 17033-360, Brazil
e-mail: kt.hoshino@gmail.com

M.L. Andersen
Department of Psychobiology, Chief of Sleep Division, Universidade Federal de São Paulo (UNIFESP), Rua Napoleao de Barros, 925, São Paulo, São Paulo, Brazil
e-mail: ml.andersen12@gmail.com

L.A. Papale
Department of Psychiatry, 6001 Research Park Blvd, Madison, WI 53719, USA
e-mail: papale@wisc.edu

T.A.F. Alvarenga
Department of Psychobiology, Universidade Federal de São Paulo (UNIFESP), São Paulo, São Paulo, Brazil

M.L. Andersen, S. Tufik (eds.), *Rodent Model as Tools in Ethical Biomedical Research*, DOI 10.1007/978-3-319-11578-8_22

researchers (Timo-Iaria et al. 1970), which helped make the rat the most utilized species in the investigation of the sleep-wake cycle for the past 30 years, particularly in studies involving structural and functional aspects of the nervous system. The selection of the rat as an adequate subject of sleep investigation occurred when it was demonstrated that it sleeps close to 13 h daily and possess all the sleep stages observed in human sleep (Timo-Iaria et al. 1970).

The investigation of the sleep-wake cycle in humans requires electrophysiological recording using cortical electrodes. The recording of oscillations in the electrical potential produced by the nervous system is generically called an oscillogram, whose most common form is the electroencephalogram (EEG). In animals, recordings are obtained by implanting electrodes either directly over the cortex (eletrocorticogram—ECoG) or implanting them within the brain. In general, ECoG in the cat, rat and other animals used in experimentation are collected through electrodes placed in bipolar derivations. Once the points over which the electrodes will be implanted are determined, implantation of the electrodes allows for the simultaneous recording of different areas of the brain and several of its motor units. The possibility of having simultaneous recordings from different areas is of great value in the investigation of the manifestations of the brain and mechanisms of the processes present during sleep and wakefulness (Timo-Iaria 1985). Figure 22.1 illustrates the positioning of implanted cortical and subcortical electrodes; two pairs of wires can be seen in the snout area near the nostrils, which are used for the detection of eye, snout and whisker movement. Electrodes in the neck record head movement.

As with human EEG, those of the rat and many other animals also reveal fast (desynchronized) and slow (synchronized) oscillations. In view of this basic distinction between frequency and amplitude of the oscillations, sleep is divided into two fundamental states called slow wave sleep (or synchronized sleep) and paradoxical sleep, or rapid eye movement (REM) sleep. These states are typically observed in cats (Ursin 1968) and rats (Hoshino and Toloi 1995; Timo-Iaria et al. 1970; Valle 1992). The slow wave sleep of synchronized sleep is manifested in rather standardized patterns known as periods, delta wave sleep and theta wave sleep. It can be verified that the EEG presents periods in which one of the periods is predominant, thus determining the definition of phases or stages within synchronized sleep.

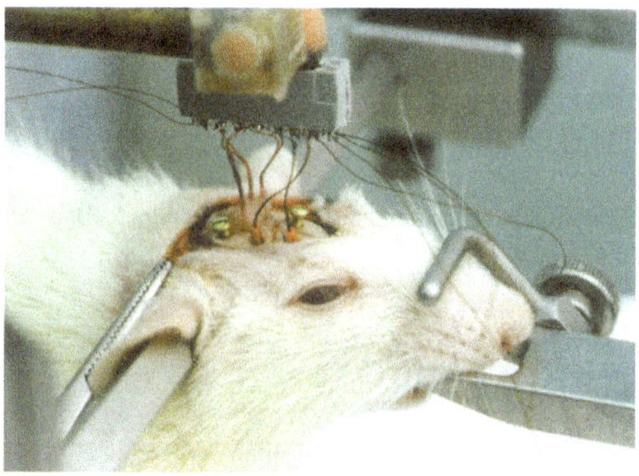

Fig. 22.1 Depiction of the surgery to implant cortical and subcortical electrodes. The socket is attached to a support on the shaft of the stereotaxic device (Andersen et al. 2001)

It is currently accepted that the sleep stages during the ECoG recordings can be determined as follows:

- Alert or active awake
- Relaxed or quiet awake
- Synchronized sleep or slow wave sleep, SS_I phase
- Synchronized sleep, SS_{II} phase
- Synchronized sleep, SS_{III} phase
- Pre-paradoxical phase
- Paradoxical phase

The occurrence of the sleep rhythm may not occur in a progressive sequence. Once in a given state of sleep, either a previous stage might be resumed or a subsequent one might come into play. This makes the classification of the evolution of sequences a description of a statistical nature. Despite this consideration, it is possible to say that the retrograde evolution of phases is common in the first sleep cycles but become more regular as sleep progresses.

The determination of the sleep structure requires assessment of the time spent by the animal in each phase of the sleep-wake cycle. For this, it is necessary to establish the phase and duration of that phase during a given recording period, a task referred to as *staging*. This is done using computer tools that establish a fixed time called a *period* in which the predominant phase can be readily identified. The sum of the periods of each phase allows for its assessment and incidence in global terms.

Recording periods of the stages may last from 1 s (Gottesman et al. 1971) to 30 s (Mistelberger et al. 1983), although the latter longer period is seldom used. The analysis of the ECoG of rats has, as of late, more often been performed by means of mathematical descriptions (Borbély et al. 1984; Ursin et al. 1989). The criteria that are adopted in sleep staging must meet certain requirements to ensure reliability and validity. Reliability is attained when staging of the classification of the sleep-wake cycle is done by either two different researchers or by the same researcher at different times with matching results. Indeed, a satisfactory degree of reliability is required for staging if one is to rely heavily on the classification, which is typically the case. Validity, in turn, is defined as meeting the criteria adopted in the classification of a given period (Neckelmann et al. 1994). In simple terms, reliability relates to how accurate the individual is when performing the assessment when adopting the criteria or gauging tools. Validity is the degree of adequacy of those criteria or gauging tools. Throughout the assessment of the different phases of the sleep-wake cycle, the information from staging in relation to its reliability and validity are relevant in the interpretation of results and merit of the work because researchers must independently classify the ECoG/electromyogram (EMG) pattern with a satisfactory degree of consistency.

Parameters that are commonly used in research with rats are (1) sleep efficiency: the percentage of sleep time throughout the recording; (2) sleep latency: the time needed to begin the first sleep episode once recording has begun. The onset of sleep is when there is total cortical synchrony and hypotonic muscles for three periods of 30 consecutive seconds; (3) paradoxical sleep latency: the time it takes for the first episode of paradoxical sleep to occur after the onset of sleep. Some studies quantify the time it takes for the occurrence of the first episode of paradoxical sleep from the beginning of a recording session; (4) total vigil time: the percentage of vigil time throughout the recording; (5) total slow wave sleep time: the percentage of synchronized sleep throughout the recording period; (6) total paradoxical sleep time: the percentage of paradoxical sleep throughout the recording period; and (7) number and average duration of the sleep and vigil episodes. Awakenings are defined as events that last at least 10–15 s with a marked modification in the ECoG and EMG pattern followed by a sleep episode. Short vigil episodes ($\leq$4 s) are important markers for sleep fragmentation (Trachsel et al. 1991).

Data collected in a manner consistent with the criteria previously mentioned allow for the quantification of the time required in each of the phases of the sleep-wake cycle as a function of the percentage of the recording time. The quantitative composition of the various sleep phases is determined in relation to the total sleep time when analysis is performed.

With regard to the phases of the sleep-wake cycle in rats, attentive awake is characterized by high frequency (30–40 Hz) and low amplitude ECoG (approximately 30 mV) followed by accentuated muscular tonus. During this vigil phase, theta waves (or theta rhythm) are also observed. Theta waves are important electrophysiological manifestations used to evaluate the level of alertness of the animal. This rhythm becomes conspicuous when there is movement of the head, eyes, snout and whiskers. According to Vanderwolf (1969), the theta rhythm is clear and regular in the hippocampus whenever the rat performs voluntary behavior (exploring the environment, assuming an upright posture), as opposed to irregular theta, which is often seen as high frequency waves that prevail in automatic behavior and reflexes such as scratching or eating (Valle 1992).

The theta waves depicted in Fig. 22.2 are present in cortical area 10 (A_{10}), in hippocampal areas CA_1 and CA_3 and in the ventrolateral nucleus of the thalamus (VLN). In A_{10} the ECoG assumes a predominantly tonic and continuous desynchronized pattern, which during the intensification of the level of alertness may increase, thereby lowering the voltage even further and increasing the frequency of the potential. The pattern indicates that the surge of activity in this area may increase substantially as a function of the level of alertness, but also that the tonus of the area remains active (Hoshino and Toloi 1995; Valle 1992; Schmidek et al. 1972, 1974; Timo-Iaria et al. 1970). Moreover, it can be observed that during the vigil episode, an intense exploratory behavior is manifested that is characterized by head, eyes, snout and whisker movement.

When evaluating Fig. 22.3, one comes to the realization that the theta waves disappear for a few seconds, giving way to delta waves that are identified by high voltage and low frequency (under 4.0 Hz). In this period, known as relaxed vigil, the level of alertness of the animal significantly decreases, as reflected by immobility and the disappearance of theta waves. The potentials tend to become synchronized in all regions that are recorded. Note that during relaxation (that is, when the level of attention falls) movement comes to a minimum. In the center of the graph, attention resumes and is expressed in the occurrence of theta waves (underlined) and movement of the head and eyes.

There has been considerable debate regarding staging in rodents that is reflected by the number of different classification criteria, particularly for synchronized sleep. The main criteria are based on the presence of sleep spindles (Timo-Iaria et al. 1970; Ursin and Larsen 1983) or the presence of slow wave activity (Borbély

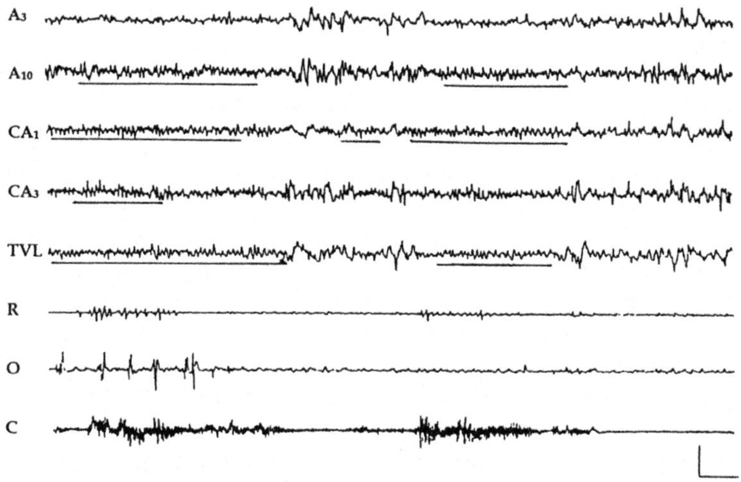

Fig. 22.2 Typical attentive vigil of the rat, characterized by theta rhythm (*underlined*) in A_{10}, in hippocampal areas CA_1 and CA_3 and in the ventral lateral nucleus of the thalamus (VLN) with movement of the snout and vibrissae, eyes and head. In cortical area 3 (A_3), there are dominantly desynchronized patterns, indicating intense alertness. Between the first and second burst of theta waves, there are high voltage delta waves, which indicate a brief period of relaxed vigil. Calibration: 100 μV and 1 s

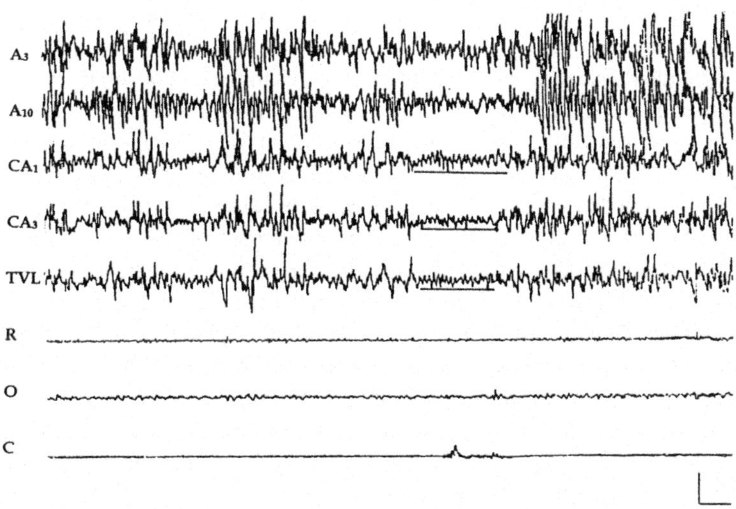

Fig. 22.3 Relaxed awake state of the same rat depicted previously. This state is characterized by two high voltage delta wave periods, separated by a short desynchronized period in A_{10} in addition to brief attentive vigil periods characterized by theta waves (*underlined*) in the hippocampus and the ventrolateral nucleus of the thalamus, during which there is a surge in eye and head movement. Calibration: 100 μV and 1 s

et al. 1984; Gottesman et al. 1971). While some authors consider synchronized sleep to be practically homogeneous, others differentiate the phases. Such studies describe the different stages and attempt to draw equivalents to human sleep, distinguishing stages with spindles, delta wave with spindles and only delta waves (Timo-Iaria et al. 1970) or with low and high frequency slow wave sleep. Here, it is important to mention that the analysis of the amplitude of waves requires computer algorithms to avert subjective interpretation.

Delta activity seems to be an indicator of depth of synchronized sleep. Once the delta band potential increases, the ability to waken the animal using acoustic stimuli becomes increasingly difficult (Neckelmann and Ursin 1993). When the rat falls asleep, the waves that indicate relaxed vigil change abruptly, and high frequency potentials become dominate (12–25 Hz). These are dubbed sleep periods (Fig. 22.4). According to the classification set forth by Timo-Iaria et al. (1970), this period is called the SS_I phase, or phase 1 of synchronized sleep.

Sleep periods of the rat resemble those of humans. During the first phase of synchronized sleep, the ECoG potentials are high voltage. These signals predominate in the frontal and parietal regions in the rat, and in some subcortical regions, the ECoG potentials are within the sigma band (10–14 Hz, Benington et al. 1994).

The SS_I phase is followed by the SS_{II} phase and is characterized during the ECoG by the presence of delta waves (frequency between 0.5 and 4Hz) coupled or not to the periods of 13 Hz oscillations (Fig. 22.5). Sleep in this phase is deeper than the preceding phase, as shown by the significant increase in the auditory signal necessary to waken the animal. As with SS_I, SS_{II} also occurs in humans and cats.

In the SS_{III} phase, the 13 Hz oscillations periods cease and only delta waves remain (Fig. 22.6). This phase is also present in humans and cats, and is the deepest form of sleep.

Pre-paradoxal sleep is classified by theta waves in the hippocampus, which is the main criteria determining desynchronized sleep. During this period, cortical waves have a synchronized morphology. Timo-Iaria et al. (1970) dubbed this pre-paradoxical sleep, which lasts approximately 30 s (Benington et al. 1994). Hoshino (1977) proposed that pre-paradoxical sleep can be considered the first stage of paradoxical sleep (PS-I) in view of: (a) phasic movement of the whiskers and other events related to paradoxical sleep occur (PS-II), such as an increase in respiratory frequency; (b) paradoxical sleep is manifested by successive ipsilateral cortical spindles when there is unilateral lesion of the mesencephalic reticular formation or a bilateral lesion when sub-anesthetic doses of sodium pentobarbital are injected in rats deprived of paradoxical sleep (Hoshino 1980) (Fig. 22.7).

Figure 22.8 depicts the ECoG patterns of paradoxical sleep, which in humans as well as in rats follows the phase of synchronized delta waves. In **a**, where the EEG pattern of several regions of the human cortex can be observed, desynchronized waves occur (high frequency and low voltage) in all recorded areas (Moruzzi 1972). This is also referred to as paradoxical sleep (Jouvet 1962, 1967, 1972, 1994). The latter denomination was adopted by Jouvet because the ECoG expression closely resembles attentive alertness, even though sleep is occurring. In **b**, the

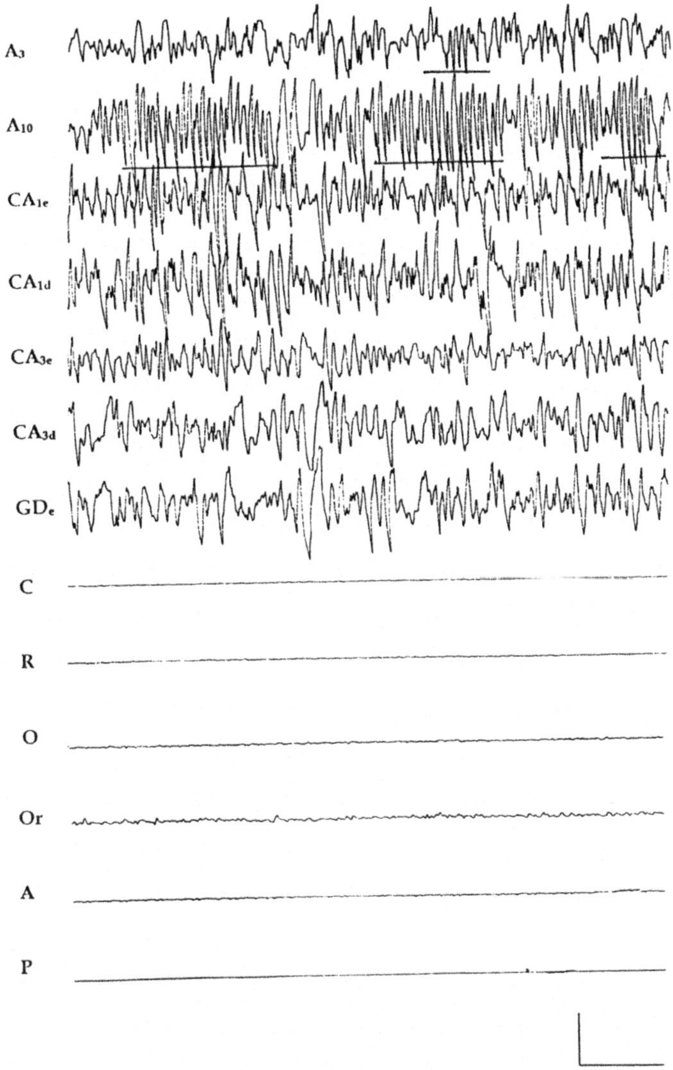

Fig. 22.4 The SS_I of synchronized sleep, characterized by the presence of well modulated periods of 13 Hz oscillations (*underlined*) in A_3 (with only one burst in A_{10}). In the other regions, the ECoG reveals waves of several frequencies and duration, including slow delta-type waves. A_{10}: area 10. A_3: area 3. CA_{1e}: hippocampal area CA_1, *left side*. CA_{1d}: hippocampal area CA_1, *right side*. CA_{3e}: hippocampus area CA_3, *left side*. CA_{3d}: hippocampal area CA_3, *right side*. DG_e: left dentate gyrus. Calibration: 100 μV and 1 s

electrophysiological manifestation of desynchronized sleep of the rat is shown. A desynchronized pattern is predominant in the more frontal regions (A_3 and especially in A_{10}), while in the hippocampus and VLN, theta waves predominate.

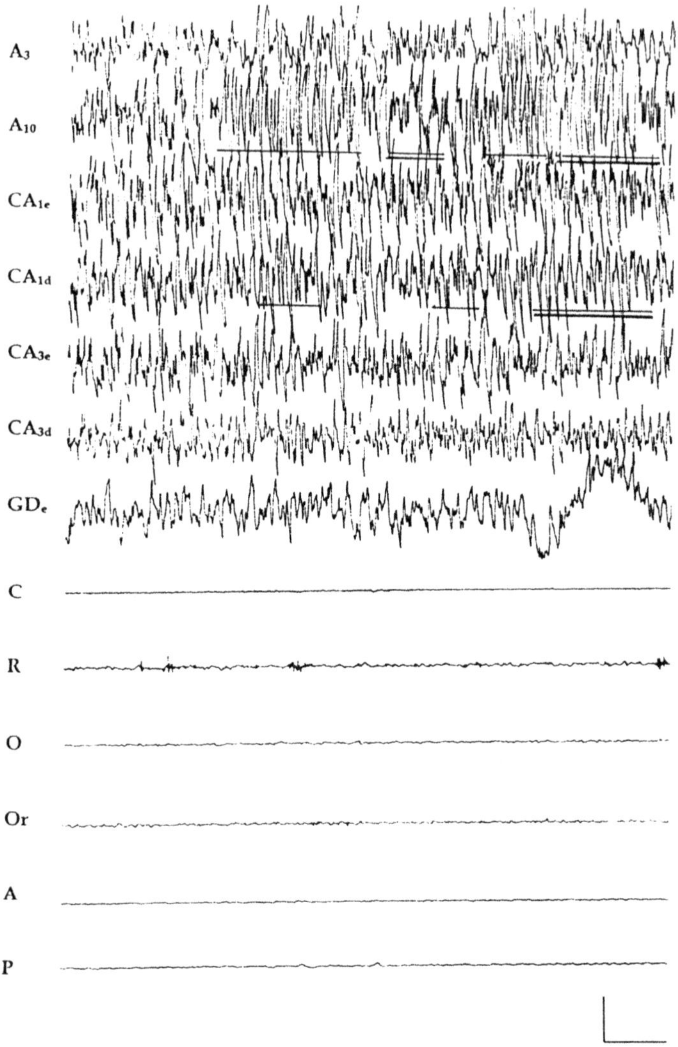

Fig. 22.5 The SS_{II} phase of synchronized sleep, with alternating theta periods (*underlined*) and delta waves (*double underlined*). Note the presence of theta periods also in the CA_1 of the hippocampus. In humans as well as in rats, an uncommon small and short movement of the face can sometimes be observed. Calibration: 100 μV and 1 s

No movement is present during paradoxical sleep. Phasic movements may, however, occur in the whiskers, eyes and head (Fig. 22.9).

In 1990, Timo-laria and collaborators verified that movement of the head during oneiric activity was predominantly in the rostrum and were taken as characteristics dream manifestation. In general, the proportion of movement prevalence of the snout in relation to that of the eyes is of 2:1. This proportion suggests the rat has

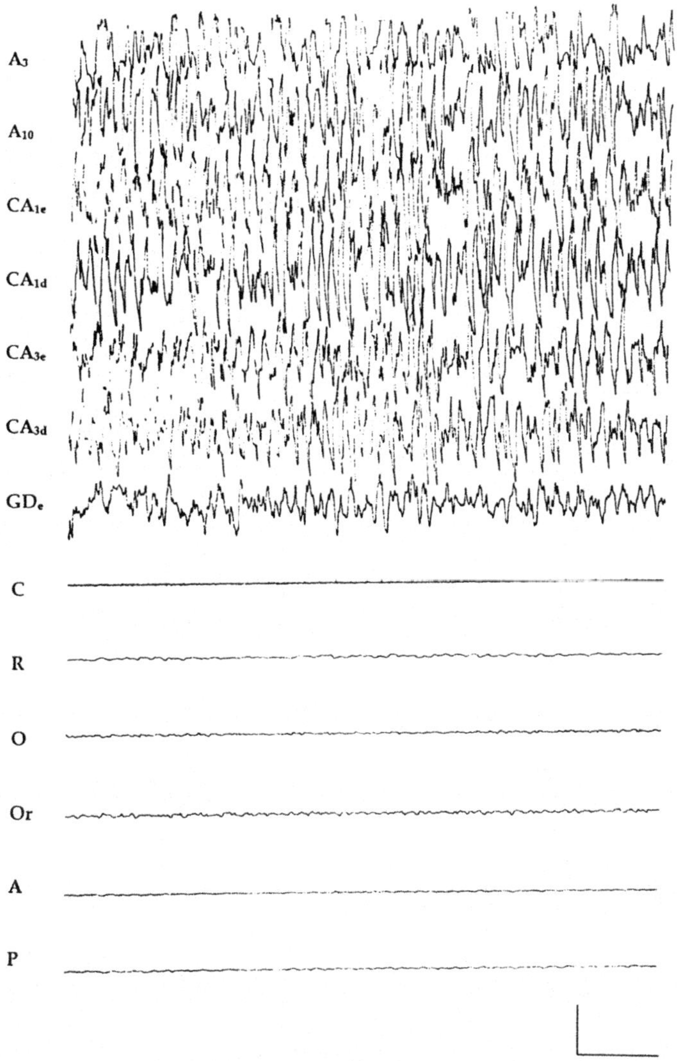

Fig. 22.6 The SS_{III} phase of synchronized sleep, characterized by the predominance of delta waves (frequency between 0.5–4Hz). Calibration: 100 μV and 1 s

more olfactory and tactile dreams (moving the snout to smell and whiskers to explore the environment mechanically) than visual dreams. Of note, one fact that indicates the occurrence of dreams without visual input is the rapid eye movement in the absence of any other body movement, which occurs at a significant frequency when the rat begins to open its eyes (Silva and Hoshino, unpublished data).

In synchronized sleep breathing and pressure become regular and slightly reduced. In paradoxical sleep, however, the most conspicuous tonic alterations (cortical

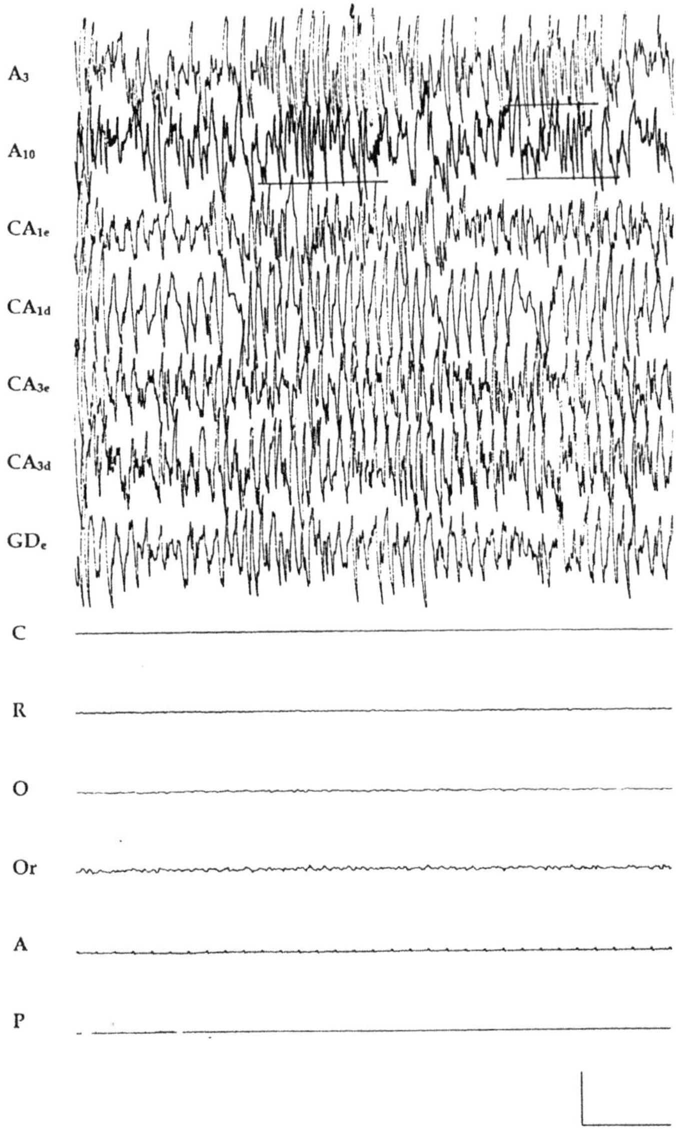

Fig. 22.7 Example of pre-paradoxical, or the intermediary phase in which cortical periods co-exist. In this episode, the frequency of oscillations is 2.5 Hz (*underlined*) and theta waves (7.5 Hz) in the hippocampus and dentate gyrus. Calibration: 100 μV and 1 s

desynchrony, muscular atonia, and hippocampal theta) and phasic (movement of the whiskers, eyes, muscular bursts, and cardio-respiratory oscillations) alterations of these parameters take place. In the rat, Junqueira and Krieger (1976) identified hypertension during paradoxical sleep. This hypertension decreases only after

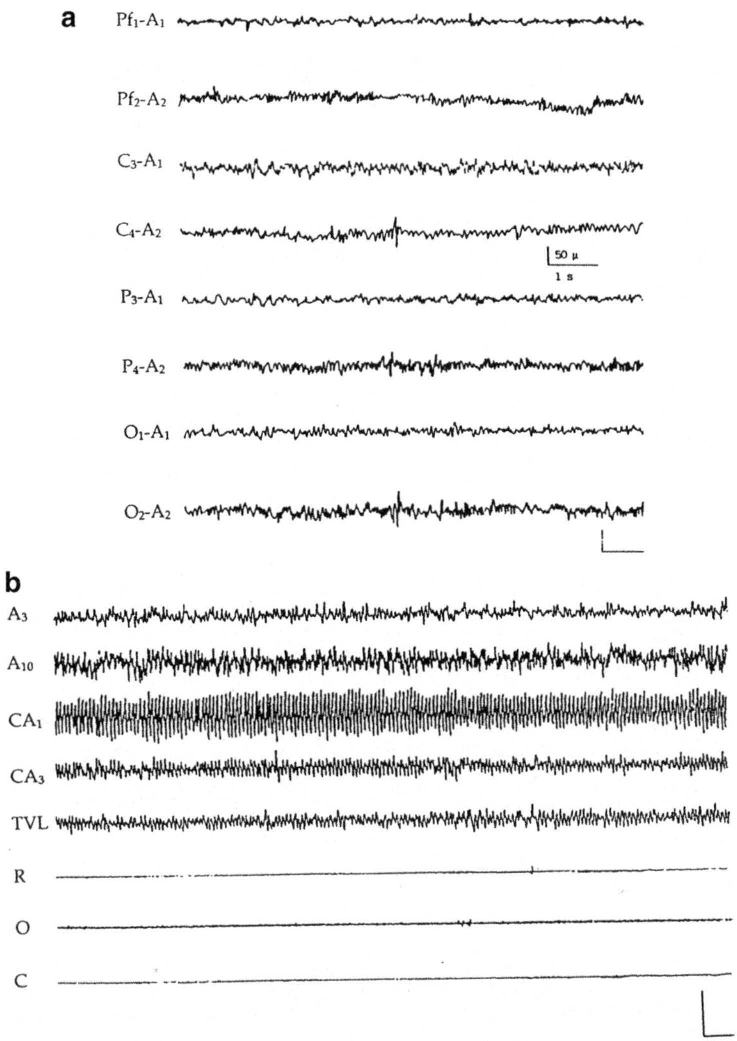

Fig. 22.8 Paradoxical sleep in humans (**a**) and rats (**b**). **a**, a generalized desynchronized pattern, extending from the pre-frontal regions (Pf_1, *left*; Pf_2, *right*); in A_1 and A_2 the electrodes are fixed to the left ear, A_1, on the right ear, A_2, up to occipital areas (O_1, *left*; O_2, *right*), including a region close to the central cleft (C_3; C_4) and the parietal area (P_3; P_4). Calibration: 50 µV and 1 s. **B**, desynchronized pattern in the frontal regions and theta waves in the hippocampus (CA_1 and CA_3) as well as the VLN, oscillating at a frequency of 6.1 Hz. Note the slight movement of the face, whiskers and eyes. Calibration: 100 µV and 1 s

sinoaortic denervation. The meaning of this discrepancy remains unclear, but it is possible that hypertension during paradoxical sleep is characteristic of rodents or even of animals that serve as prey, and that it serves to enhance defensive characteristics, such as the fall of arterial pressure, contrary to what happens in man, cat, dog,

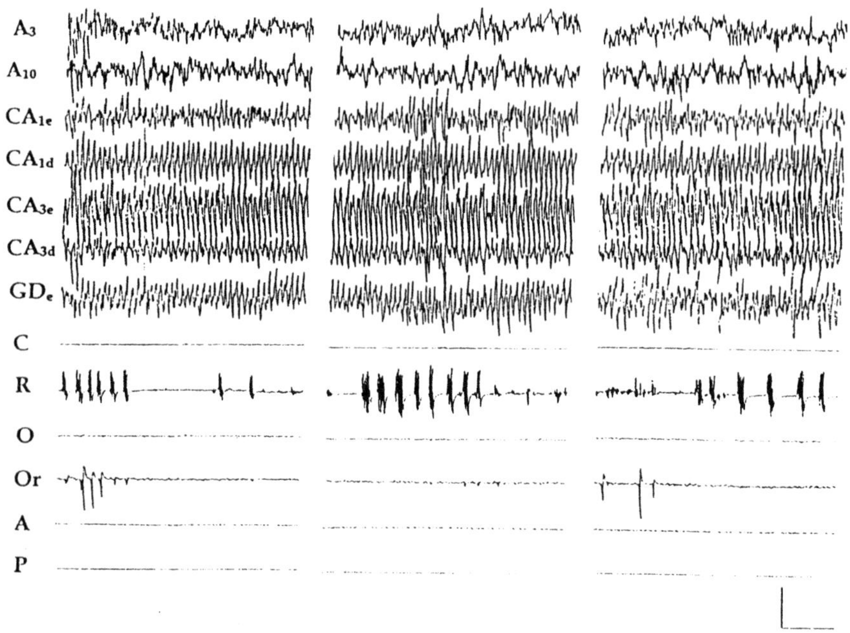

Fig. 22.9 Three periods of paradoxical sleep, with high voltage theta waves in the hippocampus and dentate gyrus and intense oneiric movement of the whiskers and ears. Here, eye movement is minimal. The movement of the whiskers is characterized by a typical pattern (fast and broad) of predominant of whisker vibration. Such bursts occur at a high frequency (8–10 s). Calibration: 100 μV and 1 s

monkey, and other predators (Valle 1992). Another explanation is that the redistribution of blood may increase peripheral resistance in the muscles (Krieger 1989).

The evolution of the sleep-wake cycle of rats is somewhat irregular at the onset of sleep. The first cycles are incomplete, although they do become regular and almost always end with a brief awakening. This is characteristic of animals that naturally serve as prey to predators (e.g., rat, mouse, rabbit, guinea pig, etc.) while predators (e.g., man, cat, dog) rarely awaken between one cycle and the other. Brief periods of wakefulness or even marked superficial sleep (microarousals), however, may also occur in humans. It is possible that such brief vigil periods are periods of olfactory exploration of the environment to perhaps identify a nearby predator. An interesting fact, and one that supports the possible predator hypothesis, is that the rat does not actually physically move at all while frequent movement of the nostrils and whiskers is recorded. The animal most likely awakens and sniffs the environment but does not move its body so as not to reveal its presence to a nearby predator. Figure 22.10 depicts such awakenings at the end of each cycle that characterizes poly-phase sleep, a pattern that requires spectral analysis to be conducted from recording segments lasting 10–20 s (Deboer et al. 2003) (Fig. 22.11).

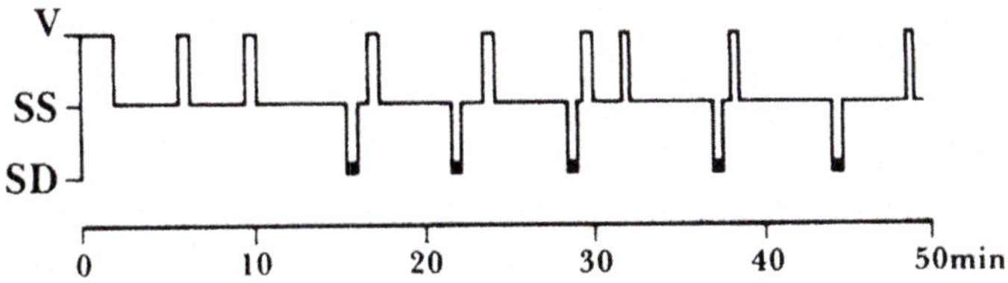

Fig. 22.10 Hypnogram of the sleep cycle of rats (Timo-Iaria 1985). The sleep cycle of the rat lasts approximately 10 min

Fig. 22.11 A rat sleeping with its vertex against the cage floor (Laboratory of Sleep of Rats of the Psychobiology Department of Universidade Federal de São Paulo/Brazil)

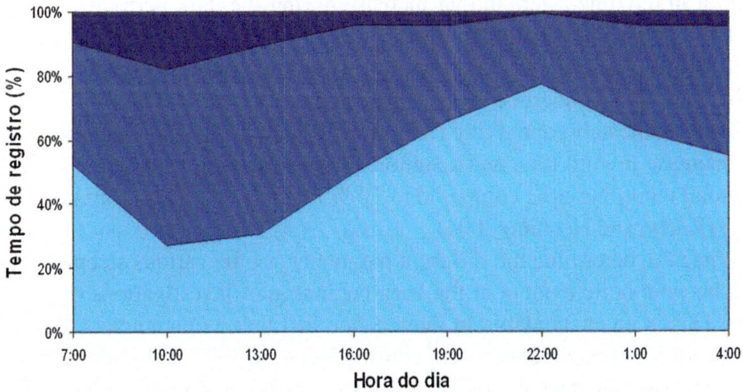

Fig. 22.12 Relative representation (percentage) of the amount of vigil time, slow waves, and paradoxical sleep in rats over a 24 h period. Awake predominates in the dark period and sleep during the light, but the rat sleeps throughout the entire light period (Department of Psychobiology Universidade Federal de São Paulo/Brazil)

The rat is an animal whose vigil period is kept during the night whereas sleep occurs during the day. Sleep behavior is characterized by the animal becoming quiet, keeping its eyes closed, and adopting different postures. It is common to sleep with its vertex against the cage floor (Fig. 22.12). Preference for one or another position is often determined by the adopted sleep posture, which is very much influenced by room temperature. Rats, like many other species, sleep very close to each other, sometimes in heaps in order to keep warm.

As depicted in Fig. 22.12, sleep also occurs in the dark period, although in a lesser proportion than during the light period. This figure illustrates the distribution of sleep states of the rat over a 24 h period. It is clear that synchronized sleep as well as desynchronized sleep lasts for less time at night than during the day. According to Van Luijtelaar and Coenen (1983), the rat sleeps about 62 % of the daytime period and for 33 % of the dark period, and much of the dark period is actually spent eating, walking, exploring the environment, interacting with other rats, mating, and nesting. Adding up all the sleep periods shows that 47.5 % of a 24 h period is spent sleeping and 17.1 % of the total sleeping time is light sleep (SS_I and SS_{II} phases), 21 % is deep sleep (SS_{III}) and 9.4 % is desynchronized sleep. Borbély and Neuhaus (1979) obtained similar proportions and Friedman et al. (1979) found slightly different values (20.5 %) light synchronized sleep, 23.8 % deep synchronized sleep, and 11 % desynchronized sleep.

According to Landis et al. (1989), rats sleep deeper in the first 3 h of a given light period and less so in the first 3 h of the dark period. At about the 12th and 15th hour of sleep, paradoxical sleep is predominant (Kleinlogel 1983a, b).

The daily structure of the rat's sleep varies according the age and gender of the animal. Paradoxical sleep occupies the largest portion of time in the first days of life. Synchronized sleep appears only afterwards, and progressively intensifies until it reaches the percentages of those of an adult. As the animal grows older, there is a reduction in daytime sleep and an increase during the dark period (Van Gool and Mirmiran 1983). The structure of female sleep is slightly different with sleep reduction at the proestrus-phase of the estrous cycle that precedes estrus, during which the graafian follicle undergoes maturation (Schwierin et al. 1998). This can promote a rebound on the subsequent day (Kleinlogel 1983a, b). During lactation there is a slight increase in vigil time and a significant reduction of paradoxical sleep (3.0 % of the total recording time compared to 10 % in control females with normal estrus cycles) (Rocha and Hoshino 2004).

The need to determine the different parameters of the various sleep structures in the rat became quite evident at the very beginning of investigations of sleep/wake cycles. This is understandable in view of the fact that its parameters allow for the quantification of fundamental events during sleep. In 1972, Schmidek and collaborators showed that despite the ventroflexion position during paradoxical sleep, animals exhibited complete muscle atonia, including cervical muscles. When room temperature was kept under 24 °C, the rat sleeps curled up and tends to stretch out entirely at temperatures closer to 30 °C. When sleep time at these different temperatures is quantified, it can be verified that the animal sleeps less at temperatures under 24 °C and over 30 °C and that alert periods are longer. Sleep reduction occurs at the expense of paradoxical sleep and the period of synchronized sleep is kept reasonably constant. This is to say that paradoxical sleep reduces in function of either very low or very high temperatures. Acclimatizing rats to low temperatures (14 °C) and subsequently having rats sleep at higher temperatures (24 °C) produced partially suppressed paradoxical sleep. This suggests that a possible mechanism exists that is related to thermo-regulation that interrupts paradoxical sleep.

The comparison of the parameters of the sleep structures has, in addition, a high heuristic value as the high population density among rats subjected to sleep deprivation led to aggressive behavior. At low temperatures, the animals tend to sleep more, keeping each other warm (Sandrin and Hoshino 1996). At high temperatures, however, there is temporary reduction of sleep time. After a few days, the vigil and rest periods tend to synchronize among rats to ensure enough high quality sleep and to inhibit aggressive behavior induced by sleep deprivation.

It can be safely said that monitoring sleep parameters in the rat will continue to be needed for the next generation of information. This information will contribute in the comprehension of the neurophysiology of sleep, the underlying mechanisms involved in sleep disturbances and perhaps the discovery of those mechanisms. Such is evident as a great number of the studies requires the comparison of the sleep structures of the experimental groups to those of their respective control groups so that the action of an independent variable can be determined. First, the conditions in which the natural life sleep structure occurs are the reference that enables the comprehension of experimenter-induced alterations. Furthermore, comparative studies of the natural sleep structure among different animals are a routine procedure in phylogenetic studies of sleep. Second, an altered sleep structure produced by an imposition may serve as reference for the determination of the effects of an unknown variable.

Analysis of the Microstructure of Sleep in Rodents

Sleep, assessed by classical analysis, appears as a cyclic process with regular alternation between slow wave sleep and paradoxical episodes that become evident in ECoG recordings. The rhythmic properties observed at a phenomenological level are also reflected in the microstructure of the sleep ECoG. Many studies have demonstrated rhythmic alterations over a recording period (Uchida et al. 1992; Aeschbach and Borbély 1993; Mann et al. 1993) that are reflected in ECoG power oscillations within distinct frequency bands. The ECoG has proven to be a reliable means of gauging brain activity, which has been found to be heavily dependent on cerebral metabolism. Studies have shown that the alterations in cerebral blood flow are closely correlated to changes in ECoG activity (Cao et al. 1998; Kudoh et al. 1997; Nagata et al. 1989; Tolonen and Sulg 1981).

Towards the end of the 1920's, neuropsychiatrist Hans Berger employed the use of electronic amplifiers in his recordings and carried out systematic investigations into the several different electric potentials generated in the neocortex. His studies resulted in the characterization of the microstructure of sleep composed of the alpha, beta, theta and gamma rhythms (Andersen et al. 2001). The action potentials that are generated in the central nervous system (CNS) or in the muscles are of very low voltage, ranging from as little as 10 μV to a few hundred μV. This is the reason why amplifiers are used to amplify the original potentials hundreds if not thousands times, so that they can be detected and recorded on paper graphically or visualized on the screen of a computer (Andersen et al. 2001).

Synchronization is characterized by neuronal action potentials of high voltage and low frequency and results from the composition of a small number of potentials in phase and/or with very little discrepancy (Adrian 1934). Conversely, desynchronization results from the composition of a high number of neuronal action potentials that occur out of phase, generating compound potentials of low voltage but of high frequency whose unitary potentials overlap at every instant (Andersen et al. 2001).

Human EEG recordings at rest (either sleep or relaxed wakefulness) reveal regular low frequency oscillations (i.e., alpha, theta, and delta rhythms). Berger demonstrated that the alpha rhythm occurs in posterior regions of the human cortex. When the person is in relaxed wakefulness and receiving little sensorial stimuli, this rhythm disappears approximately 200 ms after the eyes open (expressed today as alternative alertness) but reappears immediately (or at most a latency of approximately 200 ms) when the eyes close. This factor clearly indicates that regular oscillations are programmed into neurons by central pattern generators that oscillate at deterministic frequencies, and that these frequencies do not occur by chance. The human alpha rhythm (8–12 Hz) and the theta rhythm (4–8 Hz) are particularly important in the investigation of the deterministic property of regular neural oscillations. Their regularity, or "stationarity" as is said in systems analysis jargon, however, is kept for only very short (under 1 s) periods of time (in view of the fact that synaptic events that determine electrophysiological events are very short lasting, in the order of few milliseconds) and quickly change as a function of what the circuitry is performing at a given time. The ECoG power in the theta frequency band (4–8 Hz) during quiet waking increases during sleep deprivation, and predicts the subsequent homeostatic increase of slow wave activity during sleep in the ECoG power between 0.5 and 4 Hz. These findings indicated that theta power in waking is an ECoG variable that reflects the rise in sleep propensity.

Similar to what occurs in humans, ECoG is generated through rapid (desynchronization) and slow oscillations (delta sleep waves, and theta waves) in rats. The mechanisms that generate neuronal action potentials in the rat certainly do not differ much from what is encountered in humans, cats or other species. Adrian (1934), while engaged in one of his multiple approaches to investigating the mechanisms and origins of electrical potentials of the cerebral ganglia of roaches, observed that they look very much like human alpha waves and theta waves of other species.

Rodents exhibit a variety of waking behaviors that include active exploration and quiescence with low locomotion. These differences in behavior are reflected in the ECoG and the morphology of the waves can be carefully analyzed (Figs. 22.13 and 22.14). During active waking that is associated with voluntary movements, the ECoG is desynchronized and is dominated by fast frequencies and regular theta activity (4–8 Hz). Quiet waking, on the other hand, is characterized by a mixed pattern with slower waves (Vyazovskiy and Tobler 2005). Sleep spindles and slow waves are the hallmarks of the quiet waking ECoG (Bjorvatn et al. 1997). In rats, two different types of spindles have been described—anterior and posterior spindles (Gottesmann 1992; Terrier and Gosttesamnn 1978). The anterior spindles

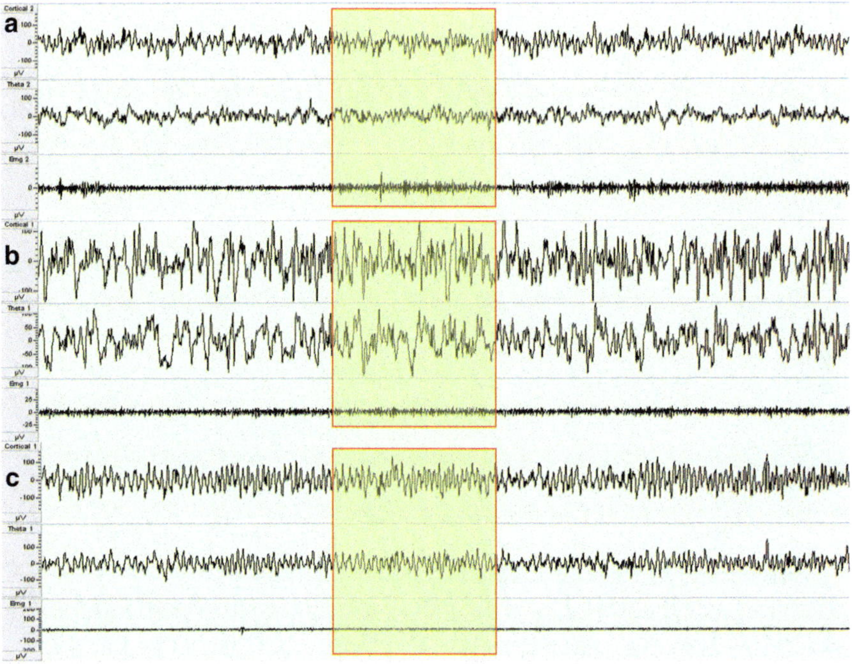

Fig. 22.13 Typical examples of polygraphic recordings. Analysis of wave morphology during wakefulness (**a**), slow wave sleep (**b**), and paradoxical sleep (**c**). Recordings and analyses were performed at the Department of Psychobiology of Universidade Federal de São Paulo/Brazil. ECoG and EMG signals were digitally collected and processed with the Somnologica/Add-life sleep recording software package (Embla Medical, Reykjavik, Iceland)

have a frequency between 5 and 15 Hz with a mean of 11.2 Hz, whereas the posterior spindles (absent in the frontal cortex) vary between 9 and 18 Hz with a mean of 12.4 Hz (Terrier and Gosttesamnn 1978). According to studies on the neurophysiological mechanisms underlying ECoG spindles and slow waves, an inverse relationship is expected (Nuñez et al. 1992; Steriade et al. 1996). Both types of rhythms seem dependent on hyperpolarization of thalamocortical neurons, with spindles occurring at membrane potentials between −55 and −65 mV, and slow wave oscillations occurring at potentials between−68 and −90 mV (Nuñez et al. 1992). The power densities for the transition stage, meaning the short-lasting sleep stage usually occurring just before paradoxical sleep (Gottesmann and Gandolfo 1986; Gottesmann 1992), is visually characterized by massive spindling as well as theta oscillations. A power spectrum analysis of the ECoG also clearly reflects this. Transition is an unusual sleep stage; the spindles may represent a sign of a deep slow wave sleep stage, whereas the low-frequency and theta rhythm may be an index of cerebral activation (Gottesmann and Gandolfo 1986; Gottesmann 1992). The theta oscillations originate in the hippocampus, and the theta peak in transition and paradoxical sleep can thus only be seen in frontoparietal and not in frontofrontal ECoG.

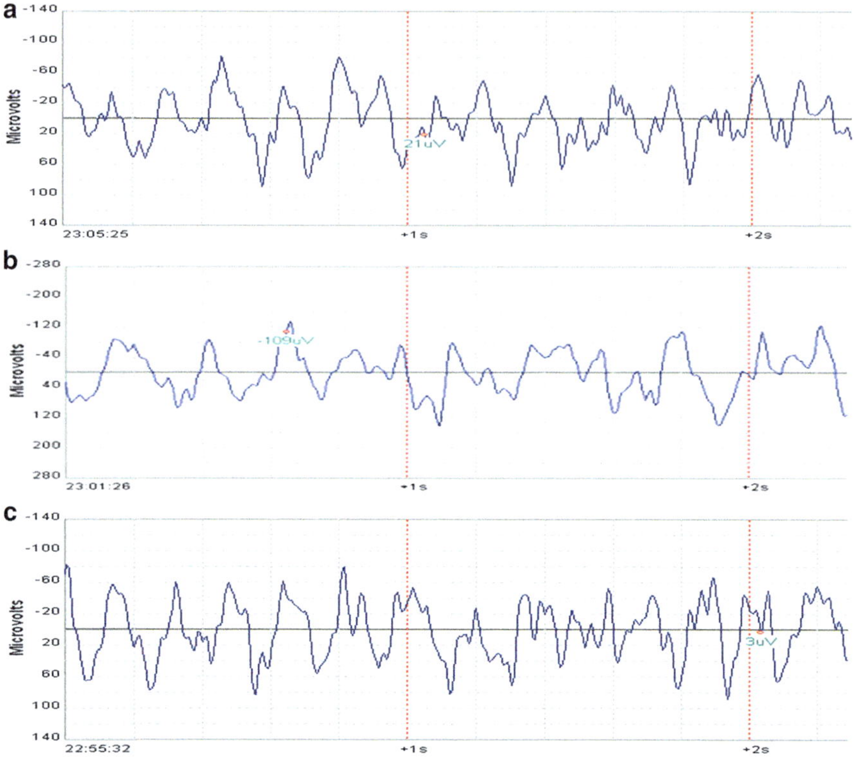

Fig. 22.14 Typical examples of polygraphic recordings. Analysis of wave morphology during wakefulness (**a**) with primarily theta waves, indicating active wakefulness in the rodent (**b**) slow wave sleep dominated by delta waves, and paradoxical sleep (**c**) that is dominated by theta waves and a lack of muscle activity. Recordings and analyses were performed at the Department of Psychobiology of Universidade Federal de São Paulo/Brazil. ECoG and EMG signals were digitally collected and processed with the Somnologica/Add-life sleep recording software package (Embla Medical, Reykjavik, Iceland)

In Wistar rats, spindles increase in number, duration, and amplitude as sleep deepens (Terrier and Gosttesamnn 1978; Gottesmann 1992). Campbell and Feinberg (1993), however, did not find a reciprocal relationship between Sprague–Dawley rats and humans. For each animal, threshold values separating the different sleep-wake stages were determined by thoroughly examining the polygraph record and visually relating scored stages with corresponding power spectrum values (Bjorvatn et al. 1997). The visual scoring criteria and the algorithm for the semiautomatic scoring had been thoroughly defined earlier (Neckelmann and Ursin 1993; Neckelmann et al. 1994). Scored stages are defined as active waking, quiet waking, light slow wave sleep, deep slow wave sleep, transition-type sleep, and paradoxical sleep.

The ECoG power is markedly reduced with a significant increase in delta activities accompanied by a decrease in beta activity, thereby producing a diffuse slow-wave

ECoG pattern (Lu et al. 2001a, b). The theta rhythm, which can be recorded from the hippocampus, is observable during wakefulness and paradoxical sleep. Some authors have divided the theta band in two frequency bands: slow theta (4–6.5 Hz) and fast theta (6.5–9 Hz) (Marrosu et al. 1997; Timofeeva and Gordon 2001). These bands would be linked to two theta generators that both depend on the arousal stage. Thus, slow theta is associated with quiet waking activities and fast theta can be observed during the active waking phase associated to voluntary movements. Mechanisms and controls of this theta rhythm are less known, but it appears to be under cholinergic control (Timofeeva and Gordon 2001). These two types of theta were observed during paradoxical sleep (Vanderwolf et al. 1977).

Spectral analysis of the ECoG can be performed by digital systems in which the sleep stages are determined. The total energies of the ECoG and EMG spectrums can be recorded in different frequency bands (delta: 0.5–4 Hz, slow theta: 4–6.5 Hz, fast theta: 6.5–9 Hz, alpha: 9–14 Hz). These energies are expressed as percentages of the total energy in the spectrum. Sleep stages can also be quantified using an automatic or manual staging method. Each laboratory has developed the criteria for the automatic sleep staging system (Fig. 22.15). For example, Crouzier et al. (2006)

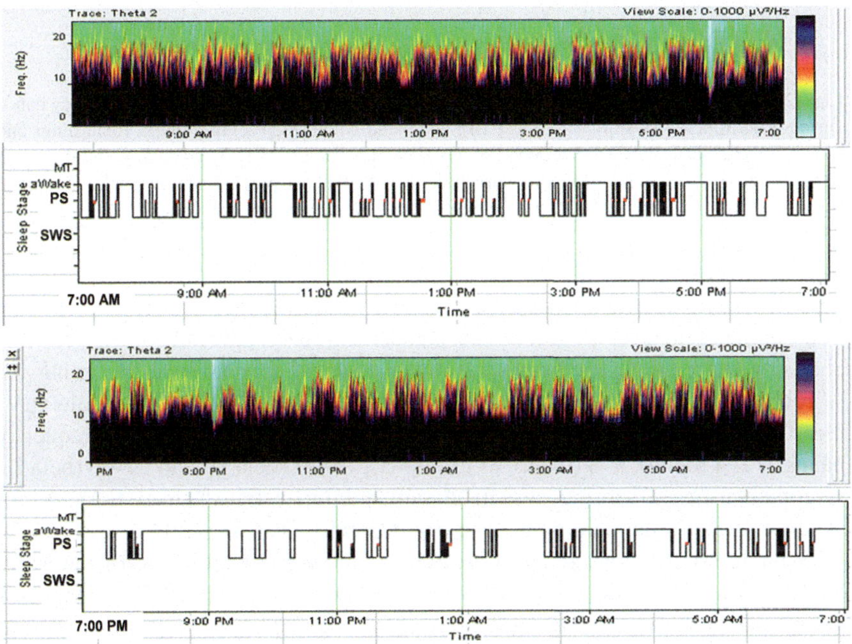

Fig. 22.15 Typical examples of the sleep-wake cycle in rodents. Cortical ECoG power spectral density in different frequency bands and corresponding hypnograms during the light (7 am–7 pm) and dark periods (7 pm–7 am). Recordings and analyses were performed at the Department of Psychobiology of Universidade Federal de São Paulo/Brazil. ECoG and EMG signals were digitally collected and processed with the Somnologica/Add-life sleep recording software package (Embla Medical, Reykjavik, Iceland) (*SWS* slow wave sleep, *PS* paradoxical sleep)

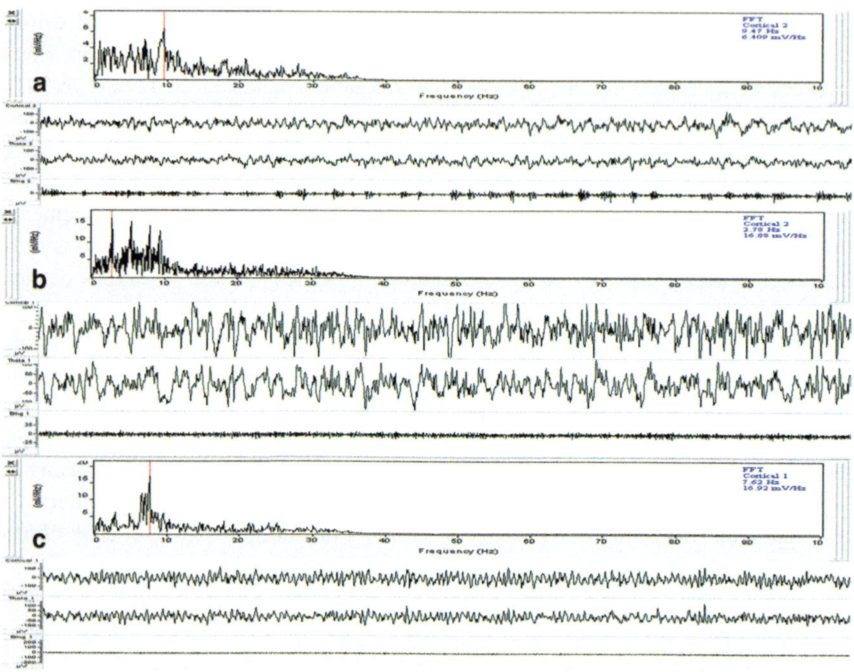

Fig. 22.16 Typical examples of polygraphic recordings. The FFT analysis for the frequency bands during wakefulness (**a**), slow wave sleep (**b**), and paradoxical sleep (**c**) in rodents. Recordings and analyses were performed at the Department of Psychobiology of Universidade Federal de São Paulo/Brazil. ECoG and EMG signals were digitally collected and processed with the Somnologica/Add-life sleep recording software package (Embla Medical, Reykjavik, Iceland)

described each stage as calculated every 5.12 s from the ECoG and EMG spectra. They defined four sleep stages: awake, paradoxical sleep, light slow wave sleep and deep slow wave sleep. The hypnogram was calculated over 3 days of recording.

Radek et al. (2004) calculated the average ECoG amplitude in microvolts (μV) for 30 s epochs using Fast Fourier Transform (FFT) analysis for the frequency bands of 1–4 (delta), 4–8 (theta), 8–13 (alpha), 13–25 (beta 1), and 25–50 (beta 2). A computer-based scoring system classified Fourier transformed ECoG/EMG states as awake, slow wave sleep, and paradoxical sleep (Fig. 22.16).

Behaviorally, the awake ECoG/EMG state corresponded to the rats actively exploring, grooming, or sitting still with eyes open. The sleeping EEG/EMG corresponded to the rats being immobile, eyes closed, and frequently in a curled position. The rats would also show brief motor twitches during the paradoxical state. The sleep data was expressed as the average time spent in either the awake, slow wave sleep, or paradoxical sleep state during the total scoring period. The ECoG global frequency band (0–30 Hz) was divided into four frequency bands: delta (0–4 Hz), theta (4–8 Hz), alpha (8–13 Hz) and beta (13–30 Hz) (Lu et al. 2001a, b).

The gamma rhythm is also a subject of intense study. Rhythmic synchronization of neural discharges in the gamma band, 30–100 Hz, is thought to provide a spatial and temporal dynamic link between cortical areas involved in functional networks (Traub et al. 1997). This frequency band is part of background neural activity that is related to a variety of non-conscious cortical processes occurring during wakefulness as well as sleep and while under anesthesia (Steriade et al. 1996).

Based on the aforementioned studies, it is fair to state that power densities differ throughout the frequency ranges between different sleep and wake stages; more studies on this theme are warranted.

The Importance of Investigating the Sleep-Wake Cycle in Rats

The use of animals in scientific experimentation and teaching activities is required for the advancement of medical care for humans and animals. The contribution animals have made in the various scientific fields has promoted the development of long-needed prophylactic measures as well as of countless effective therapies for human and animal ailments. An example of such contributions are the discoveries of insulin, tranquilizers, anesthetics, and anti-depressants. Equally relevant advances are the development of vaccines for malignant diphtheria and for poliomyelitis as well as drugs that allow for life-saving transplantation in animals. Moreover, animals have played a fundamental role in the development of novel medical treatment with the therapeutic application of antibiotics, and will most likely continue to be of great use in the validation of new medications and treatment for various diseases, possibly averting epidemic disasters. Finally, animal experimentation is essential in the comprehension of diseases like cancer, AIDS, and Alzheimer's disease.

In the 1960's, rodents became the animal of choice in the investigation of the sleep-wake cycle. As of 1970, Timo-laria and collaborators emphasized the relevance of studies conducted on rats and contributed in making the rat the species most utilized in scientific research, preferably in the research of matters involving the structural and functional mechanisms of the nervous system. The adoption of the rat in sleep research was strengthened when it was demonstrated that it, in addition to sleeping approximately 13 h in a 24 h period, possesses almost all of the sleep phases as humans (Timo-Iaria et al. 1970). From that moment, other relevant studies utilized the rat or other rodents to evaluate the sleep-wake cycle and paved the way for countless fundamental discoveries to be made in human physiology, such as the discovery of a gene for narcolepsy in rodents (Chap. 24). Besides being a fundamental discovery in itself, this aides in the study of symptoms of this particular disease.

Thus, under ethical principles, the adoption of animals (and of rodents in particular) is ever more present in various areas of scientific investigation, including neuronal, endocrinological, physiological, histological, and immunological studies, among others. These studies yield relevant information regarding the physiological mechanisms and behaviors that correspond to what is considered normal or not, helping to establish valuable patterns that guide diagnosis and making their use justifiable for scientific purposes.

After 50 years of the study of sleep, several questions remain unclear. Throughout almost 90,000 different studies regarding sleep, we are still very far from understanding its function. Questions about why we need to sleep a certain amount of hours each night to be capable of being coherent the next day are not well understood. This fascinating process occurs in each organism in a homeostatic rhythm and is extremely important for the maintenance of life.

References

Adrian ED. Electrical activity of the nervous system. Arch Neurol. 1934;32:1125–36.

Aeschbach D, Borbély AA. All-night dynamics of the human sleep EEG. J Sleep Res. 1993;2:70–81.

Andersen ML, Valle AC, Iaria CT, Tufik S. Implantação de eletrodos para o estudo eletrofisiológico do ciclo vigília-sono do rato. 1st ed. São Paulo: Universidade Federal de São Paulo-UNIFESP/EPM; 2001. 62p.

Benington JH, Kodali SK, Heller HC. Scoring transitions to REM sleep in rats based on the EEG phenomena of pre-REM sleep: an improved analysis of sleep structure. Sleep. 1994;17:28–36.

Berger H. Über das elektrenkephalogramm des menschen. Arch Psychiatr Nervenkr. 1930;87: 527–70.

Bjorvatn B, Fagerland S, Ursin R. EEG power densities (0.5–20 Hz) in different sleep-wake stages in rats. Physiol Behav. 1997;63:413–7.

Borbély AA, Neuhaus HU. Sleep-deprivation: effects on sleep and EEG in the rat. J Comp Neurol. 1979;133:71–87.

Borbély AA, Tobler I, Hanagasioglu M. Effect of sleep deprivation on sleep and EEG power spectra in the rat. Behav Brain Res. 1984;14:171–82.

Campbell IG, Feinberg I. A cortical EEG frequency with a REM specific increase in amplitude. J Neurophysiol. 1993;69:1368–71.

Cao Y, D'Olhaberriague L, Vikingstad BS, Levine SR, Welch KMA. Pilot study of functional MRI to assess cerebral activation of motor function after poststroke hemisparesis. Stroke. 1998;29:112–22.

Crouzier D, Baubichon D, Bourbon F, Testylier G. Acetylcholine release, EEG spectral analysis, sleep staging and body temperature studies: a multiparametric approach on freely moving rats. J Neurosci Methods. 2006;151:159–67.

Deboer T, Vansteensel MJ, Detari L, Meijer JH. Sleep states alter activity of suprachiasmatic nucleus neurons. Nat Neurosci. 2003;6:1086–90.

Friedman L, Bergmann BM, Rechtschaffen A. Effects of sleep deprivation on sleepness, sleep intensity, and subsequent sleep in the rat. Sleep. 1979;1:369–91.

Gottesman C, Juan de Mendonza JL, Lacoste G, Lallement B, Rodrigues L, Tasset M. Etude sur l'analyse et la quantification automatiques des differents états de veille et de sommeil chez rat. C R Acad Sci. 1971;272:301–2.

Gottesmann C. Theta rhythm: the brain stem involvement. Neurosci Biobehav Rev. 1992;16:25–30.

Gottesmann C, Gandolfo G. A massive but short lasting forebrain deafferentation during sleep in the rat and cat. Arch Ital Biol. 1986;124:257–69.

Hoshino K. Estágios de manifestação neocortical do estado dessincronizado de sono no rato. Tese. Botucatu: Universidade Estadual Paulista; 1977.

Hoshino K. Electrocortical patterns of REM-sleep in mesencephalic reticular formation lesioned rats. Rev Cienc Biomed. 1980;1:31–40.

Hoshino K, Toloi Jr G. Neocortical spindling during wakefulness in the rat. Braz J Med Biol Res. 1995;28:337–42.
Jouvet M. Recherches sur les structures nerveuses et les mécanismes reponsables des differentes phases du sommeil physiologique. Arch Ital Biol. 1962;100:125–206.
Jouvet M. Neurophysiology of the states of sleep. Physiol Rev. 1967;47:117–77.
Jouvet M. The role of monoamines and acetylcholine-containing neurons in the regulation of the sleep-waking cycle. Ergeb Physiol. 1972;64:166–307.
Jouvet M. Paradoxical sleep mechanisms. Sleep. 1994;17:7–83.
Junqueira LF, Krieger EM. Blood pressure and sleep in the rat in normotension and in neurogenic hypertension. J Physiol. 1976;259:725–35.
Kleinlogel H. The female rat's sleep during oestrous cycle. Neuropsychobiology. 1983a;10: 228–37.
Kleinlogel H. Sleep in various species of laboratory animals. Neuropsychobiology. 1983b;9:174–7.
Krieger EM. Arterial baroreceptor ressetting in hypertension (the J.W. McCubbin memorial lecture). Clin Exp Pharmacol Physiol. 1989;15(Suppl):3–17.
Kudoh M, Takahashi S, Yonezawa H. Correlation between quantitative EEG and cerebral blood flow and oxygen metabolism in patients with dementia of Alzheimer type. Rinsho Shinkeigaku. 1997;37:359–65.
Landis CA, Levine JD, Robinson CR. Decreased slow-wave and paradoxical sleep in the rat chronic pain model. Sleep. 1989;12:167–77.
Lu J, Zhang YH, Chou TC, Gaus SE, Elmquist JK, Shiromani P, Saper CB. Contrasting effects of ibotenate lesions of the paraventricular nucleus and subparaventricular zone on sleep-wake cycle and temperature regulation. J Neurosci. 2001a;21:4864–74.
Lu X-C, Williams AJ, Tortella FC. Quantative electroencephalography spectral analysis and topographic mapping in a rat model of middle cerebral artery occlusion. Neuropathol Appl Neurobiol. 2001b;27:481–95.
Mann K, Backer P, Roschke J. Dynamical properties of the sleep EEG in different frequency bands. Int J Neurosci. 1993;73:161–9.
Marrosu F, Cozzolino A, Puligheddu M, Giagheddu M, Di Chiara G. Hippocampal theta activity after systemic administration of a non-peptide delta-opioid agonist in freely-moving rats: relationship to D1 dopamine receptors. Brain Res. 1997;776:24–9.
Mistelberger RE, Bergmann BM, Waldenar W, Rechtschaffen A. Recovery sleep following sleep deprivation in intact and suprachiasmatic nuclei-lesioned rats. Sleep. 1983;6:217–33.
Moruzzi G. The sleep-waking system. Ergeb Physiol. 1972;64:1–165.
Nagata K, Tagawa K, Hiroi S, Shishido F, Uemura K. Electroencephalografic correlates of blood flow and oxygen metabolism provided by positron emission tomography in patients with cerebral infarction. Electroenceph Clin Neurophysiol. 1989;72:16–30.
Neckelmann D, Ursin R. Sleep stages and EEG power spectrum in relation to acoustical stimulus arousal threshold in the rat. Sleep. 1993;16:467–77.
Neckelmann D, Olsen OE, Fagerland S, Ursin R. The reliability and functional validity of visual and semiautomatic sleep/wake scoring in the Moll-Wistar rat. Sleep. 1994;17:120–31.
Nuñez A, Amzica F, Steriade M. Intrinsic and synaptically generated delta (1–4 Hz) rhythms in dorsal lateral geniculate neurons and their modulation by light-induced fast (30–70 Hz) events. Neuroscience. 1992;51:269–84.
Radek T, Decker MW, Jarvis MF. The adenosine kinase inhibitor ABT-702 augments EEG slow waves in rats. Brain Res. 2004;1026:74–83.
Rocha LRM, Hoshino K. Some aspects of the sleep of lactating rat dams. Hypnos Magazine 2004;1–9. http://www.flass.icb.usp.br/hypnos/artigos.html
Sandrin MFN, Hoshino K. Sono de ratos confinados em alta densidade populacional. Dissertação de mestrado. Instituto de Biociências, UNESP, Botucatu; 1996.
Schmidek WR, Hoshino K, Schmidek M, Timo-Iaria C. Influence of environmental temperature on the sleep-wakefulness cycle in the rat. Physiol Behav. 1972;8:363–71.
Schmidek WR, Timo-Iaria C, Schmidek M. Influence of loxapine on the sleep-wakefulness cycle of the rat. Pharmacol Biochem Behav. 1974;2:747–51.

Schwierin B, Borbley AA, Tobler I. Sleep homeostasis in the female rat during the estrous cycle. Brain Res. 1998;811:96–104.
Steriade M, Amzica F, Contreras D. Synchronization of fast (30–40 Hz) spontaneous cortical rhythms during brain activation. J Neurosci. 1996;16:392–417.
Terrier G, Gosttesamnn C. Study of cortical spindles during sleep in the rat. Brain Res Bull. 1978;3:701–6.
Timofeeva OA, Gordon CJ. Changes in EEG power spectra and behavioral states in rats exposed to the acetylcholinesterase inhibitor chlorpyrifos and muscarinic agonist oxotremorine. Brain Res. 2001;893:165–77.
Timo-Iaria C. O sono. Ciência Hoje. 1985;4:66–76.
Timo-Iaria C, Negrao N, Schmidek WR, Hoshino K, Lobato de Menezes CE, Leme da Rocha T. Phases and states of sleep in the rat. Physiol Behav. 1970;5:1057–62.
Timo-Iaria C, Yamashita R, Hoshino K, Melo AS. Rostrum movements in desynchronized sleep as a prevalent manifestation of dreaming activity in Wistar rats. Braz J Med Biol Res. 1990;23:617–20.
Tolonen U, Sulg IA. Comparison of quantitative EEG parameters from four different analysis techniques in evaluation of relationships between EEG and CBF in brain infarction. Electroencephalogr Clin Neurophysiol. 1981;51:177–85.
Trachsel L, Tobler I, Achermann P, Borbely AA. Sleep continuity and the REM-nonREM cycle in the rat under baseline conditions and after sleep deprivation. Physiol Behav. 1991;49:575–80.
Traub RD, Jefferys JG, Whittington MA. Simulation of gamma rhythms in networks of interneurons and pyramidal cells. J Comput Neurosci. 1997;4:141–50.
Uchida S, Maloney T, Feinberg I. Beta (20–28 Hz) and delta (0.3–3 Hz) EEGs oscillate reciprocally across NREM and REM sleep. Sleep. 1992;15:352–8.
Ursin R. The two stages of slow sleep in the cat and their relation to REM sleep. Brain Res. 1968;11:347–56.
Ursin R, Larsen M. Increased sleep following intracerebroventricular injection of the delta sep-inducing peptide in rats. Neurosci Lett. 1983;40:145–9.
Ursin R, Bjorvatn B, Sommerfelt L, Underland G. Increased waking as well as increased synchronization following administration of selective 5-HT uptake inhibitors to rats. Behav Brain Res. 1989;34:117–30.
Valle AC. Estudo comparativo de algumas manifestações equivalentes do alerta e do sono dessincronizado no rato. Dissertação. São Paulo: Universidade Federal de São Paulo/EPM; 1992.
Valle AC, Timo-Iaria C, Sameshima K, Yamashita R. Theta waves and behavioral manifestations of alertness and dreaming activity in the rat. Braz J Med Biol Res. 1992;25:745–50.
Van Gool WA, Mirmiran MA. Age-related changes in the sleep pattern of male adult rats. Brain Res. 1983;279:394–8.
Van Luijtelaar ELJM, Coenen AML. An EEG averaging technique for automated sleep-wake stage identification in the rat. Physiol Behav. 1983;33:837–41.
Vanderwolf CH. Hippocampal electrical activity and voluntary movement in the rat. Electroencephalogr Clin Neurophysiol. 1969;26:407–81.
Vanderwolf CH, Kramis R, Robinson T. Hippocampal electrical activity during waking behaviour and sleep: analyses using centrally acting drugs. Ciba Found Symp. 1977;58:199–226.
Vyazovskiy VV, Tobler I. Theta activity in the waking EEG is a marker of sleep propensity in the rat. Brain Res. 2005;1050:64–71.

Chapter 23
Narcolepsy-Cataplexy in the Rats

Christopher M. Sinton

Narcolepsy

In humans, narcolepsy is a debilitating and important sleep disorder that was first described in 1877 by Westphal, a German physician, based on his observation of a patient with sudden sleep attacks associated with symptoms of motor incapacity and aphasia. He considered it a form of epilepsy. Three years later, the French neurologist Gélineau (Fig. 23.1) described the disorder for the first time as a distinct clinical entity rather than a symptom of another condition, and proposed the term narcolepsy, meaning “sleep seizure”. In the 1880 paper, Gélineau described a patient who, from the age of 36, began to fall asleep suddenly during the day and had sudden falls or “astasia” (cataplexy), which occurred whenever he became emotional.

Narcolepsy is characterized by two main symptoms, excessive daytime sleepiness and cataplexy, and three supplementary symptoms, hypnagogic or hypnopompic hallucinations, sleep paralysis, and fragmented nocturnal sleep (Choo and Guilleminault 1998). From the perspective of work done in rodents, the most important symptom is cataplexy. A cataplectic episode has been described as an abrupt and reversible decrease or loss in muscle tone without alteration in the level of consciousness (Guilleminault and Gelb 1995). This is a defining symptom of narcolepsy, though it should be noted that a precise definition of cataplexy remains debatable as it is also described as a transition between, or a mixed state of, wakefulness and rapid eye movement (REM) sleep (Hishikawa and Shimizu 1995). Hence, patients who collapse in a position that allows sleep will typically progress from cataplexy into REM sleep (i.e., a narcoleptic episode), while others will remain conscious until the episode is over (i.e., a cataplectic episode). In fact, after the original

C.M. Sinton (✉)
Department of Medicine, Arizona Respiratory Center, University of Arizona,
1501 N. Campbell Avenue, Tuscon, AZ 85724-5030, USA
e-mail: csinto@email.arizona.edu

M.L. Andersen, S. Tufik (eds.), *Rodent Model as Tools in Ethical Biomedical Research*, DOI 10.1007/978-3-319-11578-8_23

Fig. 23.1 The French neurologist, Jean Baptiste Edouard Gélineau (1859–1906) was the first to describe narcolepsy as a clinical condition

discovery of narcolepsy in mice, and following careful study of these episodes in mice and rats, we noted that essentially all episodes in mice transition very rapidly to REM sleep (Willie et al. 2003). Indeed, these transitions are so rapid that only very rarely, and then only for a few seconds, have we been able to observe true cataplexy (i.e., wakefulness with muscle atonia) in mice. Hence for mice, the terms cataplectic episode and narcoleptic episode are interchangeable. However, to maintain consistency with the human literature on narcolepsy, we have proposed the use of the term cataplexy to describe these episodes in mice and the term narcolepsy-cataplexy to describe their condition.

Genetically Modified Mice and CNS Dysfunction

The use of genetically modified mice has provided a wealth of information about the functions of many proteins. Proteins that are expressed primarily in the central nervous system (CNS), especially as receptors or receptor ligands, have been important targets for these types of studies. The elucidation of the cellular basis of some CNS dysfunctions, for example those that manifest as psychiatric disorders, may eventually lead to this type of investigation. This will be complicated, however, by the likelihood that multiple genes, and thus multiple proteins, combined with epistatic interaction between the genes, are involved.

This means that disruption or substitution of a single gene will not produce an animal model of the disorder, but may still provide important indications about the function of the gene.

Mammalian sleep is a complex behavior with an unknown function, and the mechanisms that initiate and underlie sleep and wakefulness states have only recently begun to be understood. This followed pioneering studies over many years in which various regions in the CNS were lesioned to show the importance of the area to sleep, or the electrical activity of groups of neurons across the different sleep states was recorded to show how their activity changed. These data were subsequently combined with neuroanatomical and pharmacological studies to produce models of the changes that occur at a cellular level in the brain during sleep and wakefulness. Like other CNS dysfunctions, sleep disorders such as narcolepsy were considered as a likely consequence of complex interactions between various neurotransmitter systems, possibly caused by multiple genetic differences. Therefore, it was unexpected that a mouse lacking a single gene for the precursor of two related peptides that act as receptor ligands in the CNS was found to exhibit narcolepsy-cataplexy (Chemelli et al. 1999). This is a description of the narcoleptic mouse, the story of its discovery, and a brief review of subsequent studies in a narcoleptic rat model.

The Identification of Orexin (Hypocretin)

The discovery of the orexins (also called hypocretins) and their receptors (Sakurai et al. 1998) followed the identification of a novel cDNA sequence with the characteristic seven transmembrane signature of a G protein-coupled cell surface receptor (GPCR) that was without a known ligand (i.e., it was an "orphan" receptor). Sakurai et al. (1998) used a cell-based reporter system to identify possible peptide ligands for the receptor. In brief, these authors challenged cell lines that expressed the orphan GPCR with various tissue extracts and monitored for a transduction effect, or an activation of the cell line. They eventually found several tissue extracts that caused an increase in Ca^{2+} levels in these cells. They then purified the extracts and sequenced the resulting peptides, which were named orexin-A and orexin-B. Both of these peptides came from a single precursor, prepro-orexin. Using the prepro-orexin cRNA probe, Sakurai et al. (1998) then performed *in situ* hybridization to show the distribution of orexin in the rat brain. The results showed that orexins were uniquely concentrated around the perifornical area of the lateral hypothalamus (Fig. 23.2), an area critically important for feeding. Indeed, when orexin was subsequently administered intracerebroventricularly, it was found to stimulate food consumption in rats. Further confirmation of the importance of orexin for energy balance came from studies that showed that prepro-orexin mRNA was upregulated during fasting. In fact, it was because the peptide was able to increase feeding that it was named orexin, from the word orexigenic, i.e., that which stimulates the appetite for food.

We have described the work of Sakurai et al. (1998) in detail as it led directly to the generation of the orexin knockout mouse. However, an independent group published their work on the discovery of the orexin peptides (de Lecea et al. 1998), just prior to the publication of Sakurai et al. (1998). This group used different

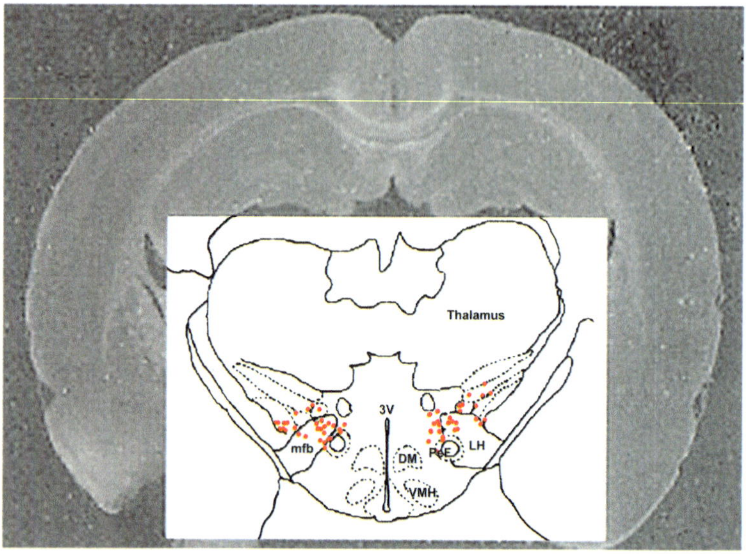

Fig. 23.2 Schematic visualization of the location of prepro-orexin mRNA-containing neurons in a coronal section of the adult rat brain. Orexin-containing cells, shown in red superimposed on anatomical structures in the hypothalamic and subthalamic areas, were closely clustered in an area around the perifornical and lateral hypothalamus. *3 V* third ventricle, *LH* lateral hypothalamus, *PeF* perifornical nucleus, *mfb* medial forebrain bundle, *DM* dorsomedial hypothalamus, *VMH* ventromedial hypothalamus

methods (1998), and they called these neuropeptides hypocretins, however the findings were essentially identical. Shortly thereafter, and while the orexin knockout mouse was being developed, Peyron et al. (1998) described projections from the orexin-containing cell bodies throughout the brain (Fig. 23.3). The projection pattern was exceptional because it was observed that orexin neurons innervated widespread areas throughout the brain, including direct projections onto most of the major nuclei that had previously been implicated in the processes of cortical activation and known to play a critical role in modulating sleep–wakefulness.

The Discovery of the Narcoleptic Phenotype

A series of subsequent experiments identified the relationship between orexin and narcolepsy. Orexin knockout mice were first generated with a null mutation induced by targeted disruption of the prepro-orexin gene (Chemelli et al. 1999). Since we already had evidence that orexin was involved in feeding behavior, we expected that these knockout mice would show differences in energy homeostasis. As part of these studies, we recorded the activity of the 14–15 week-old mice, in an open field apparatus to measure their energy expenditure from movement through a 24 h period.

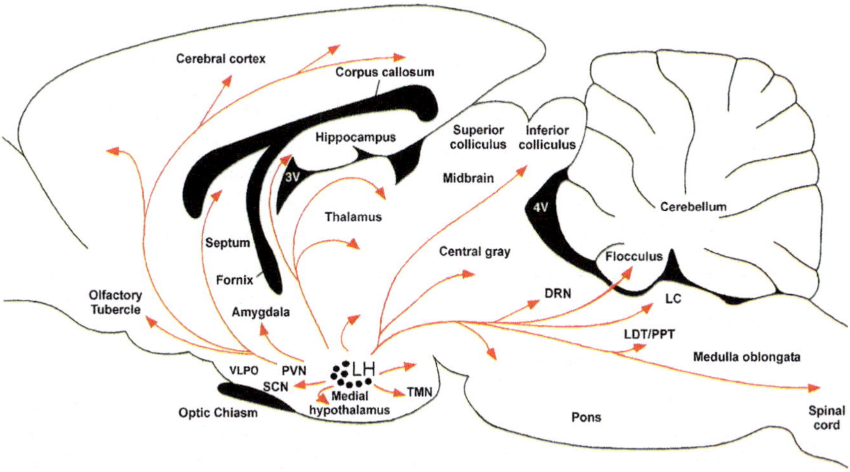

Fig. 23.3 Schematic of a sagittal section of a rat brain showing the widely distributed projection pathways from orexin-containing cells that are clustered in the lateral hypothalamus (LH) and some of the important areas that they innervate. *DRN* dorsal raphe, *LC* locus coeruleus, *LDT/PPT* latcrodorsal tegmental/pedunculopontine tegmental nuclei, *PRF* pontine reticular formation, *PVN* paraventricular nucleus, *SCN* suprachiasmatic nucleus, *TMN* tuberomamillary nucleus, *VLPO* ventrolateral preoptic area, *3 V/4 V* third/fourth ventricle

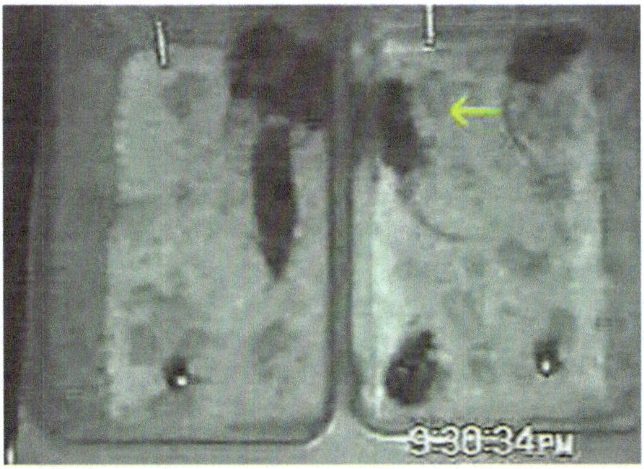

Fig. 23.4 An example frame from one of the first videotapes of orexin knockout mice under infrared light. The beginning of an episode of behavioral arrest is observed in the orexin knockout, marked by the arrow, as the mouse becomes unsteady and starts to drag its hind limbs

Surprisingly, we found that the orexin knockout mice showed periods of reduced activity during the dark phase when thc animals would normally be most active. To observe what might be happening during these periods of lowered activity, we then videotaped the mice with a camera mounted overhead under infrared light. These videotapes of the knockout mice, as compared to their wild-type littermates, demonstrated that this curious pattern of activity occurred only in the knockout mice. This activity was more unusual than expected because it was recognizable as an abrupt cessation of purposeful motor activity associated with a sudden, sustained change in posture that was maintained throughout the episode (Fig. 23.4).

Careful study of the videotapes, therefore, revealed brief periods when it looked as though a behavioral switch had been turned "off" and then "on". The knockout mice would suddenly stop moving and remain completely motionless and, for this reason, we called these periods "behavioral arrests". Interestingly, these behavioral arrests also ended abruptly, sometimes with resumption of the same purposeful motor activity that had been occurring before the cessation of movement. To see what these episodes looked like when they were first recorded, some video clips from the original publication (Chemelli et al. 1999) are available at http://www.cell.com/cgi/content/full/98/4/437/DC1. We initially thought that the orexin knockout mice were undergoing some form of epileptic seizure activity, and so we prepared to record their electroencephalograms (EEG). Before doing that, however, we also recorded some video from the side of the cage and with the camera closer to the animals. A couple of those recordings gave a clearer view of one of the mice during these episodes, and from the twitching vibrissae and slight jerking of the body, we determined that we were observing the characteristic posture of a mouse in REM sleep. Since the mouse had been awake and engaged in normal activity and then had apparently instantaneously begun an episode of REM sleep, it was apparent that the mouse might be narcoleptic. Twenty-four hour continuous EEG recording was therefore required together with associated electromyogram (EMG) activity to characterize the sleep/wakefulness behavior of the mice and to ensure that the episodes were not seizure-related.

These first EEG/EMG recordings demonstrated a complete lack of seizures while also showing an interesting disruption of sleep regulation with more fragmented episodes of sleep, particularly during the dark phase. These recordings did not reveal, however, any direct transition from wakefulness into REM sleep, despite the fact that videotapes of the mice when they were unrestricted and not attached to an EEG/EMG cable had apparently shown such transitions. It seemed likely that the cable we had been using might have been too heavy and in particular had enough inertia such that when the mouse collapsed into the REM sleep posture, it was shaken back awake by the resistance of the cable to any movement. Hence, we developed a different method of recording sleep in mice using a system that had less weight and inertia than the previously described method (Sinton et al. 1981) (Fig. 23.5).

Once the development of this new method of recording vigilance states in the mouse was completed (Fig. 23.6)—it is described in detail elsewhere in this volume—we observed direct transitions from wakefulness to REM sleep in all the orexin knockout mice in terms of the EEG/EMG signals that characterized wakefulness followed, directly, by REM sleep. We had therefore shown that these mice were indeed narcoleptic (Fig. 23.7).

Cataplectic Episodes in the Orexin Knockout Mouse

We also visually characterized these episodes of behavioral arrest objectively from the videotapes, so that we had an accurate summary of their frequency and appearance in unrestricted mice. For this, we developed specific criteria to differentiate

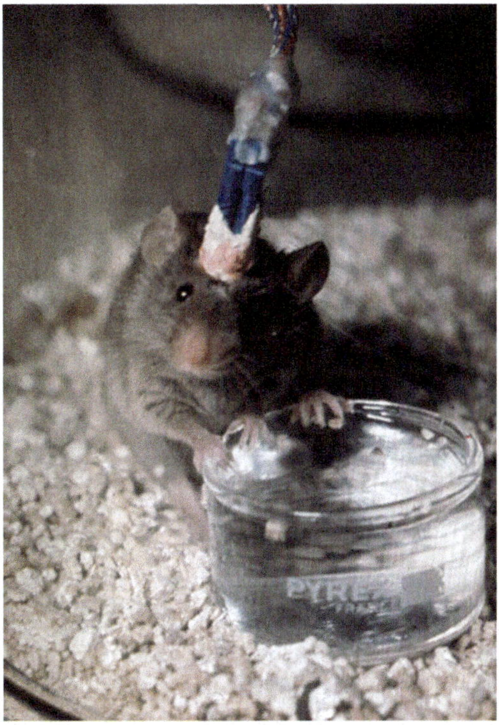

Fig. 23.5 Original methodology for implanting mice with electrodes for EEG/EMG monitoring. Although the weight of the implant and cable could be easily supported by a mouse, two important changes were necessary: lowering the overall height of the implant attached to the skull and shortening the cable (and using thinner wires). Both of these changes had the effect of reducing the inertia of the system at the point of contact with the mouse, and thus reducing the awakening effect of the cable when the mouse jerked while sleeping or collapsed suddenly into sleep

a b

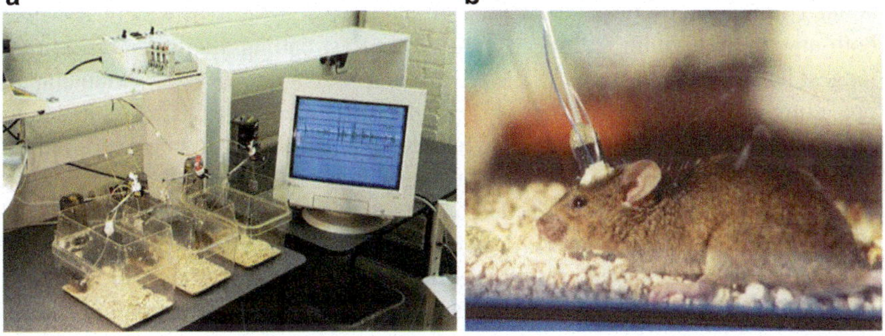

Fig. 23.6 Revised methodology for implanting mice with electrodes for EEG/EMG monitoring. Panel (**a**) shows the overall system with the mice in the recording cages and the associated computer for monitoring and archiving the signals. Panel (**b**) shows the implant. Note the size of the implant and the reduced cable thickness

them from normal quiet behavioral states with decreased overt activity, as well as from regular transitions into non-REM sleep. Independent observers used these criteria to identify episodes of behavioral arrest, but any episodes that were judged ambiguous in any aspect, by either observer, were excluded from the analysis (Chemelli et al. 1999).

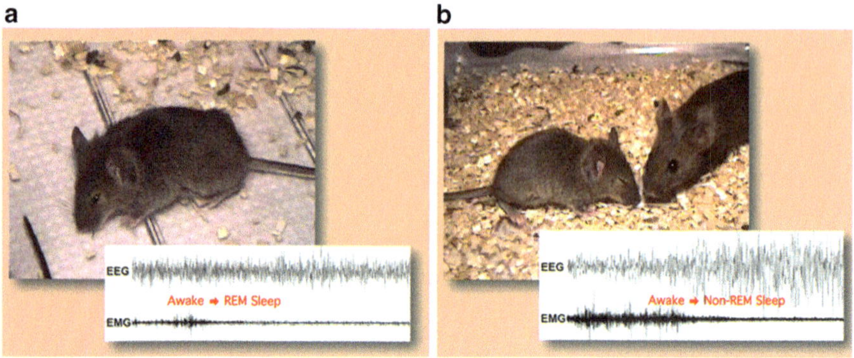

Fig. 23.7 A narcoleptic mouse (**a**) displays the characteristic direct transition from wakefulness to REM sleep as it suddenly stops moving and collapses to the floor of the cage. In panel (**b**) a normal transition from wakefulness to non-REM sleep is observed on the EEG/EMG recording, as the mouse settles gradually into a typical sleep posture after grooming and preparing the bedding on the floor of the cage

Homozygous orexin knockout mice, together with their wild-type and heterozygote littermates, were filmed for the first 4 h of the dark phase, after an initial 3 h period of acclimatization to the surroundings during the light phase. Videotapes were coded and then given to two blinded observers for scoring. Two mice were filmed in two separate open fields at once. No episodes of behavioral arrest were identified on videotapes of the wild-type or heterozygote mice, but orexin knockout mice had, on average, 17 episodes during the 4 h videotaping period with a mean duration of about 1 min per episode. The duration of single episodes varied widely, however, both among and within the mice, with the shortest ending after only 6 s and the longest lasting more than 200 s. The predominant behaviors for the 5 s immediately preceding and the 10 s following each episode were feeding, drinking, ambulating, grooming, burrowing, and climbing. Interestingly, while burrowing and climbing were often observed before, they never occurred after an episode. Conversely, while feeding and drinking rarely preceded an attack they frequently occurred immediately afterwards. Clearly observed gait ataxia lasting for a few seconds preceded about 25 % of all observed episodes. Grossly observable motor activity with side-to-side rocking, though without any change in overall posture, frequently occurred several seconds after the start of the attack. We performed close-up video studies of individual narcoleptic mice filmed from various angles to explore in greater detail the postural changes and motor activity that had been noted during the behavioral arrests. The postural changes characteristically involved sudden collapse of the head and neck with simultaneous buckling of the limb, medially, and/or laterally from the body, causing the ventral surface to fall to the cage floor at angles 45° to perpendicular to the sagittal plane. Occasionally, the mouse fell completely onto its side. Gross motor activity during attacks always resulted from episodic limb twitching that rocked the mouse about its central axis. As we had noticed on the original pilot videotape recordings, bulbar motor activity typical of REM sleep was often seen in the close-up videos during attacks, with ear twitching, eye blinking, eye movements, and chewing movements sometimes observed.

We also studied the onset and development of these episodes in male homozygote littermates with the cohorts housed in groups of four or five per cage. While mice from one set were placed into individual cages for videotaping, the other set was videotaped in the group cage, to allow for social interaction. Videotaping was conducted at weekly intervals beginning at 3–6 weeks of age. The grouped mice exhibited more episodes: 55 versus 8 at 4 weeks of age and 64 versus 26 at 6 weeks of age. Interestingly, while all mice that were group housed had attacks by 4 weeks of age, this did not occur until 6 weeks of age in the individually filmed mice. In the group-housed mice, we frequently observed chasing, tail biting, and social grooming immediately before behavioral arrest episodes. Conversely, social interaction in the group-housed setting also appeared to cut short the episodes due to stimulation from littermates. Thus the average episode duration was 45 versus 57 s at 4 weeks and 45 versus 90 s at 6 weeks in the group- versus individually-housed mice, respectively. Categorization of posture during behavioral arrests at 6 weeks revealed that in 40 % of episodes, the young mice completely fell over to their side either in the group- or individually-housed settings. In another 40 % of episodes, the mice collapsed onto their ventral surface without gross motor activity, and only in 20 % of episodes did we see the "rocking" motor activity that was frequently observed in adults.

The discovery of narcolepsy in the orexin knockout mouse was quickly accepted because an independent group had concurrently found that the autosomal recessive form of canine narcolepsy was caused by a mutation in the one of the two orexin receptor genes (the orexin-2 receptor) (Lin et al. 1999). Together, these observations firmly identified orexins as having an important function in sleep regulation and showed that the interaction of orexin with the orexin-2 receptor in the brain was essential for the normal regulation of REM sleep (Siegel 1999).

Generation of the Narcoleptic Rat

In the human, mutations of the prepro-orexin or orexin receptor genes are extremely rare (Peyron et al. 2000). In most patients with typical narcolepsy-cataplexy, undetectable to very low levels of orexin-A neuropeptide have been detected in cerebrospinal fluid. Thus, it is likely that the number of orexin-expressing neurons is greatly diminished in narcoleptic patients (Peyron et al. 2000), though the cause for this apparent neuronal degeneration remains unknown. To mimic the postnatal loss of orexin neurons that occurs in narcoleptic patients, we generated transgenic mice in which orexin neurons expressed a cytotoxic gene product under the control of the human prepro-orexin promoter (Hara et al. 2001). Expression of this transgene resulted in the degeneration of orexin-containing neurons during development and a narcoleptic phenotype, very similar to that of the orexin null mouse as described above. We then adopted this transgenic approach to express the human orexin/ataxin-3

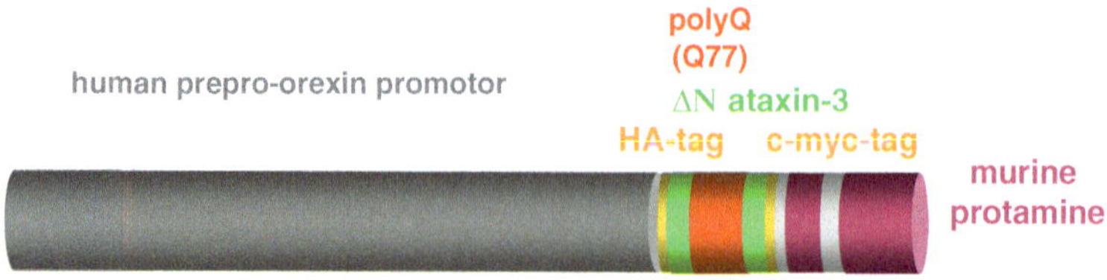

Fig. 23.8 Schematic representation of the transgene used for generating orexin/ataxin-3 rats. The orexin/ataxin-3 transgene expresses an N-terminally-truncated human ataxin-3 protein containing the toxic Q77 polyglutamine stretch under the control of the human prepro-orexin promoter. The C-terminus of the transgene contains a Myc epitope for histological examinations. The transgene was injected into pronuclei of fertilized Wistar rat eggs to generate founder animals, which were bred to generate the orexin/ataxin-3 transgenic lines

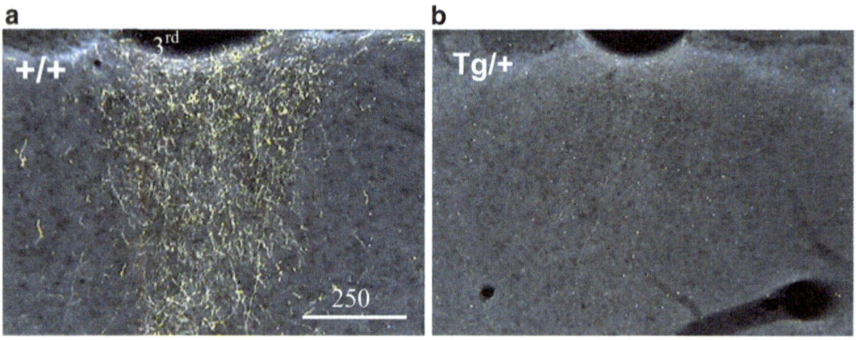

Fig. 23.9 The dense orexin-containing projections in the thalamic paraventricular nucleus of wild-type rats (+/+) were undetectable in the transgenic animals (Tg/+) at 17 weeks of age

transgene in rats (Fig. 23.8). In the transgenic animals we found that the orexin projections were essentially entirely lost (Fig. 23.9), and that these rats exhibited the narcoleptic phenotype (Beuckmann et al. 2004).

Similar to our results from the narcoleptic mice, orexin/ataxin-3 transgenic rats, when compared with wild-type rats, showed abnormalities in REM sleep as well as more fragmentation of sleep and wakefulness during the 12 h dark phase. This fragmentation was reflected by an increased number of wakefulness episodes (on average 90 versus 70 episodes/12 h in the wild-type rats). In the rat, we defined an episode of narcolepsy-cataplexy as an episode of REM sleep following wakefulness with less than 30 s of intervening non-REM sleep (Fig. 23.10). Rat vigilance state transitions tend to be less abrupt than in mice, and this definition allowed us to capture all these episodes in the narcoleptic rats and determine that their incidence was essentially zero in the wild-type rats. In fact, such episodes of behavioral arrest could be observed in all of the orexin/ataxin-3 transgenic rats, primarily during the dark phase (Fig. 23.11), whereas an episode was observed only once in a wild-type rat during a total of about 1,500 h of recording time. Orexin/ataxin-3 transgenic rats had about 4 episodes/12 h dark phase and less than 1 episode/12 h light phase. Hence, these episodes were less frequent than normal REM sleep events and also

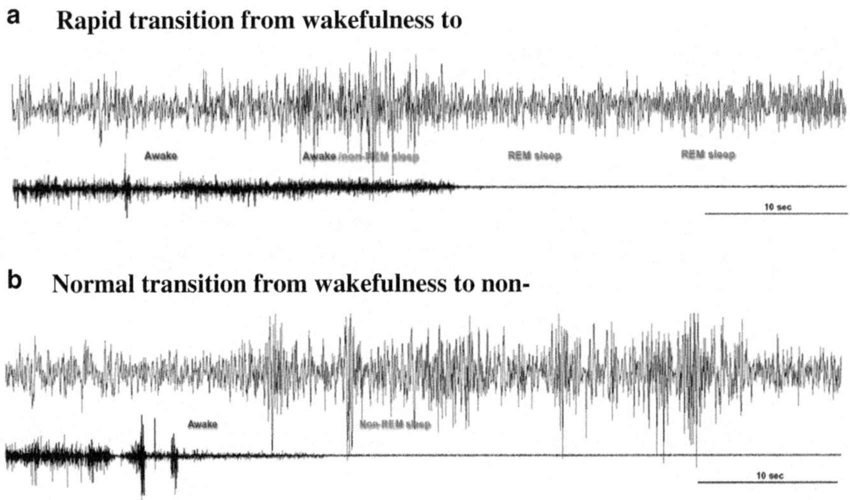

Fig. 23.10 Examples of EEG/EMG recordings from an orexin/ataxin-3 rat showing a rapid transition from wakefulness to REM sleep (**a**) and a normal progression from wakefulness to non-REM sleep (**b**) Note that in panel (**a**), about 10 s of non-REM sleep is visible on the EEG recording. This is accompanied by continuing EMG activity, before REM sleep is initiated

Fig. 23.11 Infrared videotape monitoring of a narcoleptic rat shows that an episode of behavioral arrest is initiated just as the rat starts to drink at the water bottle. The animal collapses and then suddenly resumes purposeful activity less than 60 s later

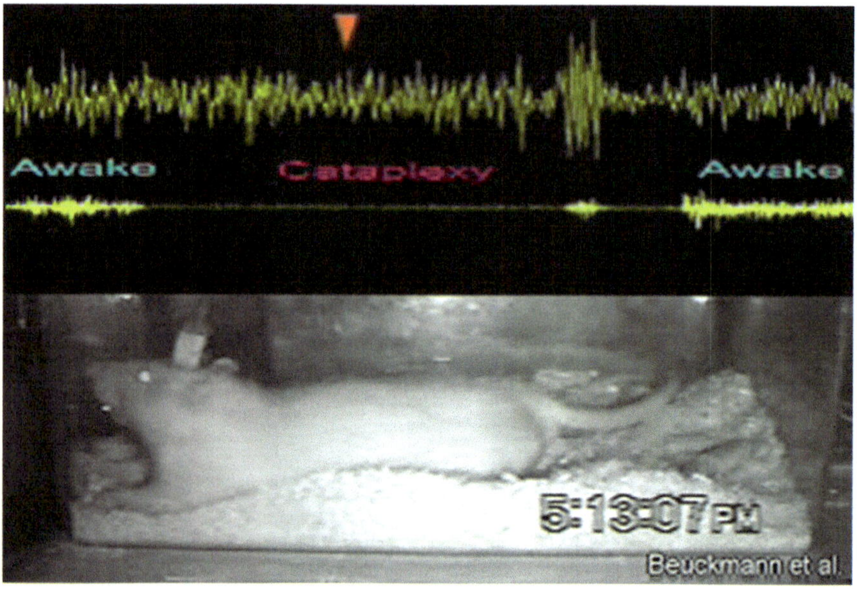

Fig. 23.12 A representative frame from the published video clip, which displays the behavior of the rat combined with a time stamp and the concurrent EEG/EMG recordings, annotated to indicate behavioral state. Here the rat, which is the same as shown in Fig. 23.11, demonstrates a cataplectic episode

less frequent than we had observed in the narcoleptic mice. However, once they started, their mean duration was not different from the duration of a normal REM sleep episode.

The behavior of these rats was also monitored for 4 h using the same infrared video recording we used with the mice. The interested reader can follow this link to view these video clips which were published as part of the first description of these rats (Beuckmann et al. 2004): http://www.jneurosci.org/cgi/content/full/24/18/4469/DC1?maxtoshow=&HITS=10&hits=10&RESULTFORMAT=&author1=beuckmann&searchid=1127338743006_10334&stored_search=&FIRSTINDEX=0&journalcode=jneuro.

We started the recording at the beginning of the dark phase, with simultaneous monitoring of the EEG and EMG signals (Fig. 23.12). Subsequent review of the video recordings showed that cataplectic episodes, as we had seen in the mouse, sometimes occurred during motivated behavior, such as ambulating or drinking. Each episode was associated with a sudden loss of muscle tone and concomitant loss of posture. However, the rocking movements along the body axis that we observed in orexin knockout mice were not present. Similar to the mice, each behavioral arrest in the transgenic rats was always abruptly terminated with the resumption of full mobility and purposeful behavior. Interestingly, in addition to the typical behavioral arrests associated with REM sleep episodes, another type of postural collapse was observed. These novel arrests could not be differentiated behaviorally from a REM sleep episode

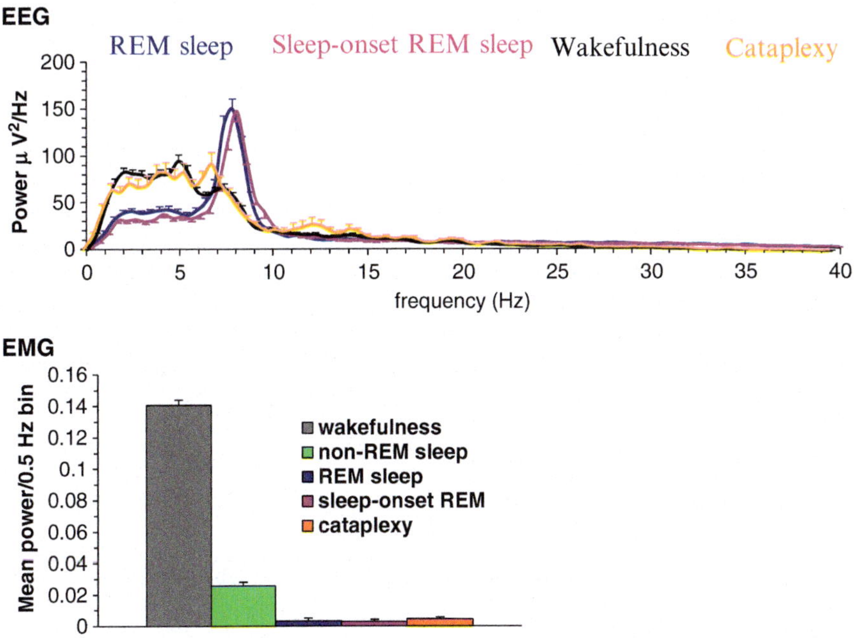

Fig. 23.13 Representative mean power spectra of the EEG and EMG signals from episodes of wakefulness, non-REM sleep, REM sleep, and cataplexy averaged across the orexin/ataxin-3 rats (N=8) after normalization. The EMG is the average integrated power for the same episodes. For the EEG, the spectra are displayed from 1 to 40 Hz. They show that episodes of cataplexy are characterized by muscle atonia combined with an EEG spectrum that is indistinguishable from that recorded during wakefulness

by video photography alone: the rat showed an identical sudden and complete loss of muscle tone and cessation of purposeful behavior. However, we used power spectral analysis of the corresponding EEG and EMG signals to show that orexin/ataxin-3 transgenic rats lost muscle tone, while the EEG maintained a frequency pattern identical to that recorded during normal wakefulness.

Our analysis of the power spectrum of the EEG signal used the standard method of fast Fourier transform (FFT) to calculate the frequency spectrum of the signal from 1 to 40 Hz in 1 Hz bins (Fig. 23.13). Those arrests that were characterized in this manner occurred both during the dark (about five episodes/12 h) and light (about three episodes/12 h) periods, but were generally shorter in duration than the behavioral arrest that began with REM sleep. As expected, muscle tone, when compared with either normal wakefulness or non-REM sleep, was essentially absent during these episodes of behavioral collapse with wakefulness-like EEG, similar to during REM sleep.

In contrast to human and canine cataplectic episodes, which are usually triggered by emotional stimuli (Aldrich 1992; Riehl et al. 1998), we could not reliably elicit behavioral arrests in the orexin/ataxin-3 transgenic rats using external emotive stimuli.

We have suggested that these episodes in the rat are equivalent to muscle atonia with consciousness in the human. Their existence therefore supports our hypothesis that in the mouse these episodes also occur but are very brief and difficult to separate from the REM sleep episodes that invariably and immediately follow in this species. Hence, narcolepsy is probably similar in all species, though the relative expression of cataplectic and narcoleptic episodes will vary. However, since we can more easily observe cataplexy separately from narcolepsy in the rat, the behavioral arrests are more concordant with the current terminology of narcolepsy-cataplexy in humans. We thus propose the use of the term cataplexy to describe behavioral arrests in the rat and narcolepsy-cataplexy to describe the condition. Clarification and potentially separate use of the terms narcolepsy and cataplexy may follow agreement and recognition of these episodes as transitional mixed states of wakefulness and REM sleep in humans (Hishikawa and Shimizu 1995), as we have found them to be in the rodent.

Summary and Conclusions

Studies continue to determine the functional role of orexin in normal animals, and results from the narcoleptic rats have been particularly important in this regard. We found that transgenic rats were less awake only during the dark, or active, period, and that this deficit in wakefulness was greater at the end of the period. This time corresponds to the timing, in humans, of maximal circadian alertness, which consolidates wakefulness at the end of the active part of the daily cycle when the homeostatic drive for sleepiness is highest (Dijk and Czeisler 1994). Our data thus support the proposal that orexin may contribute to the daily variation in alertness, particularly towards the end of the active period. We also found an apparently continuous effect of the loss of orexin on the expression of REM sleep, expressed as an increase in REM sleep throughout the normally active phase, though the REM sleep bout length remained unchanged. These results indicate that orexin has an inhibitory influence on the appearance of REM sleep, and we have hypothesized that another function of orexin is to inhibit the initiation of a REM sleep episode. This is probably most important when the brain state, including cortical activation, is most similar to REM sleep, as it is during active wakefulness. Hence, when orexin is not present, the switch into REM sleep from wakefulness is no longer inhibited and a direct transition of the narcoleptic episode becomes more likely.

Acknowledgements I am grateful for the contributions made by all who authored the studies described here and to those who advised us as this work proceeded. I particularly thank Rick Chemelli, Jon Willie, and Masashi Yanagisawa, who made these discoveries possible when they videotaped an orexin knockout mouse during the dark period and first observed what they thought might be a seizure.

References

Aldrich MS. Narcolepsy. Neurology. 1992;42:34–43.

Beuckmann CT, Sinton CM, Williams SC, Richardson JA, Hammer RE, Sakurai T, et al. Expression of a poly-glutamine-ataxin-3 transgene in orexin neurons induces narcolepsy-cataplexy in the rat. J Neurosci. 2004;24:4469–77.

Chemelli RM, Willie JT, Sinton CM, Elmquist JK, Scammell T, Lee C, et al. Narcolepsy in orexin knockout mice: molecular genetics of sleep regulation. Cell. 1999;98:437–51.

Choo KL, Guilleminault C. Narcolepsy and idiopathic hypersomnolence. Clin Chest Med. 1998; 19:169–81.

de Lecea L, Kilduff TS, Peyron C, Gao X, Foye PE, Danielson PE, et al. The hypocretins: hypothalamus-specific peptides with neuroexcitatory activity. Proc Natl Acad Sci U S A. 1998; 95:322–7.

Dijk DJ, Czeisler CA. Paradoxical timing of the circadian rhythm of sleep propensity serves to consolidate sleep and wakefulness in humans. Neurosci Lett. 1994;166:63–8.

Guilleminault C, Gelb M. Clinical aspects and features of cataplexy. Adv Neurol. 1995;67: 65–77.

Hara J, Beuckmann CT, Nambu T, Willie JT, Chemelli RM, Sinton CM, et al. Genetic ablation of orexin neurons in mice results in narcolepsy, hypophagia, and obesity. Neuron. 2001;30:345–54.

Hishikawa Y, Shimizu T. Physiology of rem sleep, cataplexy, and sleep paralysis. Adv Neurol. 1995;67:245–71.

Lin L, Faraco J, Li R, Kadotani H, Rogers W, Lin X, et al. The sleep disorder canine narcolepsy is caused by a mutation in the hypocretin (orexin) receptor 2 gene. Cell. 1999;98:365–76.

Peyron C, Faraco J, Rogers W, Ripley B, Overeem S, Charnay Y, et al. A mutation in a case of early onset narcolepsy and a generalized absence of hypocretin peptides in human narcoleptic brains. Nat Med. 2000;6:991–7.

Peyron C, Tighe DK, van den Pol AN, de Lecea L, Heller HC, Sutcliffe JG, et al. Neurons containing hypocretin (orexin) project to multiple neuronal systems. J Neurosci. 1998;18:9996–10015.

Riehl J, Nishino S, Cederberg R, Dement WC, Mignot E. Development of cataplexy in genetically narcoleptic dobermans. Exp Neurol. 1998;152:292–302.

Sakurai T, Amemiya A, Ishii M, Matsuzaki I, Chemelli RM, Tanaka H, et al. Orexins and orexin receptors: a family of hypothalamic neuropeptides and G protein-coupled receptors that regulate feeding behavior. Cell. 1998;92:1. page following 696.

Siegel JM. Narcolepsy: a key role for hypocretins (orexins). Cell. 1999;98:409–12.

Sinton CM, Valatx JL, Jouvet M. Increased sleep time in the offspring of caffeine-treated dams from two inbred strains of mice. Neurosci Lett. 1981;24:169–74.

Willie JT, Chemelli RM, Sinton CM, Tokita S, Williams SC, Kisanuki YY, et al. Distinct narcolepsy syndromes in orexin receptor-2 and orexin null mice: molecular genetic dissection of non-rem and rem sleep regulatory processes. Neuron. 2003;38:715–30.

Chapter 24
Hypoxia: Introduction of Mechanisms and Consequences

Juliana Cini Perry and Michael John Decker

Oxygen (O_2) is an important chemical element that represents approximately 21 % of the Earth's atmosphere. Until the end of the 1800s, science has known little about the effects of O_2 depletion in the body. This prompted researchers of the late nineteenth century to investigate the physiological effects of high altitude. In 1878, French physiologist Paul Bert provided the first scientific evidence that the lack of O_2 (hypoxia) caused an incomplete saturation of the blood leading to "undue hyperpnea with exercise, nausea, headaches, and great depression". Since that time, research and clinical observations have demonstrated that hypoxia, whether acute or chronic, causes certain predictable physiologic responses. These occur irrespective of whether hypoxia is induced by a pathological disease or by the environment, such as by exposure to high altitude.

Research into hypoxic mechanisms gained momentum in the early 1900s. Between 1921 and 1922, Joseph Barcroft led an expedition into the Andes to systematically collect and analyze data describing acute and chronic responses occurring during travel to high altitudes. During this expedition, at different altitudes, his group monitored alterations in ventilation, values of arterial carbon dioxide ($PaCO_2$), and arterial oxygenation (PaO_2). From these data, Barcroft observed that, at 14,000 ft, increased respiratory rate raised alveolar O_2 pressures 12–15 mm higher than would otherwise be expected.

In addition to respiratory parameters, the hematological response to high altitude hypoxia was also observed. Specifically, Barcroft and the other investigators were

J.C. Perry, Ph.D. (✉)
Technology Innovation, Rua Jaime Balão, 765, Curitiba, Paraná, CEP 80040-340, Brazil
e-mail: julianaciniperry@gmail.com

M.J. Decker, Ph.D., Diplomate
American Board of Sleep Medicine, School of Nursing, Case Western Reserve University, 10900 Euclid Ave, Cleveland Heights, OH 44106, USA
e-mail: Mdecker01@gmail.com

M.L. Andersen, S. Tufik (eds.), *Rodent Model as Tools in Ethical Biomedical Research*, DOI 10.1007/978-3-319-11578-8_24

Table 24.1 The relationship of gas values in arterial blood during acute vs. chronic hypoxia in humans

	Baseline	Acute hypoxia	Chronic hypoxia vompensated blood gas
pH	~7.4	>7.5	~7.5
PaO_2	>85 mmHg	<70 mmHg	~80 mmHg
$PaCO_2$	~40–45 mmHg	<35 mmHg	~32 mmHg
SaO_2	>93 %	<80 %	90 %
HCO_3^-	22–24 mEq/L	>22–24 mEq/L	16 mEq/L

PaO_2 partial oxygen pressure in arterial blood, *$PaCO_2$* partial carbon dioxide pressure in arterial blood, *SaO_2* arterial oxyhemoglobin saturation, *HCO_{3-}* arterial bicarbonate

subjected to blood analysis before departure from their homeland at sea level (baseline), then repeatedly during their exposure to progressively higher altitudes, and finally, during a gradual descent back to sea level. Of the analyses performed, the most intriguing was related to reticular cell counts. Barcroft reported that in contrast to high altitude natives, all members of the expedition exhibited an increase in reticular cells, blood cell precursors. The counts peaked in the first week of the subjects being at 14,000 ft and remained above normal during the entire stay. However, upon return to sea level, the number of reticular cells fell below baseline values. Barcroft showed that the number of these cells changes at different altitudes, and the cells were characterized as having a "blood breathing function" (Barcroft et al. 1923). Later, hemoglobin molecules were considered to be responsible for O_2 transportation.

In the decades following Barcroft's high altitude expedition through the Andes, much emphasis was placed on describing the respiratory and hematological responses of individuals and populations exposed to high altitude hypoxia. An early study demonstrated that acute exposure to high altitude hypoxia leads to hemoglobin O_2 desaturation, hyperventilation and decreased arterial carbon dioxide (CO_2) concentrations. Table 24.1 highlights general alterations in gas values of arterial blood induced by acute and chronic high altitude hypoxia.

In contrast with environmentally induced hypoxia, disease processes lead to a slightly different profile of arterial blood gas. The diagnosis and treatment of hypoxia are based on identifying the nature and frequency of gas exchange impairment (chronic vs. intermittent, induced either environmentally or pathologically). Some diseases lead to inadequate ventilation while others induce impairment of gas diffusion through the alveoli, resulting in reduced gas transport from the lungs to the tissues (Table 24.2). Insufficient oxygenation results in a reduction of O_2 concentration in the blood and a consequent deficiency of O_2 in the tissues. Thus, hypoxia is characterized by a reduction of O_2 concentration in arterial blood, a condition that greatly stresses all organs.

Table 24.2 Definitions for most commonly encountered types of hypoxia and clinical examples

Causes of hypoxia	Definitions	Clinical example
Hypoxemic hypoxia	Low concentrations of atmospheric O_2 (reduction in arterial pO_2)	High altitude travel
Anemic hypoxia	Low effective hemoglobin concentration	Blood loss, anemia
Shunt hypoxia	Reduction of ventilation-perfusion and gas diffusion in the alveoli	Arteriovenous malformation
		Hepatopulmonary syndrome
Cardiogenic hypoxia	Inadequate transport of O_2 by blood to the tissues due to a decrease in cardiac output	Congestive heart failure
Histotoxic hypoxia	Incapacity of the tissues to make use of O_2 (inhibition of cellular enzyme activity)	Arsenic
Hypermetabolic hypoxia	Increased ATP requirement or energy metabolism (hyperthermia)	High fever

pO_2 arterial pressure of oxygen, *ATP* Adenosine 5′-triphosphate

Hypoxia Induced by Sleep-Related Breathing Disorders in Humans

Obstructive Sleep Apnea

Sleep and breathing disorders are now seen as a major public health problem. Obstructive sleep apnea (OSA) affects 2–4 % of the adult population (i.e., apnea-hypopnea index >5 or more episodes per hour[1] and daytime sleepiness). Considering the apnea-hipopnea index 24 % of middle age males and 9 % of females have apneas during sleep. Additionally, a Brazilian community-based survey study used a probabilistic three-stage cluster sample of São Paulo has shown a high prevalence of sleep apnea.[2] OSA was observed in 32.8 % of the participants. This study is the first apnea survey of a large metropolitan area in South America, identifying a higher prevalence of OSA than previous epidemiological studies. This can be explained by the use of current techniques and clinical criteria, inclusion of older groups, and the higher prevalence of obesity in the studied population. As shown in Fig. 24.1, OSA is characterized by repetitive complete (apnea) or partial (hypopnea) obstruction of the upper airways resulting in pauses in breathing and subsequent O_2 desaturation. OSA is associated with adverse clinical consequences such as excessive daytime sleepiness, coronary artery disease, increased risk for stroke, diabetes, glucose intolerance, and academic and social under-achievements. The classic daytime manifestation of apnea is excessive sleepiness, but other symptoms such as cognitive

[1] In clinical practice, the severity of apnea is associated to the apnea-hypopnea index, defined by the number of such episodes per hour of sleep. The task force suggested apnea hypopnea index (AHI) cutpoints of 5, 15, and 30 events/hour to indicate mild, moderate, and severe levels of (Young et al. 2002).

[2] Tufik et al. (2010).

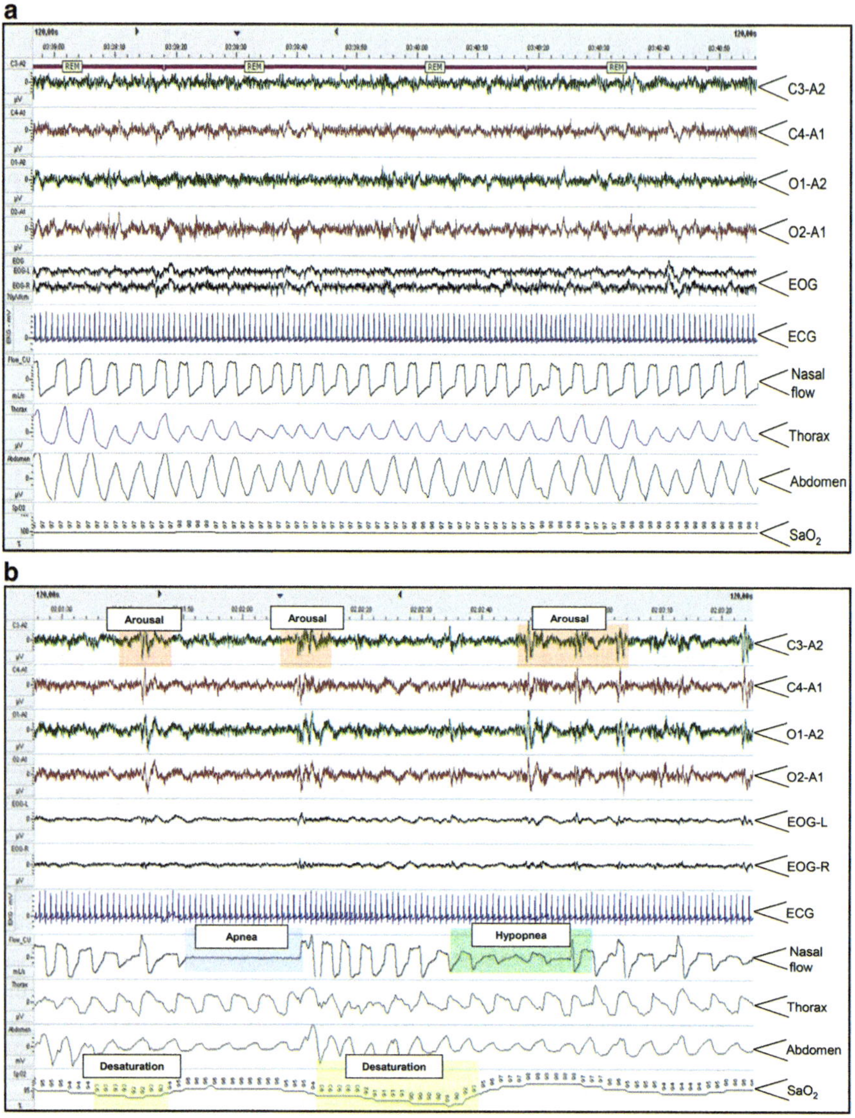

Fig. 24.1 Polysomnography in a normal subject (**a**) and an apneic patient (**b**) during 120 s. Episode of obstructive apnea and hypopnea occurring during REM sleep is demonstrated. Note the increasing ventilatory effort across the episode indicating its obstructive nature, arousal and desaturation. Definition of abbreviations: EEG (C3-A2; C4-A1; O1-A2; O2-A1): eletroencephalogram; *EOG* eletrooculogram, *ECG* eletrocardiogram, *SaO2* arterial oxyhemoglobin saturation. *Illustration*: Sleep Institute-AFIP

deficits and fatigue are commonly reported. Several neurobehavioral morbidities that greatly impact public health and the economy can be traced to obstructive sleep apnea. More direct effects of excessive daytime sleepiness are the accidents involving vehicles and labor. The situation is even more serious due to the high prevalence of obstructive sleep apnea among adults, who accounted for 800,000 obstructive sleep apnea-related motor-vehicle collisions that incurred an estimated cost of 3.4 billion dollars and claimed 1,400 lives in the year 2000 alone.[3]

The diagnosis of OSA is based on the combination of characteristic clinical features in addition to compatible findings on instrumental tests in which multiple physiologic signals are monitored simultaneously during a night of sleep. A full night polysomnography, conducted by a technologist in a sleep laboratory, is the gold standard for the diagnosis of suspected OSA. The polysomnography apparatus allows the simultaneous recording of neurophysiological and cardiorespiratory variables reflecting the quantity and quality of sleep achieved (Fig. 24.1).

Chronic Obstructive Pulmonary Disease

In Chronic Obstructive Pulmonary Disease (COPD) destruction of lung tissue leads to reduced alveolar gas exchange. COPD is characterized by extreme reduction of exhaling flow. This results in the inability to move O_2 from the environment into the blood as well as a reduction in the flow of CO_2 from the blood back into the environment. Unlike sleep apnea, hypoxia induced by lung disease leads to consistently high blood levels of CO_2 and acidosis. Hypoventilation causes the most important gas-exchange alteration in COPD patients, leading to hypercapnia (Box 1) and hypoxemia, especially during rapid-eye-movement sleep (REM), when marked respiratory muscle atonia occurs. OSA and COPD lead to the blood gas values depicted in Table 24.3.

Table 24.3 Comparative arterial blood gas values during OSA vs. COPD

	Baseline	During OSA event	COPD
pH	~7.40	<7.29	~7.32
PaO_2	>85 mmHg	<60 mmHg	55 mmHg
$PaCO_2$	~40–45 mmHg	>45 mmHg	70 mmHg
SaO_2	>93 %	<80 %	85 %
HCO_3^-	22–24 mEq/L	22–26 mEq/L	30 mEq/L

PaO_2 partial oxygen pressure in arterial blood, *$PaCO_2$* partial carbon dioxide pressure in arterial blood, *SaO_2* arterial oxyhemoglobin saturation, *HCO_{3-}* arterial bicarbonate

Hypercapnia is an excess of CO_2 in the blood. Carbon dioxide is a gaseous product of the body's metabolism and is normally expelled through the lungs. Upon first examination, it would seem that any respiratory condition that causes hypoxia would also cause hypercapnia. But hypercapnia in general only occurs in association with hypoventilation or by circulatory insufficiency.

[3] Sassani et al. (2004).

Animal Models

The use of animal models in the study of human conditions requires validation of the behavioral manifestation and an association with the major characteristics observed in the human situation. Mammalian models (e.g., the dog, pig, baboon and lamb) have been used to determine the effects of hypoxia. These models, however, have ethical limitations and incur high costs over time. Among the animal models proposed to investigate the effects of hypoxia, rodents are of particular interest. Rats and mice make for more than adequate subjects as their response to hypoxia parallels that seen in humans.

Anoxia: O_2 supply to the cell is assumed to be practically zero.
Ischemic-hypoxia: blood hypoperfusion or loss of blood flow to tissue (decrease in cardiac output) and a consequent reduction of O_2 supply to the cells.
Hypoxia: reduction in arterial pO_2 and a consequent deficiency of O_2 in the tissues.
Hypoxemia: reduction of O_2 specifically in the blood.
Hyperoxia: condition opposite to hypoxia, in which there is an excess of O_2 in body tissues or a higher than normal partial pressure of O_2.

Models of Anoxia

Approximately 2.9–9.0 infants per each 1,000 delivered experience some degree of ischemic-anoxic or prolonged anoxic insult. Perinatal brain damage is usually brought about by intrauterine asphyxia following an acute reduction of the uterine or umbilical circulation. The areas most heavily affected are the cerebral cortex and the basal ganglia. The fetus reacts to a severe lack of O_2 with activation of the sympathetic-adrenergic nervous system and a redistribution of cardiac output in favor of the central organs (brain, heart and adrenals). If the asphyxic insult persists, the fetus is unable to maintain circulatory centralization, and the cardiac output and extent of cerebral perfusion plunge.

Unlike hypoxia, which is a reduction in O_2, anoxia is defined as a complete absence of O_2. To induce perinatal anoxia in rodents, on the day of parturition, the dams are anesthetized and hysterectomized, and the isolated intact uterus is immediately immersed in a 37 °C saline bath for some minutes to induce anoxia. Following anoxia, the uterus horns are rapidly opened and the pups removed and stimulated to breathe. A variation to this approach requires that cesarean delivered pups are immediately placed for 15–20 min into a sealed chamber infused with 100 % nitrogen. In this model, the rodents are first submitted to hypoxia and later to anoxia.

Models of Ischemic-Hypoxia

In humans, disturbance of brain perfusion and oxygenation is a leading cause of perinatal brain damage. Ischemic-hypoxia insult during early fetal or neonatal stages leads to the damaging of immature neurons, resulting in behavioral and psychological dysfunctions, such as motor or learning disabilities, cerebral palsy or epilepsy. Ischemic-hypoxia models assess hypoxia associated with a reduction in blood flow (hypoperfusion). An example of such a model is the clamping of the umbilical cord in which the blood and O_2 supply are blocked in the near-term rat fetuses. One of the most frequently employed protocols combines ischemia, induced via unilateral carotid ligation, with exposure to hypoxia (<11 % atmospheric O_2) for a duration of several minutes to hours. The ischemia-hypoxia protocol described by Levine[4] and modified by Rice[5] and colleagues determines that the insult should be induced on the 7th–8th postnatal day in the pups. The animals are anesthetized and submitted to a unilateral carotid clamping. After recovery, the rats are exposed to 8 % O_2 in a humidified chamber at 36 °C for 1 or 2 h. Brain damage, seen histologically, is generally confined to the cerebral hemisphere ipsilateral to the arterial occlusion, and consists of selective neuronal death or infarction, depending on the duration of the systemic hypoxia. Unilateral ischemia-anoxia induces severe neuropathology, such as gross atrophy of the sensorimotor cortex, hippocampus, striatum, and thalamus.

Models of Hypoxia

Hypoxia Perinatal in Rodents

As a major cause of fetal brain damage with long lasting behavioral implications, hypoxia and hypoxia-ischemia have been comprehensively studied since the 1950s. In the development of an animal model of prenatal hypoxia in humans, the selection of the age at which the injury takes place is critical, as the susceptibility of neurons and glia changes considerably in the fetus and newborn.

In one experimental protocol,[6] postnatal rat pups were randomly assigned to either a (1) intermittent hypoxia-inducing chamber; (2) a normoxic chamber—exposed to only compressed air, or; (3) non-handled pups were left undisturbed with the dam. Once inside their respective chambers, rat pups were exposed to either intermittent hypoxia or normoxia during the subsequent 2 h. Rat pups were then removed from the chambers and returned to the dam, for 45 min. Pups were then returned to their

[4] Levine (1960).

[5] Rice et al. (1981).

[6] Decker et al. (2003, 2005).

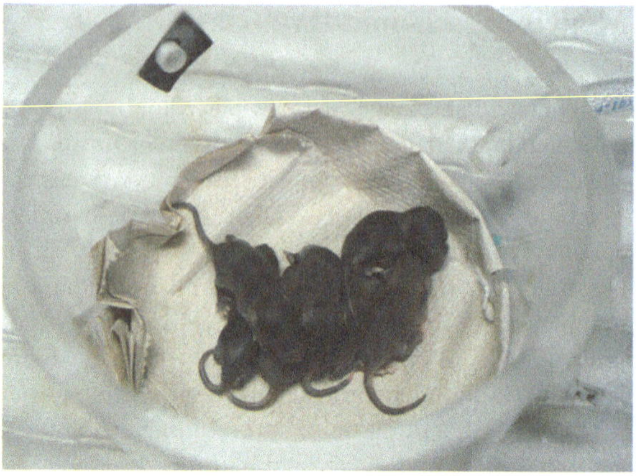

Fig. 24.2 Intermittent hypoxia insult (10 % O_2) induced on the 7th–11th postnatal day in the C57BL/6 J mice

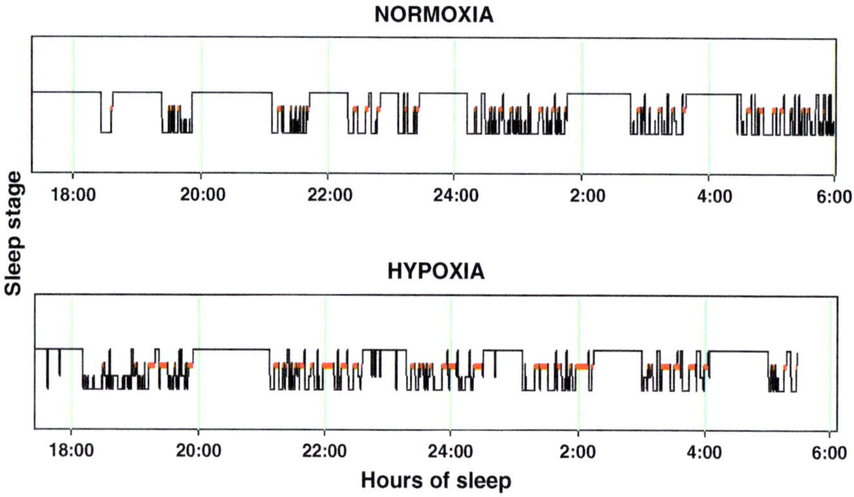

Fig. 24.3 The histograms show the sleep-wake architecture of post-hypoxia juvenile rats compared with normoxia. Neonatal hypoxia stimuli led to increased paradoxical sleep (*red*) and decreased wakefulness in juvenile rats measured during the dark phase of the circadian cycle periods (18:00–6:00)

respective chambers and again exposed to intermittent hypoxia or normoxia for an additional 2 h. Between postnatal 8 and 11 days, hypoxic pups were exposed to 2 h of intermittent hypoxic (or normoxic), followed by a 45 min feeding and grooming session, repeated three times each day (Fig. 24.2).

Behavioral consequences of neonatal intermittent hypoxia are impaired working memory, locomotor hyperactivity, and diminished levels of arousal. Post-hypoxic rats exhibited reductions in wakefulness and increased paradoxical sleep during the lights-off portion of the circadian cycle, as depicted in Fig. 24.3. Neurochemical consequences include reduced dopamine levels and hippocampal cell death.

Hypoxia in the Adult Rodent

Hypoxia is characterized by a reduction of O_2 concentration in arterial blood and a consequent deficiency of O_2 in the tissues. Sustained hypoxia protocols reproduce high altitude or chronic lung disease. The model that is commonly used to mimic sleep apnea is intermittent hypoxia. In this model, hypoxia insults lasting 10–90 s are alternated with similar periods of normoxia (Fig. 24.4).

Hypoxia is performed in a specially built chamber connected to a supply of O_2 and N_2 gas (Fig. 24.5). Inflow of O_2 and N_2 into the chamber is controlled by a computer program to produce sustained hypoxia (5–10 %) or intermittent hypoxia (cycles of room air to 5–10 % O_2).

Chronic sustained hypoxia increases wakefulness, reduces paradoxical sleep and induces slow wave sleep fragmentation. The sleep-wake architecture of rats under the influence of 15.5 % O_2 content was unchanged, compared with that of normoxic controls. One consequence of sustained hypoxia (10 % O_2) is reduced amplitude of the electrocardiogram and low percentages of slow wave and paradoxical sleep in rats. There is partial recovery of both slow wave and paradoxical sleep after 1–2 weeks of hypoxia, although none of the hypoxia values equal control values obtained before the exposure to low O_2. Studies using milder intermittent hypoxia insults have proven to be less disruptive to the sleep architecture. In contrast, the more severe intermittent hypoxia (5 % O_2) induced a subtle modification of slow wave sleep and severe and sustained paradoxical sleep deprivation during the light phase

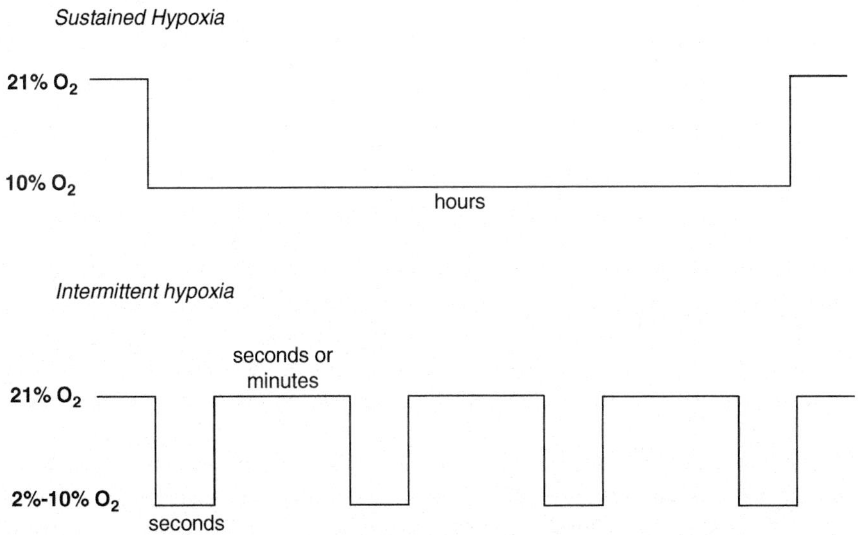

Fig. 24.4 Schematic design of sustained hypoxia and intermittent hypoxia protocols. In the sustained hypoxia episode the O_2 concentration was kept constant for a period of several hours, whereas the duration of each intermittent hypoxic episode lasted between 10 and 40 s, depending on the length of apneas, which are interspersed with normoxic periods that last several seconds or minutes

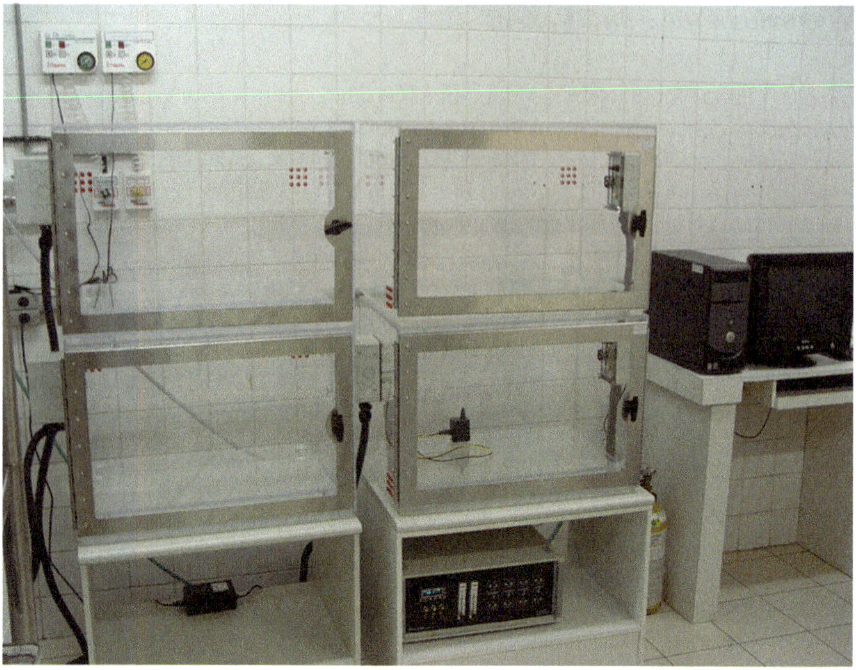

Fig. 24.5 Hypoxia chamber. Intermittent hypoxia is induced in a specially built chamber (30×20×20 in., Oxycycler model A44X0, Biospherix, Redfield, NY, USA) connected to a supply of O_2 and N_2 gas. Sensors measured O_2 concentration, CO_2 concentration (<0.01 %), humidity (40–50 %) and temperature (22–24 °C). Inflow of O_2 and N_2 into the chamber are controlled by a computer programmed to produce cycles of minutes room air to seconds 2–10 % O_2

and a substantial paradoxical sleep rebound during the dark phase in C57BL/6 J mice. The most sophisticated models have utilized either airway obstruction or the delivery of hypoxic gases with the onset of sleep and subsequent removal of the stimulus when arousal or wakefulness occurs. With hypoxia stimuli, wakefulness episodes were longer and more frequent, and paradoxical sleep was decreased and slow wave sleep episodes were shorter and more frequent. An important outcome in the comparison of sleep-wake patterns in sustained hypoxia vs. intermittent hypoxia stimuli during sleep is that the effects of both conditions on sleep were qualitatively similar. However, the effects of sustained hypoxia were of larger magnitude in slow wave sleep.

It has been proposed that hypoxia and sleep fragmentation are implicated in cardiovascular risk associated to OSA. Animal models have revealed that intermittent hypoxia is the critical stimulus underlying development of increased sympathetic activity and hypertension. These alterations can be related to several different components including augmented sympathetic nerve activity, altered function of arterial chemoreceptors, and elevated norepinephrine and dopamine concentrations.

The brain is particularly vulnerable to the effects of hypoxia, which produces extensive neuronal damage in selected regions. Intermittent hypoxia may cause an extensive lesion to susceptible brain neurons that does not always result in neuronal death, but rather is followed by a marked impairment of brain functioning, e.g., various behaviors, stress response, learning and memory. When intermittent hypoxia occurs during a critical period of brain development, it disrupts the functional integrity of the dopaminergic system and induces substantial cognitive and behavioral alterations. In adult rats, exposure to intermittent hypoxia has been associated with alterations of monoamine concentrations in the brain. Hypoxia can reduce cerebral perfusion and damage specific subsets of neurons in the cortex, basal ganglia and hipoccampus. This occurs through neurophysiological alterations in the function of ion channels, O_2 sensors, signaling pathways, neuromodulators, induction of apoptosis, or oxidative stress. In general, most neurons respond to hypoxia by decreasing their metabolic demand. Because the brain has limited O_2 reserves and a limited ability to utilize anaerobic processes, most neurons reduce their metabolic requirements by decreasing their activity.

The Canine Model

Basic research on sleep apnea using experimental animals may help to further the understanding and prevention of OSA. To establish a natural model of sleep-disordered breathing, Hendricks and colleagues[7] investigated respiration during wakefulness and sleep in the English bulldog. This breed is characterized by an abnormal upper airway anatomy, with enlargement of the soft palate and narrowing of the oropharynx. During sleep, the animals had disordered respiration and episodes of O_2 desaturation (<90 % for prolonged durations). In REM sleep, the bulldogs had episodes of both central and obstructive apnea, the latter being associated with REM movements of the rib cage and abdomen. During wakefulness, the bulldogs were hypersomnolent as evidenced by shortened sleep latency.

Yet another canine OSA model was described by Kimoff and colleagues.[8] Healthy adult dogs were prepared with a tracheostomy and with implanted electroencephalographic and nuchal electromyographic recording electrodes. A silent occlusion valve was attached to the outer end of the endotracheal tube. The electroencephalogram and electromyogram were monitored continuously by a computer that determined the leep-wake state. At a predetermined time after each sleep onset, a signal was transmitted from the computer to the valve controller, resulting in airway occlusion. When the dog aroused from sleep, the occlusion was released. The telemetry unit, measurements of ventilatory and arousal responses were obtained during daytime sleep. This canine model has been shown to reproduce the characteristic apnoea and hypersomnolence of human OSA. The advantage of this model is that sleep during long-term OSA can be compared with both normal sleep before OSA and recovery sleep after OSA.

[7] Hendricks et al. (1987).

[8] Kimoff et al. (1994).

Potential Pitfalls

Isocapnic and Arterial Blood Acidosis

The methodology employed in several reports utilizes an isocapnic, hypoxia inducing gas mixture of 10 % O_2, 3–5 % CO_2, and balance nitrogen. Some authors add CO_2 to maintain an atmospheric CO_2 pressure near 38 Torr,[9] approximating that of the arterial blood. Thus, during exposure to the isocapnic hypoxic gas mixture, as the rodent becomes hypoxemic and increases minute ventilation, the addition of CO_2 to the gas mixture sustains arterial CO_2 concentrations within the normative range of 35–45 Torr. The net result is a relatively pure hypoxic insult without corresponding hypo- or hypercapnic-induced derangement in acid–base physiology. During an apneic event, cessation of ventilation results in decreased alveolar ventilation. This increases arterial CO_2 levels and induces arterial blood acidosis. While isocapnic hypoxia minimizes potential physiologic confounds induced by altered blood CO_2 levels, it does not induce the same derangement in acid–base physiology that typically accompanies apneic events. Thus, isocapnic hypoxia has the advantage of providing insight into the pathogenic effect of hypoxia alone.

Hypocapnic and Arterial Blood Alkalosis

During the first hours of exposure to low O_2 (10 %) without a CO_2 mixture there are increases in respiration rate and pH, O_2 saturation falls to about 50 %, and decreases of pCO_2 occur, as shown in Fig. 24.6. Hypocapnia is a state in which the level of CO_2 in the

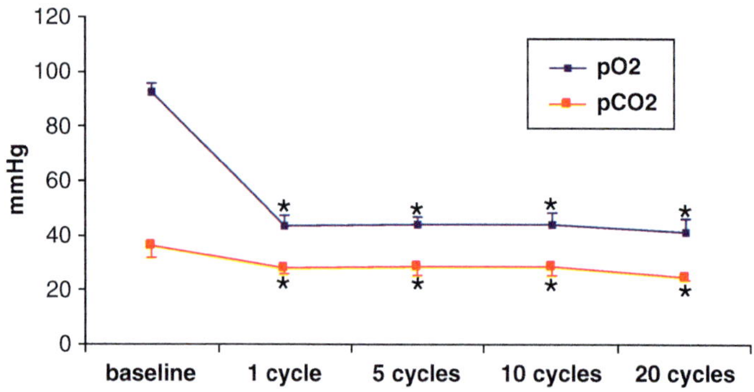

Fig. 24.6 Mean of pO_2 and pCO_2 (mmHg) in arterial blood samples collected before intermittent hypoxia and at the end of the 1st, 5th, 10th and 20th cycle of hypoxia in rats. Each cycle consisted of 2 min with 10 % O_2 followed by 2 min with 20.8 % O_2. One way ANOVA with repeated measurement followed post hoc Newman-Keuls test. $*p<0.05$. (Perry et al. 2007)

[9]The symbol Torr is a non-international system of units of pressure (1 Torr = 1 mmHg = 9.337×10^{-3} psi).

blood is lower than normal. This can result from deep or rapid breathing, known as hyperventilation. The hyperventilation of pure hypoxia (without a CO_2 mixture) is accompanied by respiratory alkalosis. The effects of hypoxia on sleep are linked to O_2 pressure rather than to secondary effects related to hypoxia and respiratory alkalosis. Furthermore, the increase of blood pressure in response to 30 days of intermittent hypoxia is most likely due to hypoxia per se rather than the addition of CO_2. However, following several weeks of chronic hypoxic exposure, the hyperventilation gradually diminishes and CO_2 levels stabilize, as was first described by Barcroft in 1921.

Future Directions

Obstructive sleep apnea is associated with fragmentation of sleep due to the repeated occurrence of end-apneic arousal throughout the night. Arousals are provoked by stimuli generated during upper airway obstruction. Mechanoreceptor stimuli produced during obstructed inspiratory efforts appear to play a major role in mediating the end-apneic arousal response. The sleep disruption resulting from repeated arousals plays a major role in the pathogenesis of most of the consequences of OSA (i.e., neuropsychiatric, respiratory, and cardiovascular) and may contribute to the progression of OSA severity. However, further studies are required to elucidate the relative contribution of sleep fragmentation versus hypoxia in producing these complications and the precise mechanisms involved. Isolated effects of hypoxia and sleep fragmentation in the animal models have been subject to extensive research. Investigating whether associations between hypoxia and sleep fragmentation exert any influence on cognitive function, or on neurochemical and cardiovascular alterations, may be useful in the construction of a more complete model to assess the prevalent syndrome in humans.

Final Considerations

Most sleep disorders result from complex interactions between genes and the environment. Modern molecular techniques are increasingly applied to determine the contribution of genes to sleep and its disorders. Studies proposed in this application will also advance our understanding of the extent to which phenotype and genotype predetermine susceptibility to intermittent apneas and hypoxia-induced cardiovascular and cognitive morbidity. Investigation of the impact produced by hypoxia in animals models contributes to the knowledge of how poorly oxygenated tissue affects the systems. Despite such effects, the cells and tissue may harbor the capacity to adapt to a limited O_2 supply and to unleash a variety of mechanisms to maintain brain integrity and functions under pathophysiological conditions. Knowledge of factors like underlying genotypic traits is necessary to not only confer vulnerability or resistance to intermittent hypoxia, but also to modulate responsiveness to therapeutic interventions aimed at restoring respiratory, cardiovascular, cognitive and metabolic dysfunction following intermittent hypoxic insults.

References

American Academy of Sleep Medicine. Sleep-related breathing disorders in adults: recommendations for syndrome definition and measurement techniques in clinical research. Sleep. 1999;22: 667–89.

American Academy of Sleep Medicine ICSD 2. Internacional Classification of Sleep Disorders. Diagnostic and coding manual. 2nd ed. Westchester: American Academy of Sleep Medicine; 2005.

Bao G, Randhawa PM, Fletcher EC. Acute blood pressure elevation during repetitive hypocapnic and eucapnic hypoxia in rats. J Appl Physiol. 1997;82:1071–8.

Barcroft J. The effect of altitude on the dissociation curve of the blood. J Physiol. 1911; XLII:44–63.

Barcroft J, Binger CA, Bock AV, Doggart JH, Forbes HS, Harrop G, et al. Observations upon the effect of high altitude on the physiological processes of the human body, carried out in the peruvian Andes, chiefly at Cerro de Pasco. Philos Trans R Soc Lond Ser B. 1923;211:351–480.

Bittencourt LRA, Marson O, Nery LE, Tufik N. Complicações cardiovasculares da síndrome da apnéia do sono obstrutiva. J Pneumol. 1998;24:311–6.

Bittencourt LR, Suchecki D, Tufik S, Peres C, Togeiro SM, Bagnato MC, et al. The variability of the apnoea-hypopnoea index. J Sleep Res. 2001;10:245–51.

Bittencourt LRA, Poyares D, Tufik S. Hipertensão arterial sistêmica e síndrome da apnéia e hipopnéia obstrutiva do sono: aspectos fisiopatológicos. Hipertensão. 2003;6:86–90.

Bittencourt LR, Silva RS, Santos RF, Pires ML, Mello MT. Excessive daytime sleepiness. Rev Bras Psiquiatr. 2005;27:16–21.

Decker MJ, Hue GE, Caudle WM, Miller GW, Keating GL, Rye DB. Episodic neonatal hypoxia evokes executive dysfunction and regionally specific alterations in markers of dopamine signaling. Neuroscience. 2003;117:417–25.

Decker MJ, Jones KA, Solomon IG, Keating GL, Rye DB. Reduced extracellular dopamine and increased responsiveness to novelty: neurochemical and behavioral sequelae of intermittent hypoxia. Sleep. 2005;28:169–76.

Findley LJ, Levinson MP, Bonnie RJ. Driving performance and automobile accidents in patients with sleep-apnea. Clin Chest Med. 1992;13:427–35.

Fletcher EC. Invited review: physiological consequences of intermittent hypoxia: systemic blood pressure. J Appl Physiol. 2001;90:1600–5.

George CF, Nickerson PW, Hanly PJ, Millar TW, Kryger MH. Sleep apnoea patients have more automobile accidents. Lancet. 1987;2:447.

Gozal D, Daniel JM, Dohanich GP. Behavioral and anatomical correlates of chronic episodic hypoxia during sleep in the rat. J Neurosci. 2001;21:2442–50.

Hamrahi H, Stephenson R, Mahamed S, Liao KS, Horner RL. Regulation of sleep-wake states in response to intermittent hypoxic stimuli applied only in sleep. J Appl Physiol. 2001;90: 2490–501.

Hendricks JC, Kline LR, Kovalski RJ, O'Brien JA, Morrison AR, Pack AI. The English bulldog: a natural model of sleep-disordered breathing. J Appl Physiol. 1987;63:1344–50.

Ip MSM, Lam B, Ng MMT, Lam WK, Tsang KWT, Lam KSL. Obstructive sleep apnea is independently associated with insulin resistance. Am J Respir Crit Care Med. 2002;165:670–6.

Kellogg RH. "La Pression barometrique ": Paul Bert's hypoxia theory and its critics. Respir Physiol. 1978;34:1–28.

Kimoff RJ, Makino H, Horner RL, Kozar LF, Lue F, Slutsky AS, et al. Canine model of obstructive sleep apnea: model description and preliminary application. J Appl Physiol. 1994;76:1810–7.

Laszy J, Sarkadi A. Hypoxia-induced sleep disturbance in rats. Sleep. 1990;13:205–17.

Lavie P, Herer P, Hoffstein V. Obstructive sleep apnoea syndrome as a risk factor for hypertension: population study. Br Med J. 2000;320:479–82.

Leger D. The cost of sleep-related accidents: a report for the National Commission on Sleep Disorders Research. Sleep. 1994;17:84–93.

Levine S. Anoxia-ischemic encephalopathy in rats. Am J Pathol. 1960;36:1–17.

Littner MR, Kushida C, Wise M, Davila DG, Morgenthaler T, Lee-Chiong T, et al. Standards of Practice Committee of the American Academy of Sleep Medicine. Practice parameters for clinical use of the multiple sleep latency test and the maintenance of wakefulness test. Sleep. 2005;28:113–21.

Mello MT, Santana MG, Souza LM, Oliveira PC, Ventura ML, Stampi C, et al. Sleep patterns and sleep-related complaints of Brazilian interstate bus drivers. Braz J Med Biol Res. 2000;33:71–7.

Miwa S, Fujiwara M, Inoue M. Effects of hypoxia on the activities of noradrenergic and dopaminergic neurons in the rat brain. J Neurochem. 1986;47:63–9.

Naegele B, Launois SH, Mazza S, Feuerstein C, Pepin JL, Levy P. Which memory processes are affected in patients with obstructive sleep apnea? An evaluation of 3 types of memory. Sleep. 2006;29:533–44.

Nieto FJ, Young TB, Lind BK, Shahar E, Samet JM, Redline S, et al. Association of sleep-disordered breathing, sleep apnea, and hypertension in a large community-based study. JAMA. 2000;283:1829–36.

Pappenheimer JR. Sleep and respiration of rats during hypoxia. J Physiol. 1977;266:191–207.

Pappenheimer JR. Hypoxia insomnia: effects of carbon monoxide and acclimatization. J Appl Physiol. 1984;57:1696–703.

Peppard PE, Young T, Palta M, Skatrud J. Prospective study of the association between sleep-disordered breathing and hypertension. N Engl J Med. 2000;342:1378–84.

Perry JC, D'Almeida V, Souza FG, Schoorlemmer GHM, Colombari E, Tufik S. Consequences of subchronic and chronic exposure to intermittent hypoxia and sleep deprivation on cardiovascular risk factors in rats. Respir Physiol Neurobiol. 2007;156:250–8.

Perry JC, D'Almeida V, Antunes IB, Tufik S. Distinct behavioral and neurochemical alterations induced by intermittent hypoxia or paradoxical sleep deprivation in rats. Prog Neuropsychopharmacol Biol Psychiatry. 2008a;32:87–94.

Perry JC, D'Almeida V, Lima MMS, Godoi F, Vital MABF, Oliveira MGM, et al. Intermittent hypoxia and sleep restriction associations: motor, cognition and neurochemical alterations. Behav Brain Res. 2008b;189:373–80.

Polotsky VY, Rubin AE, Balbir A, Dean T, Smith PL, Schwartz AR, et al. Intermittent hypoxia causes REM sleep deficits and decreases EEG delta power in NREM sleep in the C57BL/6J mouse. Sleep Med. 2006;7:7–16.

Punjabi NM, Sorkin JD, Katzel LI, Goldberg AP, Schwartz AR, Smith PL. Sleep-disordered breathing and insulin resistance in middle-aged and overweight men. Am J Respir Crit Care Med. 2002;165:677–82.

Rice JE, Vannucci RC, Brierley JB. The influence of immaturity on hypoxia-ischemic brain damage in the rat. Ann Neurol. 1981;9:131–41.

Roubein IF, Embree LJ, Jackson DW, Ordway FS. The effect of hypoxia on monoamine levels in discrete regions of aged rat brain. Neurobiol Aging. 1981;2:37–40.

Santos EH, de Mello MT, Pradella-Hallinan M, Luchesi L, Pires ML, Tufik S. Sleep and sleepiness among Brazilian shift-working bus drivers. Chronobiol Int. 2004;21:881–8.

Sassani A, Findley LJ, Kryger M, Goldlust E, George C, Davidson TM. Reducing motor-vehicle collisions, costs, and fatalities by treating obstructive sleep apnea syndrome. Sleep. 2004;27:453–8.

Tuor UI, Del Bigio MR, Chumas PD. Brain damage due to cerebral hypoxia/ischemia in the neonate: pathology and pharmacological modification. Cerebrovasc Brain Metab Rev. 1996;8: 159–93.

Veasey SC, Davis CW, Fenik P, Zhan G, Hsu YJ, Pratico D, et al. Long-term intermittent hypoxia in mice: protracted hypersomnolence with oxidative injury to sleep-wake brain regions. Sleep. 2004;27:194–201.

Young T, Palta M, Dempsey J, Skatrud J, Weber S, Badr S. The occurrence of sleep-disordered breathing among middle-aged adults. N Engl J Med. 1993;328:1230–5.

Young T, Peppard PE, Gottlieb DJ. Epidemiology of obstructive sleep apnea—a population health perspective. Am J Resp Crit Care. 2002;165:1217–39.

Chapter 25
Assessment of Motor Function in Rodents: Behavioral Models Sharing Simplicity and Multifaceted Applicability

Part 1: The Open-Field Test

Roberto Frussa-Filho, Daniela Fukue Fukushiro, Camilla de Lima Patti, Eduardo Ary Villela Marinho, Sonia Regina Kameda, and Rita de Cassia Carvalho

The whole is more than the sum of its parts. The profound complexity of biological beings cannot be reproduced in either the test tube or culture dish. Neuroscience research, in particular, critically depends on behavioral experimentation. However, while ethical and technical issues preclude the use of human subjects for many important experiments, a number of factors also limit the widespread use of non-human primates (see Jinnah et al. 2005). Thus, the majority of research on behavioral animal models has focused on other species, particularly small rodents such as rats or mice.

Within this context, the evaluation of rodents' motor activity is crucial for the correct interpretation of behavioral models of memory, anxiety, depression, aggression, addiction, etc. Conversely, emotionality, memory, and other "mental states"

R. Frussa-Filho (In Memoriam) • S.R. Kameda
Department of Pharmacology, Universidade Federal de São Paulo (UNIFESP), São Paulo, São Paulo, Brazil

D.F. Fukushiro, Ph.D. (✉)
Department of Pharmacology, Universidade Federal de São Paulo (UNIFESP), Rua Botucatu, 862, Ed. Leal Prado, 1° andar, São Paulo, São Paulo 04023062, Brazil
e-mail: danifukushiro@gmail.com

C.L. Patti, Ph.D.
Instituto de Genética e Erros Inatos do Metabolismo, Universidade Federal de São Paulo (UNIFESP), São Paulo, São Paulo, Brazil

E.A.V. Marinho
Departamento de Ciências da Saúde, Universidade Estadual de Santa Cruz, Ilhéus, Bahia, Brazil

R.d.C. Carvalho
Department of Pharmacology, Universidade Federal de São Paulo (UNIFESP), São Paulo, São Paulo, Brazil

This chapter is dedicated to Dr. João Palermo-Neto, who initiated Dr. Frussa-Filho and many other Brazilian neuroscientists into the exciting world of the open-field model.

M.L. Andersen, S. Tufik (eds.), *Rodent Model as Tools in Ethical Biomedical Research*, DOI 10.1007/978-3-319-11578-8_25

can dramatically modify specific animal models of movement disorders. Indeed, the intricate circular connection between mental states and motor function is splendidly defined by Walle Nauta's question: "What is a thought except a movement that is not connected to a motor neuron?"

Assessment of motor activity and movement disorders in rodents may, therefore, lead to an unlimited source of scientific knowledge. The aim of the following chapters is not to exhaustedly describe and discuss either the multiple methods available for assessing rodents' general activity (see Pierce and Kalivas 1997 for this purpose) or rodents' models of specific movement disorders (see Jinnah et al. 2005). These three chapters will critically discuss three rodent models of motor function that share two marked characteristics: simplicity and wide applicability. All of the to-be-discussed models (open-field, catalepsy, and orofacial dyskinesia) are simple to perform and each respective apparatus can be bought by or constructed in any laboratory in the world. From another point of view, while the first two models (open-field test—Chap. 26 and catalepsy test—Chap. 27) can be used to investigate several neuroscientific parameters beyond motor function, the last model to be discussed (orofacial dyskinesia—Chap. 28) evaluates a movement disorder shared by many different neurological diseases. These are ideal models for researchers who believe that science is built from ideas, not just money and technology.

The Open-Field Test

History

The open-field test was originally described by Hall (1934) for the study of emotionality in rats. As reviewed by Prut and Belzung (2003), the procedure consists of subjecting an animal, usually a rodent, to an environment from which escape is prevented by surrounding walls. Hall's apparatus consisted of a brightly illuminated circular arena of about 1.2 m diameter closed by a wall of 0.45 m high. Hall used defecation and urination in the above-described open-field to measure individual differences in rats' emotionality. Thenceforth, the use of this test has diversified enormously, differing, for example, in shape of the environment (circular, square, or rectangular), lighting, subjects, and mainly in the behavioral parameters quantified (see Walsh and Cummins 1976; Belzung 1999; Prut and Belzung 2003 for review). This concern notwithstanding, during the last 70 years, the open-field test has become one of the most (likely the most) widely used test in neuropsychobiology and neuropsychopharmacology. Indeed, even nowadays, studies using this very simple method have been published in the best periodicals of biological psychiatry (Abílio et al. 1999, for example), neurobiology of aging (Silva et al. 1996, for example), and neuropsychopharmacology (Chinen et al. 2006, for example).

As pointed out by Walsh and Cummins (1976), the popularity of the open-field test likely stems from the simplicity of the apparatus, the easy and rapid measurement of clearly defined behaviors, and a generally accepted interpretation of these

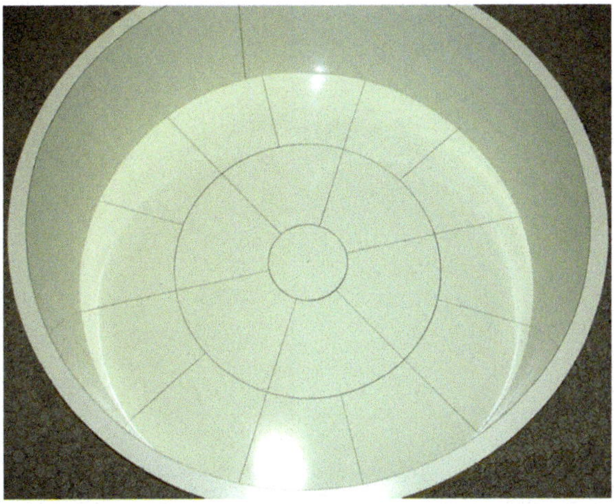

Fig. 25.1 The open-field apparatus used for rats

behaviors. In addition, certain of the measured behaviors are sensitive to a wide range of genetic, experiential, physiological, and pharmacological manipulations.

Methodology

As mentioned above, the open-field test has undergone enormous methodology diversification, the description and discussion of which is not the scope of this chapter (see Walsh and Cummins 1976; Eilam 2003; Lipkind et al. 2004 for this purpose). Thus, for the sake of illustration and practicality, a specific open-field methodology will be described here. This open-field methodology is the one used by our research group and many of the examples that we will present in this chapter on the multiple potential applications of the test will be related to this specific open-field methodology.

The apparatus consists of a wood circular arena of 97 cm in diameter closed by a wall of 32.5 cm high (rats) or 40 cm in diameter and 50 cm high (mice[1]) with an open top and floor divided into 19 squares (Figs. 25.1 and 25.2). Animals are placed individually[2] in the center of the arena and hand-operated counters and stopwatches are employed to score the following behavioral parameters, typically during the span of 5 min:

- Total ambulation frequency = number of any floor units entered;
- Peripheral square ambulation frequency = number of entrances into the floor units near the walls of the apparatus;
- Central square ambulation frequency = number of entrances into any floor unit not near the walls of the apparatus;

[1]Although paradoxical, the wall of the apparatus used for mice is higher because at times, they jump in an attempt to escape.

[2]Before (gently) handling the animals, the experimenter must always "wash" his/her hands with food pellets.

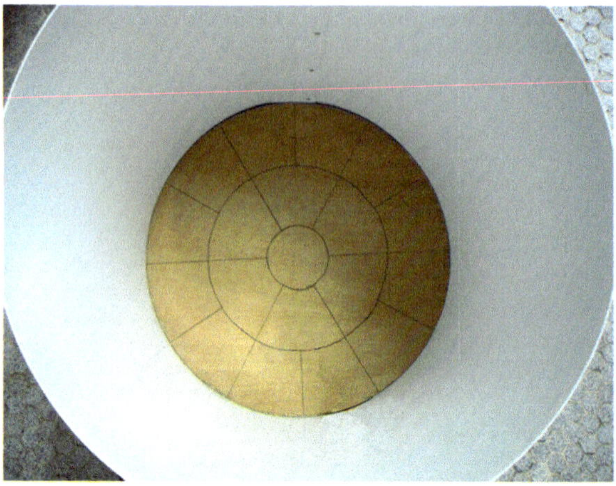

Fig. 25.2 The open-field apparatus used for mice

- Rearing frequency = number of times the animal lifts both of its forefeet from the floor. When rearing along the walls of the arena, the animal typically leans against the walls;
- Immobility duration = total seconds of lack of movement;
- Grooming duration = total seconds of rapid cleaning movements of the forefeet towards the face and/or the body. A typical complete grooming bout starts with the animal scratching its face, progressively moving down along the body, and terminating with the tip of the tail. Both complete and non-complete bouts (interrupted at some point along the body) are counted as grooming behavior.

Figures 25.3, 25.4, and 25.5 illustrate the different open-field parameters described above of both rats and mice.

The open-field apparatus must be cleaned with alcohol-water 5.0 % before placement of the animals to obviate possible bias due to clues left by previous subjects. To minimize possible effects of circadian changes on open-field behavior, experimental and control observations must be alternated, each animal being tested at the same time in each session.

Application: Evaluation of Motor Activity

Drug-Induced Inhibition of Motor Activity

The open-field test has been extensively used to evaluate motor activity. For example, in rats and mice, it has accurately detected the inhibitory motor effects of different neuroleptic drugs, such as haloperidol (Bernardi and Palermo-Neto 1979; Bernardi et al. 1981; Vital et al. 1995; Conceição and Frussa-Filho 1996; Frussa-Filho et al. 1997; Queiroz and Frussa-Filho 1997; Abílio et al. 1999, 2003b; Perry et al. 2004), sulpiride (Frussa-Filho and Palermo-Neto 1990), and droperidol (Frussa-Filho and Palermo-Neto 1991). While all of the neuroleptic drugs block D_2 dopamine receptors

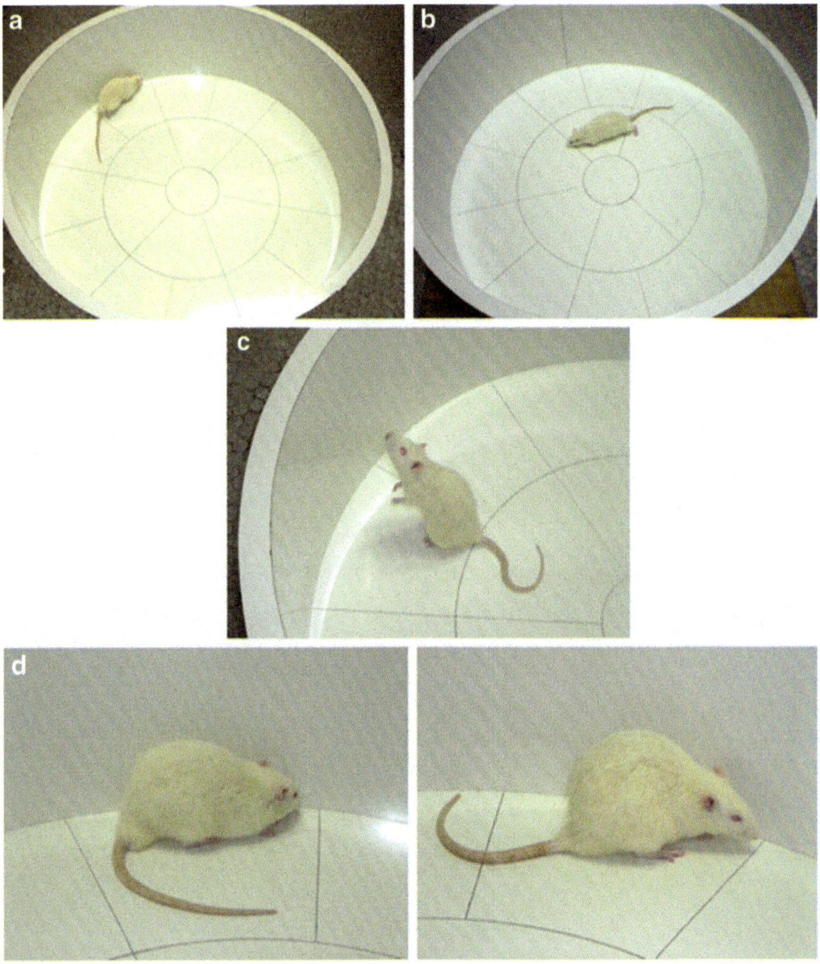

Fig. 25.3 Peripheral locomotion (**a**), central locomotion (**b**), rearing (**c**), and immobility (**d**) of a rat in the open-field

to some degree, dopaminergic transmission in both the dorsal (caudate and putamen) and ventral (nucleus accumbens) striatum has been extensively implicated in motor function (Kelly et al. 1975; Pijnenburg et al. 1975; Delfs et al. 1990; see also Mink 2007 for a recent review). In this respect, the motor inhibitory effect induced by dopamine receptor blockers devoid of antipsychotic activity was also demonstrated by rats' open-field behavior. This is the case of the D_2 antagonist metoclopramide (Frussa-Filho and Palermo-Neto 1988).

Importantly, open-field studies demonstrated that D_2 dopamine receptor blockers devoid of antipsychotic activity produced a general decrease in motor activity—and not only a hypolocomotion effect. Indeed, compared to activity chambers, which quantify only locomotion and rearing (by interruptions of horizontal and vertical photoelectric beams, respectively), open-field studies can assure a true inhibition of

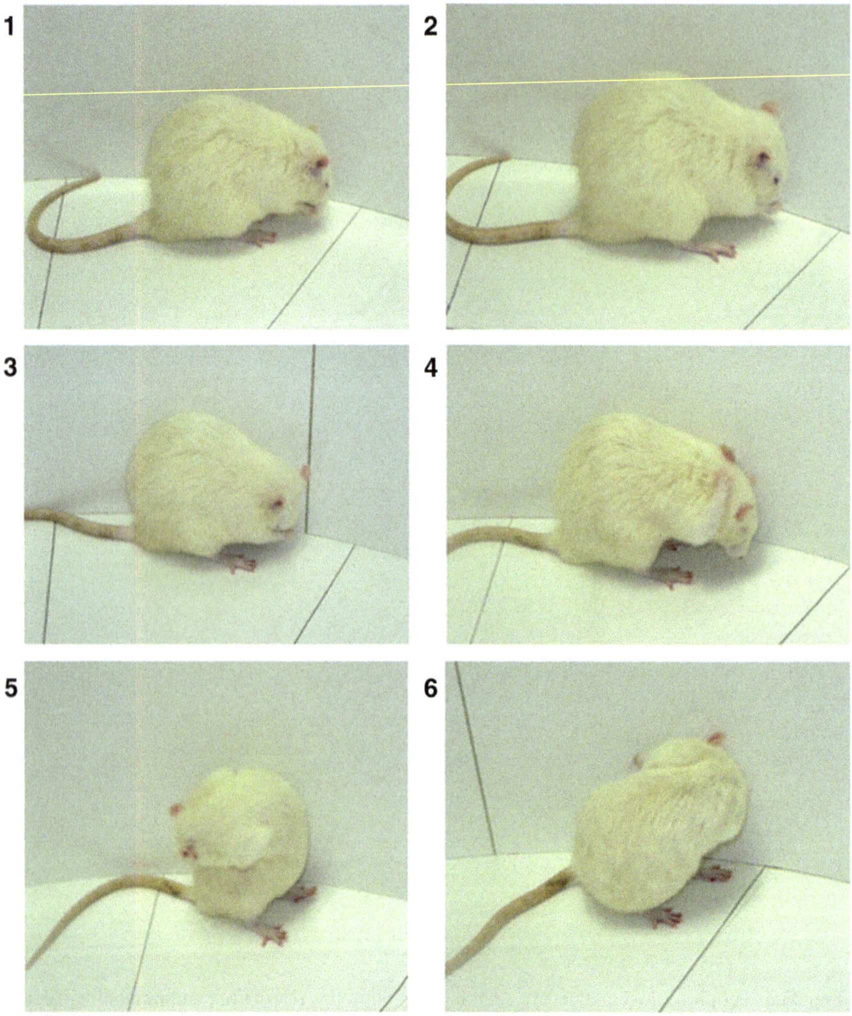

Fig. 25.4 Sequence of grooming behavior of a rat in the open-field

motor activity by detecting a concomitant decrease in grooming behavior and, most important, an increase in immobility duration. As we will later discuss in detail, the methodological advantages of the open-field apparatus can prevent misinterpretations of motor activity alterations produced, for example, by the behavioral competition phenomenon.

Besides D_2 dopamine receptor blockers, the open-field has been effective in detecting the motor depression produced by indirect dopaminergic antagonists. This is the case of reserpine, a monoamine-depleting agent. As with neuroleptics, open-field behavior of both rats and mice treated with reserpine has revealed a general inhibition of motor function. Indeed, reserpine-treated rats (Abílio et al. 2003a; Peixoto et al. 2003, 2005; Faria et al. 2005) and mice (Carvalho et al. 2003; Castro

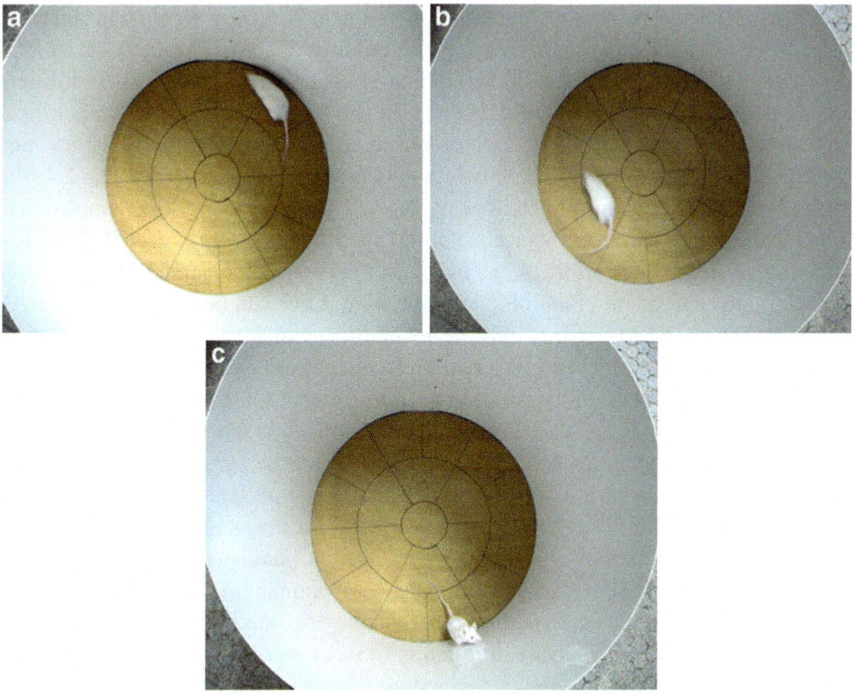

Fig. 25.5 Peripheral locomotion (**a**), central locomotion (**b**), and rearing (**c**) of a mouse in the open-field

et al. 2006) present a decrease in both locomotion and rearing frequencies and a concomitant increase in immobility duration in the open-field apparatus.

Drug-Induced Enhancement of Motor Activity: Parameter Specificity

Similar to other experimental models that evaluate motor activity, the open-field paradigm is bidirectional, with the ability to detect and quantify an ***increase*** in motor activity produced by pharmacological and non-pharmacological factors. However, a great advantage of the open-field model is its ability to differentiate a general motor activity enhancement from an increase in a specific parameter of motor function. For example, using the open-field method, we demonstrated that a putative motor stimulant dose (2 mg/kg) of amphetamine acutely administrated in mice is, in fact, only a ***locomotor*** stimulant dose because it reduces rearing frequency but does not modify immobility duration (Conceição and Frussa-Filho 1996). Similarly, we have shown, through the quantification of mice's open-field behavior, that a putative motor stimulant dose (1.8 g/kg) of ethanol produces an inhibitory effect on motor function when acutely administered. Indeed, this ethanol dose produced an increase in locomotion frequency but a concomitant decrease in rearing frequency and an ***increase*** in immobility duration (Araujo et al. 2005). In another study (Araujo et al. 2006a), we investigated the effects of three different doses of ethanol (1.4, 1.8, and 2.2 g/kg) on the parameters of mice's open-field behavior. While the three ethanol doses increased locomotion frequency and decreased rearing frequency, the immobility duration was not modified by the 1.4 g/kg dose, but was significantly increased by the 1.8 g/kg and 2.2 g/kg ethanol doses. Demonstrating the unique importance of the open-field model for accurately evaluating the motor effects of drugs, we have also recently demonstrated that the well-known motor stimulant effect of some morphine doses can be challenged. Indeed, we (Patti et al. 2005) verified that the locomotor stimulant dose of 20 mg/kg of morphine concomitantly decreased rearing frequency and grooming duration, without modifying the immobility duration of mice observed in an open field.

These results indicate that locomotion quantification does not always reflect the general effect of drugs on motor activity; this can be better investigated in the open-field model, where other motor behavioral parameters are also taken into account. The specific and at times, opposite effects of acute administration of amphetamine, as well as ethanol or morphine, on different parameters of motor function may be related to the above mentioned "behavioral competition phenomenon". Thus, for example, it could be hypothesized that the locomotor stimulant effects of morphine would influence the manifestation of other motor behaviors due to behavioral competition (e.g., the animal cannot present locomotion and rearing at the same time). However, this does not seem to be the case since doses of morphine that did not increase locomotion frequency (10 and 15 mg/kg) also decreased rearing frequency and duration of grooming. In addition, no correlation was observed between locomotion and rearing or grooming behaviors in mice acutely treated with morphine (see Patti et al. 2005). The behavioral competition hypothesis also seems improbable when considering ethanol and amphetamine. Indeed, we have demonstrated that mice "sensitized" by repeated injections of ethanol or amphetamine present a further increase in locomotion frequency and a tolerance to the inhibitory effect of these drugs on rearing frequency (Araujo et al. 2005 and unpublished data, respectively).

Evaluation of the Behavioral Competition Phenomenon

The above concerns notwithstanding, the behavioral competition phenomenon exists and can lead to serious misinterpretations of drug motor effects, particularly when automated activity chambers or more sophisticated systems of video of movement analysis and behavior recognition (see Costa et al. 2007) are used instead of the traditional open-field model. In fact, only the open-field paradigm can accurately detect the critical difference between mild stereotyped behaviors and immobility. This differentiation is particularly relevant in the case of the investigation of motor effects of dopaminergic psychostimulants, such as amphetamine, which can produce stereotyped behaviors and locomotor stimulation at the same doses. For example, using the open-field paradigm, we have recently demonstrated the phenomenon of rapid-onset behavioral sensitization to amphetamine in mice (Chinen et al. 2006). An important finding of that investigation was that the rapid-onset behavioral sensitization phenomenon was not specific for amphetamine-induced stereotypes, but was also demonstrated for amphetamine-induced hyper-locomotion in mice. This result differs from that of Kuczenski and Segal (1999a) who found that, opposite from the stereotyped behavior, the locomotor stimulant effect of amphetamine did not undergo rapid-onset sensitization in rats. Although experimental subjects differed between studies (rats versus mice), differences in the methodological quantification of stereotyped and locomotor behaviors may better account for the differences in results. Indeed, in the study of Kuczenski and Segal (1999ab), both types of behavior (locomotion and stereotypy) were quantified in the same experimental rectangular chambers. Possibly because of the chamber's lack of specificity for measuring each behavior, there was a clear-cut competition between locomotor and stereotyped behaviors such that stereotypy replaced locomotor activation. In our study, the use of a circular open-field apparatus (without corners) to specifically measure locomotor behavior as well as the use of rectangular wire mesh cages to measure stereotyped behavior allowed the expression of the locomotor-activating and the stereotyped effects of the same doses of amphetamine.

In order to illustrate the behavioral competition phenomenon and how it can lead to misinterpretations of drug effects on motor activity, we will describe an original investigation, specifically designed for this chapter.

Two experiments were performed. In each experiment, 20 three-month-old female Swiss mice, ranging in weight from 35 to 40 g, were randomly allocated to two groups: SAL and AMP. In experiment 1, the animals received an i.p. injection of saline (SAL) or 5.0 mg/kg amphetamine (AMP) and were immediately placed in the open-field apparatus. After 15 min, mice's motor activity was quantified during eight 5-min sessions, each separated by 30 min (total duration of open-field exposure: 235 min). In experiment 2, the animals received an i.p. injection of saline (SAL) or 5.0 mg/kg amphetamine (AMP) and were immediately placed in a rectangular wire mesh cage ($16 \times 30 \times 19$ cm). Fifteen minutes later, their stereotyped behavior was quantified during eight 20-s sessions, each separated by 30 min (total duration of wire mesh cage exposure: 235 min). Stereotyped behavior was quantified according to the scoring system proposed by Setler et al. (1976), with some modifications validated in our laboratory for mice. Briefly, animals' behavior was scored (ranging from 0 to 4) by an observer who was unaware of the drug treatment. The grading system was as follows: 0, asleep or stationary; 1, active; 2, active with predominantly stereotyped sniffing and rearing; 3, stereotyped sniffing with bursts of licking and/or gnawing and biting; 4, continual licking and/or gnawing of cage grids. Animals were used only once.

Acute amphetamine administration produced an initial increase in total, peripheral, and central locomotion (Figs. 25.6a–c, respectively) of mice in the 1st session observation (15 min) followed by a decrease in these parameters in the two subsequent observation sessions (45 min and 75 min). A mild increase in locomotion was observed from the 4th (105 min) to the 7th (205 min) observation sessions. Mice receiving amphetamine also presented an initial decrease in rearing frequency

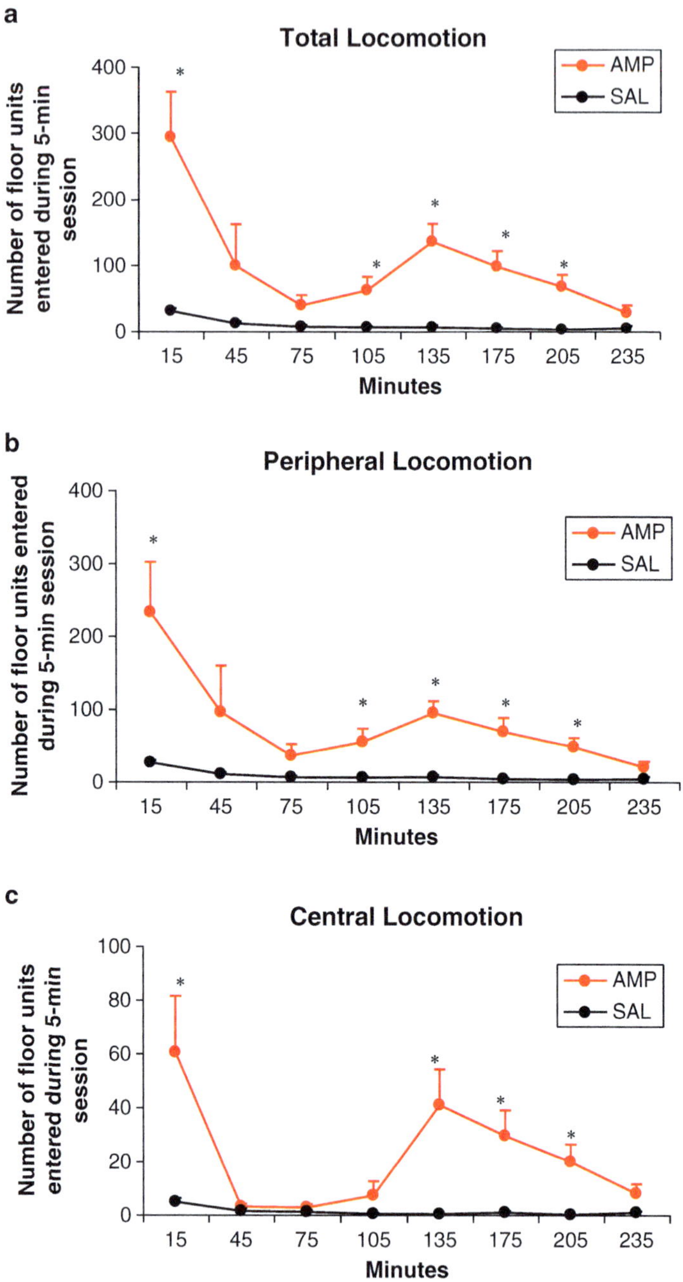

Fig. 25.6 Total (**a**), Peripheral (**b**), and Central (**c**) locomotion frequencies of mice that received an i.p. injection of saline (SAL) or 5.0 mg/kg amphetamine (AMP). Data represent means ± S.E. *T* test for independent samples. *$p < 0.05$ compared to the SAL group

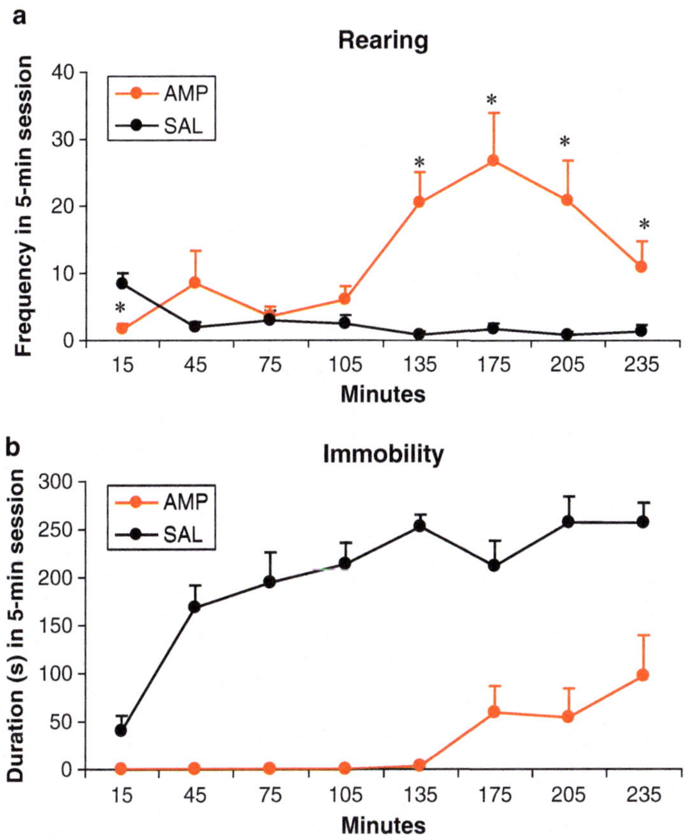

Fig. 25.7 Rearing frequency (**a**) and Immobility duration (**b**) of mice that received an i.p. injection of saline (SAL) or 5.0 mg/kg amphetamine (AMP). Data represent means ± S.E. *T* test for independent samples. *$p < 0.05$ compared to the SAL group

(15 min) and a substantial increase in this behavioral parameter from the 5th observation session (135 min) onwards (Fig. 25.7a). Immobility duration (Fig. 25.7b) of the AMP group was decreased during all of the observation sessions. Therefore, the decrease in immobility duration observed during the 2nd (45 min) and 3rd (75 min) observation sessions indicates that the concomitant absence of effects on locomotion and rearing frequencies in the AMP group when compared to the control group did not reflect an absence of enhancement in general motor function. In other words, another motor parameter should be enhanced.

Data from experiment 2 (Fig. 25.8) indicated that acute amphetamine administration produced stereotyped behaviors during all of the observation sessions. Notably, the higher scores of amphetamine-induced stereotypy in experiment 2 as well as the amphetamine-induced decrease in locomotion and rearing (when compared to the first session) in experiment 1 were observed during the same observation

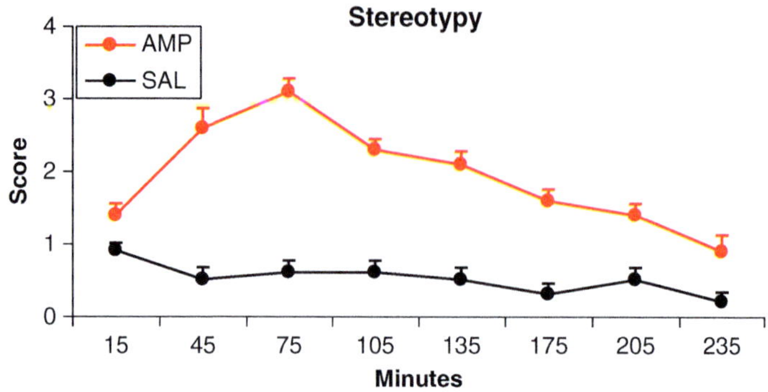

Fig. 25.8 Stereotypy scores of mice that received an i.p. injection of saline (SAL) or 5.0 mg/kg amphetamine (AMP). Data represent means ± S.E. *T* test for independent samples. $^*p < 0.05$ compared to the SAL group

sessions (2nd—45 min and 3rd—75 min). These results demonstrate that behavioral competition between stereotypy and locomotion developed following acute amphetamine administration in mice.

Taken together, the above-mentioned data demonstrates some of the advantages of the open-field apparatus over activity boxes. Within this context, the open-field apparatus seems to be more sensitive to and capable of detecting behavioral competition between locomotion and stereotypy than activity boxes. Whereas the automated program used by activity boxes may not accurately differentiate stereotyped behaviors from immobility, since the former is characterized by the replacement of normal locomotor and rearing behaviors for repetitive movements towards one specific place, quantifying immobility by direct observation in the open-field allows us to easily differentiate these behaviors. In line with this idea, following administration of a certain psychostimulant drug, a reduction in locomotion that is accompanied by a reduction in immobility of rodents in the open-field may indicate the development of behavioral competition between locomotion and stereotypy.

In addition, as previously mentioned in this chapter, most activity boxes are biased to favor stereotypy over other behaviors because of the presence of corners.

Evaluation of the Effects of Non-pharmacological Factors on Motor Activity

Furthermore, this experimental paradigm is very sensitive to and capable of detecting and quantifying bidirectional changes in motor activity produced not only by pharmacological agents, but also by physiological and environmental factors, such as aging, sleep deprivation, social isolation, and environmental enrichment (Ricardo et al. 2005; Araujo et al. 2006a; Fukushiro et al. 2007b). As for the effects of aging on rats' open-field behavior, we have demonstrated that aging produces a general inhibition of motor activity, reducing locomotion and rearing frequencies and increasing immobility duration (Silva et al. 1996; Ricardo et al. 2005).

Translational Potential

Drug-Induced Inhibition of Open-Field General Motor Activity in Rodents: Animal Models of Bradykinesia in Humans

From a clinical point of view, neuroleptic-induced decrease in the general motor activity of rats and mice tested in the open-field apparatus can be considered an animal model of neuroleptic-induced bradykinesia in humans (Hansen and Hoffman 1997; Terry et al. 2007) and reserpine-induced inhibition of rats' and mice's open-field behavior can be used as an animal model of bradykinesia in Parkinson's disease (Paulson and Stern 1997; Tadaiesky et al. 2006). Furthermore, the decreased locomotion and increased immobility in the open-field apparatus presented by 3-nitropropionic acid-treated mice has been proposed as an animal model of bradykinesia in Huntington's disease (Rosenstock et al. 2004). Indeed, while Huntington's disease is characterized by the inhibition of complex II of respiratory chain, 3-nitropropionic acid is an irreversible inhibitor of the mitochondrial complex II enzyme succinate (Palfi et al. 1996). In addition, it has been shown that Huntington's disease patients display bradykinesia throughout the entire duration of the disease (Thomson et al. 1988).

Neuroleptic Withdrawal-Induced Enhancement of Different Open-Field Motor Parameters in Rodents: Animal Models of Tardive Dyskinesia, Neuroleptic-Induced Psychosis, and Increased Responsiveness to Drugs of Abuse

The enhancement of general or parameter-specific motor activity in the open-field test has clinical implications. For example, whereas acute administration of neuroleptic drug decreases general activity in the open-field (see above), ***withdrawal*** from chronic neuroleptic treatment enhances open-field spontaneous locomotion and rearing frequencies, increases grooming frequency and duration, and decreases immobility duration in the rat (Bernardi and Palermo-Neto 1979; Bernardi et al. 1981; Frussa-Filho and Palermo-Neto 1988, 1991; Vital et al. 1995; Queiroz and Frussa-Filho 1997; Abílio et al. 1999, 2003b; Perry et al. 2004).

These effects may be a consequence of the development of central dopaminergic pathway supersensitivity (Palermo-Neto 1982; Palermo-Neto and Frussa-Filho 2001). Indeed, behavioral supersensitivity is thought to result from receptor site proliferation in mesolimbic and striatal brain tissues in response to a chronic dopamine receptor blockade (Burt et al. 1977; Muller and Seeman 1978; Fleminger et al. 1983;

Vital et al. 1998). In this way, whereas increased locomotor activity has been related to a dopaminergic hyperfunction in the mesoaccumbens system (Kelly et al. 1975; Pijnenburg et al. 1975; Delfs et al. 1990), increased rearing (Al-Khatib et al. 1995) and grooming (Neisewander et al. 1995) behaviors have been related to nigrostriatal hyperfunction.

Thus, from a clinical point of view, neuroleptic withdrawal-induced enhancement of open-field rearing and grooming parameters may be related to neuronal mechanisms underlying tardive dyskinesia since striatal dopaminergic supersensitivity has been proposed as a possible contributing factor to the development of tardive dyskinesia in schizophrenics receiving long-term neuroleptic treatment (Klawans 1973; Casey 1995; Ebadi and Srinivasan 1995; Latimer 1995).

In this regard, neuroleptic withdrawal-induced increase in open-field locomotion frequency is proposed as a physiopathological model of neuroleptic withdrawal-induced psychosis since it has been suggested that mesolimbic dopaminergic supersensitivity is related to this pathology (Chouinard et al. 1978; Chouinard and Jones 1980).

Besides increasing spontaneous open-field locomotion, neuroleptic withdrawal also potentiates the locomotor stimulant effect of various drugs of abuse in rodents. This phenomenon has been related to the neural mechanisms underlying the high prevalence of addiction in neuroleptic withdrawn schizophrenic patients (LeDuc and Mittleman 1995). Indeed, all addictive drugs cause dopamine release in the nucleus accumbens, which has been extensively associated with drug dependence (see the next topic for more details). Thus, neuroleptic-induced mesolimbic (mesoaccumbens in particular) dopaminergic supersensitivity has been proposed as a major factor contributing to the increased responsiveness to drugs such as cocaine in schizophrenic stimulant abusers chronically treated with neuroleptics (LeDuc and Mittleman 1993). In line with this rationale, we have recently verified that withdrawal from chronic neuroleptic treatment increased both cocaine-induced place preference (a well-known animal model of craving) and cocaine-induced open-field hyperlocomotion in mice (Fukushiro et al. 2007a).

Increased Open-Field Hyperlocomotion After Repeated Administration of Drugs of Abuse in Mice: An Animal Model of Drug Dependence

Most, if not all, drugs with abuse potential stimulate locomotion in rodents. As mentioned above, this locomotor stimulation has been extensively related to increased dopaminergic neurotransmission in the mesoaccumbens system (Kelly et al. 1975; Pijnenburg et al. 1975; Delfs et al. 1990), which is related to drug reward (Di Chiara and Imperato 1988; Koob 1992; Weiss et al. 1992; Self and Nestler 1995). In this way, while tolerance of many of the effects of repeated drug treatment occurs, the locomotor stimulant effect of various drugs of abuse often becomes progressively greater with repeated administration (Robinson and Becker 1986; Piazza et al. 1990; De Vries et al. 1998). This phenomenon, called behavioral sensitization, has been hypothesized to reflect neuronal adaptations that seem to be crucial to the development of drug addiction (Wise and Bozarth 1987; Robinson and Berridge 1993). Within this context, the quantification of locomotor activity of rodents through the open-field test is highly effective in detecting the behavioral sensitization phenomenon that develops after repeated administration of amphetamine (Bellot et al. 1996; Costa et al. 2001; Frussa-Filho et al. 2004; Alvarez et al. 2006; Araujo et al. 2006b; Chinen et al. 2006), ethanol (Bellot et al. 1996; Araujo et al. 2005, 2006a), morphine, cocaine, nicotine, and ecstasy (Frussa-Filho et al. unpublished data).

Application: Evaluation of Anxiety

The open-field test has become a very useful and convenient procedure for measuring not only motor activity, but also anxiety, memory, and even drug/environment conditioning in rodents. Although the main scope of this chapter is to discuss its utility for motor activity quantification, a critical analysis of its other potential experimental utilities provides a comprehensive understanding of the model, which is also necessary for the consideration of some methodological concerns related to its use.

In terms of utilizing the open-field as a paradigm for measuring anxiety, an excellent revision on this topic was recently made by Prut and Belzung (2003). As pointed out by these authors, "an increase in central locomotion or in time spent in the central part of the device without modification of total locomotion and of vertical exploration can be interpreted as an anxiolytic-like effect[3] while the contrary, that is, a decrease of these variables, is associated with anxiogenic effects." Indeed, rodents spontaneously prefer the periphery of the apparatus to activity in the central parts of the open-field. Both rats and mice walk near the open-field walls, a behavior called thigmotaxis. This natural aversion to the central parts of the open-field apparatus was first observed and experimentally used by Hall (1934). He observed that rats presented a higher locomotion frequency when they were food deprived, but not all rats ate when food was provided in the apparatus. Animals that did not eat were termed emotional. When compared to non-emotional rats, they entered the central part of the arena fewer times and exhibited higher levels of defecation.

[3] In the majority of the behavioral studies in which animal models of anxiety are used, the term "anxiolytic (or anxiogenic) effect" is avoided and replaced by the term "anxiolytic- (or anxiogenic-) ***like*** effect". We will not adopt such a procedure because in our opinion, rather than a mere scientific care, it represents a philosophical point of view. Indeed, the term "anxiolytic-like behavior" suggests that only the "special" *Homo sapiens* species are capable of being anxious, which is an anthropocentric and somewhat arrogant point of view.

Face and Construct Validity

According to Prut and Belzung (2003), anxious behavior in the open-field is triggered by two factors: individual testing (the animal is separated from its social group) and agoraphobia (as the arena is very large relative to the animal's breeding or natural environment). Based on these considerations, they propose that the open-field test satisfies both the face and the construct validity criteria as a model of human anxiety. Face validity implies that the anxiety response observed in the animal is similar to that observed in humans. In the open-field, the observed behavior is avoidance of threatening places (floor units further from the walls of the apparatus), which can also be observed in humans. Construct validity implies a similar etiology. In rodents, forced confrontation with novelty is stressful (Cigrang et al. 1986). Stress induces anxious behaviors, as it does in humans, which suggests a similar etiology and, thus, construct validity (Prut and Belzung 2003).

Predictive Validity—Benzodiazepines

Concerning the predictive validity, Prut and Belzung (2003) propose that the open-field test is sensitive to the anxiolytic effects produced by classical benzodiazepines and $5HT_{1A}$ receptor agonists. However, our personal experience with classical benzodiazepines contradicts this notion. Indeed, we have invariably verified a ***decrease*** in the percent of central locomotion after acute treatment with classical benzodiazepines (unpublished data). According to Prut and Belzung (2003), acute administration of benzodiazepine receptor full agonists primarily induces anxiolytic effects as is elicits an increase in the percent of entries into the central part of the open-field (56 % of the studies). However, the authors acknowledge that in some studies, these drugs had no effect (31 % of the studies) or even anxiogenic effects (13 %). At first sight, our personal experience contradicts the literature; however, a critical analysis of these numbers leads to the opposite conclusion. Indeed, 56 % of the studies confirming the expected effect of a classical benzodiazepine on an animal model of anxiety is, in fact, a very low percentage. For example, excluding the studies of the "one-trial tolerance phenomenon" (see File et al. 1990; Pereira et al. 1999; Frussa-Filho and Ribeiro 2002), as far as we know, there is no paper showing that a classical benzodiazepine did not produce anxiolytic effects in the elevated plus maze (see Chap. 17 for details of the model). Accordingly, as in our case, most researchers will not submit to publication data showing that a classical benzodiazepine did not elicit an anxiolytic effect in an alleged animal model of anxiety. In addition, if such a negative result is submitted, the chance of a negative editorial decision will be, undoubtedly, very high. Taken together, these arguments lead to the speculation that, in contrast to the ***published*** studies, most ***performed*** studies did not detect an anxiolytic effect of classical benzodiazepines on the open-field test.

Predictive Validity—5HT$_{1A}$ Serotonergic Receptor Agonists

Whereas the anxiolytic effects of classical benzodiazepines on the open-field test are questionable, the anxiolytic effects of 5HT$_{1A}$ receptor agonists seem to be much more conceivable. Indeed, as pointed out by Prut and Belzung (2003), partial agonists such as buspirone, gepirone, and ipsapirone were anxiolytic in 73.3 % of the studies. This can be compared to the effects of these compounds on other animal models of anxiety. For example, Belzung (2001) showed that 5-HT$_{1A}$ receptor agonists exhibited anxiolytic activity in 74 % of the preclinical studies. Thus, the ability of the open-field to detect anxiolysis of 5-HT$_{1A}$ receptor agonists seems to be the same as that of other animal models of anxiety, which renders this model suitable for the assessment of the anxiolytic activity of such compounds.

Predictive Validity—Spontaneously Hypertensive Rats (SHR)

Experiments on the behavior of spontaneously hypertensive rats (SHR) also support the use of the open-field test as an animal model of anxiety. SHR exhibit less anxiety-related behavior relative to their normotensive strain, the Wistar-Kyoto (WKY). As measured in the elevated plus-maze (EPM), the SHR enter the open arms more frequently and have an increased duration of open arm time (Gentsch et al. 1987; Berton et al. 1997; Durand et al. 1999, 2000, 2003; Pollier et al. 2000; Ferguson and Gray 2005). In this regard, Gentsch et al. (1987) demonstrated that SHR exhibited an increased central square locomotion when compared to WKY. However, the proposed "anxiolytic" characteristics of SHR were questioned in view of the significant hypoactivity presented by WKY. Indeed, McCarty and Kirby (1982) showed that in the open-field test, the measures of activity for this strain were well below the values of four other inbred normotensive or hypertensive strains. In contrast, the activity of SHR was well within the ranges of the other strains.

Within this context, we compared the anxiety levels of SHR with those of equally active normotensive control rats (Wistar EPM-1 rats) in several animal models of anxiety. Besides confirming the higher central square locomotion and increased percent entries and time in the open arms of the elevated plus-maze, we found that SHR present anxiolytic behavior in another well-known animal model of anxiety: the social interaction test (Goto et al. 1993). In line with these results, we demonstrated that in comparison with both WKY and Wistar EPM-1 rats, SHR showed the same effects in the elevated T-maze model as those observed in benzodiazepine-treated normotensive rats (Conceição et al. 1994). Our original finding that the "anxiolytic behavior" of SHR could be detected through the open-field test with the same effectiveness as other rodent models of anxiety—even when compared to equally active control strains—was confirmed by other research groups. For example, compared to the inbred rat strain Lewis (LEW), SHR show low indices of experimental anxiety when submitted to a variety of behavioral tests, such as the open-field, the elevated plus-maze, and the black–white box (Ramos et al. 1997, 1998, 2002, 2003; Izídio et al. 2005; Vendruscolo et al. 2006). These strains do not differ in their activity levels in either novel or familiar environments (Ramos et al. 1997, 1998, 1999, 2002; Ramos and Mormede 1998; Takahashi et al. 2001; Vendruscolo et al. 2003, 2006; Brüske et al. 2007).

Predictive Validity—Conclusions

Taking together the predictive validity evidence concerning the use of the open-field test as a rodent model of anxiety, the main criticism found is related to the contradictory effects of classical benzodiazepines on the model. Clearly, further studies are necessary in order to determine the optimum methodological specifications necessary to improve this aspect of the predictive validity of the model. Indeed, as pointed out in previous reviews of the open-field test, open-field behavior can be modified by a great variety of factors, such as size and shape (circular, square, rectangular) of the arena, lighting conditions, light/dark cycle, duration of the testing, animal housing conditions before testing (social, individual), diet, species, strain, sex, and age of the animals, familiarity with the apparatus (single exposure, repeated testing), etc. (Walsh and Cummins 1976; Bernardi and Palermo-Neto 1980; Choleris et al. 2001; Tou and Wade 2002; Eilam 2003; Prut and Belzung 2003; Lipkind et al. 2004). The last modifying factor mentioned, familiarity with the apparatus, is the basis of the next use of the open-field paradigm to be discussed: learning/memory.

Application: Evaluation of Learning/Memory

Learning and memory can also be evaluated in the open-field test through the quantification of open-field habituation. Shortly, memory is quantified by the decrease in general activity observed in session 2 when compared to session 1 (when the animal was first exposed to the apparatus). Since this learning/memory model is described in detail in Chap. 15, we will only discuss a few specific observations that come from the authors' own practice.

Aversion Level of the Open-Field Environment

It is important to assure a low aversive environmental condition during the open-field sessions. For example, we (Frederico et al. 1994a) have verified a complete absence of the habituation phenomenon in an elevated open-field (that is, an aversive open-field apparatus without walls and elevated 50 cm from the ground). If the open-field environment is aversive, it increases the aversive component of novelty, reducing rodents' general activity. In a second session, the animal should present a decrease in general activity due to habituation to the appealing component of novelty, but an increase in general activity because of the habituation to the aversive component of novelty. As a result, the general activity remains unchanged. Accordingly, the general activity of mice observed in the elevated open-field during the first session was significantly reduced compared to the general activity of mice tested in a traditional open-field apparatus with the same dimensions (Frederico et al. 1994b). Importantly, acute administration of the anxiolytic agent chlordiazepoxide before both sessions increased general activity in the first and decreased it in the second session, supporting the habituation phenomenon in the elevated open-field (Frederico et al. 1994a).

Duration of Sessions

As appropriately pointed out by Quillfeldt (Chap. 15), session duration may range from 2–10 min in order to assure detection of open-field habituation. Within this context, our personal practice advises the use of longer durations in the first session (10 min, for example, quantified min by min) and shorter durations in the second session (2 min, for example, also quantified min by min) due to the phenomenon of intra-session habituation. Indeed, general activity markedly decreases after the first two minutes in the open-field apparatus (see Fig. 25.9, for illustration), masking the inter-session habituation phenomenon and, consequently, memory quantification.

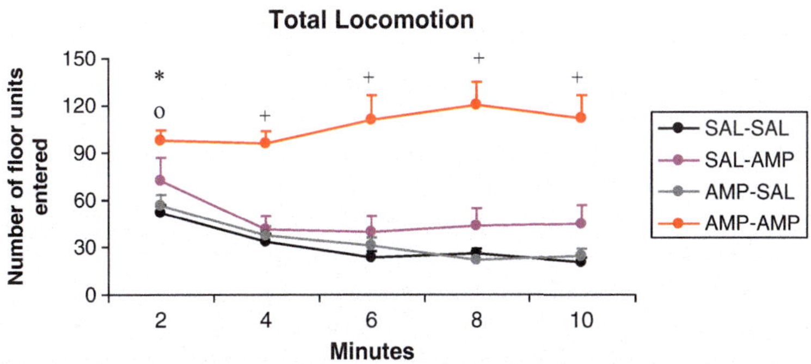

Fig. 25.9 Total locomotion frequency throughout the open-field 10-min session of mice treated with saline (SAL) or 2.0 mg/kg amphetamine (AMP) every other day for 13 days and, 7 days later, challenged with saline (SAL) or 2.0 mg/kg amphetamine (AMP). *T*-Test for independent samples. Two-way ANOVA with repeated measures revealed significant effects of time, treatment, and challenge and significant time × treatment, time × challenge, time × treatment × challenge, and treatment × challenge interactions. *T*-Test for paired samples revealed that total locomotion frequency was decreased at all time intervals with regard to the 2-min interval for all groups, except the AMP-AMP group. *$p<0.05$ when compared to the SAL-SAL group. °$p<0.05$ when compared to the AMP-SAL group. ⁺$p<0.05$ when compared to all other groups

Interestingly, while inter-session open-field habituation is a measure of memory, intra-session habituation seems to be a measure of motivation. We have demonstrated that both cocaine and amphetamine inhibit the phenomenon of intra-session open-field habituation (Araujo et al., submitted; Fukushiro et al., submitted).

Interval Between Session 1 and Session 2

As discussed by Quillfeldt (Chap. 15), the magnitude of the interval between session 1 and session 2 is critical in determining the type of memory evaluated (24 h or more—long-term memory; less than 6 h—short-term memory; up to 3 min—working memory). The point we would like to stress here is that longer intervals can be very useful for detecting long-term memory deficits, which at times, are not detected by the classical 24 h or 48 h intervals. For example, whereas we found clear-cut deficits in memory of old rats evaluated through the inhibitory avoidance task, we did not succeed in detecting aging-induced open-field habituation deficits using the 24 h interval between session 1 and session 2 (Silva et al. 1996). More recently, however, we demonstrated marked impairment of open-field habituation in old rats when the second session was performed 4 months after the first session (Frussa-Filho, unpublished data).

Open-Field Habituation Versus Drug-Induced Environmental Conditioning Versus State-Dependence: The Importance of Measuring Different Open-Field Behavioral Parameters

As with other behavioral models of memory (see Chap. 15), open-field habituation can be used to evaluate various aspects of learning/memory. When an experimental intervention (drug, for instance) is applied before session 1, the open-field model can show (during session 2) memory (habituation) alterations due to drug effects on both acquisition and consolidation. If the drug is injected directly after session 1, only consolidation is evaluated (considering a 24 h or more interval between session 1 and session 2 and drugs with short half-life). Finally, if the drug is injected directly before session 2, only recall will be evaluated (see above).

However, considering the first experimental situation (drug injection before session 1), effects on the habituation phenomenon may be misinterpreted due to both the drug-induced environmental conditioning and the state dependence phenomena. This situation is likely to occur if the drug in question is a psychostimulant such as amphetamine or cocaine. Indeed, we (Alvarez et al. 2006; Chinen et al. 2006) have demonstrated that one pairing of amphetamine-induced hyperlocomotor effect and the open-field apparatus can be enough to produce conditioned locomotion (i.e., an increase in locomotion frequency when the animal is re-exposed to the apparatus under a non-drug state when compared to session 2 locomotion of a control group that had received control solution before session 1). Thus, the environment (open-field apparatus) would act as a conditioned stimulus and the animal's hyperlocomotion, in the absence of the drug, would be the conditioned response.

At times, the magnitude of the conditioned locomotion is similar to the magnitude of the locomotion parameter presented by non-drugged animals exposed to the apparatus for the first time. Thus, it is impossible to differentiate between an amnestic effect of the drug (due to impairment of acquisition, consolidation, or both) and the development of conditioned locomotion. Interestingly, whereas an amnestic effect involves impaired memory, conditioned locomotion requires the ***remembrance*** of a previous hyperlocomotion in the environment (open-field apparatus).

Increasing the complexity of this experimental situation, the increased locomotion in session 2 should be due to a state dependence phenomenon (i.e., the retrieval of learned information requires that the animal be in a state similar to that in which the memory of this information was acquired—see Chap. 15). Indeed, the open-field apparatus could be interpreted differently under the drug and non-drug states. This new (non-drug) state might cause animals to detect and explore new characteristics of the environment. Thus, the environment would be the same but the different drug-state of the animal would impair the recall of the habituation task. The animal, therefore, would behave as if the open-field apparatus was a new environment. In other words, the rat (or mouse) had never been in that environment ***under that specific drug state***.

Within this complex context, the quantification and analysis of different open-field parameters seem to be critical to correctly interpreting the phenomenon under the behavioral data. For example, if increased locomotion in the second session is related to the environmental conditioning phenomenon, it is expected that only the parameters increased by the drug during the first session will be enhanced during the second session (performed in the absence of drug effects). Alternatively, if the enhanced locomotion frequency during session 2 is related to amnestic effects of the drug previously administered in session 1, the animal will behave as if the open-field is a new environment (that is, an increase in rearing frequency as well as a decrease in grooming and immobility durations). This would be the case for both acquisition and consolidation impairments (directly produced by the previous drug administration in session 1) and recall inhibition (due to the state dependence phenomenon). These two putative mechanisms should be differentiated by drug administration before both session 1 and session 2 to avoid the state dependence effects on recall. In this case, the habituation phenomenon must be evaluated by intra-group comparisons (session 1 versus session 2 of the same group) because of the stimulant effect of the drug. Although not an easy task, the differentiation and understanding of the drug-induced environmental conditioning phenomenon and the state dependence phenomenon are extremely important since both are related to cravings in humans (Damianopoulos and Carey 1992; Carey and Damianopoulos 1994; Hoffman et al. 1997; Hotsenpiller et al. 2002; Hotsenpiller and Wolf 2002; Patti et al. 2006).

Ethical and Final Comments

Since its development by Hall (1934), the open-field method has become one of the most widely used tools in behavioral research (see Walsh and Cummins 1976; Bernardi and Palermo-Neto 1980; Choleris et al. 2001; Prut and Belzung 2003). One of its great advantages is that its use is ethical. The open-field is a model developed for animal **observation**. Since it is used to evaluate motor function and learning/memory, it is imperative that the environmental conditions be as non-aversive as possible. As for anxiety evaluation, a more aversive environment is necessary at times but this can be achieved with small lighting and sound increases.

Novelty is one of the most important aversive stimuli in the open-field test. However, in the studies designed to evaluate motor function, previous extensive habituation is desirable. In the studies aimed to evaluate anxiety and learning/memory, novelty must be present. However, even this aversive stimulus has an appealing counterpoint, ensuring long life to this unique scientific experimental model.

References

Abílio VC, Freitas FM, Dolnikoff MS, Castrucci AML, Frussa-Filho R. Effects of continuous exposure to light on behavioral dopaminergic supersensitivity. Biol Psychiatry. 1999;45:1622–9.

Abílio VC, Araujo CC, Bergamo M, Calvente PR, D'Almeida V, Ribeiro Ride A, et al. Vitamin E attenuates reserpine-induced oral dyskinesia and striatal oxidized glutathione/reduced glutathione ratio (GSSG/GSH) enhancement in rats. Prog Neuropsychopharmacol Biol Psychiatry. 2003a;27:109–14.

Abílio VC, Vera Jr JA, Ferreira LS, Duarte CR, Martins CR, Torres-Leite D, et al. Effects of melatonin on behavioral dopaminergic supersensitivity. Life Sci. 2003b;72:3003–15.

Al-Khatib IMH, Dökmeci I, Fujiwara M. Differential role of nucleus accumbens and caudate-putamen in mediating the effect of nomifensine and methamphetamine on ambulation and rearing of rats in the open-field test. Jpn J Pharmacol. 1995;67:69–77.

Alvarez JN, Fukushiro DF, Tatsu JA, De Carvalho EP, Gandolfi AC, Tsuchiya JB, et al. Amphetamine-induced rapid-onset sensitization: role of novelty, conditioning and behavioral parameters. Pharmacol Biochem Behav. 2006;83:500–7.

Araujo NP, Camarini R, Souza-Formigoni ML, Carvalho RC, Abílio VC, Silva RH, et al. The importance of housing conditions on behavioral sensitization and tolerance to ethanol. Pharmacol Biochem Behav. 2005;82:40–5.

Araujo NP, Andersen ML, Abílio VC, Gomes DC, Carvalho RC, Silva RH, et al. Sleep deprivation abolishes the locomotor stimulant effect of ethanol in mice. Brain Res Bull. 2006a;69:332–7.

Araujo NP, Fukushiro DF, Cunha JL, Levin R, Chinen CC, Carvalho RC, et al. Drug-induced home cage conspecifics' behavior can potentiate behavioral sensitization in mice. Pharmacol Biochem Behav. 2006b;84:142–7.

Bellot RG, Camarini R, Vital MA, Palermo-Neto J, Leyton V, Frussa-Filho R. Monosialoganglioside attenuates the excitatory and behavioural sensitization effects of ethanol. Eur J Pharmacol. 1996;313:175–9.

Belzung C. Measuring exploratory behaviour. In: Crusio WE, Gerlai R, editors. Handbook of molecular genetic techniques for brain and behavior research (Techniques in the behavioral and neural sciences). Amsterdam: Elsevier; 1999. p. 739–49.

Belzung C. Rodent models of anxiety-like behaviors: are they predictive for compounds acting via non-benzodiazepine mechanisms? Curr Opin Investig Drugs. 2001;2:1108–11.

Bernardi MM, Palermo-Neto J. Effects of abrupt and gradual withdrawal from long-term haloperidol treatment on open-field behavior of rats. Psychopharmacology (Berl). 1979;65:247–50.

Bernardi MM, Palermo-Neto J. Atividade geral: conceito e medidas. Psicologia. 1980;1:43–51.

Bernardi MM, De Souza H, Palermo-Neto J. Effects of single and long-term haloperidol administration on open field behavior of rats. Psychopharmacology (Berl). 1981;73:171–5.

Berton O, Ramos A, Chaouloff F, Mormade P. Behavioral reactivity to social and nonsocial stimulations: a multivariate analysis of six inbred rat strains. Behav Genet. 1997;27:155–66.

Brüske GR, Vendruscolo LF, Ramos A. Two inbred rat strains contrasting for anxiety-related behaviors show similar levels of defensive responses to cat odor. Behav Brain Funct. 2007;3:17.

Burt DR, Creese I, Snyder SH. Antischizophrenic drugs: chronic treatment elevates dopamine receptor binding in brain. Science. 1977;196:326–8.

Carey RJ, Damianopoulos EN. Conditioned cocaine induced hyperactivity: an association with increased medial prefrontal cortex serotonin. Behav Brain Res. 1994;62:177–85.

Carvalho RC, Silva RH, Abílio VC, Barbosa PN, Frussa-Filho R. Antydiskinetic effects of risperidone on animal models of tardive dyskinesia in mice. Brain Res Bull. 2003;60:115–24.

Casey DE. Motor and mental aspects of extrapyramidal syndromes. Int Clin Psychopharmacol. 1995;3:105–14.

Castro JP, Frussa-Filho R, Fukushiro DF, Silva RH, Medrano WA, Ribeiro R, et al. Effects of baclofen on reserpine-induced vacuous chewing movements in mice. Brain Res Bull. 2006; 68:436–41.

Chinen CC, Faria RR, Frussa-Filho R. Characterization of the rapid-onset type of behavioral sensitization to amphetamine in mice: role of drug-environment conditioning. Neuropsychopharmacology. 2006;31:151–9.

Choleris E, Thomas AW, Kavaliers M, Prato FS. A detailed ethological analysis of the mouse open-field test: effects of diazepam, chlordiazepoxide and an extremely low frequency pulsed magnetic field. Neurosci Biobehav Rev. 2001;25:235–60.

Chouinard G, Jones BD, Annable L. Neuroleptic-induced supersensitivity psychosis. Am J Psychiatry. 1978;135:1409–10.

Cigrang M, Vogel E, Misslin R. Reduction of neophobia in mice following lesions of the caudate-putamen. Physiol Behav. 1986;36:25–8.

Conceição IM, Frussa-Filho R. Effects of microgram doses of haloperidol on open-field behavior in mice. Pharmacol Biochem Behav. 1996;53:833–8.

Conceição IM, Goto SH, Frussa-Filho R. Evaluation of memory in an elevated T maze: a comparison between spontaneously hypertensive, Wistar-Kyoto and Wistar EPM-1 rats. Braz J Med Biol Res. 1994;27:731–5.

Costa FG, Frussa-Filho R, Felicio LF. The neurotensin receptor antagonist, SR48692, attenuates the expression of amphetamine-induced behavioural sensitisation in mice. Eur J Pharmacol. 2001;428:97–103.

Costa FG, Frussa-Filho R, Canteras NS, Valera AG, Felicio LF. Blockade of neurotensin receptors during amphetamine discontinuation indicates individual variability. Neuropeptides. 2007;41: 83–91.

Damianopoulos EN, Carey RJ. Conditioning, habituation and behavioral reorganization factors in chronic cocaine effects. Behav Brain Res. 1992;49:149–57.

De Vries TJ, Schoffelmeer AN, Binnekade R, Mulder AH, Vanderschuren LJ. Drug-induced reinstatement of heroin- and cocaine-seeking behaviour following long-term extinction is associated with expression of behavioural sensitization. Eur J Neurosci. 1998;10:3565–71.

Delfs JM, Schreiber L, Kelley AE. Microinjection of cocaine into the nucleus accumbens elicits locomotor activation in the rat. J Neurosci. 1990;10:303–10.

Di Chiara G, Imperato A. Drugs abused by humans preferentially increase synaptic dopamine concentrations in the mesolimbic system of freely moving rats. Proc Natl Acad Sci U S A. 1988;85:5274–8.

Durand M, Berton O, Aguerre S, Edno L, Combourieu I, Mormede P, et al. Effects of repeated fluoxetine on anxiety-related behaviours, central serotonergic systems, and the corticotropic axis in SHR and WKY rats. Neuropharmacology. 1999;38:893–907.

Durand M, Aguerre S, Fernandez F, Edno L, Combourieu I, Mormede P, et al. Strain-dependent neurochemical and neuroendocrine effects of desipramine, but not fluoxetine or imipramine, in spontaneously hypertensive and Wistar-Kyoto rats. Neuropharmacology. 2000;39:2464–77.

Durand M, Mormede P, Chaouloff F. Wistar-Kyoto rats are sensitive to the hypolocomotor and anxiogenic effects of mCPP. Behav Pharmacol. 2003;14:173–7.

Ebadi M, Srinivasan SK. Pathogenesis, prevention, and treatment of neuroleptic-induced movement disorders. Pharmacol Rev. 1995;47:575–604.

Eilam D. Open-field behavior withstands drastic changes in arena size. Behav Brain Res. 2003; 142:53–62.

Faria RR, Abílio VC, Grassl C, Chinen CC, Negrão LT, De Castro JP, et al. Beneficial effects of vitamin C and vitamin E on reserpine-induced oral dyskinesia in rats: critical role of striatal catalase activity. Neuropharmacology. 2005;48:993–1001.

Ferguson SA, Gray EP. Aging effects on elevated plus maze behavior in spontaneously hypertensive, Wistar-Kyoto and Sprague-Dawley male and female rats. Physiol Behav. 2005;85: 621–8.

File SE, Mabbutt PS, Hitchcott PK. Characterization of the phenomenon of "one-trial tolerance" to the anxiolytic effect of chlordiazepoxide in the elevated plus-maze. Psychopharmacology (Berl). 1990;102:98–101.

Fleminger S, Rupniak NM, Hall MD, Jenner P, Marsden CD. Changes in apomorphine-induced stereotypy as a result of subacute neuroleptic treatment correlates with increased D-2 receptors, but not with increases in D-1 receptors. Biochem Pharmacol. 1983;32:2921–7.

Frederico PC, Nakagami A, Camargo EY, Monzillo LU, Conceição IM, Frussa-Filho R. Efeitos da novidade sobre as ações do clordiazepóxido em um campo aberto modificado. Anais da IX Reunião Anual da Federação de Sociedades de Biologia Experimental; 1994. p. 70.

Frederico PC, Nakagami A, Camargo EY, Monzillo LU, Conceição IM, Frussa-Filho R. Efeitos da altura e de espaços abertos sobre a emocionalidade de camundongos. Anais da IX Reunião Anual da Federação de Sociedades de Biologia Experimental; 1994. p. 44.

Frussa-Filho R, Palermo-Neto J. Effects of single and long-term metoclopramide administration on open field and stereotyped behavior of rats. Eur J Pharmacol. 1988;149:323–9.

Frussa-Filho R, Palermo-Neto J. Effects of single and long-term administration of sulpiride on open-field and stereotyped behavior of rats. Braz J Med Biol Res. 1990;23:463–72.

Frussa-Filho R, Palermo-Neto J. Effects of single and long-term droperidol administration on open-field and stereotyped behavior of rats. Physiol Behav. 1991;50:825–30.

Frussa-Filho R, Ribeiro RA. One-trial tolerance to the effects of chlordiazepoxide in the elevated plus-maze is not due to acquisition of a phobic avoidance of open arms during initial exposure. Life Sci. 2002;71:519–25.

Frussa-Filho R, Abílio VC, Bergamo M, Palermo-Neto J. Behavioural subsensitivity induced by long-term administration of a low dose of haloperidol to rats. J Pharm Pharmacol. 1997;49:412–5.

Frussa-Filho R, Goncalves MT, Andersen ML, Araujo NP, Chinen CC, Tufik S. Paradoxical sleep deprivation potentiates amphetamine-induced behavioural sensitization by increasing its conditioned component. Brain Res. 2004;1003:188–93.

Fukushiro DF, Alvarez JN, Tatsu JA, De Castro JP, Chinen CC, Frussa-Filho R. Haloperidol (but not ziprasidone) withdrawal enhances cocaine-induced locomotor activation and conditioned place preference in mice. Prog Neuropsychopharmacol Biol Psychiatry. 2007a;31:867–72.

Fukushiro DF, Calzavara MB, Trombin TF, Lopez GB, Abílio VC, Andersen ML, et al. Effects of environmental enrichment and paradoxical sleep deprivation on open-field behavior of amphetamine-treated mice. Physiol Behav. 2007b;92:773–9.

Gentsch C, Lichtsteiner M, Feer H. Open field and elevated plus-maze: a behavioural comparison between spontaneously hypertensive (SHR) and Wistar Kyoto (WKY) rats and the effects of chlordiazepoxide. Behav Brain Res. 1987;25:101–7.

Goto SH, Conceição IM, Ribeiro RA, Frussa-Filho R. Comparison of anxiety measured in the elevated plus-maze, open-field and social interaction tests between spontaneously hypertensive rats and Wistar EPM-1 rats. Braz J Med Biol Res. 1993;26:965–9.

Hall CS. Emotional behavior in the rat. I. Defecation and urination as measures of individual differences in emotionality. J Comp Psychol. 1934;18:385–403.

Hansen TE, Hoffman WF. Drug-induced parkinsonism. In: Yassa R, Nair NPV, Jeste DV, editors. Neuroleptic-induced movement disorders. A comprehensive survey. New York: Cambrigde University Press; 1997. p. 341–80.

Hoffman RE, Esposito R, Rosen M, Rockholz P. Recurrent personal memories during intoxication reported by patients with alcoholism. Psychol Med. 1997;27:1441–6.

Hotsenpiller G, Wolf ME. Conditioned locomotion is not correlated with behavioral sensitization to cocaine: an intra-laboratory multi-sample analysis. Neuropsychopharmacology. 2002;27: 924–9.

Hotsenpiller G, Horak BT, Wolf ME. Dissociation of conditioned locomotion and Fos induction in response to stimuli formerly paired with cocaine. Behav Neurosci. 2002;116:634–45.

Izídio GS, Spricigo Jr L, Ramos A. Genetic differences in the elevated plus-maze persist after first exposure of inbred rats to the test apparatus. Behav Process. 2005;68:129–34.

Jinnah HA, Hess EJ, Ledoux MS, Sharma N, Baxter MG, Delong MR. Rodent models for dystonia research: characteristics, evaluation, and utility. Mov Disord. 2005;20:283–92.

Kelly PH, Seviour PW, Iversen SD. Amphetamine and apomorphine responses in the rat following 6-OHDA lesions of the nucleus accumbens septi and corpus striatum. Brain Res. 1975;94: 507–22.

Klawans Jr HL. The pharmacology of tardive dyskinesias. Am J Psychiatry. 1973;130:82–6.

Koob GF. Neural mechanisms of drug reinforcement. Ann N Y Acad Sci. 1992;654:171–91.

Kuczenski R, Segal DS. Sensitization of amphetamine-induced stereotyped behaviors during the acute response. J Pharmacol Exp Ther. 1999;288:699–709.

Latimer PR. Tardive dyskinesia: a review. Can J Psychiatry. 1995;40:S49–54.

Leduc PA, Mittleman G. Interactions between chronic haloperidol treatment and cocaine in rats: an animal model of intermittent cocaine use in neuroleptic treated populations. Psychopharmacology (Berl). 1993;110:427–36.

Leduc PA, Mittleman G. Schizophrenia and psychostimulant abuse: a review and re-analysis of clinical evidence (Review). Psychopharmacology (Berl). 1995;121:407–27.

Lipkind D, Sakov A, Kafkafi N, Elmer GI, Benjamini Y, Golani I. New replicable anxiety-related measures of wall vs. center behavior of mice in the open-field. J Appl Physiol. 2004;97: 347–59.

Mccarty R, Kirby RF. Spontaneous hypertension and open-field behavior. Behav Neural Biol. 1982;34:450–2.

Mink JW. Functional organization of the basal ganglia. In: Jankovic J, Tolosa E, editors. Parkinson's disease and movement disorders. 5th ed. Philadelphia: Lippincott Williams and Wilkins; 2007. p. 1–22.

Muller P, Seeman P. Dopaminergic supersensitivity after neuroleptics: time-course and specificity. Psychopharmacology (Berl). 1978;60:1–11.

Neisewander JL, Ong A, McCgonigle P. Anatomical localization of SKF-38393-induced behaviors in rats using the irreversible monoamine receptor antagonist EEDQ. Synapse. 1995;9: 134–43.

Palermo-Neto J. Supersensitivity, drug withdrawal, and open field behavior. Psychopharmacol Bull. 1982;18:11–2.

Palermo-Neto J, Frussa-Filho R. Behavioral models of tardive dyskinesia in rodentes. In: Bolis CL, Pani L, Licinio J, editors. Dopaminergic Systems: evolution from biology to clinical aspects. Philadelphia: Lippincott Williams and Wilkins; 2001. p. 61–81.

Palfi S, Ferrante RJ, Brouillet E, Beal MF, Dolan R, Guyot MC, et al. Chronic 3-nitropropionic acid treatment in baboons replicates the cognitive and motor deficits of Huntington's disease. J Neurosci. 1996;16:3019–25.

Patti CL, Frussa-Filho R, Silva RH, Carvalho RC, Kameda SR, Takatsu-Coleman AL, et al. Behavioral characterization of morphine effects on motor activity in mice. Pharmacol Biochem Behav. 2005;81:923–7.
Patti CL, Kameda SR, Carvalho RC, Takatsu-Coleman AL, Lopez GB, Niigaki ST, et al. Effects of morphine on the plus-maze discriminative avoidance task: role of state-dependent learning. Psychopharmacology (Berl). 2006;184:1–12.
Paulson HL, Stern MB. Clinical manifestations of parkinson's disease. In: Watts RL, Koller WC, editors. Movement disorders. Neurologic principles and practice. New york: McGraw-Hill; 1997. p. 183–99.
Peixoto MF, Abílio VC, Silva RH, Frussa-Filho R. Effects of valproic acid on an animal model of tardive dyskinesia. Behav Brain Res. 2003;142:229–33.
Peixoto MF, Araujo NP, Silva RH, Castro JP, Fukushiro DF, Faria RR, et al. Effects of gabaergic drugs on reserpine-induced oral dyskinesia. Behav Brain Res. 2005;160:51–9.
Pereira JKD, Vieira RJ, Konishi CT, Ribeiro RA, Frussa-Filho R. The phenomenon of "one-trial tolerance" to the anxiolytic effect of chlordiazepoxide in the elevated plus-maze is abolished by the introduction of a motivational conflict situation. Life Sci. 1999;65:101–7.
Perry JC, Vital MA, Frussa-Filho R, Tufik S, Palermo-Neto J. Monosialoganglioside (GM1) attenuates the behavioural effects of long-term haloperidol administration in supersensitive rats. Eur Neuropsychopharmacol. 2004;14:127–33.
Piazza PV, Deminiere JM, Le Moal M, Simon H. Stress- and pharmacologically-induced behavioral sensitization increases vulnerability to acquisition of amphetamine self-administration. Brain Res. 1990;514:22–6.
Pierce RC, Kalivas PW. A circuitry model of the expression of behavioral sensitization to amphetamine-like psychostimulants. Brain Res Brain Res Rev. 1997;25:192–216.
Pijnenburg AJ, Honig WM, Van Rossum JM. Inhibition of D-amphetamine induced locomotor activity by injection of haloperidol into the nucleus accumbens of the rat. Psychopharmacologia. 1975;41:87–95.
Pollier F, Sarre S, Aguerre S, Ebinger G, Mormede P, Michotte Y, et al. Serotonin reuptake inhibition by citalopram in rat strains differing for their emotionality. Neuropsychopharmacology. 2000;22:64–76.
Prut L, Belzung C. The open field as a paradigm to measure the effects of drugs on anxiety-like behaviors: a review. Eur J Pharmacol. 2003;463:3–33.
Queiroz CM, Frussa-Filho R. Effects of buspirone on dopaminergic supersensitivity. Life Sci. 1997;61:371–82.
Ramos A, Mormede P. Stress and emotionality: a multidimensional and genetic approach. Neurosci Biobehav Rev. 1998;22:33–57.
Ramos A, Berton O, Mormède P, Chaouloff F. A multiple-test study of anxiety-related behaviours in six inbred rat strains. Behav Brain Res. 1997;85(1):57–69.
Ramos A, Mellerin Y, Mormede P, Chaouloff F. A genetic and multifactorial analysis of anxiety-related behaviours in Lewis and SHR intercrosses. Behav Brain Res. 1998;96:195–205.
Ramos A, Moisan MP, Chaouloff F, Mormede C, Mormede P. Identification of female-specific QTLs affecting an emotionality related behavior in rats. Mol Psychiatry. 1999;4:453–62.
Ramos A, Kangerski AL, Basso PF, Da Silva Santos JE, Assreuy J, Vendruscolo LF, et al. Evaluation of Lewis and SHR rat strains as a genetic model for the study of anxiety and pain. Behav Brain Res. 2002;129:113–23.
Ramos A, Correia EC, Izídio GS, Brüske GR. Genetic selection of two new rat lines displaying different levels of anxiety-related behaviors. Behav Genet. 2003;33:657–68.
Ricardo VP, Frussa-Filho R, Silva RH, Lopez GB, Patti CL, Zanier-Gomes PH, et al. Effects of social isolation on aging-induced orofacial movements in rats. Physiol Behav. 2005;86: 203–8.
Robinson TE, Becker JB. Enduring changes in brain and behavior produced by chronic amphetamine administration: a review and evaluation of animal models of amphetamine psychosis. Brain Res Rev. 1986;11:157–98.

Robinson TE, Berridge KC. The neural basis of drug craving: an incentive-sensitization theory of addiction. Brain Res Rev. 1993;18:247–91.

Rosenstock TR, Carvalho AC, Jurkiewicz A, Frussa-Filho R, Smaili SS. Mitochondrial calcium, oxidative stress and apoptosis in a neurodegenerative disease model induced by 3-nitropropionic acid. J Neurochem. 2004;88:1220–8.

Self DW, Nestler EJ. Molecular mechanisms of drug reinforcement and addiction. Annu Rev Neurosci. 1995;18:463–95.

Setler P, Sarau H, Mckenzie G. Differential attenuation of some effects of haloperidol in rats given scopolamine. Eur J Pharmacol. 1976;39:117–26.

Silva RH, Felicio LF, Nasello AG, Vital MA, Frussa-Filho R. Effect of ganglioside (GM1) on memory in senescent rats. Neurobiol Aging. 1996;17:583–6.

Tadaiesky MT, Andreatini R, Vital MA. Different effects of 7-nitroindazole in reserpine-induced hypolocomotion in two strains of mice. Eur J Pharmacol. 2006;535:199–207.

Takahashi RN, Berton O, Mormede P, Chaouloff F. Strain-dependent effects of diazepam and the 5-HT2B/2C receptor antagonist SB 206553 in spontaneously hypertensive and Lewis rats tested in the elevated plus-maze. Braz J Med Biol Res. 2001;34:675–82.

Terry Jr AV, Gearhart DA, Warner S, Hohnadel EJ, Middlemore ML, Zhang G, et al. Protracted effects of chronic oral haloperidol and risperidone on nerve growth factor, cholinergic neurons, and spatial reference learning in rats. Neuroscience. 2007;150:413–24.

Thomson PD, Berardelli A, Rothwell JC, Day BL, Dick SPR, Benecke R, et al. The coexistence of bradykinesia and chorea in Huntington's disease and its implications for theories of basal ganglia control of movement. Brain. 1988;111:223–44.

Tou JCL, Wade CE. Determinants affecting physical activity levels in animal models. Exp Biol Med. 2002;227:587–600.

Vendruscolo LF, Takahashi RN, Bruske GR, Ramos A. Evaluation of the anxiolytic-like effect of NKP608, a NK1-receptor antagonist, in two rat strains that differ in anxiety-related behaviors. Psychopharmacology (Berl). 2003;170:287–93.

Vendruscolo LF, Terenina-Rigaldie E, Raba F, Ramos A, Takahashi RN, Mormède P. A QTL on rat chromosome 7 modulates prepulse inhibition, a neuro-behavioral trait of ADHD, in a Lewis x SHR intercross. Behav Brain Funct. 2006;2:21.

Vital MA, Frussa-Filho R, Palermo-Neto J. Effects of monosialoganglioside on dopaminergic supersensitivity. Life Sci. 1995;56:2299–307.

Vital MA, Frussa-Filho R, Palermo-Neto J. Monosialoganglioside increased the in vivo affinity of D2 receptors for apomorphine in supersensitive rats. Ann N Y Acad Sci. 1998;845:428–9.

Walsh RN, Cummins RA. The open-field test: a critical review. Psychol Bull. 1976;3:482–504.

Weiss F, Hurd YL, Ungerstedt U, Markou A, Plotsky PM, Koob GF. Neurochemical correlates of cocaine and ethanol selfadministration. Ann N Y Acad Sci. 1992;654:220–41.

Wise RA, Bozarth MA. A psychomotor stimulant theory of addiction. Psychol Rev. 1987;94:469–92.

Chapter 26
Assessment of Motor Function in Rodents: Behavioral Models Sharing Simplicity and Multifaceted Applicability

Part 2: The Catalepsy Test

Roberto Frussa-Filho, Daniela Fukue Fukushiro, Camilla de Lima Patti, Cibele Cristina Chinen, Sonia Regina Kameda, and Rita de Cassia Carvalho

The Catalepsy Test

Introduction

According to the classical review by Sanberg et al. (1988), experimental catalepsy is defined as "a failure of laboratory animals to correct an externally imposed posture". Indeed, as stated by these authors, if a normal animal is placed in an unusual posture, it will change its position in a matter of seconds. Conversely, a cataleptic animal will remain in this position for a long period of time (i.e., for several minutes or more).

Cataleptic behavior may be experimentally induced in different species of rodents, such as rats (Morpurgo 1962), mice (György et al. 1969) and guinea pigs (Costall and Naylor 1975), using drugs that block dopaminergic transmission (Morpurgo 1962; Conceição and Frussa-Filho 1993; Lucas et al. 1997; Agovic et al. 2008), cholinergic agonists (Klemm 1985; Sousa et al. 2001), opioid agonists (De Ryck and

R. Frussa-Filho (In Memoriam) • S.R. Kameda
Department of Pharmacology, Universidade Federal de São Paulo (UNIFESP), São Paulo, São Paulo, Brazil

D.F. Fukushiro, Ph.D. (✉)
Department of Pharmacology, Universidade Federal de São Paulo (UNIFESP), Rua Botucatu, 862, Ed. Leal Prado, 1° andar, São Paulo, São Paulo 04023062, Brazil
e-mail: danifukushiro@gmail.com

C.L. Patti, Ph.D.
Instituto de Genética e Erros Inatos do Metabolismo, Universidade Federal de São Paulo (UNIFESP), São Paulo, São Paulo, Brazil

C.C. Chinen • R.d.C. Carvalho
Department of Pharmacology, Universidade Federal de São Paulo (UNIFESP), São Paulo, São Paulo, Brazil

M.L. Andersen, S. Tufik (eds.), *Rodent Model as Tools in Ethical Biomedical Research*, DOI 10.1007/978-3-319-11578-8_26

Teitelbaum 1984; Zarrindast et al. 2002) or cannabinoids (Gough and Olley 1978; Rácz et al. 2008). Some authors have used the term *catatonia* to define morphine-induced catalepsy in order to differentiate it from neuroleptic-induced catalepsy (De Ryck and Teitelbaum 1984). More rarely, the term catatonia has been used to define neuroleptic-induced catalepsy (De Sousa-Moreira et al. 1982). Nevertheless, the majority of studies in the literature have used the term catalepsy indiscriminately (Sanberg et al. 1988).

Methodology

Different laboratories have used distinct tests to measure catalepsy in rodents, such as the cork test (Lloyd et al. 1981), the grid test (Fuenmayor and Vogt 1979), the parallel bars test (Amir et al. 1981), the wood block test (Shipley et al. 1981) and the inclined screen test (Barnes et al. 1990). The behavioral catalepsy test that is employed most commonly is undoubtedly the horizontal bar test, which was originally described by Kuschinsky and Hornykiewicz (1972). Briefly, during this test, the animal's forepaws are placed on a horizontal bar that is situated a few centimeters above a workbench, and the amount of time that the animal remains in this imposed posture is quantified. An excellent description of the methodological variations of this test can be found in Sanberg et al. (1988). The specific methodology of the catalepsy test used for mice in our laboratory is described below in order to provide a detailed example of the utilization of this animal model.

The apparatus used in our laboratory consists of a horizontal glass bar, 0.5 cm in diameter and 30 cm in length, that is fixed at a height of 4 cm above the working surface. In accordance with recommendations provided by multiple reports (Barnes et al. 1990; Dijk et al. 1991; Rocha et al. 1997), a mouse performs three trials of a single catalepsy determination at one time point. The sum (or average) of these trials enables minimization of the stress-induced data variability produced by the handling procedure as compared to use of a single trial per determination. In each trial, the animal's forepaws are placed on the bar and the duration of catalepsy is timed from the instant that the animal is placed on the bar to the instant that it removes both forepaws from the bar or climbs over the bar with the hind limbs. Considering the sum of the three trials, a cut-off of 19 min is used in each determination. This long cut-off time is employed because short cut-off values have been suggested to obscure modifications in the evolution of the cataleptic behavior during chronic drug treatment regimens (Sanberg et al. 1988). Depending on the pharmacokinetic profile of the catalepsy-induced agent, the latency of the first catalepsy determination (composed of three trials, as explained above) and the number of catalepsy determinations (separated by 19 min, as explained above) will vary. As suggested by Undie and Friedman (1988) and Lipska et al. (1995), the total amount of time that the animal remains in a cataleptic posture (i.e., the sum of the durations of the catalepsy determinations) is calculated to provide the cumulative catalepsy time for each animal. The cumulative catalepsy time (in seconds) is then transformed to a logarithmic value (ln) in order to normalize the data, as proposed by Ferré et al. (1990) and this is the value used to perform statistical analysis.

Applicability: Reserpine-, 6-OHDA- and Rotenone-Induced Catalepsy in Rodents as Animal Models of Rigidity in Parkinson's Disease

Parkinson's Disease

Parkinson's disease, also known as idiopathic Parkinson syndrome, was originally described in 1817 by James Parkinson in *An Essay on the Shaking Palsy* as a pathology characterized by "involuntary tremulous motion, with lessened muscular power, in parts not in action and even when supported; with a propensity to bend the trunk forwards, and to pass from a walking to a running pace: the senses and intellect being uninjured". Four decades later, Jean-Martin Charcot added rigidity to Parkinson's clinical description and named the syndrome Parkinson's disease.

Currently, Parkinson's disease is considered a progressive neurodegenerative motor disorder characterized by a classic triad of cardinal symptoms: resting tremor, bradykinesia (slowness of movement) and rigidity, combined with a loss of normal postural control (see McAuley 2003 and Lane and Dunnett 2008 for review). Dopaminergic neuronal degeneration in the substantia nigra with consequent depletion of striatal dopamine and the presence of eosinophilic intracytoplasmic inclusions (Lewy bodies) in surviving cells are characteristic features of this disease (McAuley 2003; Klockgether 2004; Rao et al. 2006). The specific molecular mechanisms involved in these alterations are still unclear. Multiple factors are likely responsible for the observed neurodegeneration (Olanow and Tatton 1999; Díaz et al. 2001; Nehru et al. 2008). One of the most important molecular alterations related to the progressive loss of dopaminergic neurons in substantia nigra pars compacta in Parkinson's disease seems to be mitochondrial complex-1 inhibition (Sherer et al. 2003; Schmidt and Alam 2006).

The diagnosis of Parkinson's disease is based on presentation of the cardinal symptoms by the patient (Rao et al. 2006), which only appear when over 70 % of the originally present dopamine content is lost (Bernheimer et al. 1973; Riederer and Wuketich 1976). Among the above-mentioned cardinal symptoms of Parkinson's disease, rigidity is of particular importance to this chapter. Rigidity is defined as an increased resistance of a joint to passive movement of a limb segment (Klockgether 2004). It may be constant or intermittent throughout the movement. In addition, it may be generalized (i.e., it may involve a limb as well as axial muscles). Rigidity is more marked when the patient is standing than seated and it may be potentiated by stress, anxiety and chiefly contra-lateral contractions (Delwaide 2001). Rigidity may contribute to the impairment experienced by parkinsonian patients by interfering with voluntary movement, particularly its speed of execution, since more strength is needed to move a joint in this condition (Delwaide 2001).

Treatment of Parkinson's disease is mainly pharmacological, including the administration of drugs that enhance the activity of the nigrostriatal dopaminergic pathway. Dopamine agonists (such as bromocriptine, pergolide, pramipexole and ropinirole) and L-DOPA, which is the precursor of dopamine, are the most commonly used drugs in patients with mild and more severe symptoms of the disease, respectively (Rao et al. 2006). L-DOPA is particularly effective in treating bradykinesia and rigidity (Miyasaki et al. 2002), but is less effective in controlling symptoms like speech, postural reflex and gait disturbance (Rao et al. 2006). It is usually combined with carbidopa in order to increase its bioavailability and avoid the peripheral adverse effects of dopamine (Rao et al. 2006). Other drugs used in the treatment of Parkinson's disease include anticholinergic agents (use of these agents are limited by their low effectiveness and adverse effects), MAO-B and COMT inhibitors (Rao et al. 2006) and antiglutamatergic drugs (Klockgether 2004).

Since Parkinson's disease does not spontaneously arise in animals, animal models to study this pathology are developed by methodological approaches that produce some or most of its characteristic features. Among these approaches is the administration of agents such as reserpine, 6-hydroxydopamine (6-OHDA), 1-methyl-4-phenyl-1,2,3,6-tetrahydropyridine (MPTP), rotenone, high doses of methamphetamine and iron. Below, we will describe the reserpine, 6-OHDA and rotenone models, with a focus on the cataleptic behavior produced by administration of these drugs, since only limited evidence has demonstrated catalepsy after administration of the other indicated agents.

Reserpine-Induced Catalepsy

It has been established that reserpine depletes vesicle stores of dopamine and other neurotransmitters, such as adrenaline, noradrenaline, histamine and serotonin (Gerlach and Riederer 1996). As a result of dopamine depletion, it produces a state that is mildly similar to that observed in Parkinson's disease, although in the absence of the specific degeneration processes and Lewy body formation. Indeed, as previously revised by Gerlach and Riederer (1996), the behavioral effects of reserpine administration in rats include a general reduction in motor activity (bradykinesia and catalepsy) and the development of tremors. Therefore, the reserpine model seems to be a phenomenological rather than a pathophysiological model of Parkinson's disease, since it reproduces the cardinal symptoms and, to some degree, the neurochemical alterations of this syndrome in animals, in the absence of the progressive degeneration of dopaminergic neurons.

Importantly, while reserpine-induced inhibition of rodent motor activity in the open-field has been used as a model of bradykinesia (see Chap. 26), reserpine-induced catalepsy could be proposed as an animal model of rigidity in Parkinson's disease. In this way, from a translational point of view, the failure of animals to correct an externally imposed posture (catalepsy) could reflect resistance to a movement, which is characteristic of Parkinson's rigidity. Catalepsy has previously been suggested as an indicator of Parkinson's rigidity in animals (Dutra et al. 2002).

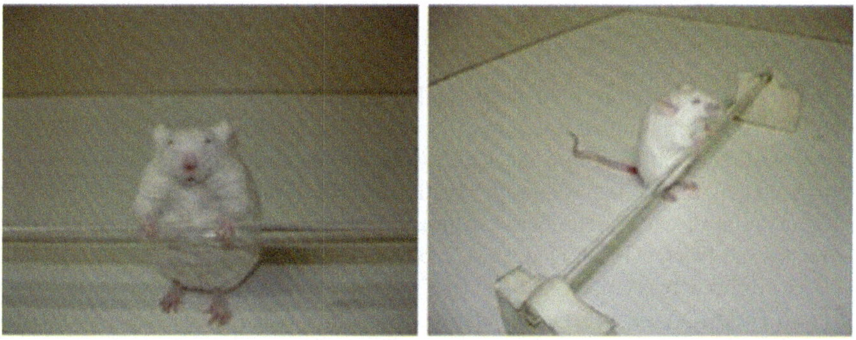

Fig. 26.1 Reserpine-induced catalepsy in a mouse

In fact, Lorenc-Koci et al. (1996) demonstrated that the electromyographic (EMG) recordings during catalepsy in rats were similar to those observed during muscle rigidity in Parkinson's.

It should be noted, however, that it is very difficult to distinguish bradykinesia from rigidity in animals, since both of these symptoms are related to motor inhibition. Therefore, reserpine-induced catalepsy could mimic both of these behaviors. This concern notwithstanding, the broad use of reserpine-induced inhibition of motor activity in the open-field as a classical animal model for Parkinson's disease-associated bradykinesia (see Chap. 26), suggests that the reserpine-induced catalepsy model is a suitable model for rigidity. To better illustrate this model, reserpine-induced catalepsy in a mouse is shown in Fig. 26.1.

Data from the literature indicate that drugs used in the treatment of Parkinson's rigidity (and also bradykinesia) and drugs with similar mechanisms of action reduce reserpine-induced catalepsy in animals. Previous studies have demonstrated the reversal of reserpine-induced catalepsy by the administration of L-DOPA (Johnson et al. 1976; Namba et al. 1980, 1981), dopamine agonists such as bromocriptine and apomorphine (Johnson et al. 1976), SKF 38393—a selective D_1 agonist—and quinpirole—a selective D_2 agonist (Hubbard and Trugman 1993) and cabergoline—a relatively novel D_2 agonist (Miyagi et al. 1996), as well as the NMDA glutamatergic antagonist MK-801 (Dutra et al. 2002) in rodents.

Due to the transient and non-selective characteristics of neurotransmitter depletion promoted by the reserpine model, it no longer in common use. A modification of this model has been introduced that combines reserpine with a-methyl-p-tyrosine (a dopamine synthesis inhibitor) administration in order to achieve increased specificity (specific reduction in dopamine transmission) and, consequently, more similarities with the clinical aspects of Parkinson's disease (Gerlach and Riederer 1996). In recent years, the administration of 6-OHDA to rodents and MPTP to non-human primates have been the most commonly used animal models of Parkinson's disease.

6-OHDA-Induced Catalepsy

Systemic administration of the toxin 6-OHDA produces no effects on the brains of animals because it is unable to cross the blood–brain barrier. However, when this toxin is directly infused into the lateral ventricle or brain structures (such as the substantia nigra, medial forebrain bundle and striatum), it produces a selective degeneration of catecholaminergic neurons (Ungerstedt 1968; Bloom et al. 1969; Uretsky and Iversen 1970; Díaz et al. 2001; Sarkaki et al. 2008).

As reviewed by Gerlach and Riederer (1996), the mechanism of the neurotoxic action of 6-OHDA seems to be related to the production of free radicals. Thus, 6-OHDA is apparently selectively taken up into catecholaminergic neurons (due to its structural similarity to catecholamines), where it generates hydrogen peroxide and subsequently hydroxyl radicals, leading to cell death.

After bilateral injections of 6-OHDA into the nigrostriatal pathway, the degeneration of dopamine neurons and the consequent depletion of dopamine are so severe that the animals are unable to feed or hydrate without assistance and most of them die (Lane and Dunnett 2008). Due to the disadvantages of this model (intensive animal care and loss of many animals), several authors have adopted an alternative approach in which they bilaterally inject 6-OHDA specifically into the medial forebrain bundle (Hayakawa et al. 1999; Cannon et al. 2007; Datla et al. 2007; Strome et al. 2007). This technique produces specific depletion of noradrenaline in the hypothalamus and of dopamine in the striatum (Smith and Young 1974), but does not incapacitate the animal. This approach has been reported to result in a reduction of locomotor activity and induction of catalepsy and tremor in rodents (Jolicoeur et al. 1991; Hayakawa et al. 1999), making it useful for the study of Parkinson's disease. In contrast to the reserpine model, the 6-OHDA bilateral lesion model is not only a phenomenological, but also a pathophysiological animal model of Parkinson's disease since it produces similar symptoms to those observed clinically and leads to neuronal degeneration.

Another alternative that overcomes the problems observed with bilateral injections of 6-OHDA is the unilateral injection of 6-OHDA into the striatum (Andén et al. 1966; Ungerstedt and Arbuthnott 1970) or the substantia nigra (During et al. 1992; Sarkaki et al. 2008). This approach has also been widely used as an animal model of Parkinson's disease. This technique produces degeneration of dopamine neurons and depletion of dopamine in the striatum on the same side as the lesion (Gerlach and Riederer 1996), resulting in both catalepsy behavior (Sarkaki et al. 2008) and dopamine agonist-induced turning behavior (Schwarting and Huston 1996). Although this latter behavioral model (turning behavior) does not imitate the symptoms of Parkinson's disease, it has proven to be very useful for detecting clinically effective anti-parkinsonian drugs (Ungerstedt 1971).

While 6-OHDA-induced turning behavior has received substantial attention in the literature (Metz et al. 2005; Pinna et al. 2006; Tronci et al. 2007; Liu et al. 2008), the relevance of 6-OHDA-induced catalepsy to this topic is clear, since, as suggested for reserpine-induced catalepsy, catalepsy induced by 6-OHDA bilateral or unilateral lesions is an animal model of Parkinson's disease rigidity.

Furthermore, the same drugs that are used to treat Parkinson's rigidity (and other symptoms, such as bradykinesia) reverse the catalepsy induced by 6-OHDA lesions. Hayakawa et al. (1999) demonstrated that the appearance of catalepsy induced by bilateral infusion of 6-OHDA into the medial forebrain bundle in rats was antagonized by the systemic administration of L-DOPA (10 and 20 mg/kg, i.p.), the dopamine receptor agonist talipexole (0.02 and 0.05 mg/kg, i.p.) and the adenosine receptor antagonist theophylline (10 and 20 mg/kg, i.p.). Regarding this last result, data from the literature suggest that adenosine (2A) antagonists may confer a benefit to parkinsonian patients by modulating dopamine transmission (Bara-Jimenez et al. 2003). In another study, Díaz et al. (2001) verified that administration of the dopamine agonist apomorphine (0.5 mg/kg, i.p.) decreased the duration of catalepsy induced by infusion of 6-OHDA into the ventricles of rats. In this respect, parenterally administered apomorphine has been clinically used for the treatment of Parkinson's disease. It is particularly useful for patients experiencing a sudden, unexpected and resistant "off" period (Goetz et al. 2005; Rao et al. 2006). Indeed, L-DOPA-treated patients may experience a "wearing-off" effect, which is characterized by a shorter duration of benefit from each L-DOPA dose, causing worsening of parkinsonian symptoms (see Rao et al. 2006).

Some authors have demonstrated that 6-OHDA-induced catalepsy may intensify with repeated testing, but only when the tests are conducted in the same environmental context (Klein and Schmidt 2003; Srinivasan and Schmidt 2004). In this regard, context-dependent intensification of parkinsonian symptoms (for instance, rigidity) might also occur in Parkinson's disease (Klein and Schmidt 2003).

Rotenone-Induced Catalepsy

Mitochondrial complex-1 inhibition seems to be closely related to the progressive loss of dopaminergic neurons in substantia nigra pars compacta in Parkinson's disease (Sherer et al. 2003; Schmidt and Alam 2006). Rotenone is a potent specific inhibitor of mitochondrial complex-1 that crosses both the blood–brain barrier and the cell membrane easily because of its lipophilic structure (Betarbet et al. 2000). It is the most potent rotenoid, which is a family of natural cytotoxic compounds extracted from various parts of *Leguminosa* plants. It selectively destroys dopaminergic neurons that produce Parkinson's disease-like behavioral dysfunction (Alam et al. 2004; Sherer et al. 2007). A rotenone-induced behavioral effect in rodents is catalepsy. Similar to reserpine- and 6-OHDA-induced catalepsy, rotenone-induced catalepsy can be considered an animal model of Parkinson's disease rigidity. Nehru et al. 2008 demonstrated that rotenone at a dose of 2 mg/kg for 21 days (but not for 14 days) produced cataleptic behavior in rats. This long latency for the development of catalepsy in this model is probably related to the progressive magnitude of neurodegeneration and suggests rotenone-induced catalepsy in rodents as a very interesting phenomenological and pathophysiological animal model of Parkinson's disease rigidity.

Applicability: Neuroleptic-Induced Catalepsy in Rodents as an Animal Model of Neuroleptic-Induced Parkinsonism

Neuroleptic-Induced Parkinsonism

Parkinsonism is one of the most common side-effects of antipsychotic drugs. Although many studies have focused on other neuroleptic-induced syndromes due to their apparent irreversibility, parkinsonism presents the greatest morbidity (Hansen and Hoffman 1997).

Clinically, all of the cardinal signs present in idiopathic Parkinson's disease occur in neuroleptic-induced parkinsonism (Hall et al. 1956; Goetz and Klawans 1981; Friedman 1992). Thus, the symptoms of neuroleptic-induced parkinsonism may be grouped into five categories: bradykinesia, rigidity, tremor, loss of postural reflexes and a miscellaneous category, including symptoms of seborrheic dermatitis, hypophonia (soft and monotonous speech), micrographia, hypersalivation and hypomimia (lack of facial expression) (Hansen and Hoffman 1997). Although all of the symptoms of idiopathic Parkinson's disease occur in drug-induced parkinsonism, their comparative frequencies may differ. Specifically, whereas rigidity and tremor occur with the same frequency in patients with idiopathic Parkinson's disease, rigidity is more common than tremor in patients with neuroleptic-induced parkinsonism (Hausner 1983; Rozzini et al. 1985).

Concerning epidemiology, Hansen and Hoffman (1997) analyzed 23 studies and found the overall rate of occurrence of neuroleptic-induced parkinsonism to be 28.2 %, with 3,630 cases identified among 12,886 patients at risk. These values are questionable because different studies presented rates of occurrence that were extremely variable (from 10 to 75 %).

Treatment strategies for neuroleptic-induced parkinsonism include modifications in the dosage or type of neuroleptic as well as the introduction of other medications to treat the side-effect. Concerning the latter possibility, the use of anticholinergic agents, both for the treatment and prophylaxis of neuroleptic-induced parkinsonism, has been discussed in previous reviews (McEvoy 1983; Keepers and Casey 1986). In general, anticholinergic agents seem to be effective in the treatment of neuroleptic-induced parkinsonism, but are less effective in the prophylaxis of this pathology (Keepers and Casey 1986). According to Hansen and Hoffman (1997), although anticholinergic drugs have clear antiparkinsonian effects, the extent of their efficacy may be limited and they may be associated with additional side-effects. Complete remission of neuroleptic-induced parkinsonism following the inclusion of anticholinergic drugs in a patient's treatment regimen is rare.

Many physicians do not treat parkinsonism induced by neuroleptics due to the limitations cited above for the use of anticholinergic agents and especially because they expect patients to develop spontaneous tolerance to the pathology (Cahan and Parrish 1960; DiMascio and Demirgian 1970; Ayd 1971; Klett and Caffey 1972). However, short- and long-term studies have indicated that tolerance should not be expected in many cases. Indeed, no evidence of tolerance was found when the severity of neuroleptic-induced parkinsonism was evaluated 6 weeks (Simpson 1970) or 3 months after treatment with neuroleptics (Kruse 1960). Some previous reports have indicated that this syndrome can be aggravated over time. Thus, among 33 patients studied for 3–11 years, neuroleptic-induced parkinsonism worsened by 20–30 % among patients maintained under stable or increasing dosages of neuroleptics (Casey et al. 1986).

In fact, neuroleptic-induced parkinsonism may persist for prolonged periods of time after withdrawal from the antipsychotic drug. At first, studies indicated that the syndrome persisted for only 10–60 days after neuroleptic withdrawal (Ayd 1961; Demars 1966). More recently, studies have reported greater persistence of this syndrome, particularly in the elderly. Stephen and Williamson (1984) verified that the overall time of remission of drug-induced parkinsonism was 7 weeks in a group of 48 elderly patients (76 years old, in average); however, one of them did not recover until 36 weeks post-treatment. In another study, Crane (1976) reported that in 5 of 12 patients, the syndrome persisted for 6 months to 2 years after neuroleptic withdrawal. Nevertheless, the possible pre-existence of idiopathic Parkinson's disease that was previously "masked" by neuroleptic-induced parkinsonism should be considered in cases where the syndrome persists for a very long period of time in the elderly (Hansen and Hoffman 1997).

Neuroleptic-Induced Catalepsy

Whereas experimental catalepsy has been considered a useful tool for the study of neurobiological mechanisms of Parkinson's disease (Sanberg et al. 1988), catalepsy induced by dopamine receptor blockers has been recognized as an animal model of neuroleptic-induced parkinsonism (Janssen et al. 1965; De Graaf and Korf 1986; Sanberg et al. 1988; Gras et al. 2008). Indeed, much evidence has associated catalepsy induced by dopaminergic blockers with a reduction in nigrostriatal dopaminergic activity (Fot et al. 1970; Sanberg 1980; Calderon et al. 1988), similar to that observed in neuroleptic-induced parkinsonism (see Hansen and Hoffman 1997). For instance, the classical Sanberg (1980) study verified that catalepsy induced by haloperidol was almost completely abolished in rats previously treated with an intrastriatal injection of kainic acid, but not in rats previously submitted to cortical ablation. The author concluded that the cataleptic effects of haloperidol were mediated by post-synaptic dopamine receptors localized in striatal neurons. In parallel, neuroleptics with little inhibitory effect on the nigrostriatal dopaminergic system, such as thioridazine, sulpiride and clozapine, are less effective in producing catalepsy in rodents and have been associated with a lower incidence of neuroleptic-induced parkinsonism in humans (Honma and Fukushima 1976, 1978; Sanberg et al. 1988). Finally, it is of note that, similar to what was previously described in the literature for neuroleptic-induced parkinsonism, there are reports indicating tolerance, sensitization or lack of alteration of rodents' cataleptic behavior following repeated administration of neuroleptics (Barnes et al. 1990; Frussa-Filho et al. 1992).

The study of factors that may modify the evolution of catalepsy induced by repeated treatment with dopaminergic blockers is relevant, since it can contribute to understand the plasticity of dopaminergic transmission as well as the evolution of neuroleptic-induced parkinsonism. As mentioned above, different studies have reported a decrease (tolerance), an enhancement (sensitization) or no alteration of catalepsy during long-tem administration of neuroleptics.

Specifically, the development of tolerance to neuroleptic-induced cataleptic behavior was verified by Asper et al. (1973), Gessa and Tacliamonte (1975) and Ezrin-Waters and Seeman (1977) in rats. Curiously, Campbell and Baldessarini (1981) reported tolerance to the *duration* of catalepsy, but sensitization to the emergence of this behavior (i.e., a decrease in the *latency* to the beginning of this parameter), after 5 months of haloperidol i.p. administration in rats. Conversely, sensitization to the duration of catalepsy was observed by Barnes et al. (1990) following daily or weekly administration of different doses of haloperidol for 22 days. Furthermore, Antelman et al. (1986) verified that a single injection of haloperidol enhances the cataleptic behavior of rodents in response to a subsequent injection of this neuroleptic administered 15 or 22 (but not 8) days later. Finally, concerning studies that report no effects of repeated treatment with neuroleptics on the cataleptic behavior of rodents, György et al. (1969) verified that long-term administration of haloperidol produced neither tolerance nor sensitization to the cataleptic behavior of rats. In addition, no alteration in the cataleptic behavior of rats that were orally treated with 10 mg/kg flupentixol or 10 mg/kg haloperidol for 12 consecutive days was observed by Nielsen (1974). Puri and Lal (1974) also failed to observe modifications in the percentage of rats that presented catalepsy after a 12-day treatment with increasing doses (5–20 mg/day) of haloperidol. In line with these latter studies, we verified that daily administration of 2 mg/kg haloperidol for 30 days did not alter the duration of catalepsy in rats (Frussa-Filho et al. 1992). This lack of modifications in catalepsy was observed in both old and socially isolated animals.

Several methodological differences may explain the discrepancies mentioned above. First, the effects of repeated administration of neuroleptics on the evolution of experimental catalepsy should be noted. Although the same neuroleptic (haloperidol) was used in all investigations, the studies differed with regard to the dose, schedule and duration of treatment and the administration pathway. Furthermore, other differences related to the type of apparatus used to measure catalepsy, the parameters considered to quantify this behavior and the moment at which quantification of the cataleptic behavior was conducted (i.e., several times during the treatment or a single time over different periods after the end of treatment) may have contributed to the contradictory results found in the literature.

Studies examining the influence of methodological variables on the evolution of experimental catalepsy are rare. Indeed, despite the above-mentioned study performed in our laboratory, which demonstrated the absence of effects of age and housing conditions on the evolution of catalepsy induced by haloperidol in rats (Frussa-Filho et al. 1992), only two other studies in this field have been published (Ezrin-Waters and Seeman 1977; Barnes et al. 1990). Concerning the Ezrin-Waters and Seeman's investigation (1977), tolerance to catalepsy following long-term oral administration of haloperidol (0.75 and 1.5 mg/kg) was observed with different doses and administration pathways employed during the challenge session. Barnes et al. (1990) investigated the effects of dose (0.05, 0.5, 1.5 and 5 mg/kg), the frequency of administration (daily or weekly), the frequency of tests (repeated or a single test) and the apparatus (horizontal bar or inclined screen) on the evolution of catalepsy induced by haloperidol. The analysis of the results revealed that, although no experimental variable produced a definitive pattern of alteration in catalepsy over time, all of the factors influenced this behavior when considered together. Thus, tolerance was observed only for the animals treated daily with 1.5 mg/kg haloperidol and tested repeatedly on the horizontal bar. On the other hand, sensitization was observed for the animals treated daily or weekly with different doses of haloperidol and after repeated tests or a single test either on the horizontal bar or on the inclined screen. Nevertheless, sensitization developed more vigorously following weekly administration of haloperidol and with repeated tests using the horizontal bar.

Concerning sensitization, different hypotheses have been formulated in order to explain the enhancement of catalepsy following repeated neuroleptic administration. Within this context, the possible influence of repeated exposure of an animal to the catalepsy test in the evolution of the cataleptic behavior has been the cause of extensive investigation and controversy. Sanberg et al. (1980) and Brown and Handley (1980) have suggested that the phenomenon may be related to a type of tonic immobility that can be progressively elicited in rodents after just a few seconds of retention (Klemm 1971; Bures et al. 1976; Amir et al. 1981). On the other hand, Iwata et al. (1989) have suggested that the phenomenon may be related to learning processes. These possibilities will be discussed in more detail elsewhere in this chapter.

Regarding the mechanisms that have been proposed to explain tolerance to the cataleptic behavior following repeated administration of neuroleptics, the development of post-synaptic D_2 dopamine receptor supersensitivity in the striatum has been suggested (Ezrin-Waters and Seeman 1977; Korf 1988). Indeed, striatal dopaminergic supersensitivity produced by long-term administration of dopaminergic blockers has been demonstrated electrophysiologically (Skirboll and Bunney 1979), biochemically (Gianutsos et al. 1975; Burt et al. 1977; Vital et al. 1998) and behaviorally. Concerning behavioral models, several studies have demonstrated that apomorphine-induced stereotypy is significantly increased after withdrawal from not only haloperidol (Tarsy and Baldessarini 1974; Bernardi and Palermo Neto 1979; Bernardi et al. 1981; Vital et al. 1995; Queiroz and Frussa-Filho 1999; Abílio et al. 1999, 2003, 2004; Perry et al. 2004), but also from metoclopramide (Frussa-Filho and Palermo-Neto 1988), sulpiride (Frussa-Filho and Palermo-Neto 1990) and droperidol (Frussa-Filho and Palermo-Neto 1991) treatment.

Finally, in addition to the previously mentioned factors (training, supersensitivity and conditioning), the time interval between the administration of an injections seems to be crucial to the evolution of cataleptic behavior induced by neuroleptics. Thus, Masuda et al. (1982) verified that, whereas daily administration of different doses of haloperidol produced tolerance to the cataleptic effect of the drug in mice, administration of haloperidol every 2 or 4 days produced robust sensitization to this behavior.

Applicability: SCH 23390-Induced Catalepsy as an Animal Model for Evaluation of Dopaminergic Transmission Mediated by D_1 Dopamine Receptors

SCH 23390-Induced Catalepsy Versus Neuroleptic-Induced Catalepsy

Soon after identification of the antipsychotic effect of chlorpromazine, 53 years ago, recognition of dopamine D_2 receptor antagonism as a common denominator among pharmacologically diverse antipsychotics resulted in the exploration of a large number of selective D_2 antagonists. Indeed, all conventional neuroleptics that are effective for the treatment of schizophrenia are mixed D_1/D_2 antagonists or selective D_2 antagonists (see Hess et al. 1986; Waddington 2000). As a consequence, the blockade of D_2 dopamine receptors was suggested as the mechanism of antipsychotic action of these drugs (Creese et al. 1976; Seeman et al. 1976). The ability of these conventional neuroleptics to produce catalepsy showed a positive correlation with their ability to improve psychotic symptoms. Furthermore, the ability of these drugs to induce catalepsy and promote antipsychotic effects showed a positive correlation with their ability to antagonize D_2 receptors (Ezrin-Waters et al. 1976; Owen et al. 1978; Seeman 1981; Laduron 1983; Creese et al. 1983).

For many years, catalepsy was specifically and exclusively associated with the blockade of D_2 receptors (see Klemm and Block 1988); however, this notion was contradicted by introduction of the selective D_1 antagonist SCH 23390 (SCH) in the beginning of the 1980's (Iorio et al. 1983; Hyntel 1983, 1984). In the first study in this field, Iorio et al. (1983) failed to verify catalepsy after the administration of SCH, in the following year, Christensen et al. (1984) demonstrated the cataleptic action of this D_1 antagonist. This effect elicited by SCH was confirmed in rats (Hess et al. 1986; Schmidt et al. 2005) and mice (Klemm and Block 1988. Meschler et al. 2000) in subsequent studies using different catalepsy tests (Morelli and Di Chiara 1985). Hess et al. (1986) that the in vivo induction of catalepsy by SCH is due to the blockade of D_1 receptors.

SCH-induced catalepsy and neuroleptic-induced catalepsy share similarities but also present important differences. Calderon et al. (1988) demonstrated that catalepsy induced by SCH or haloperidol is abolished by a quinolinic acid-induced striatal lesion. In this regard, catalepsy induced either by D_1 or D_2 blockers is mainly mediated by striatal dopamine receptors (Calderon et al. 1988; Costall and Naylor 1975; Ossowska et al. 1990) that are different efferent striatal pathways (Ögren and Fuxe 1988). A study conducted by Wardas et al. (1995) demonstrated that the intrastriatal administration of a D_1 antagonist produces dose-dependent experimental catalepsy. Ossowska et al. (1990) verified that bilateral injection of the D_1 antagonist SCH, the preferentially D_2 antagonist haloperidol or the selective D_2 antagonist sulpiride into the ventro-rostral striatum induces dose-dependent catalepsy. However, the injection of these drugs into the dorso-rostral striatum was ineffective Notwithstanding these observations, several studies have demonstrated the cataleptic effects of D_1 antagonists after their administration not only into the caudate-putamen (Hauber et al. 2001; Wardas et al. 1995; Ossowska et al. 1990; Rowlett et al. 1991), but also in the subthalamic nucleus (Hauber 1998) the globus pallidus (Hauber and Lutz 1999) and the nucleus accumbens (Ossowska et al. 1990). In the above-mentioned study of Wardas et al. (1995), equivalent doses of SCH produced similar catalepsy when administered either into the nucleus accumbens or the striatum; however, administration of a 100-fold higher dose of sulpiride into the nucleus accumbens was necessary to produce similar catalepsy to that observed after administration of this antagonist into the striatum. This result suggests that, in the nucleus accumbens, D_1, but not D_2, dopamine receptors play an important role in catalepsy induced by dopaminergic antagonists. Interestingly, haloperidol produced similar effects when administered into both brain structures, probably due to its mixed action on dopaminergic receptors. Thus, it seems that several brain structures may be involved in catalepsy induced by the D_1 antagonist SCH.

Different authors have suggested that catalepsy induced by SCH is phenomenologically distinct from catalepsy induced by neuroleptics (Morelli and Di Chiara 1985; Undie and Friedman 1988). Whereas SCH-induced catalepsy develops rapidly (i.e., presents short latency to occurrence) and persists for a short period of time, neuroleptic-induced catalepsy develops gradually (i.e., exhibits longer latency to occurrence) and persists for longer periods of time (Ögren and Fuxe 1988; Ushijima et al. 1995). Another difference between the cataleptic responses induced by D_1 or D_2 blockade was demonstrated by Ögren and Fuxe (1988). These authors verified that the benzodiazepine receptor antagonist RO 15-1788 reduced catalepsy induced by raclopride, but not by SCH. They proposed that SCH-induced catalepsy is predominantly mediated by nigrothalamic GABAergic pathways, while neuroleptic-induced catalepsy is predominantly mediated by striatopallidal GABAergic pathways.

In addition, both catalepsy induced by SCH and catalepsy induced by neuroleptics are influenced by the cholinergic system. Within this context, catalepsy induced by haloperidol (Klemm 1985; Sayers et al. 1976; Setler et al. 1976; Arnt et al. 1986), fluphenazine, spiroperidol (Undie and Friedman 1988), droperidol (Arnt and Christensen 1981) or SCH (Morelli and Di Chiara 1985; Undie and Friedman 1988; Arnt et al. 1986) was attenuated by atropine or scopolamine (Morelli and Di Chiara 1985; Arnt and Christensen 1981; Setler et al. 1976; Arnt et al. 1986). However, only SCH-induced catalepsy was potentiated by the cholinergic agonist pilocarpine (Undie and Friedman 1988; Klemm 1985). Thus, the authors suggested that, although the cataleptic responses produced by either D_1 or D_2 antagonists are influenced by dopamine/acetylcholine interactions, different subtypes of muscarinic cholinergic receptors seem to participate in each response.

In addition to providing a functional correlate of dopamine D_1 receptor antagonism (Mizuki et al. 1996), SCH-induced catalepsy is also a useful model for investigation of the functional interactions between D_1- and D_2-like dopamine receptors (see Klemm 1993). Several studies in the literature have suggested a complex interaction between catalepsy responses induced by SCH and neuroleptics. Different authors have verified that SCH-induced catalepsy is antagonized by selective D_2

agonists in rats (Meller et al. 1985; Morelli and Di Chiara 1985) and mice (Zarrindast and Habibi-Moini 1991). Conversely, it should be noted that Klemm and Block (1988) verified an increase in catalepsy induced by SCH following administration of the D_2 agonist bromocriptine in mice tested on the inclined screen.

Interestingly, data relating to the effects of the selective D_1 agonist SKF 38393 (SKF) on catalepsy induced by D_2 antagonists have been contradictory. Zarrindast and Habibi-Moini (1991) verified that SKF (8 mg/kg, i.p.) attenuates sulpiride-induced catalepsy in mice submitted to the bar test. Conversely, Ögren and Fuxe (1988) reported that 5 mg/kg SKF (s.c.) potentiated catalepsy produced by raclopride in rats tested on the inclined screen test.

The interaction between cataleptic behaviors produced by SCH and D_2 antagonists can be demonstrated by their synergy. Klemm and Block (1988) observed that both SCH (1 mg/kg) and the D_2 antagonist molindone (10 mg/kg) produce cataleptic responses of similar magnitude in mice submitted to the inclined screen test. However, the concomitant administration of these drugs produced a cataleptic response with a magnitude that was higher than the sum of the cataleptic responses induced by each drug alone. This synergistic effect of SCH and D_2 antagonists on cataleptic behavior was also observed in rats by Wanibuchi and Usuda (1990).

Evolution of SCH-Induced Catalepsy

Given that: (1) blockade of D_1 receptors produces catalepsy, (2) many classical neuroleptics predominantly antagonize D_2 receptors, but may also block D_1 receptors and (3) that there is a close interaction between D_1 and D_2 receptors in the development of cataleptic behavior, the participation of D_1 receptors in neuroleptic-induced parkinsonism has been suggested (Morelli and Di Chiara 1985). Investigation of the evolution of cataleptic behavior during long-term administration of SCH may help to elucidate the participation of D_1 receptors in the above-mentioned contradictory results related to the evolution of both neuroleptic-induced catalepsy and parkinsonism.

Hess et al. (1986) were the first scientists to study the evolution of catalepsy during long-term administration of SCH. Using the bar test, these authors quantified the cataleptic behavior of rats during the 1st, 3rd, 5th, 7th, 14th an 21st days of the daily treatment with 0.5 mg/kg SCH or 0.2 mg/kg of the D_2 antagonist spipirone. In this study, tolerance to the cataleptic effect of spipirone, but not of SCH, developed. However, these results should be interpreted with caution because of the limited cut-off time used by the authors (120 s). In fact, while on the 1st day of treatment the catalepsy value for SCH-treated rats was 107 ± 30 s, catalepsy reached a maximum value of 120 ± 0 s on the other observation days. This methodology failure prevented evaluation of the sensitization phenomenon in this investigation. Interestingly, Hess et al. (1986) reported that SCH-treated animals presented an increase in the number of D_1 (but not D_2) receptors as well as enhancement of spontaneous locomotion and SKF- induced locomotion and stereotypy during withdrawal. Such supersensitivity of D_1 receptors should have facilitated the development of tolerance to the catalepsy induced by SCH; nonetheless, it did not occur.

In another study, Hess et al. (1988) demonstrated the development of tolerance to the cataleptic behavior induced by spipirone in rats using the bar test. Similar to the findings of their previous study, neither tolerance nor sensitization was observed after long-term administration of SCH for 21 days. Again, the interpretation of these results may have been compromised because maximum values of catalepsy were reached 20 to 40 min after injection of the drug (peak of action of SCH) due to the short cut-off that was used. Although long-term treatment with SCH did not produce tolerance to the cataleptic behavior elicited by a subsequent injection of SCH, it produced tolerance to the catalepsy elicited by a subsequent injection of spipirone (cross-tolerance). Long-term treatment with spipirone produced tolerance to its own cataleptic behavior, but not to the catalepsy elicited by SCH. Long-term co-treatment with SCH and spipirone did not produce tolerance to the cataleptic effect of these drugs and, again, the detection of a possible sensitization was compromised by the maximum values of catalepsy observed following the 1st day of observation.

Lappalainen et al. (1989) reported that chronic treatment of rats with a low dose regimen (0.1 mg/kg) of SCH 23390 did not lead to the development of tolerance to the cataleptic response in the bar test. In addition, although the cataleptic effect of a higher dose regimen (0.5 mg/kg) of SCH 23390 was significantly reduced after a 6-day treatment, subacute tolerance was gradually reversed and was no longer significant after 12 and 18 days. Different functional mechanisms were proposed to explain the lack of tolerance to SCH 23390- induced catalepsy in these three studies. Possible involvement of a learning process due to repeated testing was not experimentally considered by Hess et al. (1986, 1988) or Lappalainen et al. (1989) despite the fact that the animals were repeatedly tested for catalepsy during the course of the experiments.

In order to investigate this possibility, we examined the effects of chronic SCH 23390 treatment on the cataleptic behavior of mice tested repeatedly over time or only once on the horizontal bar (Chinen and Frussa-Filho 1999). To functionally evaluate the sensitivity of D_1-like dopamine receptors after chronic SCH 23390 treatment, the effects of a challenge injection of the D_1-like dopamine receptor agonist SKF 38393 on grooming behavior were quantified. Interestingly, the data showed that mice that were repeatedly exposed to the bar test during repeated SCH treatments presented strong sensitization to the cataleptic behavior induced by this dopamine D_1 blocker. Conversely, a trend toward tolerance was observed in mice submitted to a single test at the end of the repeated treatment regimen with the dopamine D_1 antagonist. Importantly, sensitization to the cataleptic behavior in the multiple test SCH 23390-treated group was observed in spite of the development of a D_1-like dopamine receptor supersensitivity, which was suggested by an increase in SKF 38393-induced grooming behavior. Indeed, this D_1-like dopamine receptor supersensitivity phenomenon would be expected to induce tolerance and not sensitization to SCH-induced catalepsy.

Collectively, the above-described results concerning the evolution of neuroleptic- or SCH-induced catalepsy demonstrate that they do not always reflect the status of nigrostriatal dopaminergic function. A possible explanation for this flaw may stem from the fact that the catalepsy behavior of a rodent is subject to conditioning and other types of learning processes. These conditioning and learning processes may also contribute to the evolution of neuroleptic-induced parkinsonism (St Jean et al. 1964a, b). This conditioned or "learned" catalepsy in rodents can also be applied to the study of depression.

Applicability: Conditioned and Learned Catalepsy as Models of a Nocebo Effect and Depression, Respectively

The possible influence of conditioning or other types of learning processes in the expression of cataleptic behavior was originally suggested by Stanley and Glick (1976). These authors verified that rats repeatedly tested on the horizontal bar after an acute injection of haloperidol presented higher catalepsy scores than animals that received the same treatment but were tested for catalepsy only once. This phenomenon was later replicated in both rats and mice (Brown and Handley 1980; Hillegaart et al. 1987; Iwata et al. 1989; Ferré et al. 1990).

Two main hypotheses can be raised in order to explain this enhanced catalepsy observed in neuroleptic-treated rodents that are repeatedly tested on the horizontal bar. First, Pavlovian conditioning between the rigidity state of the animals and the procedure of placing them on the bar may have been occurred. Second, this increased catalepsy could be the consequence of a learning process that is independent of neuroleptic-induced rigidity. More specifically, when placed repeatedly on a bar with an imposed posture, the animals may "learn" to maintain such posture over long periods of time.

Consistent experimental evidence corroborates both possibilities, which are not necessarily exclusive. In addition, these possibilities may be related to different applications of the catalepsy test.

Conditioned Catalepsy as an Animal Model of the Nocebo-Effect

As mentioned above, conditioned catalepsy in neuroleptic- (or other dopamine receptor blocker-) treated rodents could be the result of Pavlovian conditioning between the rigidity produced by these drugs and the procedure of repeatedly placing the animals on the bar. In this case, the dopamine receptor blocker-induced rigidity would be the unconditioned stimulus, the cataleptic response on the bar would be the unconditioned response, the procedure of placing the animal on the bar would be the conditioned stimulus and the increase in the cataleptic response after repeated testing would be the conditioned response. This hypothesis is supported by the demonstration that the rigidity induced by neuroleptic drugs or SCH could be conditioned to the injection procedure as well (Chinen and Frussa-Filho 1999). In that study, we found that the catalepsy duration (quantified only once at the end of the pharmacological treatment) was longer in mice that were treated chronically with SCH and challenged with saline (group SCH-SAL) compared to mice submitted to an identical long-term treatment with SCH 23390 but not challenged with any injection (group SCH-NI) as well as compared to mice chronically treated and challenged with saline (group SAL-SAL). Since the cataleptic behavior of the SCH-NI, the SAL-SAL and the SAL-NI groups did not differ (i.e., the challenge saline injection did not produce a cataleptic effect per se, nor did SCH pre-treatment per se), these data strongly suggest the development of a conditioning phenomenon between SCH-induced rigidity and the injection procedure. The same pattern of results were obtained when the classical neuroleptic haloperidol or the selective D_2 dopamine receptor blocker metoclopramide were used instead of SCH (Chinen and Frussa-Filho 1999).

Irrespective of the conditioned stimulus (injection procedure or the procedure of placing the animal on the cataleptic bar), neuroleptic-induced conditioned catalepsy in rodents can be used as a new and simple animal model of the placebo (or, in such a case, nocebo) effect (see Chinen and Frussa-Filho 1999). This should be of interest because, although the view of placebo as a conditioned response is not new, previous analyses have remained at a descriptive rather than an experimental level (see Suchman and Ader 1992).

From a clinical point of view, since catalepsy is a well accepted model of neuroleptic-induced parkinsonism, it is tempting to speculate whether these conditioning phenomena to neuroleptic- (or SCH-) induced rigidity—quantified through the catalepsy test—are related to the development of tolerance in neuroleptic-induced parkinsonism in humans (see above). Interestingly, as stated earlier, psychological factors related to the placebo effect may play a fundamental role in the treatment of neuroleptic-induced parkinsonism (St Jean et al. 1964a, b).

Learned Catalepsy as an Animal Model of Depression

Although consistent evidence supports the existence of Pavlovian conditioning of drug-induced rigidity in enhancement of the catalepsy by test repetition (or even by repeated drug injections), it has been demonstrated that the test repetition-induced increase in catalepsy could be the consequence of a learning process independent of drug-induced rigidity. Indeed, test repetition-induced catalepsy may develop in drug-naive rodents. This behavioral phenomenon was firstly reported by Sanberg et al. (1980). These authors demonstrated that control rats (injected with saline or not injected) showed an increased cataleptic response with repeated testing (the animals were submitted to 4 bar exposures/day for 4 day-sessions that were separated by 48 h). This finding was corroborated by other studies that used both

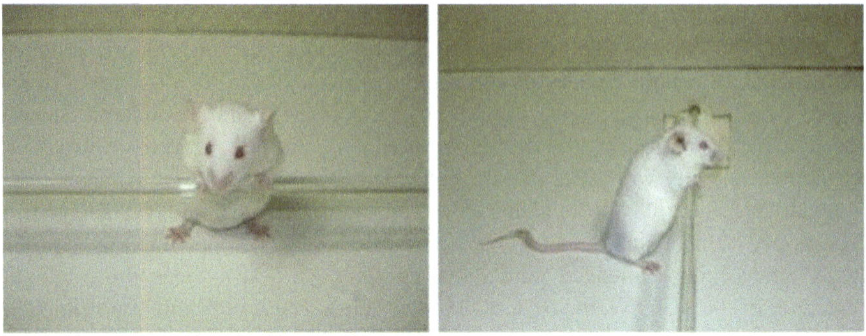

Fig. 26.2 Learned catalepsy in mice

rats and mice (Brown and Handley 1980; Ferré et al. 1990; Chinen and Frussa-Filho 1999) (Fig. 26.2).

The participation of learning processes in catalepsy has been elegantly demonstrated by Iwata et al. (1989). These authors verified that whereas cataleptic behavior was increased when mice were repeatedly trained (first tests) and challenged (last test) with either the forepaws or the hind limbs on the horizontal bar, this increase in cataleptic behavior did not occur when training was conducted with the forepaws on the bar and challenge with the hind limbs on the apparatus.

According to Ferré et al. (1990), the animal would "learn" to maintain the imposed abnormal posture in order to avoid handling when both of its forepaws touch the floor. Indeed, handling has been considered a stressor (Axelrod and Reisine 1984). Within this context, conditioned catalepsy would be very similar to immobility behavior quantified in Porsolt's forced swimming test (Porsolt et al. 1977). Indeed, in the classical Porsolt's test, rats or mice are placed in an inescapable receptacle filled with water and the duration of immobility is used as a depressive behavioral index. In other words, when the animal "gives up" escaping, it becomes immobile, which is a behavior that is counteracted by classical antidepressant agents. Similarly, leaned catalepsy occurs when an animal "gives up" moving from the bar.

The efficacy of antidepressant agents in modifying catalepsy induced by neuroleptics has been documented with rather contradictory results (Al-Khatib et al. 1989; Silva et al. 1990; Khisti et al. 1997; Pires et al. 1998, 2005). This concern notwithstanding, studies investigating the effects of antidepressants on "learned" catalepsy (i.e., repeated testing-induced catalepsy in drug-naïve animals) are rare. Here, we describe data from our laboratory as well as the other two relevant studies, all of which show that learned catalepsy can be used as a model of depression in rodents.

Fundaró (1998) induced learned catalepsy by repeatedly pinching mice at the scruff and, immediately after each pinch, placing them on parallel bars. This author verified that female mice presented a more rapid occurrence and a prolonged duration of learned catalepsy when compared to male controls. This finding is consistent with data from the literature showing that, in humans, women tend to have an earlier age-at-onset for depression, exhibit more frequent depressive episodes and have a greater number of depressive symptoms than men (Smith et al. 2008). Importantly, in the study of Fundaró (1998) the antidepressants imipramine (20 mg/kg, i.p.), amitriptiline (30 mg/kg, i.p.) and fluoxetine (20 mg/kg, i.p.) augmented the number of pinches/bar exposures necessary to induce catalepsy and reduced the total duration of catalepsy. More recently, using a very similar protocol, Tikhonova et al. (2006) demonstrated that chronic treatment with the antidepressant imipramine reduced the severity of learned catalepsy in NPK mice. This mouse strain is predisposed to this type of catalepsy and—interestingly—presents many depression-like behavioral features.

Importantly, while Fundaró (1998) demonstrated the inhibitory effects of antidepressants acutely administered in the learned catalepsy phenomenon, Tikhonova et al. (2006) demonstrated such an inhibitory effect after withdrawal from repeated antidepressant treatment (3 days after the end of antidepressant treatment). Based upon this result, we examined whether repeated antidepressant treatment would effectively attenuate learned catalepsy *during* drug treatment (which would more closely resemble the clinical approach). Thus, we performed an experiment (unpublished data) in which 30 Swiss female mice (3 months old, 20–30 g) were randomly allocated into 3 groups of 10 animals each: VEH, FLU5 and FLU10. The animals were treated daily with an i.p. injection of vehicle solution, 5 or 10 mg/kg fluoxetine, respectively, for 13 days. On 6th, 8th, 10th, 12th and 14th days of treatment, the mice were placed on the horizontal bar 60 min before receiving their treatment, in order to quantify their cataleptic behavior (catalepsy quantification sessions 1, 2, 3, 4 and 5, respectively). In order to induce the learned catalepsy phenomenon, the catalepsy behavior quantification method described in the beginning of the chapter was modified. Briefly, each mouse was submitted to 15 bar exposures, which a 5-min intertrial period and cut-off of 250 s. Figure 26.3 illustrates the results of this

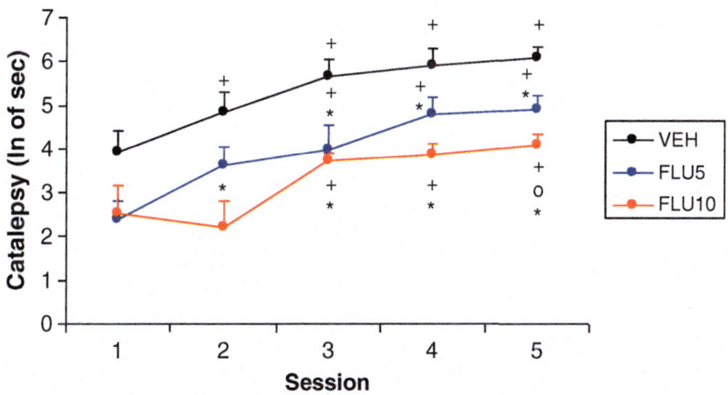

Fig. 26.3 Evolution of learned catalepsy in mice submitted to repeated tests on the horizontal bar during treatment with vehicle or fluoxetine. The animals were treated daily with an injection of vehicle (VEH), or 5 (FLU5) or 10 (FLU10) mg/kg fluoxetine for 13 days. They were tested for catalepsy behavior on days 6 (session1), 8 (session 2), 10 (session 3), 12 (session 4) and 14 (session 5) of treatment. ANOVA followed by Duncan's test and *T* test for paired samples. *$p<0.05$ compared to the VEH group in the same session; °$p<0.05$ compared to the FLU5 group in the same session; ⁺$p<0.05$ compared to session 1

experiment. All of the groups presented a progressive increase in the duration of catalepsy during the sessions, indicating the development of learned catalepsy for all animals. However, the development of learned catalepsy seemed to be retarded by fluoxetine treatment, since only the VEH group presented an increase in catalepsy duration after session 2, compared to session 1 (in the two fluoxetine groups, a significant increase in catalepsy duration was verified from session 3 onwards). Notably, when compared to the VEH group, the animals treated with fluoxetine exhibited a decrease in catalepsy duration during most of the sessions (from session 3–5 for the FLU5 group and from session 2–5 for the FLU10 group). In addition, the FLU10 group presented a decrease in catalepsy duration when compared to the FLU5 group in session 5 (14th day of treatment).

These data show that repeated treatment with the antidepressant fluoxetine retarded the evolution and shortened the duration of learned catalepsy in mice, emphasizing the importance of this test as an animal model of depression.

Ethical and Final Comments

The catalepsy test has been demonstrated to be a valuable tool for the study of neurobiological and pathogenic mechanisms as well as potential therapeutic approaches to treat symptoms of a number of human diseases. Besides its importance in scientific studies, the catalepsy test (specifically the horizontal bar test) is a very cheap and ethical model. This test can be performed using only a piece of cylindrical glass and two small boxes that support the glass.

From an ethical standpoint, as an open-field test (see Chap. 26), the catalepsy test was developed for animal ***observation***. Therefore, environmental conditions must be minimally averse in order to avoid the influence of stress on the behavior of the animals. In addition, success of the experiment requires that the animals be *gently* placed on the bar.

References

Abílio VC, Freitas FM, Dolnikoff MS, Castrucci AML, Frussa-Filho R. Effects of continuous exposure to light on behavioral dopaminergic supersensitivity. Biol Psychiatry. 1999;45:1622–9.

Abílio VC, Vera Jr JA, Ferreira LS, Duarte CR, Martins CR, Torres-Leite D, et al. Effects of melatonin on behavioral dopaminergic supersensitivity. Life Sci. 2003;7:3003–15.

Abílio VC, Silva RH, Carvalho RC, Grassl C, Calzavara MB, Registro S, et al. Important role of striatal catalase in aging- and reserpine-induced oral dyskinesia. Neuropharmacology. 2004; 47:263–72.

Agovic MS, Yablonsky-Alter E, Lidsky TI, Banerjee SP. Mechanisms for metoclopramide-mediated sensitization and haloperidol-induced catalepsy in rats. Eur J Pharmacol. 2008;587:181–6.

Alam M, Mayarhofer A, Schmidt WJ. The neurobehavioral changes induced by bilateral rotenone lesion in medial forebrain bundle of rats are reversed by L-DOPA. Behav Brain Res. 2004;151:117–24.

Al-Khatib IM, Fujiwara M, Ueki S. Relative importance of the dopaminergic system in haloperidol-catalepsy and the anticataleptic effect of antidepressants and methamphetamine in rats. Pharmacol Biochem Behav. 1989;33:93–7.

Amir S, Brown ZW, Amit Z, Ornstein K. Body pinch induces long lasting cataleptic like immobility in mice: behavioral characterization and the effect of naloxone. Life Sci. 1981;28:1189–94.

Andén NE, Dahlström A, Fuxe K, Larsson K. Functional role of the nigro-neostriatal dopamine neurons. Acta Pharmacol Toxicol. 1966;24:263–74.

Antelman SM, Kocan D, Edwards DJ, Knopf S, Perel JM, Stiller R. Behavioral effects of a single neuroleptic treatment grow with the passage of time. Brain Res. 1986;385:58–67.

Arnt J, Christensen AV. Differential reversal by scopolamine and THIP of the antiestereotypic and cataleptic effects of neuroleptics. Eur J Pharmacol. 1981;69:107–11.

Arnt J, Hytell J, Bach-Lauritsen T. Further studies of the mechanism behind scopolamine-induced reversal of antistereotypic and cataleptogenic effects of neuroleptics in rats. Acta Pharmacol Toxicol. 1986;59:319–24.

Asper H, Baggiolini M, Burki HR, Lauener H, Ruch W, Stille G. Tolerance phenomena with neuroleptics catalepsy, apomorphine stereotypies and striatal dopamine metabolism in the rat after single and repeated administration of loxapine and haloperidol. Eur J Pharmacol. 1973;22: 287–94.

Axelrod J, Reisine TD. Stress hormones: their interaction and regulation. Science. 1984;224:452–9.

Ayd Jr FJ. A survey of drug-induced extrapyramidal reactions. JAMA. 1961;175:1054–60.

Ayd FJ. Neuroleptics an antiparkinson drugs. Int Drug Ther Newslett. 1971;6:33–6.

Bara-Jimenez W, Sherzai A, Dimitrova T, Favit A, Bibbiani F, Gillespie M, et al. Adenosine A(2A) receptor antagonist treatment of Parkinson's disease. Neurology. 2003;61:293–6.

Barnes DE, Robinson B, Csernansky JG, Bellows EP. Sensitization versus tolerance to haloperidol-induced catalepsy: multiple determinants. Pharmacol Biochem Behav. 1990;36: 883–7.

Bernardi MM, Palermo Neto J. Effects of abrupt and gradual withdrawal from long-term haloperidol treatment on open-field of rats. Psychopharmacology (Berl). 1979;65:247–50.

Bernardi MM, De Souza H, Palermo Neto J. Effects of single and long-term haloperidol administration on open field behavior of rats. Psychopharmacology (Berl). 1981;73:171–5.

Bernheimer H, Birkmayer W, Hornykiewicz O, Jellinger K, Seitelberger F. Brain dopamine and the syndromes of Parkinson and Huntington. Clinical, morphological and neurochemical correlations. J Neurol Sci. 1973;20:415–55.

Betarbet R, Sherer TB, Mackenzie G, Garcia-Osuna M, Panov AV, Greenamyre JT. Chronic systemic pesticide exposure reproduces features of Parkinson's disease. Nat Neurosci. 2000;3:1301–6.

Bloom FE, Algeri S, Groppetti A, Revuelta A, Costa E. Lesions of central norepinephrine terminals with 6-OH-dopamine: biochemistry and fine structure. Science. 1969;166:1284–6.

Brown J, Handley SL. The development of catalepsy in drug-free mice on repeated testing. Neuropharmacology. 1980;19:675–8.

Bures J, Buresova O, Huston JP. Techniques and basic experiments for the study of brain and behavior. Amsterdam: Elsevier; 1976.

Burt DR, Creese I, Snyder SH. Antischizophrenic drugs: chronic treatment elevates dopamine receptor binding in brain. Science. 1977;196:326–8.

Cahan RB, Parrish DD. Reversibility of drug-induced parkinsonism. Am J Psychiatry. 1960; 116:1022–3.

Calderon SF, Sanberg PR, Norman AB. Quinolinic acid lesions of rat striatum abolish D1- and D2-dopamine receptor-mediated catalepsy. Brain Res. 1988;450:403–7.

Campbell A, Baldessarini RJ. Tolerance to behavioral effects of haloperidol. Life Sci. 1981;29: 1341–6.

Cannon JR, Hua Y, Richardson RJ, Xi G, Keep RF, Schallert T. The effect of thrombin on a 6-hydroxydopamine model of Parkinson's disease depends on timing. Behav Brain Res. 2007; 183:161–8.

Casey DE, Povlsen UJ, Meidahl B, Gerlach J. Neuroleptic-induced tardive dyskinesia and parkinsonism: changes during several years of continuing treatment. Psychopharmacol Bull. 1986;22:250–3.
Chinen CC, Frussa-Filho R. Conditioning to injection procedures and repeated testing increase SCH 23390-induced catalepsy in mice. Neuropsychopharmacology. 1999;21:670–7.
Christensen AV, Arnt J, Hyttel J, Larsen JJ, Svendesen O. Pharmacological effects of a specific dopamine D1 antagonist SCH 23390 in comparison with neuroleptics. Life Sci. 1984;34: 1529–40.
Conceição IM, Frussa-Filho R. Effects of a single administration of buspirone on catalepsy, yawning and stereotypy in rats. Braz J Med Biol Res. 1993;26:71–4.
Costall B, Naylor RJ. Neuroleptic antagonism of dyskinetic phenomena. Eur J Pharmacol. 1975;33:301–12.
Crane J. Psychiatry at a mission hospital in South Africa. Ulster Med J. 1976;45:73–5.
Creese I, Burt DR, Snyder SH. Dopamine receptor binding predicts clinical and pharmacological potencies of antischizophrenic drugs. Science. 1976;192:481–3.
Creese I, Sibley DR, Hamblin MW, Leff SE. The classification of dopamine receptors: relationship to radioligand binding. Annu Rev Neurosci. 1983;6:43–71.
Datla KP, Zbarsky V, Rai D, Parkar S, Osakabe N, Aruoma OI, Dexter DT. Short-term supplementation with plant extracts rich in flavonoids protect nigrostriatal dopaminergic neurons in a rat model of Parkinson's disease. J Am Coll Nutr. 2007;26:341–9.
De Graaf CJ, Korf J. Conditional tolerance to haloperidol-induced catalepsy is not caused by striatal dopamine receptor supersensitivity. Psychopharmacology (Berl). 1986;90:54–7.
De Ryck M, Teitelbaum P. Morphine catalepsy as an adaptive reflex state in rats. Behav Neurosci. 1984;98:243–61.
De Sousa-Moreira LF, Pinheiro MC, Masur J. Catatonic behavior induced by haloperidol, increased by retesting and elicited without drug in rats. Pharmacology. 1982;25:1–5.
Delwaide PJ. Parkinsonian rigidity. Funct Neurol. 2001;16:147–56.
Demars JPCA. Neuromuscular effects of long-term phenothiazine medication, eletroconvulsive therapy and leucotomy. J Nerv Ment Dis. 1966;143:73–9.
Dimascio A, Demirgian E. Antiparkinson drug overuse. Psychosomatics. 1970;11:596–601.
During MJ, Freese A, Deutch AY, Kibat PG, Sabel BA, Langer R, Roth RH. Biochemical and behavioral recovery in a rodent model of Parkinson's disease following stereotactic implantation of dopamine-containing liposomes. Exp Neurol. 1992;115:193–9.
Dutra RC, Andreazza AP, Andreatini R, Tufik S, Vital MA. Behavioral effects of MK-801 on reserpine-treated mice. Prog Neuropsychopharmacol Biol Psychiatry. 2002;26:487–95.
Ezrin-Waters C, Seeman P. Tolerance of haloperidol catalepsy. Eur J Pharmacol. 1977;41:321–7.
Ezrin-Waters C, Muller P, Seeman P. Catalepsy induced by morphine or haloperidol effects of apomorphine and anticholinergic drugs. Can J Physiol Pharmacol. 1976;54:516.
Ferré S, Guix T, Prat G, Jane F, Casas M. Is experimental catalepsy properly measured? Pharmacol Biochem Behav. 1990;35:753–7.
Fot R, Randrup A, Pakkenberg H. Lesions in corpus striatum and cortex of rats brain and the effect on pharmacologically induced stereotyped, aggressive and cataleptic behavior. Psychopharmacology (Berl). 1970;18:346–56.
Friedman JH. "Rubral" tremor induced by a neuroleptic drug. Mov Disord. 1992;7:281–2.
Frussa-Filho R, Palermo-Neto J. Effects of single and long-term metoclopramide administration on open-field and stereotyped behavior of rats. Eur J Pharmacol. 1988;149:323–9.
Frussa-Filho R, Palermo-Neto J. Effects of single and long-term administration of sulpiride on open-field and stereotyped behavior of rats. Braz J Med Biol Res. 1990;23:463–72.
Frussa-Filho R, Palermo-Neto J. Effects of single and long-term droperidol administration on open-field and stereotyped behavior of rats. Physiol Behav. 1991;50:825–30.
Frussa-Filho R, Otoboni JR, Uema FT, Palermo-Neto J. Effects of age and isolation on the evolution of catalepsy during chronic haloperidol treatment. Braz J Med Biol Res. 1992;25:925–8.
Fuenmayor LD, Vogt M. The influence of cerebral 5-hydroxytryptamine on catalepsy induced by brain-amine depleting neuroleptics or by cholinomimetics. Br J Pharmacol. 1979;67:309–18.

Fundaró A. Pinch-induced catalepsy in mice: a useful model to investigate antidepressant or anxiolytic drugs. Prog Neuropsychopharmacol Biol Psychiatry. 1998;22:147–58.

Gerlach M, Riederer P. Animal models of Parkinson's disease: an empirical comparison with the phenomenology of the disease in man. J Neural Transm. 1996;103:987–1041.

Gessa GL, Tacliamonte A. Effect of methadone and dextromoramide on dopamine metabolism: comparison with haloperidol and amphetamine. Neuropharmacology. 1975;14:913.

Gianutsos G, Hynes MD, Lal H. Enhancement of apomorphine-induced inhibition of striatal dopamine-turnover following chronic haloperidol. Biochem Pharmacol. 1975;24:581–2.

Goetz CG, Klawans HL. Drug-induced extrapyramidal disorders—a neuropsychiatric interface. J Clin Psychopharmacol. 1981;1:297–303.

Gough AL, Olley JE. Catalepsy induced by intrastriatal injections of delta9-THC and 11-OH-delta9-THC in the rat. Neuropharmacology. 1978;17:137–44.

Gras C, Amilhon B, Lepicard EM, Poirel O, Vinatier J, Herbin M, et al. The vesicular glutamate transporter VGLUT3 synergizes striatal acetylcholine tone. Nat Neurosci. 2008;11:292–300.

György L, Pfeifer KA, Hajtman B. Modification of certain central nervous effects of haloperidol during long-term treatment in the mouse and rat. Psychopharmacology (Berl). 1969;16: 223–33.

Hall RA, Jackson RB, Swain JM. Neurotoxic reactions resulting from chlorpromazine administration. JAMA. 1956;161:214–8.

Hansen TE, Hoffman WF. Drug-induced Parkinsonism. In: Yassa R, Vasavan Nair MP, Jeste DV, editors. Neuroleptic-induced movement disorders. New York: Cambridge University Press; 1997. p. 341–80.

Hausner RS. Neuroleptic-induced parkinsonism and Parkinson's disease: differential diagnosis and treatment. J Clin Psychiatry. 1983;44:13–6.

Hayakawa T, Sugimoto Y, Chen Z, Fujii Y, Kamei C. Effects of anti-parkinsonian drugs on neurobehavioural changes induced by bilateral 6-hydroxydopamine lesions in rats. Clin Exp Pharmacol Physiol. 1999;26:421–5.

Hess EJ, Albers LJ, Le H, Creese I. Effects of chronic SCH 23390 treatment on the biochemical and behavioral properties of D1 and D2 dopamine receptors: potentiated behavioral responses to a D2 dopamine agonist after selective D1 dopamine receptor upregulation. J Pharmacol Exp Ther. 1986;238:846–54.

Hess EJ, Norman AB, Creese I. Chronic treatment with dopamine receptor antagonists: behavioral and pharmacologic effects on D1 and D2 dopamine receptors. J Neurosci. 1988;8:2361–70.

Hillegaart V, Ahlenius S, Magnusson O, Fowler CJ. Repeated testing of rats markedly enhances the duration of effects induced by haloperidol on treadmill locomotion, catalepsy, and a conditioned avoidance response. Pharmacol Biochem Behav. 1987;27:159–64.

Honma T, Fukushima H. Correlation between catalepsy and dopamine decrease in the rat striatum induced by neuroleptics. Neuropharmacology. 1976;15:601–7.

Honma T, Fukushima H. Effects of bilateral lesions in the striatum or nucleus accumbens on the cataleptogenic activity of neuroleptics in rats. Jpn J Pharmacol. 1978;28:231–8.

Hubbard CA, Trugman JM. Reversal of reserpine-induced catalepsy by selective D1 and D2 dopamine agonists. Mov Disord. 1993;8:473–8.

Hyntel J. SCH 23390—The first selective dopamine D1 antagonist. Eur J Pharmacol. 1983; 91:153–4.

Hyntel J. Functional evidence for selective dopamine D1 receptor blockade by SCH 23390. Neuropharmacology. 1984;23:1395–401.

Iorio LC, Barnett A, Leitz FG, Houser VP, Korduba CA. SCH 23390 a potential benzazepine antipsychotic with unique interactions on dopaminergic systems. J Pharmacol Exp Ther. 1983;226:462–8.

Iwata SI, Izumi K, Shimizu T, Fukuda T. Effects of repeated testing on the incidence of haloperidol-induced catalepsy in mice. Pharmacol Biochem Behav. 1989;33:705–7.

Janssen PA, Niemegeers CJ, Schellekens KH. Is it possible to predict the clinical effects of neuroleptic drugs (major tranquillizers) from animal data? Part 1 "neuroleptic activity spectra" for rats. Arzneimittelforschung. 1965;15:104–17.

Johnson AM, Loew DM, Vigouret JM. Stimulant properties of bromocriptine on central dopamine receptors in comparison to apomorphine, (+)-amphetamine and L DOPA. Br J Pharmacol. 1976;56:59–68.
Jolicoeur FB, Rivcst R, Drumheller A. Hypokinesia, rigidity, and tremor induced by hypothalamic 6-OHDA lesions in the rat. Brain Res Bull. 1991;26:317–20.
Keepers GA, Casey DE. Clinical management of acute neuroleptic-induced extrapyramidal syndromes, Review. Curr Psychiatr Ther. 1986;23:139–57.
Khisti RT, Mandhane SN, Chopde CT. Haloperidol-induced catalepsy: a model for screening antidepressants effective in treatment of depression with Parkinson's disease. Indian J Exp Biol. 1997;35:1297–301.
Klein A, Schmidt WJ. Catalepsy intensifies context-dependently irrespective of whether it is induced by intermittent or chronic dopamine deficiency. Behav Pharmacol. 2003;14:49–53.
Klemm WR. Neurophysiologic studies of the immobility reflex (animal hypnosis). In: Ehrenpreis S, Solnitzky OC, editors. Neurosciences research, vol. 4. London: Academic; 1971. p. 165–212.
Klemm WR. Evidence for a cholinergic role in haloperidol-induced catalepsy. Psychopharmacology (Berl). 1985;85:139–42.
Klemm WR. The catalepsy of blocked dopaminergic receptors. Psychopharmacology (Berl). 1993;111:251–3.
Klemm WR, Block H. D1 and D2 receptor blockade have additive cataleptic effects in mice, but receptor effects may interact in opposite ways. Pharmacol Biochem Behav. 1988;29:223–9.
Klett CJ, Caffey Jr E. Evaluating the long-term need for antiparkinson drugs by chronic schizophrenics. Arch Gen Psychiatry. 1972;26:374–9.
Klockgether T. Parkinson's disease: clinical aspects. Cell Tissue Res. 2004;318:115–20.
Korf J. Striatal dopamine receptor supersensitivity is not the (exclusive) cause of behavioural tolerance to long-term haloperidol treatment. Psychopharmacology (Berl). 1988;95:144–5.
Kruse W. Treatment of drug-induced extrapyramidal symptoms. (A comparative study of three antiparkinson agents). Dis Nerv Syst. 1960;21:79–81.
Kuschinsky K, Hornykiewicz O. Morphine catalepsy in the rat: relation to striatal dopamine metabolism. Eur J Pharmacol. 1972;19:119–22.
Laduron PM. Dopamine sensitive adenylate ciclase as a receptor site. In: Kebabian JW, Kaiser C, editors. Dopamine receptors. ACS Symp Ser 224. Washington, DC: American Chemical Society; 1983.
Lane E, Dunnett S. Animal models of Parkinson's disease and L-dopa induced dyskinesia: how close are we to the clinic? Psychopharmacology (Berl). 2008;199:303–12.
Lipska BK, Jaskiw GE, Braun AR, Weinberger DR. Prefrontal cortical and hippocampal modulation of haloperidol-induced catalepsy and apomorphine-induced stereotypic behaviors in the rat. Biol Psychiatry. 1995;38:255–62.
Liu CQ, Hu DN, Liu FX, Chen Z, Luo JH. Apomorphine-induced turning behavior in 6-hydroxydopamine lesioned rats is increased by histidine and decreased by histidine decarboxylase, histamine H1 and H2 receptor antagonists, and an H3 receptor agonist. Pharmacol Biochem Behav. 2008;90:325–30.
Lloyd KG, Willigens MT, Worms P. Cortical lesions differently affect neuroleptic-and non-neuroleptic induced catalepsy in rats. Br J Pharmacol. 1981;54:821P.
Lorenc-Koci E, Wolfarth S, Ossowska K. Haloperidol-increased muscle tone in rats as a model of parkinsonian rigidity. Exp Brain Res. 1996;109:268–76.
Lucas G, Bonhomme N, De Deurwaerdère P, Le Moal M, Spampinato U. 8-OH-DPAT, a 5-HT1A agonist and ritanserin, a 5-HT2A/C antagonist, reverse haloperidol-induced catalepsy in rats independently of striatal dopamine release. Psychopharmacology (Berl). 1997;131:57–63.
Masuda Y, Murai S, Itoh T. Tolerance and reverse tolerance to haloperidol catalepsy induced by the difference of administration interval in mice. Jpn J Pharmacol. 1982;32:1186–8.
Mcauley JH. The physiological basis of clinical deficits in Parkinson's disease. Prog Neurobiol. 2003;69:27–48.
McEvoy JP. The clinical use of anticholinergic drugs as treatment for extrapyramidal side effects of neuroleptic drugs. J Clin Psychopharmacol. 1983;3:288–302.

Meller E, Kuga S, Friedhoff AJ, Goldstein M. Selective D2 dopamine receptor agonists prevent cataleps induced by SCH 23390, a selective D1 antagonist. Life Sci. 1985;36:1857–64.
Meschler JP, Conley TJ, Howlett AC. Cannabinoid and dopamine interaction in rodent brain: effects on locomotor activity. Pharmacol Biochem Behav. 2000;67:567–73.
Metz GA, Tse A, Ballermann M, Smith LK, Fouad K. The unilateral 6-OHDA rat model of Parkinson's disease revisited: an electromyographic and behavioural analysis. Eur J Neurosci. 2005;22:735–44.
Miller R, Wickens JR, Beninger RJ. Dopanime D1 and D2 receptors in relation to reward and performance: a case for the D1 receptor as a primary site of therapeutic action of neuroleptic drugs. Prog Neurobiol. 1990;34:143–83.
Miyagi M, Arai N, Taya F, Itoh F, Komatsu Y, Kojima M, Isaji M. Effect of cabergoline, a long-acting dopamine D2 agonist, on reserpine-treated rodents. Biol Pharm Bull. 1996;19: 1499–502.
Miyasaki JM, Martin W, Suchowersky O, Weiner WJ, Lang AE. Practice parameter: initiation of treatment for Parkinson's disease: an evidence-based review: report of the Quality Standards Subcommittee of the American Academy of Neurology. Neurology. 2002;58:11–7.
Mizuki Y, Ushijima I, Yamada M. Effects of chronic methamphetamine on SCH 23390- or haloperidol-induced catalepsy, and effects of coadministration of SCH23390 or haloperidol in mice. Pharmacol Biochem Behav. 1996;53:437–40.
Morelli M, Di Chiara G. Catalepsy induced by SCH 23390 in rats. Eur J Pharmacol. 1985;117:179–85.
Morpurgo C. Effects of antiparkinson drugs on a phenothiazine induced catatonic reaction. Arch Int Pharmacodyn Ther. 1962;1(137):84–90.
Namba MM, Quock RM, Malone MH. Narcotic antagonist potentiation of L-DOPA in the reversal of reserpine induced catalepsy. Proc West Pharmacol Soc. 1980;23:285–9.
Namba MM, Quock RM, Malone MH. Effects of narcotic antagonists on L-dopa reversal of reserpine-induced catalepsy and blepharoptosis in mice. Life Sci. 1981;28:1629–36.
Nehru B, Verma R, Khanna P, Sharma SK. Behavioral alterations in rotenone model of Parkinson's disease: attenuation by co-treatment of centrophenoxine. Brain Res. 2008;1201:122–7.
Nielsen IM. Pharmacological vs. clinical physiognomy of neuroleptics, with special reference to their sedative and antipsychotic effects. Acta Psychiatr Belg. 1974;74:473–84.
Nowak K, Welsch-Kunze S, Kuschinsky K. Conditioned tolerance to haloperidol- and droperidol-induced catalepsy. Naunyn Schmiedebergs Arch Pharmacol. 1988;337:385–91.
Ögren SO, Fuxe K. D1 and D2 receptor antagonist induce catalepsy via different efferent. Striatal pathways. Neurosci Lett. 1988;85:333–8.
Olanow CW, Tatton WG. Etiology and pathogenesis of Parkinson's disease. Annu Rev Neurosci. 1999;22:123–44.
Owen F, Crow TJ, Poulter M, Cross AJ, Longden A, Riley GJ. Increased dopamine receptor sensitivity in schizophrenia. Lancet. 1978;II:223–6.
Perry JC, Vital MA, Frussa-Filho R, Tufik S, Palermo-Neto J. Monosialoganglioside (GM1) attenuates the behavioural effects of long-term haloperidol administration in supersensitive rats. Eur Neuropsychopharmacol. 2004;14:127–33.
Pinna A, Pontis S, Morelli M. Expression of dyskinetic movements and turning behaviour in subchronic L-DOPA 6-hydroxydopamine-treated rats is influenced by the testing environment. Behav Brain Res. 2006;171:175–8.
Pires JG, Fonseca FC, Woelffel AB, Futuro-Neto HA. Evidence of interaction between fluoxetine and isosorbide dinitrate on neuroleptic-induced catalepsy in mice. Braz J Med Biol Res. 1998;31:417–20.
Pires JG, Bonikovski V, Futuro-Neto HA. Acute effects of selective serotonin reuptake inhibitors on neuroleptic-induced catalepsy in mice. Braz J Med Biol Res. 2005;38:1867–72.
Porsolt RD, Le Pichon M, Jalfre M. Depression: a new animal model sensitive to antidepressant treatments. Nature. 1977;266:730–2.
Poulos C, Hinson R. Pavlovian conditional tolerance to haloperidol catalepsy: evidence of dynamic adaptation in the dopaminergic system. Science. 1982;218:491–2.

Puri SK, Lal H. Tolerance to the behavioral and neurochemical effects of haloperidol and morphine in rats chronically treated with morphine or haloperidol. Naunyn Schmiedebergs Arch Pharmacol. 1974;282:155–70.
Queiroz CM, Frussa-Filho R. Effects of buspirone on an animal model of tardive dyskinesia. Prog Neuropsychopharmacol Biol Psychiatry. 1999;23:1405–18.
Rácz I, Bilkei-Gorzo A, Markert A, Stamer F, Göthert M, Zimmer A. Anandamide effects on 5-HT(3) receptors in vivo. Eur J Pharmacol. 2008;596:98–101.
Rao SS, Hofmann LA, Shakil A. Parkinson's disease: diagnosis and treatment. Am Fam Physician. 2006;74:2046–54.
Riederer P, Wuketich S. Time course of nigrostriatal degeneration in parkinson's disease. A detailed study of influential factors in human brain amine analysis. J Neural Transm. 1976;38:277–301.
Rodríguez-Díaz M, Abdala P, Barroso-Chinea P, Obeso J, González-Hernández T. Motor behavioural changes after intracerebroventricular injection of 6-hydroxydopamine in the rat: an animal model of Parkinson's disease. Behav Brain Res. 2001;122:79–92.
Rozzini R, Missale C, Gadola M. Drug-induced parkinsonism. Lancet. 1985;1:113.
Sanberg PR. Haloperidol-induced catalepsy is mediated by postsynaptic dopamine receptors. Nature. 1980;284:472–3.
Sanberg PR, Pisa M, Faulks IJ, Fibiger HC. Experimental influences on catalepsy. Psychopharmacology (Berl). 1980;69:225–6.
Sanberg PR, Bunsey MD, Giordano M, Norman AB. The catalepsy test: its ups and downs. Behav Neurosci. 1988;102:748–59.
Sarkaki A, Badavi M, Hoseiny N, Gharibnaseri MK, Rahim F. Postmenopausal effects of intrastriatal estrogen on catalepsy and pallidal electroencephalogram in an animal model of parkinson's disease. Neuroscience. 2008;154:940–5.
Sayers AC, Burki HR, Ruch W, Asper H. Anticholinergic properties of antipsychotic drugs and their relation to extrapyramidal side-effects. Psychopharmacology (Berl). 1976;51:15–22.
Schmidt WJ, Alam M. Controversies on new animal models of Parkinson's disease pro and con: the rotenone model of Parkinson's disease (PD). J Neural Transm Suppl. 2006;70:273–6.
Schmidt A, Vogel R, Rutledge SJ, Opas EE, Rodan GA, Friedman E. Cross-talk between an activator of nuclear receptors-mediated transcription and the D1 dopamine receptor signaling pathway. Pharmacol Biochem Behav. 2005;80:379–85.
Schwarting RK, Huston JP. The unilateral 6-hydroxydopamine lesion model in behavioral brain research. Analysis of functional deficits, recovery and treatments. Prog Neurobiol. 1996;50:275–331.
Seeman P. Brain dopamine receptors. Pharmacol Rev. 1981;32:229–313.
Seeman P, Lee T, Chau-Wong M, Wong K. Antipsychotic drug doses and neuroleptic/dopamine receptors. Nature. 1976;261:717–9.
Setler P, Sarau H, McKenzie G. Differential attenuation of some effects of haloperidol in rats given scopolamine. Eur J Pharmacol. 1976;39:117–26.
Sherer TB, Betarberbet R, Tasta CM, Seo BB, Richardson JR, Kim JH, et al. Mechanism of toxicity in rotenone models of Parkinson's disease. J Neurosci. 2003;23:10756–64.
Sherer TB, Richardson JR, Testa CM, Seo BB, Panov AV, Yagi T, Matsuno-Yagi A, Miller GW, Greenamyre JT. Mechanism of toxicity of pesticides acting at complex 1: relevance to environmental etiologies of Parkinson's disease. J Neurochem. 2007;100:1469–79.
Shipley JE, Rowland N, Antelman SM, Buggy J, Edwards DJ, Shapiro AP. Increased amphetamine stereotypy and longer haloperidol catalepsy in spontaneously hypertensive rats. Life Sci. 1981;28:745–53.
Silva SR, Futuro-Neto HA, Pires JG. Inhibition of chlorpromazine-induced catalepsy by the 5-HT-1A ligands pindolol and buspirone in mice. Braz J Med Biol Res. 1990;23:869–71.
Simpson GM. Controlled studies of antiparkinsonism agents in the treatment of drug-induced extrapyramidal symptoms. Acta Psychiatr Scand Suppl. 1970;212:44–51.
Skirboll S, Bunney BS. The effects of acute and chronic haloperidol treatment on spontaneuously firing neurons in the caudate nucleous of the cat. Life Sci. 1979;25:1419–34.
Smith GP, Young RC. A new experimental model of hypokinesia. Adv Neurol. 1974;5:427–32.

Smith DJ, Kyle S, Forty L, Cooper C, Walters J, Russell E, et al. Differences in depressive symptom profile between males and females. J Affect Disord. 2008;108:279–84.

Sousa FC, Gomes PB, Noronha EC, Macêdo DS, Vasconcelos SM, Fonteles MM, Viana GS. Effects of dopaminergic and cholinergic interactions on rat behavior. Life Sci. 2001;69:2419–28.

Srinivasan J, Schmidt WJ. Intensification of cataleptic response in 6-hydroxydopamine-induced neurodegeneration of substantia nigra is not dependent on the degree of dopamine depletion. Synapse. 2004;51:213–8.

St Jean A, Donald MW, Ban TA. Interchangeability of antiparkinson medication. Am J Psychiatry. 1964a;120:1189–90.

St Jean A, Donald MW, Ban TA. Uses and abuses of antiparkinsonian medication. Am J Psychiatry. 1964b;120:801–3.

Stanley ME, Glick SD. Interaction of drug effects with testing procedures in the measurement of catalepsy. Neuropharmacology. 1976;15:393–4.

Stephen PJ, Williamson J. Drug-induced parkinsonism in the elderly. Lancet. 1984;2:1082–3.

Strome EM, Zis AP, Doudet DJ. Electroconvulsive shock enhances striatal dopamine D1 and D3 receptor binding and improves motor performance in 6-OHDA-lesioned rats. J Psychiatry Neurosci. 2007;32:193–202.

Sutton MA, Beninger RJ. Psychopharmacology of conditioned reward: evidence for a rewarding signal at D1-like dopamine receptors. Psychopharmacology (Berl). 1999;144:95–110.

Tarsy D, Baldessarini RJ. Behavioural supersensitivity to apomorphine following chronic treatment with drugs which interfere with the synaptic function of catecholamines. Neuropharmacology. 1974;13:927–40.

Tikhonova MA, Lebedeva VV, Kulikov AV, Bazovkina DV, Popova NK. Effect of imipramine on the behavior and cerebral 5-HT1A serotonin receptors in mice genetically predisposed to catalepsy. Bull Exp Biol Med. 2006;141:48–50.

Tronci E, Simola N, Borsini F, Schintu N, Frau L, Carminati P, Morelli M. Characterization of the antiparkinsonian effects of the new adenosine A2A receptor antagonist ST1535: acute and subchronic studies in rats. Eur J Pharmacol. 2007;566:94–102.

Undie AS, Friedman E. Differences in the cataleptogenic actions of SCH 23390 and selected classical neuroleptics. Psychopharmacology (Berl). 1988;96:311–6.

Ungerstedt U. 6-Hydroxy-dopamine induced degeneration of central monoamine neurons. Eur J Pharmacol. 1968;5:107–10.

Ungerstedt U. Postsynaptic supersensitivity after 6-hydroxy-dopamine induced degeneration of the nigro-striatal dopamine system. Acta Physiol Scand Suppl. 1971;367:69–93.

Ungerstedt U, Arbuthnott GW. Quantitative recording of rotational behavior in rats after 6-hydroxy-dopamine lesions of the nigrostriatal dopamine system. Brain Res. 1970;24:485–93.

Uretsky NJ, Iversen LL. Effects of 6-hydroxydopamine on catecholamine containing neurones in the rat brain. J Neurochem. 1970;17:269–78.

Ushijima I, Mizuki Y, Yamada M. Development of tolerance and reverse tolerance to haloperidol— and SCH 23390- induced cataleptic effects during withdrawal periods after long-term treatment. Pharmacol Biochem Behav. 1995;50:259–64.

Vital MABF, Frussa-Filho R, Palermo-Neto J. Effects of monosialoganglioside on dopaminergic supersensitivity. Life Sci. 1995;56:2299–307.

Vital MA, Flório JC, Frussa-Filho R, De Lucia R, Tufik S, Palermo-Neto J. Effects of haloperidol and GM1 ganglioside treatment on striatal D2 receptor binding and dopamine turnover. Life Sci. 1998;62:1161–9.

Waddington JL. New antipsychotic drugs: preclinical evaluation and clinical profiles in the treatment of schizophrenia. In: Reveley MA, Deakin JFW, editors. The psychopharmacology of schizophrenia. London: Arnold; 2000. p. 225–50.

Wanibuchi F, Usuda S. Synergistic effects between D1 and D2 dopamine antagonist on catalepsy in rats. Psychopharmacology (Berl). 1990;102:339–42.

Zarrindast MR, Habibi-Moini S. Blockade of both D1 and D2 receptors may induce catalepsy in mice. Gen Pharmacol. 1991;22:1023–221.

Zarrindast MR, Samadi P, Haeri-Rohani A, Moazami N, Shafizadeh M. Nicotine potentiation of morphine-induced catalepsy in mice. Pharmacol Biochem Behav. 2002;72:197–202.

Chapter 27
Assessment of Motor Function in Rodents: Behavioral Models Sharing Simplicity and Multifaceted Applicability

Part 3: Orofacial Dyskinesia

Roberto Frussa-Filho, Daniela Fukue Fukushiro, Camilla de Lima Patti, Sonia Regina Kameda, Patrícia Helena Zanier Gomes, and Rita de Cassia Carvalho

Introduction

Orofacial dyskinesia is an important motor phenomenon with implications for a series of neuropathological conditions including L-DOPA treatment of Parkinson's disease (Jenner 2004), Huntington's disease (Verhagen Metman et al. 2002), and natural aging (Blanchet et al. 2004). Among these disorders, one of the most critical is tardive dyskinesia—the most serious iatrogenic movement disorder—which is, by definition, a late-onset adverse effect of prolonged administration of neuroleptic drugs (Sachdev 2000; Llorca et al. 2002). The aim of the present chapter is to discuss the rodent behavioral models that have been used to investigate the above mentioned neuropathological conditions.

R. Frussa-Filho (In Memoriam) • S.R. Kameda
Department of Pharmacology, Universidade Federal de São Paulo (UNIFESP), São Paulo, São Paulo, Brazil

D.F. Fukushiro, Ph.D. (✉)
Department of Pharmacology, Universidade Federal de São Paulo (UNIFESP), Rua Botucatu, 862, Ed. Leal Prado, 1° andar, São Paulo, São Paulo 04023062, Brazil
e-mail: danifukushiro@gmail.com

C.L. Patti, Ph.D.
Instituto de Genética e Erros Inatos do Metabolismo, Universidade Federal de São Paulo (UNIFESP), São Paulo, São Paulo, Brazil

P.H.Z. Gomes
Universidade Federal do Triângulo Mineiro, Uberaba, Minas Gerais, Brazil

R.d.C. Carvalho
Department of Pharmacology, Universidade Federal de São Paulo (UNIFESP), São Paulo, São Paulo, Brazil

M.L. Andersen, S. Tufik (eds.), *Rodent Model as Tools in Ethical Biomedical Research*, DOI 10.1007/978-3-319-11578-8_27

Fig. 27.1 Tongue protrusion in the rat

Methodology

In both rats and mice, orofacial dyskinesia is evaluated through the quantification of three cardinal behavioral parameters: tongue protrusions, vacuous chewing frequencies and the duration of facial musculature twitching. These parameters are usually measured by direct observation, although more sophisticated recording methodologies have already been proposed (see Levy and Ellison 1987; Levy et al. 1987). For practical reasons, a specific methodology of orofacial dyskinesia measurement will be described here.

To quantify the occurrence of oral dyskinesia, rats (or mice) are placed individually in observation cages (16 cm × 30 cm × 19 cm) without food and water. Hand-operated counters are employed to score tongue protrusion and vacuous chewing frequencies and stopwatches are used to quantify the duration of facial musculature twitching. Tongue protrusion (Fig. 27.1) is operationally defined as a visible extension of the tongue outside of the mouth and not directed at anything. Individual tongue protrusions during a bout of oral dyskinesia are each preceded by visible retraction of the tongue (Neisewander et al. 1996). Vacuous chewing movements (Fig. 27.2) are referred to as single mouth openings in the vertical plane not directed toward physical material (Bergamo et al. 1997). Twitching of the facial musculature is defined as the sum of the duration of the rapid, rhythmic, cheek movements that occurred independently of jaw movements, often resulting in the appearance of the eyeball bouncing up and down, or in and out, and lasting 3 s or more (Tran et al. 1998; Abílio et al. 2002). If tongue protrusions, vacuous chewing movements or twitching of the facial musculature occur during a period of grooming, they must not be taken into account. The incidence of these parameters is measured continuously, usually for 10 min. Mirrors are placed under the floor and behind the back wall of the cage to permit observation of oral dyskinesia when the animal is faced away from the observer. To avoid differences in the behavior of experimental and control groups of rats due to circadian changes, experimental and control observations must be alternated. The observations should be made by observers who are blind to the animals' group

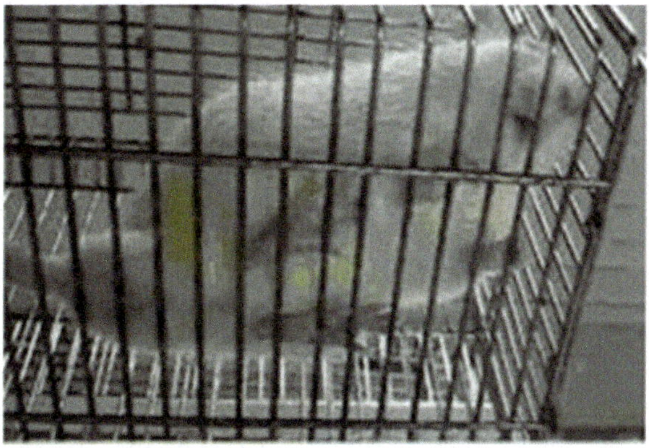

Fig. 27.2 Vacuous chewing movements the rat

assignment. We have observed that these observation criteria are not subjective, since excellent inter-observer agreement was found (Pearson's correlation = 0.98 for each parameter—Abílio et al. 2002).

It is important to highlight that while in some papers the three orofacial parameters are quantified (Bergamo et al. 1997; Queiroz et al. 1998; Queiroz and Frussa-Filho 1999), in others only two (Abílio et al. 2002; Peixoto et al. 2003; Faria et al. 2005) or even one (generally vacuous chewing movements—Egan et al. 1996; Carvalho et al. 2003; Turrone et al. 2003) are measured. Since these three orofacial parameters cannot be simultaneously expressed, the possibility that they compete with each other cannot be discarded. In order to avoid possible masked alterations due to this behavioral competition, we (Ricardo et al. 2005) have recently proposed another parameter, defined as total orofacial movements (TOM). The use of total orofacial movements provides a more accurate idea of alterations in orofacial movements taken as a whole. This parameter was calculated as the sum of vacuous chewing movements and tongue protrusion frequency. Since there is a great difference in the magnitude of each parameter, we multiplied the tongue protrusion frequency by a correction factor calculated by the division of the mean of vacuous chewing movements and tongue protrusion frequencies of the control group in each experiment. The TOM parameter was calculated for each animal.

Applicability: Rodent Oral Dyskinesia as a Model of L-DOPA-Induced Dyskinesia in Patients with Parkinson's Disease

Although it was first made available in the 1960s, the dopamine precursor 3,4-dihydroxyphenyl-l-alanine (L-DOPA) remains the most effective noninvasive symptomatic relief for Parkinson's disease sufferers (see Pavón et al. 2006; Lane and Dunnett 2008). Perhaps the greatest single factor compromising long-term L-DOPA treatment is the development of abnormal involuntary movements collectively known as dyskinesia (Marsden 1994; Jenner 2004). Dyskinesias are manifested as a wide range of motor symptoms ranging from orofacial movements to abnormal movements of the extremities, similar to ballism or chorea and dystonic postures (Jenner 2004; Obeso et al. 2000, 2004). Generally, dyskinesia occurs when the dopamine concentration in the brain is the highest and is therefore called peak-dose dyskinesia (Olanow et al. 2004), but it can also occur throughout the "on" time period (the period of response of L-DOPA treatment). Within 5 years of starting L-DOPA treatment, 40 % of patients will develop dyskinesia and by 10 years this figure reaches 90 %, and can be a major limiting factor in the progression of dopaminergic therapy (Ahlskog and Muenter 2001). The appearance of dyskinesia represents a challenge to Parkinsonian therapy because it can be severe enough to warrant reducing the L-DOPA dose below optimal therapeutic levels (Pavón et al. 2006).

As recently reviewed by Putterman et al. (2007), in the last few decades, there has been steady advancement in using the hemi-parkinsonian rat to investigate L-DOPA-induced dyskinesia. In this model of Parkinson's disease, rats receive a unilateral intracerebral injection of 6-hydroxydopamine (6-OHDA) that causes ipsilateral destruction of dopamine-containing neurons (Ungerstedt and Arbuthnott 1970). Because of chronic hemidepletion of dopamine, rats will rotate contralateral to the lesion following injections of direct dopamine agonists such as apomorphine (because of the development of compensatory dopamine receptor supersensitivity in the depleted striatum), and they rotate in the ipsilateral direction in response to the indirect dopamine agonist amphetamine (Ungerstedt 1971a). Interestingly, in animals with unilateral 6-OHDA lesions, the behavioral effects of L-DOPA are similar to those of the direct dopamine receptor agonist apomorphine. Thus, the systemic administration of L-DOPA (e.g., 100 mg/kg) acutely induces intense contraversive turning (Ungerstedt 1971b). As pointed out by Schwarting and Huston (1996a, b), once inside the brain, the decarboxylation of L-DOPA to dopamine can be mediated through enzymes of residual dopamine neurons. However, such decarboxylation can also occur despite total loss of dopamine neurons. Therefore, it is assumed that the necessary enzyme is also located on non-dopaminergic elements, like serotonergic neurons and glial cells. If dopamine is formed from exogenous L-DOPA via these sources, it can then stimulate dopamine receptors (similar to apomorphine); however, in contrast to apomorphine, dopamine can be quickly inactivated by uptake into dopaminergic cells, which explains why L-DOPA is behaviorally usually rather inconspicuous in animals with intact DA neurons. Furthermore, it shows that the behavioral action of L-DOPA in the lesion model is not only dependent on DA receptor-mediated supersensitivity, but also on the loss of DA reuptake sites (Wachtel and Abercrombie 1994).

Because dopamine agonists are clinically useful in the treatment of Parkinson's disease, the induction of contralateral circling might be interpreted as a therapeutic, anti-parkinsonian effect, but clearly, prolonged rotatory behavior is not as normal or purposeful as one might expect from a "therapeutic" action. Moreover, repeated administration of a dopamine agonist is well known to cause behavioral sensitization that augments circling behavior (Morelli et al. 1989). These considerations have prompted the suggestion that agonist-induced circling behavior may more accurately model dyskinesia than a therapeutic anti-parkinsonian effect (Henry et al. 1998; Konitsiotis and Tsironis 2006). Nevertheless, it is also clear that repetitive circling in a rat is a poor model of the complex dyskinetic movements that characterize L-DOPA-induced dyskinesia in humans. Thus, the interpretation of circling behavior in the hemi-parkinsonian rat is controversial.

Within this context, the quantification of the rotational behavior in 6-OHDA lesioned rats was supplemented by the analysis of fore and hind limbs, jaw and torso movements that can be associated with dyskinesia in patients. These behaviors, which develop over time with repeated L-DOPA administration, were first characterized in rats in 1998, by Cenci et al. In their landmark study, one and a half to two months after the unilateral 6-OHDA lesion, the rats were allocated to two groups. One group began treatment with L-DOPA methyl ester (8 mg/kg, i.p.) plus benserazide (15 mg/kg, i.p.), administered as one daily injection for 20 days. Another group of 6-OHDA-lesioned rats, which received daily injections of physiological saline, served as controls. For quantification of the abnormal involuntary movements induced by L-DOPA, rats were observed individually every 20 min for 3 h (9 monitoring periods of 1 min each), following a daily dose. Movements were recognized as dyskinetic when they fulfilled the following criteria: 1-affected the side of the body contralateral to the lesion; 2-were repetitive, purposeless and not ascribable to any normal behavioral pattern. Abnormal involuntary movements were classified based on their topographic distribution into four subtypes:

(A) locomotive dyskinesia, i.e., increased locomotion with contralateral side bias;
(B) axial dystonia, i.e., contralateral twisted posturing of the neck and upper body;
(C) orolingual dyskinesia, i.e., stereotyped jaw movements and contralateral tongue protrusion;
(D) forelimb dyskinesia, i.e., repetitive rhythmic jerks or dystonic posturing of the contralateral forelimb, and/or grabbing movement of the contralateral paw.

For each of these four subtypes, each rat was scored on a scale from 0 to 4: 0: absent; 1 = occasional; 2 = frequent; 3 = continuous but interrupted by sensory distraction; 4 = continuous, severe, not interrupted by sensory distraction. Rats that mainly showed locomotion and circling in response to L-DOPA could also be scored on the other dyskinesia subtypes by lifting them up. Conversely, rats that did not spontaneously express locomotive dyskinesia could be scored on this subtype as well by lifting up their tail and hind limbs, which forced them to contact the floor with their forepaws, and elicited locomotion. The maximum score in each session was thus 144 (maximum score per observation = 16; number of observations per session = 9).

The demonstration that a very similar L-DOPA-induced dyskinesia could also be reliably quantified in 6-OHDA lesioned mice was independent and concomitantly provided by two research groups (Lundblad et al. 2004; Pavón et al. 2006). Both studies used very similar scales to evaluate the abnormal involuntary movements, since both scales were based on that originally proposed for rats by Cenci et al. (1998). As emphasized by Pavón et al. (2006), an important methodological difference was the schedule of L-DOPA administrations. While dyskinesia was triggered in 6-OHDA-lesioned mice by escalating doses of L-DOPA in the study by Lundblad et al. (2004), Pavón et al. (2006) adopted a paradigm that is more representative of human therapeutic L-DOPA use: chronic intermittent delivery of a constant dose of L-DOPA (25 mg/kg, intraperitoneal (IP), twice a day, for 25 days).

An important difference between rats or mice previously treated with 6-OHDA is an unexpected decline in the dyskinesia severity during the course of chronic L-DOPA treatment in mice (Lundblad et al. 2005). This concern notwithstanding, experiments performed in mice were very effective in providing a pharmacological validation of the model. Indeed, Lundblad et al. (2005) demonstrated a marked inhibition of dyskinesia when amantadine, buspirone and riluzole (compounds that have been shown to alleviate L-DOPA-induced dyskinesia in Parkinsonian patients) were acutely co-administered with a challenge L-DOPA injection in mice previously submitted to the model's procedure. On the other hand, Putterman et al. (2007) showed in rats that, as is the case of the clinical situation, L-DOPA dose and the level of striatal depletion are significant potentiating variables in the L-DOPA-induced dyskinesia in the 6-OHDA-lesioned animal model.

Although oral dyskinesia is not the only component of the L-DOPA-induced dyskinesia in 6-OHDA-lesioned rats and mice (and have not been quantitatively evaluated as described in the methodology section of this chapter), it seems to be especially important. Indeed, in an elegant study (Winkler et al. 2002), Cenci's research group demonstrated that opposite to the locomotive dyskinesia and the axial dystonia, the oral and the forelimb dyskinesias of rats submitted to the model were more specifically related to the denervation of the ventrolateral caudate-putamen. This topography reproduces the pattern of striatal DA denervation which is typical of Parkinson's disease (Kish et al. 1988), since the lateral caudate-putamen in the rat is functionally equivalent to the human putamen, whereas the medial caudate-putamen (and the underlying nucleus accumbens) are part of associative and limbic-related circuits (for review see Flaherty and Graybiel 1994).

Applicability: 3-Nitropropionic Acid-Induced Oral Dyskinesia in Mice as a Model of Orofacial Dyskinesia in Huntington's Disease

Huntington's disease is a hereditary autosomal dominant neurodegenerative disorder caused by an expansion of CAG repeats in chromosome 4 (Gusella et al. 1983), that during its development presents motor symptoms, psychic disorders and cognitive deficits as the most common signs. Huntington's disease, like many neurodegenerative processes, is associated with changes in Ca^{2+} homeostasis and the production of reactive oxygen species (Petersén et al. 1999).

Concerning the motor symptoms, while one of the earliest motor signs of Huntington's disease is the progressive appearance of orofacial dyskinesia (Brouillet et al. 1999), Chorea in Huntington's disease and in the L-DOPA-induced dyskinesias of Parkinson's disease may be clinically indistinguishable (Verhagen Metman et al. 2002).

3-nitropropionic acid, an irreversible inhibitor of the mitochondrial complex II enzyme succinate dehydrogenase, is responsible for oxidizing succinate to fumarate

(Palfi et al. 1996). The inhibition of this complex seems to be related to neuronal death, anatomic and neurochemical changes similar to those occurring in Huntington's disease (Beal et al. 1993; Brouillet et al. 1995).

Within this context, we have recently verified that mice repeatedly treated with 3-nitropropionic acid developed a marked oral dyskinesia characterized by an increased frequency of vacuous chewing movements (Rosenstock et al. 2004). This oral dyskinesia was associated with a decrease in locomotion frequency and an increase in immobility duration evaluated in an open-field. Importantly, it has been shown that Huntington's disease patients display bradykinesia throughout the entire duration of the disease (Thomson et al. 1988). In addition, all these motor alterations were attenuated or abolished when the animals were treated with vitamin E. In brief, mice were allocated randomly to three groups of seven to eight animals each. The animals received an intraperitoneal injection of 20 mg/kg vitamin E (VE) or control solution (CS) once daily for 9 days. From days 5 to 9 after the beginning of the treatment, 20 mg/kg 3-nitropropionic acid (3NP) or control solution was intraperitoneally injected once daily 1 h after vitamin E or control solution. Thus, the three groups of animals were as follows: CS-CS, CS-3NP, and VE-3NP. On day 10, 24 h after the last injection of their respective treatments, animals were observed for open-field behavior and orofacial movement quantification, as described above. It should be noted that vitamin E was repeatedly administered (5 days) prior to 3NP treatment because earlier studies have stressed the importance of such a repeated previous administration of this agent in order to produce its antioxidant effects in vivo (Gattaz et al. 1993; Abílio et al. 2003). Importantly, we (Rosenstock et al. 2004) also verified that besides abolishing the 3-nitropropionic acid-induced oral dyskinesia, and hypolocomotion, vitamin E also antagonized the effects of mitochondrial calcium dysfunction produced by 3-nitropropionic acid (which were associated with cellular oxidative stress and apoptosis).

These data are consistent with studies showing that both Huntington's disease brains and Huntington's disease mouse models present evidence of oxidative stress and ROS accumulation (Tabrizi et al. 2000), and the co-treatment with free radical scavengers results in a decrease in the oxidative damage in striatal and cortical synaptosomes (La Fontaine et al. 2000).

More recently, we compared the effects of 3NP on the development of orofacial dyskinesia and on succinate dehydrogenase activity in young and old mice (Rosenstock et al. 2009). Treatment with 3NP (5, 10, 15 or 20 mg/kg once a day, for 4 days) induced vacuous chewing movements in young mice. Old mice presented an increase in the basal level of orofacial movement that was not potentiated by any dose of 3NP. Histochemical analyses showed that old mice presented an increase in the succinate dehydrogenase activity. Finally, 3NP induced a decrease in SDH activity at both ages. We suggested that the 3NP-induced vacuous chewing movements in young mice are related to the inhibition of succinate dehydrogenase activity. In parallel, an enhancement in the basal activity of succinate dehydrogenase could be related to the absence of a further increase in the vacuous chewing movements presented by old mice treated with 3NP.

Applicability: Neuroleptic- or Reserpine-Induced Oral Dyskinesia in Rodents as a Model of Tardive Dyskinesia

Neuroleptics, one of the most widely prescribed groups of psychotropic drugs, have found widespread use not only for psychosis, but also in the treatment of agitation and affective disorders among adult, pediatric, and elderly patient populations as well (Worrel et al. 2000; Cooper et al. 2004; Kasckow et al. 2004; Masan 2004; McIntyre et al. 2004; Tariot et al. 2004). Notwithstanding, neuroleptic treatment has not proved to be an unmixed blessing. Neuroleptics are associated with troublesome adverse side effects, of which movement disorders are the most serious in terms of frequency, persistence, and overall impact on the well-being of patients and caregivers. Indeed, typical antipsychotics such as haloperidol and chlorpromazine have a propensity to cause both early onset motor side effects such as parkinsonism, and the late-emerging tardive dyskinesia (Muscettola et al. 1999; Tarsy et al. 2002; Janno et al. 2004; Kane 2004). Despite the advent of newer antipsychotic drugs with reduced liability to inducing these extrapyramidal motor disorders, such side effects remain a clinical concern, especially in vulnerable patient populations such as the elderly (Wolf et al. 2005).

Tardive dyskinesia is considered the most serious of the iatrogenic movement disorders (Nair and Jeste 1997). It is characterized by repetitive involuntary movements, usually involving the mouth, face, and tongue, and sometimes limb and trunk musculature. This movement disorder usually persists for months after the neuroleptic has been stopped and may be irreversible (Casey 1985; Marsden 1985).

Age is the single most frequently implicated risk factor for tardive dyskinesia, increasing both the risk of developing tardive dyskinesia and the severity and persistence of the condition (Kane et al. 1985; Wolfarth and Ossowska 1989). Within this context, the occurrence of spontaneous oral dyskinesia in elderly people who have never undergone treatment with neuroleptics has been frequently reported (Wolfarth and Ossowska 1989). The similarity between tardive dyskinesia and aging-induced dyskinesia is so remarkable that it is impossible to differentiate them on the basis of behavioral symptoms (Gerlach 1985). Accordingly, it has been suggested that long-term treatment with neuroleptics does not "cause" oral dyskinesia, but rather such treatment, by interacting with some substrate of brain aging, accelerates the emergence of a syndrome that can occur spontaneously in old age (Waddington 1990). In this respect, whereas the enhancement of oxidative stress is a well-known feature of brain aging (Cadenas and Davies 2000), neuroleptics, by blocking dopamine autoreceptors, cause a secondary increase in dopamine turnover, which may lead to increased formation of dopamine quinones and hydrogen peroxide through monoamine oxidase activity (Lohr 1991). The theory that an increase in free radical byproducts from dopamine metabolism underlies the pathophysiology of tardive dyskinesia has received considerable interest (Lohr 1991; Casey 1995; Andreassen and Jorgensen 2000; Abílio et al. 2004; Faria et al. 2005; Bishnoi et al. 2008a, b, c; Teixeira et al. 2008). Nevertheless, it has been previously suggested that a compensatory nigrostriatal dopaminergic supersensitivity induced by long-term neuroleptic treatment leads to the development of tardive dyskinesia (see Palermo-Neto and Frussa-Filho 2001 for review). However, this hypothesis has a fundamental flaw: whereas aging is the most implicated risk factor for tardive dyskinesia (Saltz et al. 1997), dopaminergic function declines with aging (Roth and Joseph 1994). In addition, old animals have a diminished capacity to develop behavioral supersensitivity and dopaminoceptor up-regulation after chronic treatment with neuroleptics (Waddington and Gamble 1980).

Orofacial dyskinesia (especially vacuous chewing movements) induced by long-term neuroleptic administration in rats has been the most extensively used phenomenological animal model of tardive dyskinesia (see Waddington 1990, 1997 and Palermo-Neto and Frussa-Filho 2001 for reviews, and Bishnoi et al. 2007a,b; Colpo et al. 2007; Samad et al. 2007; Bishnoi et al. 2007a,b for recent papers). In this context, haloperidol has been the most used neuroleptic drug and has been repeatedly (for at least 15 days) administered (once or twice daily) by an intraperitoneal (Naidu et al. 2002; Singh et al. 2003; Rogoza et al. 2004; Bishnoi et al. 2007a, b, 2008a, b), or subcutaneous (Burger et al. 2005) route. In addition, this neuroleptic has also been administered orally in drinking water (Sadan et al. 2005) or by the intramuscular route (Fachinetto et al. 2005). In this regard, a single intramuscular dose of haloperidol decanoate was able to produce oral dyskinesia in rats from the 7th to at least the 28th day after drug administration (Burger et al. 2006).

Besides neuroleptic-induced oral dyskinesia, Neisewander et al. (1994) have suggested that reserpine-induced oral dyskinesia may provide a new animal model of tardive dyskinesia. Indeed, rats treated with this monoamine-depleting agent for at least 3 days develop orofacial dyskinesia characterized by tongue protrusion, twitching of the facial musculature and vacuous chewing movements (Neisewander et al. 1991a, b, 1994, 1996; Bergamo et al. 1997; Vital et al. 1997; Queiroz et al. 1998; Queiroz and Frussa-Filho 1999; Abílio et al. 2002, 2004; Faria et al. 2005). In this regard, although reserpine is not classified as a neuroleptic, it has been used as an antipsychotic agent and has been associated with the development of tardive dyskinesia (Uhrbrand and Faurbye 1960). This reserpine-induced oral dyskinesia in rats also has other features that are consistent with tardive dyskinesia, including persistence following termination of administration and dose-dependent blockade by a D2-selection antagonist (Neisewander et al. 1991a, b). Furthermore, whereas at high doses of reserpine the response appears 3 days after a single injection of the drug (Sussman et al. 1997), at low doses the response is not evident until approximately 6–8 weeks (Neisewander et al. 1994), consistent with the protracted development of tardive dyskinesia in humans (Gerlach and Casey 1988). As with tardive dyskinesia (Wolfarth and Ossowska 1989), reserpine-induced oral dyskinesia is exacerbated by dopamine agonists like amphetamine and is mediated by the nigrostriatal dopaminergic system (Neisewander et al. 1996). Importantly, while age is one of the most frequently implicated risk factors for tardive dyskinesia, increasing both the persistence and severity of the condition (Wolfarth and ossowska 1989; Neisewander et al. 1996), we have recently verified that when compared to control adult rats, the significant increase in oral dyskinesia induced by reserpine treatment was more persistent in the older rats than in the adult animals (Bergamo et al. 1997).

More recently, studies from our laboratory have extended both haloperidol- and reserpine-induced oral dyskinesia models to mice. Indeed, using the frequency of vacuous chewing movements (since this parameter was the most reliable measure of the orofacial dyskinesia in this species), we verified that as with rats, haloperidol- and reserpine-induced oral dyskinesias in mice were conspicuous, persistent, were attenuated by dopamine D2 antagonists and were much more severe in older animals (Silva et al. 2002; Carvalho et al. 2003; Araujo et al. 2004; Castro et al. 2006 and submitted data). Figure 27.3 shows the effects of haloperidol decanoate administration on vacuous chewing movements in mice.

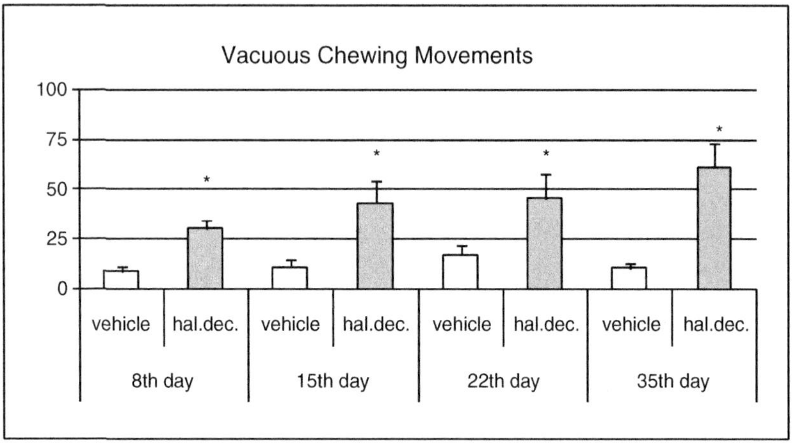

* $p<0.05$ related to vehicle (T-test).

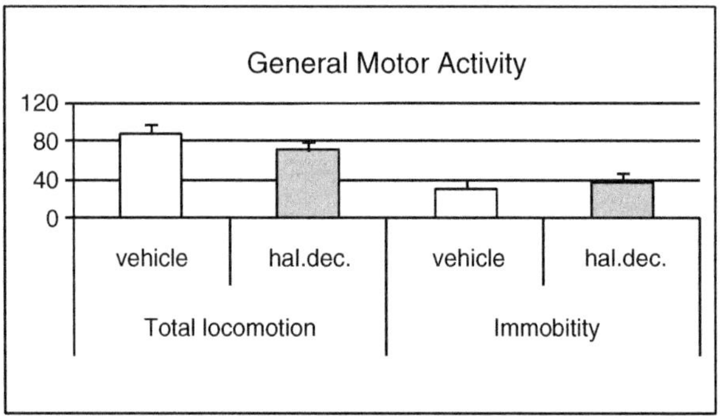

Fig. 27.3 Twenty five female mice received two subcutaneous injections of haloperidol decanoate (Janssen-Cilag) or its vehicle (gergilim oil) on days 1 and 8. The vacuous chewing movements of the mice were measured on days 8 (immediately before the second injection), 15, 22 and 35 of treatment (Panel a). Haloperidol-treated mice presented an increase in vacuous chewing movements when compared to the vehicle-treated group from the 8th day of observation onwards. On the 35th day, paired *T*-test analysis revealed that the haloperidol-treated group presented an enhanced number of vacuous chewing movements than in the 8th day. This is interesting because even without another haloperidol injection (35th day is 27 days after the last injection), vacuous chewing movements are additive. It is not related to a motor activity, since, on the 36th day, the general motor activity was analyzed in an open field and neither their general locomotion nor their immobility differed (Panel b)

Studies using haloperidol- or reserpine induced oral dyskinesia in rats or mice have strengthened the role of striatal oxidative stress in the pathophysiology of tardive dyskinesia. Indeed, several antioxidant agents have proven to be able to attenuate or reverse oral dyskinesia in such animal models. This is the case of GM1 ganglioside (Vital et al. 1997), melatonin (Abílio et al. 2002; Naidu et al. 2003a), vitamine E (Abílio et al. 2003; Faria et al. 2005), quercetin (Naidu et al. 2003b; Naidu et al. 2004), ebselen (Burger et al. 2003), diphenyl diselenide (Burger et al. 2004), vitamin C (Faria et al. 2005), spirulina maxima (Thaakur and Jyothi 2007), lazaroid (Bishnoi et al. 2007c) and curcumin (Bishnoi et al. 2008c). This very extensive series of putative therapeutic agents for tardive dyskinesia is a good example of the huge potential applicability of these very simple animal models of movement disorders. This potential applicability supports their indisputable importance in providing basic evidence about the pathophysiological mechanisms underlying such disorders.

Ethical Comments

Rodent models of orofacial dyskinesia have been developed for animal observation. Therefore, environmental conditions must be the least aversive possible to avoid the influence of stress on the behavior of the animals. Notwithstanding, since these motor disorders may be very long-lasting, the animals should always be sacrificed immediately after the end of the experiment.

References

Abílio VC, Vera Jr JA, Ferreira LS, Duarte CR, Carvalho RC, Grassl C, et al. Effects of melatonin on orofacial movements in rats. Psychopharmacology (Berl). 2002;161:340–7.

Abílio VC, Araújo CCS, Bergamo M, Calvacante PRV, D'Almeida V, Ribeiro RA, et al. Vitamin E attenuates reserpine-induced oral dyskinesia and striatal oxidized glutathione/reduced glutathione ratio (GSSG/GSH) enhancement in rats. Prog Neuropsychopharmacol Biol Psychiatry. 2003;27:109–14.

Abílio VC, Silva RH, Carvalho RC, Grassl C, Calzavara MB, Registro S, et al. Important role of striatal catalase in aging- and reserpine-induced oral dyskinesia. Neuropharmacology. 2004; 47:263–72.

Ahlskog JE, Muenter MD. Frequency of levodopa-related dyskinesias and motor fluctuations as estimated from the cumulative literature. Mov Disord. 2001;16:448–58.

Andreassen AO, Jorgensen HA. Neurotoxicity associated with neuroleptic-induced oral dyskinesia in rats. Prog Neurobiol. 2000;61:525–41.

Araujo NP, Abílio VC, Silva RH, Pereira RC, Carvalho RC, Gonzalez C, et al. Effects of topiramate on oral dyskinesia induced by reserpine. Brain Res Bull. 2004;64:331–7.

Beal MF, Hyman BT, Koroshetz W. Do defects in mitochondrial energy metabolism underlie the pathology of neurodegenerative diseases? Trends Neurosci. 1993;16:125–31.

Bergamo M, Abílio VC, Queiroz CM, Barbosa-Júnior HN, Abdanur LR, Frussa-Filho R. Effects of age on a new animal model of tardive dyskinesia. Neurobiol Aging. 1997;18:623–9.

Bishnoi M, Chopra K, Kulkarni SK. Protective effect of rutin, a polyphenolic flavonoid against haloperidol-induced orofacial dyskinesia and associated behavioural, biochemical and neurochemical changes. Fundam Clin Pharmacol. 2007a;21:521–9.

Bishnoi M, Chopra K, Kulkarni SK. Theophylline, adenosine receptor antagonist prevents behavioral, biochemical and neurochemical changes associated with an animal model of tardive dyskinesia. Pharmacol Rep. 2007b;59:181–91.

Bishnoi M, Chopra K, Kulkarni SK. U-74500A (lazaroid), a 21-aminosteroid attenuates neuroleptic-induced orofacial dyskinesia. Methods Find Exp Clin Pharmacol. 2007c;29:601–5.

Bishnoi M, Chopra K, Kulkarni SK. Co-administration of nitric oxide (NO) donors prevents haloperidol-induced orofacial dyskinesia, oxidative damage and change in striatal dopamine levels. Pharmacol Biochem Behav. 2008a;91:423–9.

Bishnoi M, Chopra K, Kulkarni SK. Protective effect of Curcumin, the active principle of turmeric (Curcuma longa) in haloperidol-induced orofacial dyskinesia and associated behavioural, biochemical and neurochemical changes in rat brain. Pharmacol Biochem Behav. 2008b;88:511–22.

Bishnoi M, Chopra K, Kulkarni SK. Protective effect of L-type calcium channel blockers against haloperidol-induced orofacial dyskinesia: a behavioural, biochemical and neurochemical study. Neurochem Res. 2008c;33:1869–80.

Blanchet PJ, Abdillahi O, Beauvais C, Rompré PH, Lavigne GJ. Prevalence of spontaneous oral dyskinesia in the elderly: a reappraisal. Mov Disord. 2004;19:892–6.

Brouillet E, Hantraye P, Ferrante RJ, Dolan R, Leroy-Willig A, Kowall NW, Beal MF. Chronic mitochondrial energy impairment produces selective striatal degeneration and abnormal choreiform movements in primates. Proc Natl Acad Sci U S A. 1995;92:7105–9.

Brouillet E, Condé F, Beal MF, Hantraye P. Replicating Huntington's disease phenotype in experimental animals. Prog Neurobiol. 1999;59:427–68.

Burger ME, Alves A, Callegari L, Athayde FR, Nogueira CW, Zeni G, Rocha JB. Ebselen attenuates reserpine-induced orofacial dyskinesia and oxidative stress in rat striatum. Prog Neuropsychopharmacol Biol Psychiatry. 2003;27:135–40.

Burger M, Fachinetto R, Calegari L, Paixão MW, Braga AL, Rocha JB. Effects of age on reserpine-induced orofacial dyskinesia and possible protection of diphenyl diselenide. Brain Res Bull. 2004;64:339–45.

Burger ME, Fachinetto R, Zeni G, Rocha JB. Ebselen attenuates haloperidol-induced orofacial dyskinesia and oxidative stress in rat brain. Pharmacol Biochem Behav. 2005;81:608–15.

Burger ME, Fachinetto R, Wagner C, Perottoni J, Pereira RP, Zeni G, Rocha JB. Effects of diphenyl-diselenide on orofacial dyskinesia model in rats. Brain Res Bull. 2006;70:165–70.

Cadenas E, Davies KJA. Mitochondrial free radical generation oxidative stress and aging. Free Radic Biol Med. 2000;29:222–30.

Carvalho RC, Silva RH, Abílio VC, Barbosa PN, Frussa-Filho R. Antidyskinetic effects of risperidone on animal models of tardive dyskinesia in mice. Brain Res Bull. 2003;60:115–24.

Casey DE. Tardive dyskinesia reversible and irreversible. In: Casey DE, Chase TN, Christensen AV, Gerlach J, editors. Dyskinesia: research and treatment. Berlin: Springer; 1985. p. 88–98.

Casey DE. Tardive dyskinesia—pathophysiology. In: Bloom FE, Kupfer DJ, editors. Psychopharmacology: the fourth generation of progress. New York: Raven; 1995. p. 1497–502.

Castro JP, Frussa-Filho R, Fukushiro DF, Silva RH, Medrano WA, Ribeiro R, et al. Effects of baclofen on reserpine-induced vacuous chewing movements in mice. Brain Res Bull. 2006;68:436–41.

Cenci MA, Lee CS, Björklund A. L-DOPA-induced dyskinesia in the rat is associated with striatal overexpression of prodynorphin- and glutamic acid decarboxylase mRNA. Eur J Neurosci. 1998;10:2694–706.

Colpo G, Trevisol F, Teixeira AM, Fachinetto R, Pereira RP, Athayde ML, et al. Ilex paraguariensis hás antioxidant potential and attenuates haloperidol-induced orofacial dyskinesia and memory dysfunction in rats. Neurotox Res. 2007;12:171–80.

Cooper WO, Hickson GB, Fuchs C, Arbogast PG, Ray WA. New users of antipsychotic medications among children enrolled in TennCare. Arch Pediatr Adolesc Med. 2004;158:753–9.

Egan MF, Hurd Y, Ferguson J, Bachus SE, Hamid EH, Hyde TM. Pharmacological and neurochemical differences between acute and tardive vacuous chewing movements induced by haloperidol. Psychopharmacology (Berl). 1996;127:337–45.

Fachinetto R, Burger ME, Wagner C, Wondracek DC, Brito VB, Nogueira CW, et al. High fat diet increases the incidence of orofacial dyskinesia and oxidative stress in specific brain regions of rats. Pharmacol Biochem Behav. 2005;81:585–92.

Faria RR, Abílio VC, Grassl C, Chinen CC, Negrão LT, de Castro JP, et al. Beneficial effects of vitamin C and vitamin E on reserpine-induced oral dyskinesia in rats: critical role of striatal catalase activity. Neuropharmacology. 2005;48:993–1001.

Flaherty AW, Graybiel AM. Anatomy of the basal ganglia. In: Marsden CD, Fahn S, editors. Movement disorders, vol. 3. Cambridge: Butterworth-Heinemann; 1994.

Gattaz WF, Emrich A, Behrens S. Vitamin E attenuates the development of haloperidol-induced dopaminergic hypersensitivity in rats: possible implications for tardive dyskinesia. J Neural Transm. 1993;92:197–201.

Gerlach J. Pathophysiological mechanisms underlying tardive dyskinesia. In: Casey DE, Chase TN, Christensen AV, Gerlach J, editors. Dyskinesia: research and treatment. Berlin: Springer; 1985. p. 99–104.

Gerlach J, Casey DE. Tardive dyskinesia. Acta Psychiatr Scand. 1988;77:369–78.

Gusella JF, Wexler NS, Conneally PM, Naylor SL, Anderson MA, Tanzi RE, et al. A polymorphic DNA marker genetically linked to Huntington's disease. Nature. 1983;306:234–8.

Henry B, Crossman AR, Brotchie JM. Characterization of enhanced behavioral responses to L-DOPA following repeated administration in the 6-hydroxydopamine-lesioned rat model of Parkinson's disease. Exp Neurol. 1998;151:334–42.

Janno S, Holi M, Tuisku K, Wahlbeck K. Prevalence of neuroleptic-induced movement disorders in chronic schizophrenia inpatients. Am J Psychiatry. 2004;161:160–3.

Jenner P. Avoidance of dyskinesia: preclinical evidence for continuous dopaminergic stimulation. Neurology. 2004;62:S47–55.

Kane JM. Tardive dyskinesia rates with atypical antipsychotics in adults: prevalence and incidence. J Clin Psychiatry. 2004;65:16–20.

Kane JM, Woerner M, Lieberman J. Tardive dyskinesia: prevalence, incidence and risk factors. In: Casey DE, Chase TN, Christensen AV, Gerlach J, editors. Dyskinesia: research and treatment. Berlin: Springer; 1985. p. 72–8.

Kasckow JW, Mulchahey JJ, Mohamed S. The use of novel antipsychotics in the older patient with neurodegenerative disorders in the long-term care setting. J Am Med Dir Assoc. 2004;5: 242–8.

Kish SJ, Shannak K, Hornykiewicz O. Uneven pattern of dopamine loss in the striatum of patients with idiopathic Parkinson's disease. Pathophysiologic and clinical implications. N Engl J Med. 1988;318:876–80.

Konitsiotis S, Tsironis C. Levodopa-induced dyskinesia and rotational behavior in hemiparkinsonian rats: independent features or components of the same phenomenon? Behav Brain Res. 2006;170:337–41.

La Fontaine MA, Geddes JW, Banks A, Butterfield DA. Effects of endogenous and exogenous antioxidants on 3-nitropropionic acid-induced in vivo oxidative stress and striatal lesions: insight into Huntington's disease. J Neurochem. 2000;75:1709–15.

Levy AD, Ellison GD. Interaction between chronic amphetamine and neuroleptic treatments on oral behavior in rats. Psychopharmacology (Berl). 1987;93:218–22.

Levy AD, See RE, Levin ED, Ellison GD. Neuroleptic-induced oral movements in rats: methodological issues. Life Sci. 1987;41:1499–506.

Llorca PM, Chereau I, Bayle FJ, Lancon C. Tardive dyskinesia and antipsychotics: a review. Eur Psychiatry. 2002;17:129–38.

Lohr JB. Oxygen free radicals and neuropsychiatric illness. Arch Gen Psychiatry. 1991;48: 1097–106.

Lundblad M, Picconi B, Lindgren H, Cenci MA. A model of L-DOPA-induced dyskinesia in 6-hydroxydopamine lesioned mice: relation to motor and cellular parameters of nigrostriatal function. Neurobiol Dis. 2004;16:110–23.

Lundblad M, Usiello A, Carta M, Håkansson K, Fisone G, Cenci MA. Pharmacological validation of a mouse model of l-DOPA-induced dyskinesia. Exp Neurol. 2005;194:66–75.

Marsden CD. Is tardive dyskinesia a unique disorder? In: Casey DE, Chase TN, Christensen AV, Gerlach J, editors. Dyskinesia: research and treatment. Berlin: Springer; 1985. p. 67–71.
Marsden CD. Problems with long-term levodopa therapy for Parkinson's disease. Clin Neuropharmacol. 1994;2:S32–44.
Masan PS. Atypical antipsychotics in the treatment of affective symptoms: a review. Ann Clin Psychiatry. 2004;16:3–13.
Mcintyre RS, Mancini DA, Lin P, Jordan J. Treating bipolar disorder. Evidence-based guidelines for family medicine. Can Fam Physician. 2004;50:388–94.
Morelli M, Fenu S, Garau L, Di Chiara G. Time and dose dependence of the 'priming' of the expression of dopamine receptor supersensitivity. Eur J Pharmacol. 1989;162:329–35.
Muscettola G, Barbato G, Pampallona S, Casiello M, Bollini P. Extrapyramidal syndromes in neuroleptic-treated patients: prevalence, risk factors, and association with tardive dyskinesia. J Clin Psychopharmacol. 1999;19:203–8.
Naidu PS, Singh A, Kulkarni SK. Carvedilol attenuates neuroleptic-induced orofacial dyskinesia: possible antioxidant mechanisms. Br J Pharmacol. 2002;136:193–200.
Naidu PS, Singh A, Kaur P, Sandhir R, Kulkarni SK. Possible mechanism of action in melatonin attenuation of haloperidol-induced orofacial dyskinesia. Pharmacol Biochem Behav. 2003a;74:641–8.
Naidu PS, Singh A, Kulkarni SK. Quercetin, a bioflavonoid, attenuates haloperidol-induced orofacial dyskinesia. Neuropharmacology. 2003b;44:1100–6.
Naidu PS, Singh A, Kulkarni SK. Reversal of reserpine-induced orofacial dyskinesia and cognitive dysfunction by quercetin. Pharmacology. 2004;70:59–67.
Nair NPV, Jeste DV. Preface. In: Yassa R, Nair NPV, Jeste DV, editors. Neuroleptic-induced movement disorders. New York: Cambridge University Press; 1997. p. 15–7.
Neisewander JL, Lucki I, McGonigle P. Behavioral and neurochemical effects of chronic administration of reserpine and SKF-38393 in rats. J Pharmacol Exp Ther. 1991a;257:850–60.
Neisewander JL, Lucki I, McGonigle P. Neurochemical changes associated with the persistence of spontaneous oral dyskinesia in rats following chronic reserpine treatment. Brain Res. 1991b;558:27–35.
Neisewander JL, Castañeda E, Davis DA. Dose-dependent differences in the development of reserpine-induced oral dyskinesia in rats: support for a model of tardive dyskinesia. Psychopharmacology (Berl). 1994;116:79–84.
Neisewander JL, Castañeda E, Davis DA, Elson HJ, Sussman AN. Effects of amphetamine and 6-hydroxydopamine lesions on reserpine-induced oral dyskinesia. Eur J Pharmacol. 1996;305:13–21.
Obeso JA, Rodríguez-Oroz MC, Rodríguez M, Lanciego JL, Artieda J, Gonzalo N, Olanow CW. Pathophysiology of the basal ganglia in Parkinson's disease. Trends Neurosci. 2000; 23:S8–19.
Obeso JA, Rodriguez-Oroz M, Marin C, Alonso F, Zamarbide I, Lanciego JL, Rodriguez-Diaz M. The origin of motor fluctuations in Parkinson's disease: importance of dopaminergic innervation and basal ganglia circuits. Neurology. 2004;62:S17–30.
Olanow CW, Damier P, Goetz CG, Mueller T, Nutt J, Rascol O, et al. Multicenter, open-label, trial of sarizotan in Parkinson disease patients with levodopa-induced dyskinesias (the SPLENDID Study). Clin Neuropharmacol. 2004;27:58–62.
Palermo-Neto J, Frussa-Filho R. Behavioural models of tardive dyskinesia in rodents: a chronological review. In: Bolis CL, Licino CL, editors. Dopaminergic system: its evolution from biology to clinical aspects. Geneva: AIREN (International Association for Research and Training in Neurosciences) and World Health Organization; 2001. p. 61–81.
Palfi S, Ferrante RJ, Brouillet E, Beal MF, Dolan R, Guyot MC, et al. Chronic 3-nitropropionic acid treatment in baboons replicates the cognitive and motor deficits of Huntington's disease. J Neurosci. 1996;16:3019–25.
Pavón N, Martín AB, Mendialdua A, Moratalla R. ERK phosphorylation and FosB expression are associated with L-DOPA-induced dyskinesia in hemiparkinsonian mice. Biol Psychiatry. 2006;59:64–74.

Peixoto MF, Abílio VC, Silva RH, Frussa-Filho R. Effects of valproic acid on an animal model of tardive dyskinesia. Behav Brain Res. 2003;142:229–33.
Petersén A, Mani K, Brundin P. Recent advances on the pathogenesis of Huntington's disease. Exp Neurol. 1999;157:1–18.
Putterman DB, Munhall AC, Kozell LB, Belknap JK, Johnson SW. Evaluation of levodopa dose and magnitude of dopamine depletion as risk factors for levodopa-induced dyskinesia in a rat model of Parkinson's disease. J Pharmacol Exp Ther. 2007;323:277–84.
Queiroz CM, Frussa-Filho R. Effects of buspirone on an animal model of tardive dyskinesia. Prog Neuropsychopharmacol Biol Psychiatry. 1999;23:1405–18.
Queiroz CM, Piovezan RD, Frussa-Filho R. Reserpine does not induce orofacial dyskinesia in spontaneously hypertensive rats. Eur J Pharmacol. 1998;356:105–8.
Ricardo VP, Frussa-Filho R, Silva RH, Lopez GB, Patti CL, Zanier-Gomes PH, et al. Effects of social isolation on aging-induced orofacial movements in rats. Physiol Behav. 2005;86:203–8.
Rogoza RM, Fairfax DF, Henry P, N-Marandi S, Khan RF, Gupta SK, et al. Electron spin resonance spectroscopy reveals alpha-phenyl-N-tert-butylnitrone spin-traps free radicals in rat striatum and prevents haloperidol-induced vacuous chewing movements in the rat model of human tardive dyskinesia. Synapse. 2004;54:156–63.
Rosenstock TR, Carvalho AC, Jurkiewicz A, Frussa-Filho R, Smaili SS. Mitochondrial calcium, oxidative stress and apoptosis in a neurodegenerative disease model induced by 3-nitropropionic acid. J Neurochem. 2004;88:1220–8.
Rosenstock TR, Abílio VC, Frussa-Filho R, Kiyomoto BH, Smaili SS. Old mice present increased levels of succinate dehydrogenase activity and lower vulnerability to dyskinetic effects of 3-nitropropionic acid. Pharmacol Biochem Behav. 2009;91:327–32.
Roth GS, Joseph JA. Cellular and molecular mechanisms of impaired dopaminergic function during aging. Ann N Y Acad Sci. 1994;719:129–35.
Sachdev OS. The current status of tardive dyskinesia. Aust N Z J Psychiatry. 2000;34:355–69.
Sadan O, Bahat-Stromza M, Gilgun-Sherki Y, Atlas D, Melamed E, Offen D. A novel brain-targeted antioxidant (AD4) attenuates haloperidol-induced abnormal movement in rats: implications for tardive dyskinesia. Clin Neuropharmacol. 2005;28:285–8.
Saltz BL, Kane JM, Woerner M, Lieberman J. Aging and tardive dyskinesia. In: Yassa R, Nair PV, Jeste DV, editors. Neuroleptic-induced movement disorders. New York: Cambridge University Press; 1997. p. 13–25.
Samad N, Khan A, Perveen T, Haider S, Abdul Haleem M, Haleem DJ. Increase in the effectiveness of somatodendritic 5-HT-1A receptors in a rat model of tardive dyskinesia. Acta Neurobiol Exp. 2007;67:389–97.
Schwarting RK, Huston JP. The unilateral 6-hydroxydopamine lesion model in behavioral brain research. Analysis of functional deficits, recovery and treatments. Prog Neurobiol. 1996a;50:275–331.
Schwarting RK, Huston JP. Unilateral 6-hydroxydopamine lesions of meso-striatal dopamine neurons and their physiological sequelae. Prog Neurobiol. 1996b;49:215–66.
Silva RH, Abílio VC, Torres-Leite D, Bergamo M, Chinen CC, Claro FT, et al. Concomitant development of oral dyskinesia and memory deficits in reserpine-treated male and female mice. Behav Brain Res. 2002;132:171–7.
Singh A, Naidu PS, Kulkarni SK. Possible antioxidant and neuroprotective mechanisms of FK506 in attenuating haloperidol-increased orofacial dyskinesia. Eur J Pharmacol. 2003;477:87–94.
Sussman AN, Tran-Nguyen LTL, Neisewander JL. Acute reserpine administration elicits long-term spontaneous oral dyskinesia. Eur J Pharmacol. 1997;337:157–60.
Tabrizi SJ, Workman J, Hart PE, Mangiarini L, Mahal A, Bates G, et al. Mitochondrial dysfunction and free radical damage in the Huntington R6/2 transgenic mouse. Ann Neurol. 2000;47:80–6.
Tariot PN, Profenno LA, Ismail MS. Efficacy of atypical antipsychotics in elderly patients with dementia. J Clin Psychiatry. 2004;65:11–5.

Tarsy D, Baldessarini RJ, Tarazi FI. Effects of newer antipsychotics on extrapyramidal function. CNS Drugs. 2002;16:23–45.
Teixeira AM, Trevizol F, Colpo G, Garcia SC, Charão M, Pereira RP, et al. Influence of chronic exercise on reserpine-induced oxidative stress in rats: behavioral and antioxidant evaluations. Pharmacol Biochem Behav. 2008;88:465–72.
Thaakur SR, Jyothi B. Effect of spirulina maxima on the haloperidol-induced tardive dyskinesia and oxidative stress in rats. J Neural Transm. 2007;114:1217–25.
Thomson PD, Berardelli A, Rothwell JC, Day BL, Dick SPR, Benecke R, et al. The coexistence of bradykinesia and chorea in Huntington's disease and its implications for theories of basal ganglia control of movement. Brain. 1988;111:223–44.
Tran TT, De Costa BR, Matsumoto RR. Microinjection of sigma ligands into cranial nerve nuclei produces vacuous chewing in rats. Psychopharmacology (Berl). 1998;137:191–200.
Turrone P, Remington G, Kapur S, Nobrega JN. Differential effects of within-day continuous vs. transient dopamine D2 receptor occupancy in the development of vacuous chewing movements (VCMs) in rats. Neuropsychopharmacology. 2003;28:1433–9.
Uhrbrand L, Faurbye A. Psychopharmacology (Berl). 1960;1:408–18.
Ungerstedt U. Postsynaptic supersensitivity after 6-hydroxy-dopamine induced degeneration of the nigro-striatal dopamine system. Acta Physiol Scand Suppl. 1971a;367:69–93.
Ungerstedt U. Striatal dopamine release after amphetamine or nerve degeneration revealed by rotational behaviour. Acta Physiol Scand Suppl. 1971b;367:49–68.
Ungerstedt U, Arbuthnott GW. Quantitative recording of rotational behavior in rats after 6-hydroxy-dopamine lesions of the nigrostriatal dopamine system. Brain Res. 1970;24:485–93.
Verhagen Metman L, Morris MJ, Farmer C, Gillespie M, Mosby K, Wuu J, et al. Huntington's disease: a randomized, controlled trial using the NMDA-antagonist amantadine. Neurology. 2002;59:694–9.
Vital MABF, Frussa-Filho R, Palermo-Neto J. Effects of monosialoganglioside GM1 on dopaminergic supersensitivity. Life Sci. 1997;56:2299–307.
Wachtel SR, Abercrombie ED. L-3,4-dihydroxyphenylalanine-induced dopamine release in the striatum of intact and 6-hydroxydopamine-treated rats: differential effects of monoamine oxidase A and B inhibitors. J Neurochem. 1994;63:108–17.
Waddington JL. Spontaneous orofacial movements induced in rodents by very long-term neuroleptic drug administration: phenomenology pathophysiology and putative relationship to tardive dyskinesia. Psychopharmacology (Berl). 1990;101:431–47.
Waddington JL. Rodent and other animal models of tardive dyskinesia during long-term neuroleptic-drug administration: controversies and implications for the clinical syndrome. In: Yassa R, Nair NPV, Jeste DV, editors. Neuroleptic-induced movement disorders. New York: Cambridge University Press; 1997. p. 225–37.
Waddington JL, Gamble SJ. Differential effects of ageing on distincts features of apomorphine stereotypy in the adult rat. Neurosci Lett. 1980;20:95–9.
Winkler C, Kirik D, Björklund A, Cenci MA. L-DOPA-induced dyskinesia in the intrastriatal 6-hydroxydopamine model of parkinson's disease: relation to motor and cellular parameters of nigrostriatal function. Neurobiol Dis. 2002;10:165–86.
Wolf WA, Bieganski GJ, Guillen V, Mignon L. Enhanced 5-HT2C receptor signaling is associated with haloperidol-induced "early onset" vacuous chewing in rats: implications for antipsychotic drug therapy. Psychopharmacology (Berl). 2005;182:84–94.
Wolfarth S, Ossowska K. Can supersensitivity of rodents to dopamine be regarded as a model of tardive dyskinesia? Prog Neuropsychopharmacol Biol Psychiatry. 1989;13:789–840.
Worrel JA, Marken PA, Beckman SE, Ruehter VL. Atypical antipsychotic agents: a critical review. Am J Health Syst Pharm. 2000;57:238–55.

Chapter 28
Experimental Diet Models in the Investigation of Obesity

Ana R. Dâmaso, Fernanda Oliveira Duarte, Marcela Sene-Fiorese, Marla Simone Jovenasso Manzoni, Elizeu Antônio Rossi, Nadia Carla Cheik, Ricardo Luis Fernandes Guerra, and Ana Claudia Garcia de Oliveira Duarte

Introduction

Several experimental models of dietary manipulation have been created for the purpose of investigating chronic-degenerative diseases. In general, such models allow for the assessment of the qualitative and quantitative alterations of nutrients and substances contained in a food regimen. Recent studies have made use of alternative

A.R. Dâmaso, (✉)
Universidade Federal de São Paulo (UNIFESP), São Paulo, São Paulo, Brazil
e-mail: ana.damaso@unifesp.br

F.O. Duarte
Departamento de Educação Física e Motricidade Humana, Universidade Federal de São Carlos, São Carlos, São Paulo, Brazil

M. Sene-Fiorese
Institute of Physics of São Carlos, Universidade of São Paulo, São Carlos, São Paulo, Brazil

M.S.J. Manzoni • E.A. Rossi
Departamento de Alimentos e Nutrição, Universidade Estadual Paulista (UNESP), Araraquara, São Paulo, Brazil

N.C. Cheik
Instituto de Ciências Biomédicas, Universidade Federal de Uberlândia, Uberlândia, Minas Gerais, Brazil

R.L.F. Guerra
Department of Human Movement Sciences, Universidade Federal de São Paulo (UNIFESP), São Paulo, São Paulo, Brazil

A.C.G.d. Oliveira Duarte
Departamento de Educação Física e Motricidade Humana - DEFMH, Universidade Federal de São Carlos, São Carlos, São Paulo, Brazil

M.L. Andersen, S. Tufik (eds.), *Rodent Model as Tools in Ethical Biomedical Research*, DOI 10.1007/978-3-319-11578-8_28

forms of dietary manipulation that have resulted in the experimental development of diseases such as diabetes, dyslipidemia, and obesity (Duarte et al. 2003; Estadella et al. 2004; Manzoni et al. 2005; Cheik 2005; Guerra 2005).

Palatable Hyperlipidic Diet

The chronic treatment of a *palatable hyperlipidic diet* leads to an elevation in body fat in rats. Such is associated with hyperleptinemia, or leptin resistance, and a palatable diet may cause hyperphagia, infertility, susceptibility to diabetes, a reduction in basal metabolism, an increase in adipose tissue, and the onset of obesity. A palatable diet may also promote hypercorticosteronemia, normoglycemia, and hyperinsulinemia, suggesting insulin resistance without an increase in food intake. In addition, recent studies verified the occurrence of hypertrophy of adipocytes and a reduction in gene expression of adiponectin, suggesting an atherogenic effect of this diet (Estadella et al. 2004; Estadella 2005). Also, leptin resistance as a consequence of exogenous obesity has been verified to promote a reduction in the secretion of GH and IGF-1 (Kopchick and Andry 2000; Zhou et al. 1998).

Although a palatable hyperlipidic diet has proven to promote a reduction in food intake, it does cause obesity (Gaíva et al. 2001). After being weaned, rats fed a palatable hyperlipidic diet, in consecutive alternated cycles, with a food restriction demonstrated an increase in body weight, a reduction in energy consumption, an increase in fat deposits, and the development of obesity (Kretschemer et al. 2005). In addition to these metabolic alterations, lipid-rich diets, especially those plentiful in saturated fat acids, exert detrimental effects on the learning abilities and memory of rats. Such adverse effects may be related to the development of insulin resistance, intolerance to glucose, and transport of this substrate in the brain (Duelli et al. 2000; Estadella 2005). Other studies have demonstrated the existence of alterations in the components of energy balance, resulting in greater metabolic efficiency of hyperlipidic diets when compared to balanced diets or to those that are rich in carbohydrates (Westerterp-Platenga 2004).

The effects caused by palatable hyperlipidic diets on animal metabolism may be influenced by factors such as age, body mass, treatment time, and diet composition (Gaíva et al. 2001; Leonhardt and Langhans 2004).

The Hypercholesterolemic Diet

The increase observed in plasmatic concentrations of cholesterol (hypercholesterolemia) is the main risk factor associated with the development of cardiovascular diseases (Abd El-Gawad et al. 2005). It is believed that a relevant function of adipose tissue is the storage of triglycerides, which would exert a tamponing effect by mediating not only serum triglyceride concentration but also hypercholesterolemia, as this is the main storage tissue of cholesterol (Krause and Hartman 1984). The ingestion of cholesterol through the diet (exogenous cholesterol) seems to be closely associated with its storage in adipose tissue. It has been verified in humans that copious ingestion of diets rich in cholesterol seems to be directly related to increases in cholesterol deposits in adipose tissue, even when plasmatic concentrations are not elevated.

Among the main effects of cholesterol-enriched diets observed in experimental protocols using laboratory animals are the increase in serum concentration of total cholesterol, LDL-cholesterol, TG, reduction in HDL, as well as increases in fat liver, suggesting the development of non-alcoholic fatty liver diseases (Abd El-Gawad et al. 2005; Islam et al. 2005; Chen et al. 2003; Feoli et al. 2003; Rossi et al. 2000).

In regards to adiposity, Zulet et al. (1999) reported a reduction in central adipose tissue in rats fed a hypercholesterolemic diet as compared to animals that were fed a normocaloric diet. In our laboratory, we have, for the past ten years, elected to work with two specific types of diet:

(a) A palatable hyperlipidic diet, which has proven its worth in the investigation of exogenous obesity. This model also seems to serve well in the examination of metabolic and hormonal alterations associated with diabetes type II and dyslipidemia.
(b) A hypercholesterolemic diet was purposely adopted for the investigation of dyslipidemia and it has also proven to be a valuable experimental model in the study of animal obesity.

Experimental Models for the Investigation of the Effects of a Palatable Hyperlipidic Diet

Our laboratory opted to make use of a diet composition that was suitable for experimental conditions with or without exercise or dietary restriction so as to allow for the observation of any possible therapeutic property that could reverse or control diseases. The aim of the changes in diet composition was to investigate the development and treatment of obesity and hypercholesterolemia, while assessing exercise regimes in the control and treatment of these diseases. The palatable hyperlipidic diet adopted in this investigation was composed of: 15 g of regular commercial chow, 10 g of roasted peanuts, 10 g of milk chocolate, and 5 g of corn-starch biscuits. These were ground, mixed, and made into pellets that totaled 21.04 KJ/g. In comparison, a normocaloric diet contains 17.03 KJ/g (Estadella 2001; Duarte 2001; Duarte et al. 2003; Estadella et al. 2004). The treatment period may range from 3 to 15 weeks. Animal food was provided ad libitum. There were no gastrointestinal disturbances that might impact the results.

Experimental Model for the Investigation of Adiposity and Hormonal Secretion

In a recent investigation carried out in our laboratory, we assessed the effects of a palatable hyperlipidic diet on relative weight, the area of the adipocytes, and the plasmatic concentrations of insulin, leptin, and ghrelin in adult male rats fed the diet

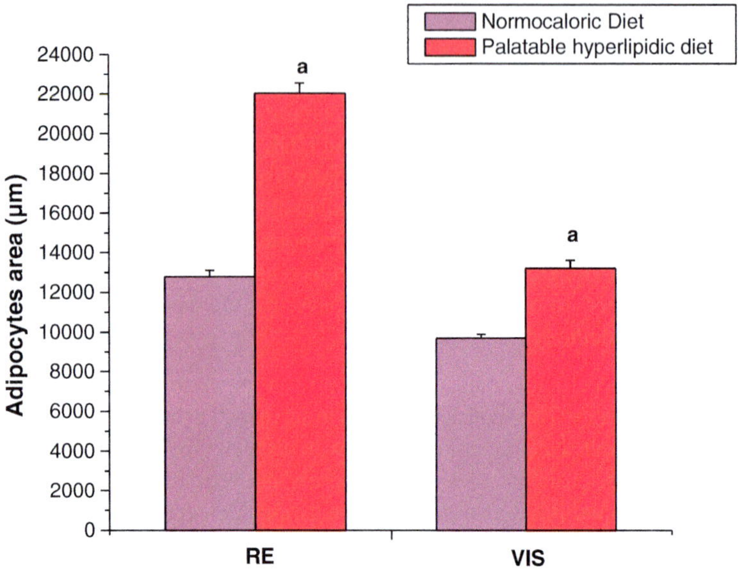

Fig. 28.1 Adipocytes area (μm) of rats fed different diets. Values are expressed as Mean ± Standard deviation (n = 8/group). A significant difference exists among the distinct letters ([a]versus N; $p \leq 0.05$) in the Tukey-Kramer test. *N* Normocaloric Diet, *RE* central retroperitoneal adipose tissue, *VIS* visceral epididymal adipose tissue

described above for 8 weeks (Cheik 2005). We observed a significant weight increase in the white adipose tissue of the central (RET) and visceral (EPI) adipose tissues of the rats fed the palatable hyperlipidic diet when compared to controls. The hypertrophy of adipocytes, which was verified in the group that ingested the higher amount of lipids (Fig. 28.1), was associated with the accumulation of fat, indicating the development of exogenous obesity.

The ingestion of a palatable hyperlipidic diet promoted a significant increase in plasmatic concentration of leptin (Fig. 28.2), without altering food ingestion. This suggests a reduction in the sensitivity to leptin by dietary induction. Hyperleptinemia is associated with these two factors: hyperinsulinemia (Fig. 28.3) and an area increase in white RET and VIS white adipose tissue (Fig. 28.1) of rats that belong to the groups fed a palatable hyperlipidic diet.

While analyzing the plasmatic concentrations of the hormone ghrelin (Fig. 28.2), a reduction of 14 % was observed in the group that ingested the hyperlipidic diet; this was the group that exhibited the highest mass of fat. This inverse association may be interpreted as an adaptive physiological response that fails in the attempt to re-establish the appropriate energy balance. Results produced herein demonstrate that relevant hormonal and metabolic alterations, reflected in the disruption of energetic homeostasis, were produced by dietary manipulation.

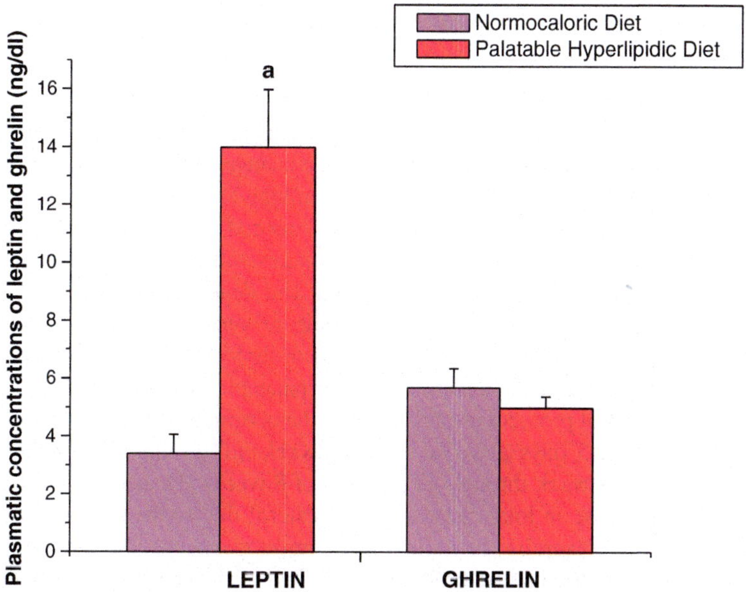

Fig. 28.2 Plasmatic concentrations of leptin and ghrelin (*ng*/dl) of rats fed different diets. Values are expressed as Mean ± Standard deviation (n = 8/group). A significant difference exists among the distinct letters ([a]versus N; $p \leq 0.05$) in the Tukey-Kramer test. *N* Normocaloric diet

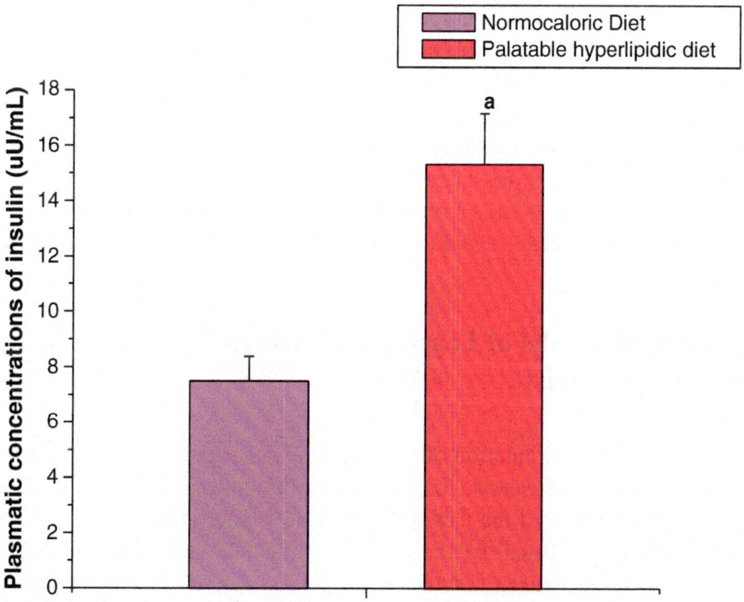

Fig. 28.3 Plasmatic concentrations of insulin (μU/mL) of rats fed different diets. Values are expressed as Mean ± Standard deviation (n = 8/group). A significant difference exists among the distinct letters ([a]versus N; $p \leq 0.05$) in the Tukey-Kramer test. *N* Normocaloric diet

The Experimental Model of Weekend Exercise Versus Continuing Exercise

Professional competitiveness and the advent of technological facilities that make our lives ever more comfortable have led to dramatic changes that are commonly associated with a sedentary lifestyle and an increase in high fat food consumption. As the practice of exercise has become restricted to a weekend activity, the term "weekend athlete" has surfaced, suggesting the need to understand the extent to which occasional exercise promotes the control of weight and degenerative chronic diseases. For this purpose, the current study sought to compare weekend exercise with continuing exercise in the development of obesity and dyslipidemia when subjects (in this study, rats) were submitted to a normocaloric or palatable hyperlipidic diet.

The experimental protocol established that the continuing exercise group would swim 90 min/day, 5 days a week, and the "weekend" group, 90 min/day for 2 consecutive days a week throughout a period of 8 weeks (Duarte et al. 2003). We observed an increase in total circulating cholesterol due to the palatable hyperlipidic diet, and that exercise, be it continuous or on weekends, did not reduce this variable. Contrary to common belief, weekend exercise seems to have exerted an exacerbating increase on the circulating HDL, however, to reduce the concentration of triglycerides. Some studies have demonstrated that the increase observed in the lipoprotein HDL occurs as a function of acceleration in the transference of free cholesterol, mediated by an increment in the activity of the enzyme cholesterol lectin acyl transferase (LCAT), which speeds the reverse transport of cholesterol (Giannini 1998). In this experimental protocol, we observed an increase in the HDL cholesterol and a reduction in the triglycerides in response to two modes of exercise while on a hyperlipidic diet, which may indicate that the combination exerts a shielding effect against the development of atherogenesis.

Exercise was effective in the control of dyslipidemia in the animals that were fed a normocaloric diet, even when the physical challenge was only on weekends. This shows that physical strain, regardless of its frequency and when associated with a balanced diet, may prevent the onset of chronic degenerative diseases.

Experimental Model of Fractioned Exercise Versus Continuous Exercise

This study contains a simulation of an applied experimental model that stimulates the increase of physical exercise for the purpose of controlling chronic diseases and by the recommendation of the CDC/ACMS (Pate 1995). This model has been successfully deployed in Brazil by the non-governmental organization CELAFISCS, Programa Agita São Paulo (Matsudo et al. 2002), whose mission statement is to foster the will to exercise increasingly through life (Whitehead 2002). CELAFISCS preaches that the ideal period of physical exercise is within the range of 60–90 min. In this model, adult rats swam 90 min, in a fractioned (3 × 30 min, 4 h interval) and continuous manner (5 times—90 min) each week for 8 weeks. The partial results of

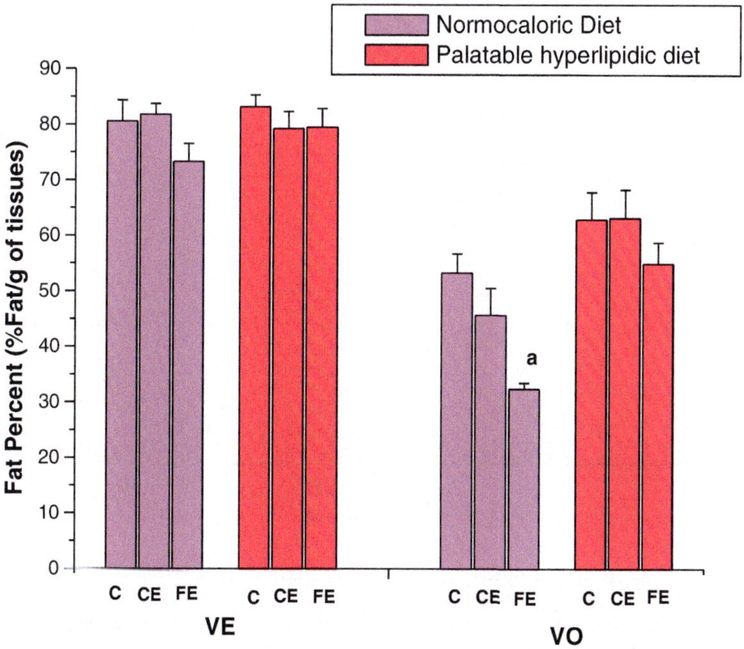

Fig. 28.4 Fat percent of epididymal tissues (VE) and omental (VO) visceral white adipose tissues in rats fed different diets. Values are expressed as Mean ± Standard error (n = 8/group). Distinct letters vary significantly ([a]versus N; $p \leq 0.05$) in the Tukey-Kramer test. *N* Normocaloric diet, *C* Control, *CE* Continuous exercise, *FE* Fractioned exercise

this study are described in the figures below (Sene-Fiorese et al. 2005). As shown in Fig. 28.4, the fat percent of the epididymal (visceral) white adipose tissue (VE) did not undergo a significant alteration from the training or the diet. The omental (VO) visceral white adipose tissue, however, was significantly reduced in the fractioned exercise group when this group was fed a normocaloric diet. This is an important effect since once this tissue is insulin resistant, it can lead to the development of Diabetes Mellitus type II.

The hyperlipidic diet promoted an increase in fat liver when compared to a normocaloric diet, a non-desired effect since such values may lead to the development of non-alcoholic fatty liver diseases. In contrast, both training modalities reduced fat liver in rats fed a palatable hyperlipidic diet (Fig. 28.5).

Experimental Model for Prolonged Ingestion of a Palatable Hyperlipidic Diet

This model aimed to determine whether a palatable hyperlipidic diet would produce diabetes mellitus II through the secretion of insulin by isolated islets, plasmatic insulin, and fast glycemia. In this study, the animals were fed a hyperlipidic diet for 15 weeks after weaning.

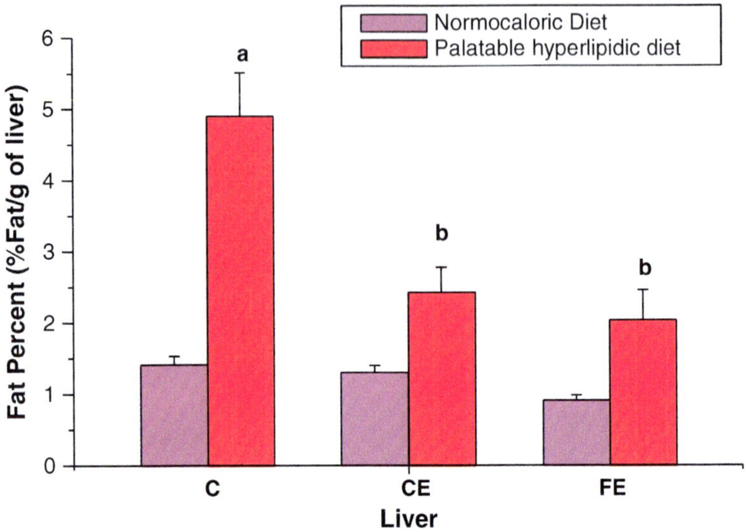

Fig. 28.5 Fat liver of rats fed different diets. Values are expressed as Mean ± Standard error (n=8/group). Distinct letters vary significantly ([a]versus N; $p \leq 0.05$) in the Tukey-Kramer test. *N* Normocaloric, *HP* Palatable hyperlipidic diet, *C* Control, *CE* Continuous exercise, *FE* Fractioned exercise

Insulin secretion by isolated islets increased in rats fed a palatable hyperlipidic diet for 15 weeks, indicating a possible adaptation to a growing demand of insulin in physiological glucose concentrations (G 8.3 mM). The alteration in demand was not enough to modify plasmatic concentrations of insulin at fast, whose stationary values might be attributable to morphological alterations of pancreatic beta cells.

Experimental Model for the Investigation of the Effect of Food Restriction

The investigation of obesity and its influence on lipid and endocrine metabolism has been the focal point of many recent studies. With the increase in obesity prevalence, some food restriction diets have claimed to perform miracles. These are engaged in as a strategic attempt to quickly lose weight, but almost always without any sort of medical supervision and without taking into consideration the potentially detrimental effects that might seriously impact one's health. Thus, it is necessary to examine data on the effects that such food restriction programs exert on the metabolisms previously mentioned (lipids and endocrine) and the accumulation of central (retroperitoneal) and visceral fat (omental and epididymal) as well as the health consequences.

In this experimental model, exogenous obese male rats were submitted to two types of restricted diets: severe restriction (SR, under 50 %) and moderate restriction (MR, under 25 %) for 5 weeks. The calculation of the amount of food that was offered to the restricted group was made based on the amount offered to the corresponding control group (Duarte et al. 2005).

Our results showed that the hyperlipidic diet promoted a greater accumulation of fat in all white adipose tissues that were analyzed in comparison to a normocaloric diet. Thus, the ingestion of large amounts of lipids and sugar increases the fat deposition and promotes the development of obesity. The increase of visceral omental fat is directly associated with the development of dyslipidemia, diabetes, and cardiovascular conditions that increase mortality. The adoption of the food restricted diet promoted the loss of central (retroperitoneal) and visceral (epididymal and omental) white adipose tissues, and the reduction in adiposity was significantly higher after severe compared to moderate restriction.

Based on the results produced by this experimental model, we suggest that engaging in a severe food restriction is not the most adequate strategy in the control of obesity and chronic degenerative diseases as this regimen led to an extreme loss of fat, a condition that may favor the development of other diseases and detrimental alterations in lipid metabolism such as reduction of HDL cholesterol and total circulating cholesterol concentrations.

In contrast, the moderate food restriction program proved to be the most effective in the control of obesity and dyslipidemia as the loss of adipose tissue was more gradual, even when animals were offered a palatable hyperlipidic diet.

Experimental Models for the Investigation of the Effects of a Hypercholesterolemic Diet

The hypercholesterolemic diet (H) that was adopted by our laboratory was standardized by Rossi (2000) and consisted of the incorporation of cholesterol and colic acid to the normocaloric chow as described below. To a batch of regular chow, add 1 % (p/p) of Sigma® C8503 cholesterol and 0.25 % (p/p) of Sigma® C1254 colic acid, incorporating these two ingredients to the chow by the nebulization method, which dissolves the components in stabilized diethyl with 10 ppm of BHT (Carlo Erba®) and absolute ethanol (Merck®), respectively. Each component is sprinkled over the chow within a hood and then left for 12 h to allow for the complete evaporation of the solvents. Subsequently, the chow is packed in opaque plastic bags to bar light and stored under refrigeration during the entire period of the experiment, which lasted between 3 and 8 weeks.

Experimental Model with Young Rats: The Effects of a Hypercholesterolemic Diet on Adiposity

Studies have demonstrated that cholesterol contained in food plays a regulatory role in the lipid metabolism of white adipose tissue (Islam et al. 2005; Le Lay et al. 2001); thus, adiposity of young (21 days) male rats was examined after an 8 week period in

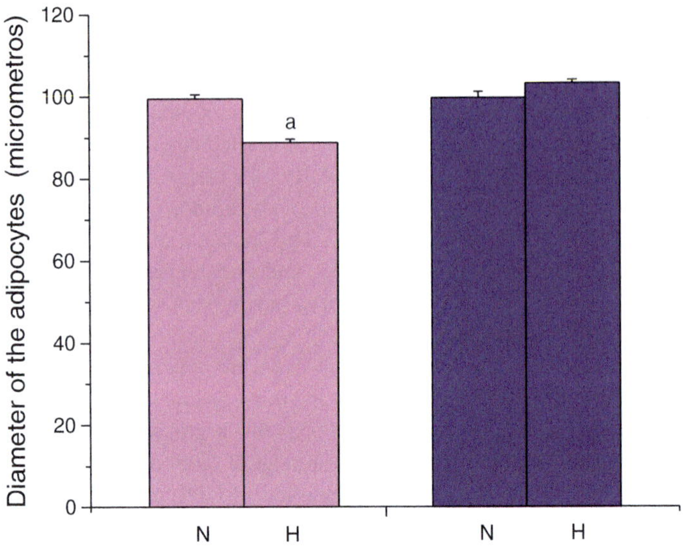

Fig. 28.6 Diameter of the adipocytes of visceral adipose tissue (magenta) and central adipose tissue (*violet*) in rats fed different diets. Values are expressed as Mean ± Standard error (n = 5/group). Distinct letters vary significantly ([a]versus N; $p \leq 0.05$) in the Kolmogorov and Smirnov test. *N* Normocaloric diet, *H* Hypercholesterolemic diet

which the animals were fed a hypercholesterolemic diet (Manzoni et al. 2005). Results suggested that this type of diet can promote a reduction in the diameter and number of adipose cells.

The effect of a diet on adiposity in our experiments seems to be correlated with plasmatic triglyceride concentrations (Fig. 28.6) as a significant reduction of these was observed in animals fed the hypercholesterolemic diet when compared to animals fed the normocaloric diet.

Cholesterol from the diet promoted alterations of the lipid metabolism and adiposity in young rats soon after weaning. The excessive ingestion of cholesterol led to lipotrophic effects on adipose tissue, which according to Frayn (2002), are associated with the development of insulin resistance.

Experimental Model With a Probiotic Soy Yogurt

We have examined the effects of a probiotic soy yogurt on dyslipidemia (Cheik 2002). Adult male rats were fed the hypercholesterolemic diet along with the soy yogurt for 8 weeks. The yogurt supplement (1 mL/rat/day) led to a significant reduction in the total cholesterol serum concentrations and triglycerides, but did not promote any meaningful alterations in the HDL cholesterol fraction. The HDL cholesterol fraction significantly increased in animals fed the normocaloric diet with the yogurt supplement when compared to the group of animals fed the normocaloric diet only.

The soy yogurt helped in the control of dyslipidemia through the reduction of total cholesterol and TG concentrations in rats fed the hypercholesterolemic diet, and also promoted an increase of HDL cholesterol in normocholesterolemic rats.

Experimental Model with Exercise for 5 or 2 Consecutive Days per Week

The results presented below were obtained from a study conducted by Guerra (2005), who compared the effect of exercise (swimming) over a period of 5 (T5) or 2 (T2) consecutive days per week on adult male rats fed the normocaloric or hypercholesterolemic diets in an attempt to identify whether different moderate continuous exercise frequencies (90 min), for a period of 8 weeks, would promote adaptation of the area of adipocytes, lipid parameters, and metabolic pathway.

Hypercholesterolemic diet significantly increased total cholesterol and plasmatic triglycerides when compared to the normocaloric diet. Such alterations show the effectiveness of this diet in promoting dyslipidemia. Although not statistically significant, all exercised groups on both diets presented an increase in HDL cholesterol, ranging from 22.5 to 26.2 %, in relation to the respective sedentary groups.

In regards to the type of exercise, both training frequencies significantly reduced the circulating concentration of cholesterol and triglycerides in hypercholesterolemic rats. These results, however, were not observed in the groups fed the normocaloric diet, demonstrating the higher efficacy of exercise in hypercholesterolemic exercised rats (5 or 2 day/week) in both periods. Our results showed that in response to the ingestion of a hypercholesterolemic diet, there was a significant increase in the area of adipocytes of the central retroperitoneal (CR) adipose tissue when compared to the normocaloric diet; however, the same did not occur in the area of visceral epididymal (VE) adipose tissue. In response to 5× week training, a significant reduction in the area of central and visceral white adipose tissues was observed, and this reduction was more accentuated in the central adipose tissue in rats fed the hypercholesterolemic diet.

In general, and considering the variations of the animals' diet content, a significant reduction was observed in the area of central and visceral adipocytes due to continuous 5× and 2×/week exercise. It is noteworthy to mention that the effects of continuous 5×/week exercise are more accentuated than those observed in animals under 2×/week training (Fig. 28.7).

Both training frequencies were also statistically effective in reducing the rate of lipid synthesis in visceral adipose tissue of hypercholesterolemic rats. In trained rats fed a normocaloric diet, the same tissue presented a percentage reduction in lipogenesis as compared to the respective control group. The importance of exercise (5 and 2 consecutive days per week) in the reduction of lipid synthesis in such tissues of animals has, thus, been demonstrated in this experimental protocol (Fig. 28.8).

It is, thus, noted that a hypercholesterolemic diet and continuous moderate exercise play an important role in metabolism by modifying lipid pathways as well

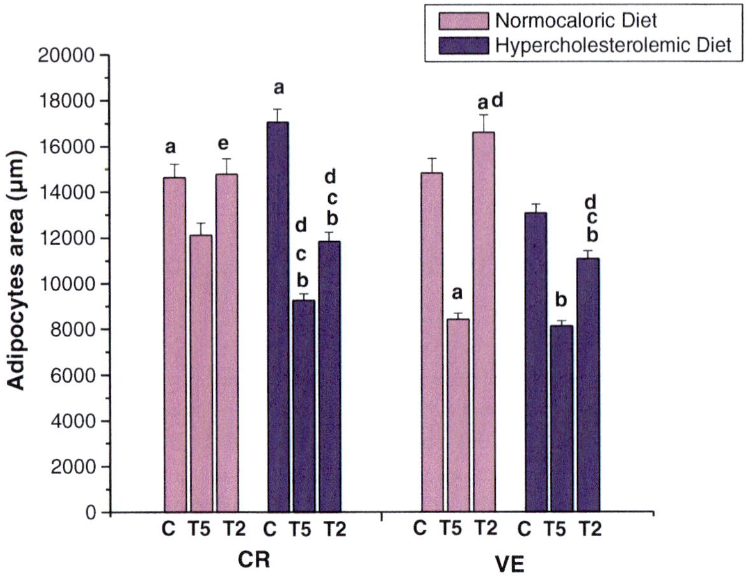

Fig. 28.7 Adipocytes area of rats fed different diets. Values are expressed as Mean ± Standard error (n = 6/group, number of areas = 300/group). Distinct letters vary significantly ([a]versus N; [b]versus H; [c]same type of exercise; [d]different exercise; $p \leq 0.05$) in the Tukey-Kramer test. *N* Normocaloric diet, *H* Hypercholesterolemic diet; *C* Control; *T5* Trained 5×/week, *T2* Trained 2×/week; *CR* Retroperitoneal adipose tissue, *VE* Visceral epididymal adipose tissue

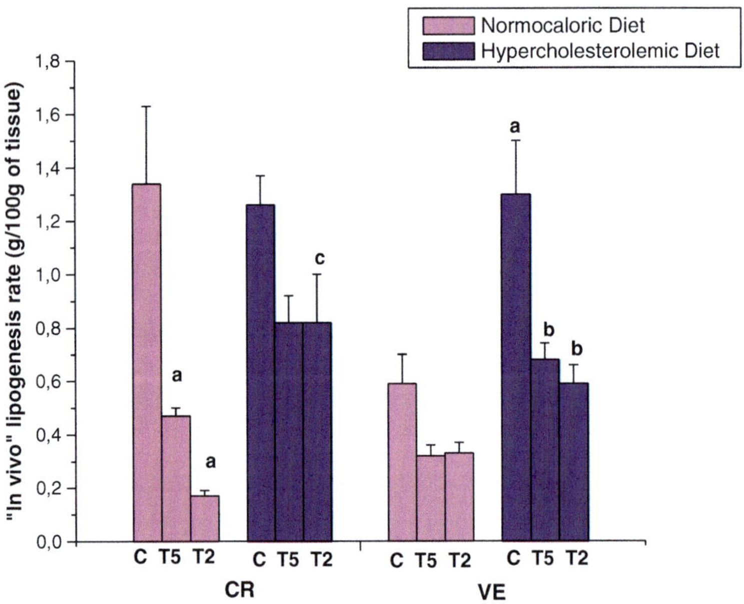

Fig. 28.8 "In vivo" *lipogenesis* rate in rats fed different diets. Values are expressed as Mean ± Standard error (n = 6/group). Distinct letters vary significantly ([a]versus N; [b]versus H; [c]same type of exercise; [d]different exercise; $p \leq 0.05$) in the Tukey-Kramer test. *N* Normocaloric diet; *H* Hypercholesterolemic diet, *C* Control; *T5* Trained 5×/week, *T2* Trained 2×/week, *CR* Central Retroperitoneal adipose tissue, *VE* Visceral epididymal adipose tissue

as the area of adipocytes in adult rats. Guerra (2005) demonstrated the efficacy of a hypercholesterolemic diet in the increase of cholesterol and triglycerides. Both exercise frequencies (5 and 2 consecutive days/week for 8 weeks) promoted important benefits, especially in rats fed a hypercholesterolemic diet, such as reduction of total cholesterol and plasmatic triglycerides, lipid synthesis of retroperitoneal and epididymal white adipose tissues, reduction of the adipocytes area, and percentage increases in the rate of lipolysis. Notwithstanding, the effects of exercise 5 consecutive times a week proved to be more evident that those resulting from exercise 2 consecutive times a week, but the latter also proved to be beneficial, especially in a hypercholesterolemic scenario. These results indicate that both protocols might contribute to the prevention and control of some chronic degenerative diseases such as hypercholesterolemia, dyslipidemias, and obesity.

Final Consideration

Different diets promoted adaptations or specific alterations in lipid metabolism and in the adiposity of the animals, and these are considered age-dependent. These diets led to significant endocrine and metabolic alterations and contributed to the development of exogenous obesity. Therefore, these experimental diets, along with some types of exercise, are the best models for understanding the mechanisms related to obesity and co-morbidities.

References

Abd El-Gawad IA, El-Sayed EM, Hafez SA, El-Zeini HM, Saleh FA. The hypocholesterolemic effect of milk yogurt and soy-yogurt containing bifidobacteria in rats fed on a cholesterol-enriched diet. Int Dairy J. 2005;15:37–44.

Brooks SP, Lampi BJ. Effect of dietary fat on whole body fat acid synthesis in weanling rats. J Nutr Biochem. 1999;10:291–8.

Cheik NC. Effects of physical training associated to fermented soy yogurt on the dislipidemia and obesity control in adult male rats feed to normocaloric and hypercholesterolemic diet. Master Thesis. PPG-CF/ UFSCar; 2002.

Cheik NC. Effects of the different nutritional interventions and exercise on hormonal regulation of food ingestion and the lipid metabolism in rats. Doctor Thesis. PPG-CF/ UFSCar; 2005.

Chen JR, Chiou SF, Suetsuna K, Yang HY, Yang SC. Lipid metabolism in hipercholesterolemic rats affected by feeding cholesterol-free diets containing different amounts of non-dialyzed soybean protein fraction. Nutrition. 2003;19:676–80.

Duarte FO. Metabolic adaptations of the two moderate swimming training, continuous and intermittent, in rats feed normocaloric or hypercholesterolemic diet. Master Thesis. UFSCar/PPGCF; 2001, p. 120.

Duarte FO, Sene MO, Oishi JC, Bidinotto P, Perez SEA, Moraes G, et al. Weekend exercise promote dislipidemia control in adult male rats feed with high fat diet. Braz J Physiotheraphy. 2003;7:229–35.

Duarte FO, Sene-Fiorese M, Zambon L, Botaro R, Freitas LF, Catelli DS, et al. Effects of food restriction on adiposity in rats with exogenous obesity. V international congress on physical education and human motricity. Motriz-J Phys Educ—UNESP. 2005;11:S62.

Duarte ACGO, Fonseca DF, Manzoni MSJ, Soave CF, Sene-Fiorese M, Dâmaso AR, et al. Hiperlipidic palatable diet promote obesity, but not change secretory capacity of insulin in rats. J Nutr. 2006;19:230–8.
Duelli R, Maurer MH, Staudt R, Heiland S, Duembgen L, Kuschinsky W. Increase cerebral glucose utilization and decreased glucose transporter Glut 1 during chronic hyperglycemia in rat brain. Brain Res. 2000;858:338–47.
Estadella D. Effects of cafeteria diet and the alternated diet cycles on lipid metabolism in sedentary and exercised rats. São Paulo, Master Thesis—Universidade Federal de São Paulo; 2001, p. 81.
Estadella D, Oyama LM, Dâmaso AR, Ribeiro EB, Oller do Nascimento CM. Effect of palatable hyperlipidic diet on lipid metabolism of sedentary and exercised rats. Nutrition. 2004;20:218–24.
Feoli AM, Roehrig C, Rotta LN, Kruger AH, Souza KB, Kessler AM, et al. Serum and liver lipids in rats and chicks fed with diets containing different oils. Nutrition. 2003;19:789–93.
Frayn KN. Adipose tissue as a buffer for daily lipid flux. Dilabetologia. 2002;45:1201–10.
Gaíva MH, Couto RC, Oyama LM, Couto GE, Silveira VL, Ribeiro EB, et al. Polynsaturated fatty acid-rich diets: effect on adipose tissue metabolism in rats. Br J Nutr. 2001;86:371–7.
Giannini SD. Aterosclerose and dislipidemia. Clinical & therapeutic: practical fundament. Ed. BG Editora e Produções Culturais Ltda; 1998. pp. 1–158.
Guerra RLF. Effects of continuous moderate exercise (5 or 2 days/week) on lipid metabolism and adipocyte area in adult normocaloric and hypercholesterolemic rats. São Carlos, Doctor Thesis—Universidade Federal de São Carlos; 2005, p. 92.
Islam KK, Knight BL, Frayn KN, Patel DD, Gibbons GF. Deficiency of PPAR-gama disturbs the response of lipogenic flux and of lipogenic and cholesterogenic gene expression to dietary cholesterol in mouse white adipose tissue. Biochem Biophys Acta. 2005;1734:259–68.
Kopchick JJ, Andry JM. Growth hormone (GH), GH receptor, and signal transduction. Mol Genet Metab. 2000;71:293–314.
Krause BR, Hartman AD. Adipose tissue and cholesterol metabolism. J Lipid Res. 1984;25:97–110.
Kretschemer BD, Schelling P, Beier N, Liebscher C, Treutel S, Kruger N, et al. Modulatory role of food, feeding regime and physical exercise on body weight and insulin resistance. Life Sci. 2005;76:1553–73.
Le Lay S, Krief S, Farnier C, Lefrère I, Le Liepvre X, Bazin R, et al. Cholesterol, a cell size-dependent signal that regulates glucose metabolism and gene expression in adipocytes. J Biol Chem. 2001;276:16904–10.
Leonhardt M, Langhans W. Fat acid oxidation and control of food intake. Physiol Behav. 2004;83:645–51.
Manzoni MSJ, Rossi EA, Carlos IZ, Vendramini RC, Duarte ACGO, Dâmaso AR. The fermented soy product supplemented with isoflavones affected fat depots in juvenile rats. Nutrition. 2005;21:1018–24.
Matsudo V, Matsudo S, Andrade D, Araújo T, Andrade E, Oliveira LC, et al. Promotion of physical activity in developing country: the Agita São Paulo experience. Public Health Nutr. 2002;5:253–61.
Ranier G, Gaudreau P, Hajjad H, Deslauriers N, Houde-Nadeau M, Brazeau P. Decreased pituitary growth hormone response to growth hormone-releasing factor in cafeteria-fed rats: dietary and obesity effects. Neuroendocrinology. 1990;52:284–90.
Rossi EA. Development and biological evaluation of the hypercholesterolemic potentital of the one new probiotic soy product Araraquara. Brasil: Universidade Estadual Paulista "Júlio de Mesquita Filho", Thesis of the Livre Docência; 2000, p. 150.
Rossi EA, Vendramini RC, Carlos IZ, de Oliveira MG, de Valdez GF. Effects of a novel fermented soy product on the serum lipids of hypercholesterolemic rabbits. Braz J Cardiol. 2000;74:213–6.
Sene-Fiorese M, Duarte FO, Zambon L, Botaro R, Freitas LF, Catelli DS, et al. Effects of continuous and fractioned exercise on adiposity in rats. V international congress on physical education and human motricity. Motriz-J Phys Educ—UNESP. 2005;11:S72.
Westerterp-Platenga MS. Fat intake and energy-balance effects. Physiol Behav. 2004;83:579–85.
Zhou X, De Schepper J, De Craemer D, Delhase M, Gys G, Smitz J, et al. Pituitary growth hormone release and gene expression in cafeteria-diet-induce obese rats. J Endrocrinol. 1998;59:165–72.